CO_2 Capture, Utilization, and Sequestration Strategies

CRC PRESS Series on Sustainable Energy Strategies

by Yatish T. Shah

CO_2 Capture, Utilization, and Sequestration Strategies

Hybrid Energy Systems: Strategy for Industrial Decarbonization

Hybrid Power: Generation, Storage, and Grids

Modular Systems for Energy Usage Management

Modular Systems for Energy and Fuel Recovery and Conversion

Thermal Energy: Sources, Recovery, and Applications

Chemical Energy from Natural and Synthetic Gas

Other related books by Yatish T. Shah

Energy and Fuel Systems Integration

Water for Energy and Fuel Production

Biofuels and Bioenergy: Processes and Technologies

For more information on this series, please visit: https://www.routledge.com/Sustainable-Energy-Strategies/book-series/CRCSES

Series Preface

While fossil fuels (coal, oil, and gas) were the dominant sources of energy during the last century, since the beginning of the twenty-first century, an exclusive dependence on fossil fuels is believed to be a nonsustainable strategy due to (a) their environmental impacts, (b) their nonrenewable nature, and (c) their dependence on the local politics of the major providers. The world has also recognized that there are in fact ten sources of energy: coal, oil, gas, biomass, waste, nuclear, solar, geothermal, wind, and water. These can generate our required chemical/biological, mechanical, electrical, and thermal energy needs. A new paradigm has been to explore greater roles of renewable and nuclear energy in the energy mix to make energy supply more sustainable and environmentally friendly. The adopted strategy has been to replace fossil energy by renewable and nuclear energy as rapidly as possible. While fossil energy still remains dominant in the energy mix, by itself, it cannot be a sustainable source of energy for the long future.

Along with exploring all ten sources of energy, sustainable energy strategies must consider five parameters: (a) availability of raw materials and accessibility of product market, (b) safety and environmental protection associated with the energy system, (c) technical viability of the energy system on the commercial scale, (d) affordable economics, and (e) market potential of a given energy option in the changing global environment. There are numerous examples substantiating the importance of each of these parameters for energy sustainability. For example, biomass or waste may not be easily available for a large-scale power system making a very large-scale biomass/waste power system (like a coal or natural gas power plant) unsustainable. Similarly, an electrical grid to transfer power to a remote area or onshore needs from a remote offshore operation may not be possible. Concerns of safety and environmental protection (due to emissions of carbon dioxide) limit the use of nuclear and coal-driven power plants. Many energy systems can be successful at laboratory or pilot scales, but may not be workable at commercial scales. Hydrogen production using a thermochemical cycle is one example. Many energy systems are as yet economically prohibitive. The devices to generate electricity from heat such as thermoelectric and thermophotovoltaic systems are still very expensive for commercial use. Large-scale solar and wind energy systems require huge upfront capital investments, which may not be possible in some parts of the world. Finally, energy systems cannot be viable without market potential for the product. Gasoline production systems were not viable until the internal combustion engine for the automobile was invented. Power generation from wind or solar energy requires guaranteed markets for electricity. Thus, these five parameters collectively form a framework for sustainable energy strategies.

It should also be noted that the sustainability of a given energy system can change with time. For example, coal-fueled power plants became unsustainable due to their impact on the environment. These power plants are now being replaced by gas-driven power plants. New technology and new market forces can also change sustainability of the energy system. For example, successful commercial developments of fuel cells and electric cars can make the use of internal combustion engines redundant in the vehicle industry. While an energy system can become unsustainable due to changes in parameters, outlined above, over time, it can regain sustainability by adopting strategies to address the changes in these five parameters. New energy systems must consider long-term sustainability with changing world dynamics and possibilities of new energy options.

Sustainable energy strategies must also consider the location of the energy system. On the one hand, fossil and nuclear energy are high-density energies and they are best suited for centralized operations in an urban area, while on the other hand, renewable energies are of low density and they are well suited for distributed operations in rural and remote areas. Solar energy may be less affordable in locations far away from the equator. Offshore wind energy may not be sustainable if the distance from shore is too great for energy transport. Sustainable strategies for one country may be quite different from another depending on their resource (raw material) availability and local market potential. The current transformation from fossil energy to green energy is often prohibited by required infrastructure and the total cost of transformation. Local politics and social acceptance also play an important role. Nuclear energy is more acceptable in France than in any other country.

Sustainable energy strategies can also be size dependent. Biomass and waste can serve local communities well at a smaller scale. As mentioned before, the large-scale plants can be unsustainable because of limitations on raw materials. New energy devices that operate well at micro- and nanoscales may not be possible on a large scale. In recent years, nanotechnology has significantly affected the energy industry. New developments in nanotechnology should also be a part of sustainable energy strategies. While larger nuclear plants are considered to be the most cost-effective for power generation in an urban environment, smaller modular nuclear reactors can be the more sustainable choice for distributed cogeneration processes. Recent advances in thermoelectric generators due to advances in nanomaterials are an example of a size-dependent sustainable energy strategy. A modular approach for energy systems is more sustainable at smaller scale than for a very large scale. Generally, a modular approach is not considered as a sustainable strategy for a very large, centralized energy system.

Finally, choosing a sustainable energy system is a game of options. New options are created by either improving the existing system or creating an innovative option through new ideas and their commercial development. For example, a coal-driven power plant can be made more sustainable by using very cost-effective carbon capture technologies. Since sustainability is time, location, and size dependent, sustainable strategies should follow local needs and markets. In short, sustainable energy strategies must consider all ten sources and a framework of five stated parameters under which they can be made workable for local conditions. A revolution in technology (like nuclear fusion) can, however, have global and local impacts on sustainable energy strategies.

The CRC Press Series on Sustainable Energy Strategies will focus on novel ideas that will promote different energy sources sustainable for long term within the framework of the five parameters outlined above. Strategies can include both improvement in existing technologies and the development of new technologies.

Series Editor,
Yatish T. Shah

CO_2 Capture, Utilization, and Sequestration Strategies

Yatish T. Shah

CRC Press
Taylor & Francis Group
Boca Raton London New York

CRC Press is an imprint of the
Taylor & Francis Group, an **informa** business

First edition published 2022
by CRC Press
6000 Broken Sound Parkway NW, Suite 300, Boca Raton, FL 33487-2742

and by CRC Press
2 Park Square, Milton Park, Abingdon, Oxon, OX14 4RN

CRC Press is an imprint of Taylor & Francis Group, LLC

Library of Congress Cataloging-in-Publication Data
Names: Shah, Yatish T., author.
Title: CO_2 capture, utilization, and sequestration strategies / Yatish T. Shah.
Description: First edition. | Boca Raton, FL : CRC Press, 2022. |
Series: Sustainable energy strategies | Includes bibliographical references and index. |
Summary: "Offering practical treatment strategies for CO_2 emission generated from various energy-related sources, this book emphasizes carbon capture, utilization, and sequestration (CCUS) with special focus on methods for each component of the strategy. While other books mostly focus on CCS strategy for CO_2, this book details the technologies available for utilization of CO_2, showing how it can be a valuable renewable source for chemicals, materials, fuels, and power instead of a waste material damaging the environment. This book is a valuable reference for readers in academia, industry, and government organizations seeking a guide to effective CCUS processes, technologies, and applications"— Provided by publisher.
Identifiers: LCCN 2021036378 (print) | LCCN 2021036379 (ebook) |
ISBN 9781032124803 (hbk) | ISBN 9781032135052 (pbk) | ISBN 9781003229575 (ebk)
Subjects: LCSH: Carbon sequestration.
Classification: LCC TD885.5.C3 S4656 2022 (print) | LCC TD885.5.C3 (ebook) |
DDC 628.5/32—dc23/eng/20211006
LC record available at https://lccn.loc.gov/2021036378
LC ebook record available at https://lccn.loc.gov/2021036379

ISBN: 978-1-032-12480-3 (hbk)
ISBN: 978-1-032-13505-2 (pbk)
ISBN: 978-1-003-22957-5 (ebk)

DOI: 10.1201/9781003229575

Typeset in Times
by codeMantra

Contents

Preface

Carbon emission in the environment is a major societal issue because it is closely linked to the global warming and resulting in negative impacts on weather patterns and sea level rise. Carbon dioxide is the single largest source of carbon emission caused by human activities. As demands for energy have increased globally, CO_2 levels have risen sharply, from pre-industrial levels of 280 ppm a century ago to over 400 ppm since 2013. These levels also include emissions generated from other, non-energy sources, including steel, aluminum, and cement production; fermentation; chemical production; and other industrial sources. Continued economic and population growth drive the increase in CO_2 emissions; and the levels of emissions, if left unchecked, are projected to exceed 530 ppm by 2100. Goals for stabilizing CO_2 levels at $\leq$450 ppm were set in 2007 to avoid serious impacts to the environment and health. To meet these goals, however, requires (a) prevention strategy which includes substantial improvements in energy efficiency, increased deployment of renewable and nuclear energy, and the development of new technologies for mitigating CO_2 emissions arising from the use of fossil fuels and other anthropogenic sources; and (b) treatment strategy which includes capture, concentrate and either store (temporary or permanent) or utilize CO_2 streams coming from various point sources. This strategy is also called CCUS strategy. Numerous books are written on carbon dioxide sequestration strategy. This is in essence a cost strategy. The present book mainly focuses on carbon dioxide utilization strategies which are revenue based.

Carbon dioxide is a very stable molecule and requires novel technologies to capture, concentrate and either sequester or utilize for valuable chemicals, materials, fuels or power generation. In 2014, the global carbon emissions from fossil-derived fuel combustion were estimated to be 9,855 million metric tons of carbon, or nearly 36 gigatonnes (Gt) of CO_2. About a quarter of these emissions come from mobile sources, such as automobiles and other forms of transportation. Much of the remainder comes from point sources—plants used for generation of electricity and heat for industry, business, and home use, as well as industrial production processes. It is these so-called point sources that are the focus of technologies for mitigating global CO_2 emissions. Efforts are being made to reduce carbon emission from vehicles by replacing internal combustion engine-driven vehicles with hybrid and electric vehicles.

While approximately 10,000 tera grams (Tg) of waste gas carbon streams is emitted globally each year, which represents a large volume of potential inputs for carbon treatment technologies, these gaseous waste streams are heterogeneous in their composition, are emitted from a wide range of geographically distributed sources, and are not always easily transported from their sources to locations where they can be processed. This heterogeneity poses challenges for carbon dioxide treatment. Sources of carbon dioxide include fossil fuel combustion, natural gas systems, coal mining, waste incineration, etc. Energy-related activities are the primary sources of U.S. anthropogenic greenhouse gas emissions, accounted for 83.8% of total greenhouse gas emissions on a carbon dioxide (CO_2) equivalent basis in 2016.

In general, treatment strategies for CO_2 work in three parts, capture and concentrate and/or purify if needed, utilize (directly or through conversion) or store for short or long terms. In the present book, we will address all three steps of treatment strategies. In the past, major efforts have been made for the capture of CO_2 emission streams from power plants and industries, concentrate the streams and sequester them underground or under the sea water. Methods were also adopted to add pre- and post-combustion or oxy-combustion treatments to reduce CO_2 or concentrate emission. The strategies for CO_2 concentration, transport and storage are discussed in Chapter 2. While storage strategy is effective in preventing CO_2 emission to the environment, it is basically a cost strategy and for power plants, it can end up increasing the cost of electricity by 50%–80%. Since, at present time, CO_2 utilization is significantly lower than generation, storage will remain as a part of overall CO_2 emission treatment strategy for a near future.

In recent years, CO_2 utilization strategy (CCUS) has become very important. CO_2 can be utilized either directly or it can be converted to useful chemicals and fuels by different methods of conversion. Various applications for direct utilization of CO_2 streams are described in Chapter 2. Indirect utilization of CO_2 is more complex and several strategies that are currently being examined are described Chapters 3–7. For both direct and indirect utilizations of CO_2, it is important to utilize carbon dioxide emission locally, where ever possible, in order to avoid expensive transportation. For example, in recent years, conversion of CO_2 to power using on-site molten carbonate fuel cell technology is being pursued by ExxonMobil and Fuel Cell Energy Corp. Carbon dioxide conversion to construction materials by carbonation technology applied to several types of waste materials can also be done near the sites where waste materials are generated.

Chapter 3 examines viability of mineral carbonation process. In recent years, carbon dioxide is converted to carbonates by mineral carbonation of natural and synthetic Ca- and Mg-bearing materials and industrial wastes containing calcium, magnesium and silicate oxides, some of which can be used to produce components of construction materials. Mineralization involves reaction of minerals (mostly calcium or magnesium silicates) with CO_2 to give inert carbonates. The reaction to form carbonates itself requires no energy inputs and actually releases heat, although significant energy is typically required to generate the requisite feed minerals. When possible, the use of industrial waste is favored. The current bottleneck, however, for viable mineral carbonation processes on an industrial scale is the reaction rate of carbonation. The chapter also examines the role of mineral carbonation in the productions of synthetic aggregates, cement and concrete for construction industry. New formulations of materials such as concrete will require testing and property validation before being accepted by users and regulators for the market.

Chapter 4 examines various types of biological treatments to convert carbon dioxide into useful biochemicals, biomaterials and biofuels. Carbon dioxide can be biologically converted to biofuels by green algae, microalgae, cyanobacteria, chemolithotrophs, and bio-electrochemical systems. In recent years, microbial electro-synthesis (MES) has captured significant attention. The chapter also discusses anaerobic and gas fermentations to convert CO_2 to syngas. Six major pathways for CO_2 fixation and conversion are also discussed. Several factors have expanded the repertoire of biobased products that can be synthesized directly from CO_2, including the large number of CO_2-utilizing microorganisms, genetic modification of microorganisms, and tailoring enzymatic/protein properties through protein engineering. Biological utilization has a large range of potential uses in the development of commercial products, including various biofuels, chemicals, and fertilizers. However, biological utilization rates and scalability remain challenges with significant potentials for the future.

Chapter 5 examines the role of heterogeneous catalysis for thermal and electrochemical conversion of carbon dioxide to useful chemical and fuels. The use of fuel cells as part of CCUS strategy is also discussed. Carbon dioxide can be chemically converted to alcohols, acids, hydrocarbons, polymer precursors, carbon monoxide, and carbon nanotubes. Carbon dioxide can also be converted to liquid fuels, fertilizer, polymers, and secondary chemicals. These conversions require catalysts to overcome kinetic barriers. Because carbon in CO_2 is in its most highly oxidized form, many of the resulting reactions are reductions, either through the addition of hydrogen or electrons. Catalysts are critical not only for making the transformation possible, but also for reducing the energy inputs to (ideally) the minimum amount dictated by the thermodynamics of the transformation, and discovery of appropriate catalysts and development of energy-efficient processes are current bottlenecks. However, combination of C1 and C2 carbon productions from CO_2 followed by the Fischer-Tropsch synthesis provides a fertile avenue for productions of higher hydrocarbons, alcohols, acids, and fuels. The chapter also examines in detail electrocatalysis process to produce a variety of chemicals like CO, methane, methanol, formaldehyde, formic acid, ethylene, ethanol, propanol, etc. New immobilized catalysts on the electrodes are being constantly evaluated. Finally, the chapter examines the role of high-temperature molten carbonate fuel cell for conversion of CO_2 to power, heat, hydrogen, and other valuable chemicals.

Chapter 6 examines the role of solar energy-based thermal, photo-thermal, and photo catalytic conversion of CO_2 to valuable chemicals and fuels. Homogeneous, heterogeneous, and immobilized homogeneous catalysts are considered. The use of renewable fuel like solar energy takes special meaning because it makes CO_2 conversion as a part of green technology. Photo-electro catalysis and novel PV/EC concept are also examined in detail. The latter approach has a strong commercial potential. While many ideas presented in this chapter are not yet successful at large scale, they present great potential for future success.

Finally, Chapter 7 presents a rapidly growing concept of plasma-activated catalysis for CO_2 conversion to many valuable fuels and chemicals. The chapter examines the roles of non-thermal plasma like DBD and warm plasmas like MW and GA and GAP on plasma-activated catalysis for four sets of reactions; CO_2 dissociation to produce CO; CO_2 hydrogenation to produce CO by reverse water gas shift reaction, to produce methane, or to produce methanol; and artificial photosynthesis ($CO_2 + H_2O$) to produce syngas, and dry reforming of methane to produce a variety of chemicals and fuels. The effectiveness of plasma-activated catalysis is examined in detail. While this approach is not quite ready for commercialization, it offers many positive features like flexibility, ease of operation at the industrial scale, convenient operating conditions, ease of scaleup, etc. The approach has a bright potential for converting CO_2 to useful products.

A number of studies have attempted to estimate the market for carbon utilization products, and one published study has reported the future market could be as high as $800 billion by 2030, utilizing 7 billion metric tons of carbon dioxide per year. Carbon dioxide is fundamentally a low-value, low-energy waste gas, which is often available in large quantities in single locations. If treatment strategies outlined above are successful, not only CO_2 emission to the environment will be prevented, but CO_2 can also become a very valuable raw material for chemicals, fuels, and power.

The present book should be valuable as a reference material for all researchers in academia, industry, and government organizations. It could also be a text material for the course on treatment of carbon emission.

Author

Yatish T. Shah received his B.Sc. in chemical engineering from the University of Michigan, Ann Arbor, USA, and MS and Sc.D. in chemical engineering from the Massachusetts Institute of Technology, Cambridge, USA. He has more than 40 years of academic and industrial experience in energy-related areas. He was chairman of the Department of Chemical and Petroleum Engineering at the University of Pittsburgh, Pennsylvania, USA; dean of the College of Engineering at the University of Tulsa, Oklahoma, USA, and Drexel University, Philadelphia, Pennsylvania, USA; chief research officer at Clemson University, South Carolina, USA; and provost at Missouri University of Science and Technology, Rolla, USA, the University of Central Missouri, Warrensburg, USA, and Norfolk State University, Virginia, USA. He was also a visiting scholar at University of Cambridge, UK, and a visiting professor at the University of California, Berkley, USA, and Institut für Technische Chemie I der Universität Erlangen, Nürnberg, Germany. Dr. Shah has previously written twelve books related to energy, nine of which are under "Sustainable Energy Strategies" book series by Taylor and Francis of which he is the editor. He has also published more than 250 refereed reviews, book chapters, and research technical publications in the areas of energy, environment, and reaction engineering. He is an active consultant to numerous industries and government organizations in the energy areas.

1 Sources of Carbon Dioxide Emission and Possible Treatment Strategies

1.1 INTRODUCTION

Carbon emission in the environment is a major societal issue because it is closely linked to the global warming and resulting in negative impacts on weather patterns and sea level rise. Two major carbon compounds emitted into the atmosphere by human activities and other natural causes are carbon dioxide and methane [1]. While methane can be more problematic greenhouse gas (24–130 times more harmful than carbon dioxide), it is emitted in a lesser volume and it eventually gets converted to carbon dioxide by atmospheric reactions. Once captured, the utilization of methane is much less challenging because methane is very reactive and can be easily used for power or the productions of other chemicals and fuels. Sources of methane are not all human made; a significant amount of methane is produced from animal manure, agricultural sources, leakage of methane hydrates, etc. Methane leakage from oil and gas industry (commonly called associated gas or gas flares), from landfill gas, or from other biomass waste treatment facilities is being captured and put to useful utilization. Since the major issue with methane emission is its capture, this book will not address the treatment of methane emission.

The emission of carbon dioxide is more human made. As demands for energy have increased globally, CO_2 levels have risen sharply, from pre-industrial levels of 280 ppm a century ago to over 400 ppm since 2013. These levels also include emissions generated from other, nonenergy sources, including steel, aluminum, and cement production; fermentation; chemical production; and other industrial sources. Continued economic and population growth drives the increase in CO_2 emissions, and the levels of emissions, if left unchecked, are projected to exceed 530 ppm by 2100. Goals for stabilizing CO_2 levels at ≤450 ppm were set in 2007 to avoid serious impacts on the environment and health [2]. To meet these goals, however, requires substantial improvements in energy efficiency, increased deployment of renewable and nuclear energy, and the development of new technologies for mitigating CO_2 emissions arising from the use of fossil fuels and other anthropogenic sources [3].

In 2014, the global carbon emissions from fossil-derived fuel combustion were estimated to be 9,855 million metric tons of carbon, or nearly 36 gigatons (Gt) of CO_2 [4]. About a quarter of these emissions come from mobile sources, such as

DOI: 10.1201/9781003229575-1

automobiles and other forms of transportation. Much of the remainder comes from point sources—plants used for generation of electricity and heat for industry, business, and home use, as well as industrial production processes. It is these so-called point sources that are the focus of technologies for mitigating global CO_2 emissions. Efforts are being made to reduce carbon emission from vehicles by replacing internal combustion engine-driven vehicles with hybrid and electric vehicles.

Carbon dioxide is a very stable molecule, and it is emitted at a significantly large scale. According to a national academy report [1], approximately 10,000 teragrams (Tg) of waste gas carbon is emitted globally each year (see Table 1.1), representing a large volume of potential inputs for carbon treatment technologies. However, these gaseous waste streams are heterogeneous in their composition, are emitted from a wide range of geographically distributed sources, and are not always easily transported from their sources to locations where they can be processed. This heterogeneity poses challenges for carbon dioxide treatment. As shown in Figure 1.1, sources of GHG include fossil fuel combustion, natural gas systems, coal mining, waste incineration, etc. Energy-related activities are the primary sources of U.S. anthropogenic greenhouse gas emissions, accounting for 83.8% of total greenhouse gas emissions on a carbon dioxide (CO_2) equivalent basis in 2016. Energy-related CO_2 emissions alone constituted 78.9% of national emissions from all sources on a CO_2 equivalent basis, while the non-CO_2 emissions from energy-related activities represent a much smaller portion of total national emissions (4.9% collectively). Emissions from fossil fuel combustion comprise the vast majority of energy-related emissions, with CO_2 being the primary gas emitted (see Figure 1.1). Globally, approximately 32,294 million metric tons (MMT) of CO_2 were added to the atmosphere through the combustion of fossil fuels in 2015, of which the United States accounted for approximately 15%.

Among approaches to reducing CO_2 emissions are three processes: carbon capture, utilization, and storage (CCUS). Capture processes are designed to remove and concentrate CO_2 from gas streams in preparation for utilization or storage. There are two types of CO_2 utilization: direct and indirect. In direct utilization, CO_2 is used for a variety of purposes like dry ice, enhance oil recovery, fire extinguishers, etc. Indirect utilization involves the conversion of CO_2 to higher-value products (e.g., chemicals, fuels, and plastics) or to stable products for long-term storage (e.g., minerals). CO_2 can also be utilized to generate power. Storage processes are

TABLE 1.1
Chronological Increase in Carbon Emission [1]

Year	Carbon Emission (Tg)
1900	About 600
1942	About 1,200
1956	About 2,000
1966	About 3,000
1980	About 5,000
2003	About 7,000
2014	About 9,900

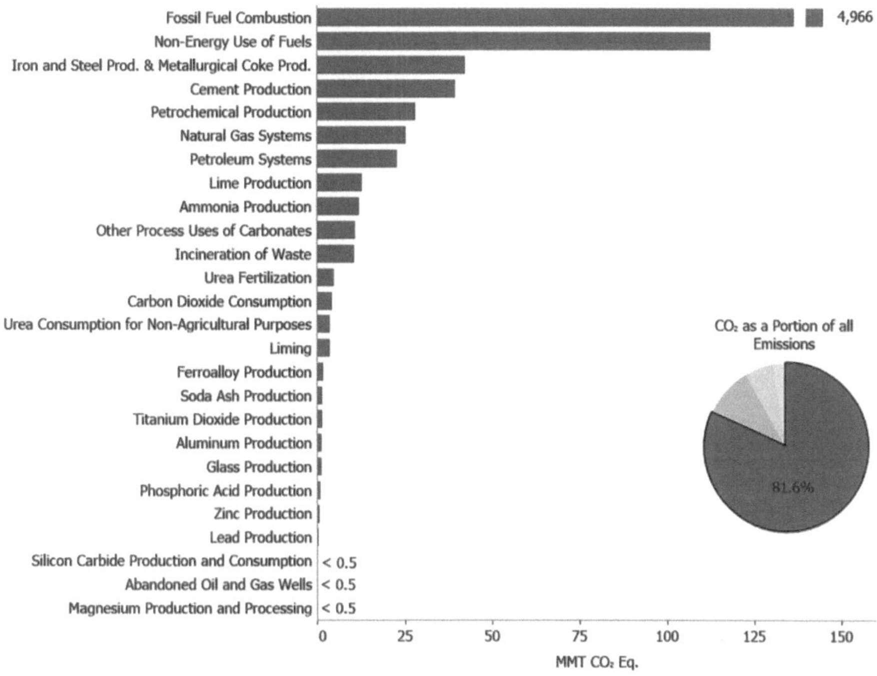

FIGURE 1.1 Sources for greenhouse gas in 2016 (MMT CO_2 Eq.) [5].

[1] Estimates are presented in units of million metric tons of carbon dioxide equivalent (MMT CO_2 Eq.), which weigh each gas by its global warming potential, or GWP, value. See the section on global warming potentials in the Executive Summary.

[2] Global CO_2 emissions from fossil fuel combustion were taken from International Energy Agency *CO_2 Emissions from Fossil Fuels Combustion – Highlights*, https://www.iea.org/publications/freepublications/publication/co2-emissions-from-fuel- combustion-highlights-2017.html, IEA (2017).

designed to store CO_2 safely and prevent its reentering the atmosphere. Sometimes temporary storage is needed to supply CO_2 continuously for downstream conversion operations. CCUS processes have been deployed at various sites for a number of years, proving that the technologies of capture, utilization, and storage are commercially available at an industrial scale. Nevertheless, the vast quantities of emissions and the associated costs for CCUS demand that new technologies be developed to make CCUS processes more efficient and economical if the goals for atmospheric CO_2 stabilization are to be met. Since utilization strategy is a revenue-based strategy, its expansion can make CO_2 to be a renewable and green source for the productions of chemicals, fuels, and power.

1.1.1 Sources of Carbon Dioxide Emission

Carbon dioxide is the most abundant greenhouse gas from anthropogenic sources, and in the United States, 98.5% of this contribution comes from the burning of fossil fuels [3]. The fuel-switching strategy for power plants from coal to natural gas was suggested because natural gas emits less carbon dioxide per unit of energy generated and because carbon dioxide contributes more to global warming than all other greenhouse gases combined. According to National Academy report [1], counter to the global trend, carbon dioxide emissions have been decreasing in the United States, from an estimated peak of 5.38 billion metric tons of carbon dioxide in 2004 (after accounting for uptake by forestry and land-use change) to an estimated 4.56 billion metric tons of carbon dioxide in 2016 (approximately 14% of the world total of 35 billion metric tons of carbon dioxide, or approximately 10,000 Tg of carbon) [5]. Trends in carbon dioxide emissions result from long- and short-term drivers, including population and economic growth, market trends, technological changes, and fuel choices. As shown in Figure 1.2, in 2016, carbon dioxide emissions in the United States come almost exclusively from the combustion of carbonaceous fuels in five major sectors: electricity generation, transportation, industrial processes and fuel use, residential fuel use, and commercial fuel use. The key categories for all GHG emission during 2016 are illustrated in Figure 1.3. Sources of carbon dioxide range from highly concentrated by-products of chemical manufacturing to relatively dilute flue gas streams from power plants with contaminants that can be problematic for carbon treatment technologies.

Figure 1.4 shows U.S. energy consumption by sources. My previous two books [6,7] examined various methods for prevention of carbon dioxide emission from these sources. These books illustrated that the best prevention strategies for the reduction of carbon dioxide are (a) to reduce the use of fossil fuels by increasing the efficiency of energy conversion processes and (b) to use hybrid energy systems to increase the contribution of renewable energy in the overall energy mix. Unfortunately, in spite of the best efforts, some emissions of carbon dioxide in the environment are inevitable.

In 2016, the analysis of world energy-associated carbon dioxide emissions with the consumption of fossil fuels indicated that coal was the largest source of carbon dioxide (13.5 billion metric tons), followed by liquid fuels (12.0 billion metric tons) and natural gas (8.0 billion metric tons) (see Figure 1.5) [5,8–10]. In addition, the global CO_2 emissions from fossil fuels combustion increased from nearly zero in 1870 to 32.0 billion metric tons in 2016 and are expected to increase up to 42 billion metric

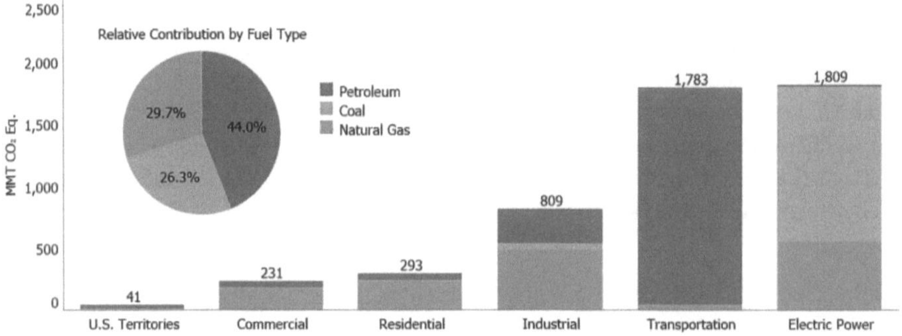

FIGURE 1.2 2016 CO_2 emissions from fossil fuel combustion by sector and fuel type (MMT CO_2 Eq.) [5].

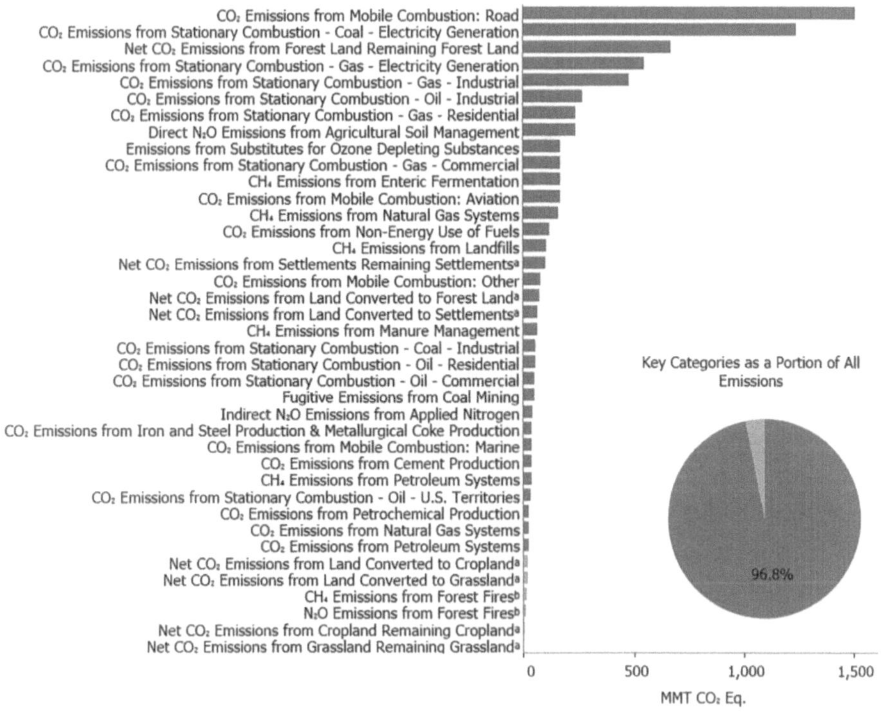

FIGURE 1.3 2016 key categories (MMT CO_2 Eq.) [5].

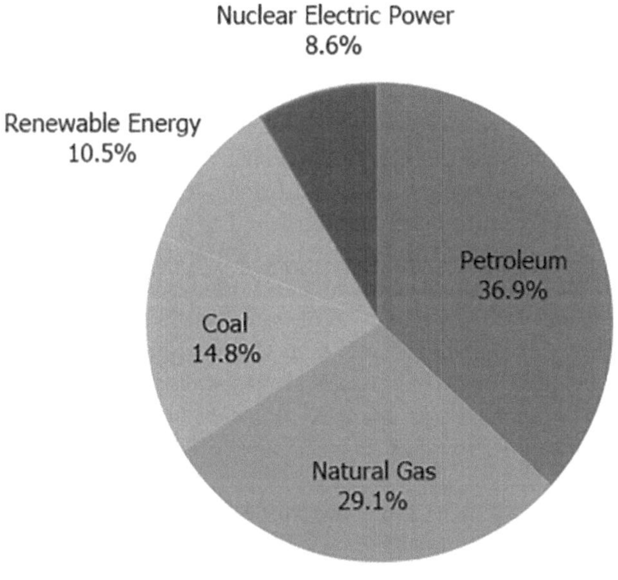

FIGURE 1.4 2016 U.S. energy consumption by energy source (percent) [5].

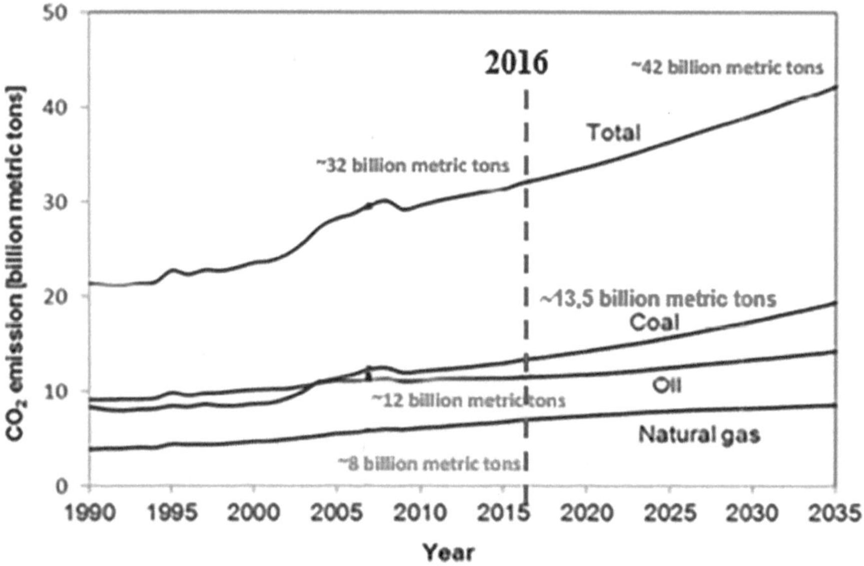

FIGURE 1.5 The world energy-associated carbon dioxide emissions by the type of fuel [8].

tons in 2035 [5]. In order to avoid CO_2 emissions, different measures such as recovery, removal, and storage have been proposed. However, carbon capture, transportation and storage also need large amounts of energy. For these reasons, the utilization of carbon dioxide in chemical conversion processes may become an important option for sustainable development, mitigation of carbon emissions, and reversal of the global warming process.

CO_2 contributed to about 77% of the world's greenhouse gas emissions (excluding water vapor) into the atmosphere in 2004 [5,8–10]. Many industrial manufacturing processes release carbon dioxide to the atmosphere, e.g. from the manufacturing of gas synthesis and combustion processes. Thus, the vented CO_2 molecule causes an increased concentration in the atmosphere. Global warming is caused by this accumulation of CO_2 in the atmosphere. These emissions should be monitored if the problem of global warming needs to be controlled.

Two main sources of CO_2 emissions within the industrial sector are manufacturing processes of industrial products where CO_2 is obtained as a by-product and from energy supply by the combustion of fossil fuels. Valorization of CO_2 emissions represents a main stake of competitiveness for large CO_2 emitters like the cement industry. Despite increasing investments in low-carbon energy alternatives and the progressive introduction of renewable energy resources, the consumption of fossil fuels is still assumed to significantly grow due to increasing worldwide energy and electricity demands caused partially by the emergence of new economic powers in the world. The overall carbon footprint of the predicted demand would exceed 40 Gt/year of CO_2 emissions by 2030 in comparison to 30 Gt/year in 2010 [5,8–10]. From a CO_2 valorization standpoint, scientific and industrial authorities are urged to suggest new schemes for producing and recycling carbon from fossil fuels for the energy market and chemical industry, in order to contribute to the reduction of environmental impacts. CO_2 transformation into added-value products will promote the transition from carbon fossil sources to low-carbon foot-print ones.

To this end, besides the power generation industry, the cement industry (large emitters of CO_2), represent a nonexploited source of carbon. In 2009, the world-wide production of cement was 3,048 Mt with 1.5–2 Gt of CO_2 emissions representing roughly 5% of anthropogenic global CO_2 emissions. Cement production processes emit large quantity of CO_2 (0,723 tCO_2 per ton of cement) coming from two main stages:

1. during pre-heating, the major part of the decomposition of limestone $CaCO_3$ into lime CaO and CO_2 named "calcination" is responsible for 60%–65% of total CO_2 emissions,
2. the clinkerization step involves the combustion of various fuels which generates additional CO_2 emissions (35%–40%).

The flue gas produced in a cement plant is released to atmosphere at 80°C and 110°C at atmospheric pressure through only one chimney stack, and such chimney vents all CO_2 emissions from the plant. The flow rate and concentration (15%–25%) of the CO_2 at the chimney stack are relatively constant over time compared to fossil fuel power plants. Actually, cement plants generally run at full load, facilitating the capture step. The associated flow rate for cement work usually ranges from 200,000 to 500,000 Nm^3/h.

For most of the recycling processes, the CO_2 flow has to be purified and concentrated. After the capture stage, if the concentration of CO_2 is higher than 90%, it is ready for geological storage or to be recycled into value-added chemicals. More effective conversion of carbon dioxide to chemicals and fuels is a potential solution to reduce its emissions.

According to EPA, key findings from the 1990 to 2018 U.S. Inventory include:

- In 2018, U.S. greenhouse gas emissions totaled 6,677 million metric tons of carbon dioxide equivalents, or 5,903 million metric tons of carbon dioxide equivalents after accounting for sequestration from the land sector.
- Emissions increased from 2017 to 2018 by 3.1% (after accounting for sequestration from the land sector). This increase was largely driven by an increase in emissions from fossil fuel combustion, which was a result of multiple factors, including more electricity use due to greater heating and cooling needs due to a colder winter and hotter summer in 2018 in comparison to 2017.
- Greenhouse gas emissions in 2018 (after accounting for sequestration from the land sector) were 10.2% below 2005 levels.

1.2 PHYSICAL AND CHEMICAL PROPERTIES OF CO_2 AND THERMODYNAMIC LIMITATIONS FOR ITS CONVERSION

As the structure of a carbon dioxide molecule is linear, it is a thermodynamically stable molecule with a measured bond strength of D = 532 kJ/mol [6,11]. The various physical and chemical properties of carbon dioxide are presented in Table 1.2. As shown in the table, the heat of formation (ΔH°) and the Gibbs free energy of

TABLE 1.2
Physical and Chemical Properties of Carbon Dioxide [8]

Property	Value and Unit
Heat of formation at 25°C	−393.5 kJ/mol
Entropy of formation at 25°C	213.6 J/K.mol
Gibbs free energy of formation at 25°C	−394.3 kJ/mol
Sublimation point at 1 atm	−78.5°C
Triple point at 5.1 atm	−56.5°C
Critical temperature	31.04°C
Critical pressure	72.85 atm
Critical density	0.468 g/cm³
Gas density at 0°C and 1 atm	1.976 g/L
Liquid density at 0°C and 1 atm	928 g/L
Solid density	1,560 g/L
Specific volume at 1 atm and 21°C	0.546 m³/kg
Latent heat of vaporization	353.4 J/g
At the triple point (−78.5°C) at 0°C	231.3 J/g
Viscosity at 25°C and 1 atm	0.015 cp
Solubility in water at 0°C and 1 atm	0.3346 g CO_2/100 g-H_2O
At 25°C and 1 atm	0.1449 g CO_2/100 g-H_2O

formation (ΔG°) of carbon dioxide are the two most important properties. Therefore, the ΔH° and ΔG° values are the most important criteria in order to estimate the thermodynamic feasibility of a reaction.

An analysis of the Gibbs free energy of the exothermic hydrogenation of CO_2 -indicates that the majority of the related reactions are thermodynamically unfavorable. Indeed, since the ΔG° values are more positive than the corresponding ΔH° values, they are less favorable. As a consequence, only a few reactions have both negative ΔG° and ΔH° values. Values of $\Delta G < 0$ either correspond to the reaction of hydrogenation or to reactions with products containing C-O bonds. Favorable values of ΔG in the hydrogenation reaction are related to the formation of water. As hydrogen must be produced at the cost of the input energy, the CO_2 mitigation cannot be achieved by any of these reactions [6,12]. Table 1.3 summarizes the values of the enthalpy and the Gibbs free energy, calculated by the ASPEN software [12–16], for the exothermic reaction in the CO_2 hydrogenations [8,12].

Hydrogenation reactions of CO_2 with a $\Delta H > 0$ can be executed [12–16]. However, reactions with highly positive ΔG° values are not favorable as the ones shown in Table 1.4 [8].

Often CO_2 conversion is accompanied by the production of CO. The reaction enthalpies for the production of the same product from either CO or CO_2 are comparable [12]. However, in most cases, CO is favored compared to CO_2 [12–16].

The reactivity of the carbon dioxide molecule is low due to its linear symmetry and the overall nonpolar nature of the molecule with a presence of the π-electron density of the double bonds and the lone pairs of electrons on the oxygen atoms and

TABLE 1.3
Values of the Enthalpy (ΔH < 0) and the Gibbs Free Energy for the Exothermic Reaction in the CO_2 Hydrogenations [8]

Reactions in the CO_2 Hydrogenations	Enthalpy (ΔH°) ΔH < 0 (kJ/mol)	Gibbs Free Energy (ΔG°) (kJ/mol)
$CO_2(g) + H_2(g) \rightarrow HCOOH(l)$	−31.0	+34.3
$CO_2(g) + 2H_2(g) \rightarrow HCHO(g) + H_2O(l)$	−11.7	+46.6
$CO_2(g) + 3H_2(g) \rightarrow CH_3OH(l) + H_2O(l)$	−137.8	−10.7
$CO_2(g) + 4H_2(g) \rightarrow CH_4(g) + 2H_2O(l)$	−259.9	−132.4
$2CO_2(g) + H_2(g) \rightarrow (COOH)2(l)$	−39.3	+85.3
$2CO_2(g) + 6H_2(g) \rightarrow CH_3OCH_3(g) + 3H_2O(l)$	−264.9	−38.0
$CO_2(g) + H_2 + CH_3OH(l) \rightarrow HCOOCH_3(l) + H_2O(l)$	−31.8	−25.8
$CO_2(g) + H_2 + CH_3OH(l) \rightarrow CH_3COOH(l) + H_2O(l)$	−135.4	−63.6
$CO_2(g) + 3H_2(g) + CH_3OH(l) \rightarrow C_2H_5OH(l) + 2H_2O(l)$	−221.6	−88.9
$CO_2(g) + H_2(g) + NH_3(g) \rightarrow HCONH_2(l) + H_2O(l)$	−103.0	+7.2
$CO_2(g) + CH_4(g) \rightarrow CH_3COOH(l)$	−13.3	+58.1
$CO_2(g) + CH_4(g) + H_2(g) \rightarrow CH_3CHO(l) + H_2O(l)$	−14.6	+74.4
$CO_2(g) + CH_4(g) + 2CO_2(g) \rightarrow (CH_3)2CO(l) + H_2O(l)$	−70.5	+51.2
$CO_2(g) + C_2H_2(g) + H_2(g) \rightarrow CH_2 = CHCOOH(l)$	−223.6	−115.0
$CO_2(g) + C_2H_4(g) \rightarrow CH_2 = CHCOOH(l)$	−49.1	+26.2
$CO_2(g) + C_2H_4(g) + H_2(g) \rightarrow C_2H_5COOH(l)$	−166.6	−56.6
$CO_2(g) + C_2H_4(g) + 2H_2(g) \rightarrow C_2H_5CHO + H_2O(l)$	−171.1	−44.4
$CO_2(g) + C_6H_6(l) \rightarrow C_6H_5COOH(l)$	−21.6	+30.5
$CO_2(g) + C_6H_5OH(l) \rightarrow mC_6H_4(OH)COOH(l)$	−6.6	+46.9

TABLE 1.4
Values of the Enthalpy (ΔH > 0) and the Gibbs Free Energy for the Hydrogenation Reactions of CO_2 [8]

Reactions in the CO_2 Hydrogenations	Enthalpy (ΔH°) ΔH > 0 (kJ/mol)	Gibbs Free Energy (ΔG°) (kJ/mol)
$CO_2(g) + CH_2 = CH_2(g) \rightarrow CH_2CH_2O(l) + CO(g)$	+152.9	+177.3
$CO_2(g) + C(s) \rightarrow 2CO(g)$	+172.6	+119.9
$3CO_2(g) + CH_4(g) \rightarrow 4CO(g) + 2H_2O(l)$	+235.1	+209.2
$CO_2(g) + CH_4(g) \rightarrow 2CO(g) + 2H_2(g)$	+247.5	+170.8
$CO_2(g) + 2CH_4(g) \rightarrow C_2H_6(g) + CO(g) + H_2O(l)$	+58.8	+88.0
$2CO_2(g) + 2CH_4(g) \rightarrow C_2H_4(g) + 2CO(g) + 2H_2O(l)$	+189.7	+208.3
$CO_2(g) + C_2H_4(g) \rightarrow C_2H_4O(g) + CO(g)$	+178.0	+176.0

also the electrophilic carbon atom [11,17,18]. Since the CO_2 molecule is a very stable one, a lot of energy must generally be supplied to trigger the desired transformation. Reactions on carbon dioxide often require high temperatures and active catalysts, with energy coming from electricity or from photons [11,17,18]. Thus, the reactions involving carbon dioxide are generally endothermic, and therefore, consume energy. For instance, considering reactions for the steam reforming of methane and CO_2 reforming of methane, the latter requires about 20% more energy input when compared to steam reforming. However, both reactions are useful for industrial applications as they produce synthesis gas with different H_2/CO molar ratios.

When carbon dioxide is used as a single reactant, it requires more energy. However, since its Gibbs free energy is −394.4 kJ/mol, it becomes thermodynamically more feasible when carbon dioxide is used with another reactant that has a higher Gibbs free energy. For example, methane, carbon (graphite), and hydrogen are some co-reactants that have higher (less negative) Gibbs energy. As an example, if we consider the dissociation of carbon dioxide to carbon monoxide where CO_2 is used as a single reactant and reduction of CO_2 by H_2 where CO_2 is used as a co-reactant, in the latter case both Gibbs free energy and heat of reaction are lower compared to the former case [12–14].

$$CO_2 \rightarrow CO + \tfrac{1}{2}O_2 \quad \Delta H° = +293\,\text{kJ/mol}, \Delta G° = +257\,\text{kJ/mol} \tag{1.1}$$

$$CO_2 + H_2 \rightarrow CO(g) + H_2O(g) \quad \Delta H° = +51\,\text{kJ/mol}, \Delta G° = +28\,\text{kJ/mol} \tag{1.2}$$

1.3 CHALLENGES FOR TREATMENT OF CARBON DIOXIDE EMISSION

Besides thermodynamic limitations for conversion of CO_2 to other chemicals and fuels, there are various other challenges in the implementation of treatment technologies to carbon dioxide emission. The data reported by EPA and national academy report indicate that carbon dioxide emissions from various sources vary in magnitude and composition. For example, the level and composition of carbon dioxide from fossil fuel combustion in 2016 were 4,966 Tg (MMT) with 12%–15% CO_2, from cement, iron/steel, and glass productions were 82.9 Tg (MMT) with 20%–35% CO_2, from ammonia manufacturing process 12.2 Tg (MMT) with greater than 98% CO_2, and from natural gas production 25.5 Tg (MMT) with 3%–4% CO_2. Such large variations of level and compositions of CO_2 waste streams from different sources make a uniform strategy for the treatment very difficult.

About one-third of U.S. carbon dioxide emissions come from electric power plants that burn fossil fuels and generate waste gases at high rates. U.S. power plant emissions totaled 1,800 Tg (MMT) in 2016 [5]. Concentrations of carbon dioxide in the flue gases from electric power plants are typically in the range of 12–15 mol% for coal-fired plants and 3–4 mol% for natural gas–powered plants [19]. While these power plants are distributed throughout the United States, they are disproportionately concentrated near major population centers. Furthermore, carbon dioxide coming from power plants often contains various other impurities such as nitrogen oxides, sulfur oxides, heavy metals, fly ash, and other species that need to be removed for many treatment technologies.

TABLE 1.5
U.S. Merchant CO_2 Supply and Demand by Sectors [5]

Merchant CO_2 Supply – Total 14.3 MMT	
Source	**Percentage (%)**
Ethanol	31
Ammonia	21
Hydrogen	21
Natural gas wells	22
Natural gas processing	3
Post combustion	1
Others	1
Merchant CO_2 Demand – Total 1.1 MMT	
End Use	**Percentage (%)**
Food processing	58
Carbonated beverages	18
Chemical processing	7
Metal fabrication	4
Agriculture	1
Other	12

Unlike carbon dioxide emitted from power plants, carbon dioxide emitted from manufacturing facilities contain higher concentrations of carbon dioxide and fewer contaminants. The chemical industry is the largest source of industrial CO_2 emissions in the United States (Figure 1.3). These include waste streams from hydrogen and ammonia manufacturing facilities, which produce particularly concentrated sources of CO_2. Another source of highly concentrated industrial CO_2 is from biofuel processing facilities, particularly ethanol fermentation plants. Cement, steel, and glass manufacturing facilities also emit significant volumes of carbon dioxide; in 2016, these facilities emitted 39.4, 1.2, and 6.9 Tg (MMT) of carbon dioxide, respectively. Cement production represents the second largest source of industrial carbon dioxide emissions in the United States. While emissions from glass production have stayed relatively constant over the past 15 years, emissions from steel production have decreased due to technological improvements and increased scrap steel utilization. The supply and demand of carbon dioxide from manufacturing industries are illustrated in Table 1.5. As shown, there is more than ten-fold difference between the supply and demand of carbon dioxide. The excess carbon dioxide is emitted to the environment unless it is further treated.

In developing a suitable technology to treat carbon dioxide emission, there are several challenges that need to overcome. As mentioned above, the carbon dioxide contained in the waste streams is rarely available in pure form, and the CO_2 composition of the waste gas stream as well as the level and nature of other impurities present in these streams vary widely, depending on its source. Each technology described in this book has its own restrictions on allowable concentrations of carbon dioxide and the level and chemical nature of other impurities. Some carbon treatment processes require purified streams. For example, carbon dioxide used in food products and food processing has high purity requirements. On the other hand, some carbon treatment technologies can valorize species other than carbon dioxide in waste gases, making the impurity of the waste gas stream a benefit. For example, LanzaTech uses microbes that convert carbon-rich waste gases containing carbon monoxide, hydrogen, carbon dioxide, methane, and other species into a variety of products. The microbes use carbon monoxide as an energy source; this carbon monoxide would otherwise need to be treated as an air pollutant, and this avoided cost increases the economic benefit of the carbon utilization technology. Some biological treatments can tolerate impurities like Sox, NOx, etc., while they do not do well in the presence of heavy metals like in flue gas from coal power plants.

The size of waste stream flow can also be an issue with some treatment technology. Since the effects of impurities on the adopted treatment technology are generally not known, nature of purification and the required feed preparation steps can be challenging. In addition, most current carbon treatment activities take an opportunistic approach to accessing waste streams, rather than a systematic approach to match waste streams with the treatment processes for which they are best suited. The most difficult carbon emission to capture is the one in transportation industry because it is highly distributed and difficult to treat in situ. Psarras et al. [20] mapped potential industrial CO_2 sources with potential utilization locations to calculate transportation and other costs of utilization processes at different locations and with different concentration needs. A coordinated approach requires better integration between separation and purification targets and information about the waste stream and processes to be used. In short, the treatment strategy needs to be custom built and optimized for the given CO_2 waste stream.

1.4 TREATMENT STRATEGIES FOR CO_2

In general, treatment strategies for CO_2 work in three parts: capture, concentrate and/or purify if needed, and utilize (directly or through conversion) or store for long term. In the present book, we will address all three steps of treatment strategies. In the past, major efforts have been made for the capture of CO_2 emission from power plants and industries, concentrate the stream, and sequester it underground or under the seawater. Methods were also adopted to add pre- and post-combustion or oxy-combustion treatments to reduce CO_2 emission. The strategies for CO_2 concentration, transport, and storage are discussed in detail in Chapter 2. While effective, this strategy turned out to be expensive particularly for the treatment of carbon dioxide coming from the power industry. The transportation cost for carbon dioxide for sequestration was significant. Generally, they ended up increasing the cost of electricity by 50%–80%.

In recent years, CO_2 utilization strategy (CCUS) has become very important. CO_2 can be utilized either directly or it can be converted to useful chemicals and fuels by different methods of conversion. Various applications for direct utilization of CO_2 streams are described in Chapter 2. Indirect utilization of CO_2 is more complex and several strategies that are currently being examined are described in Chapters 3–7. For both direct and indirect utilizations of CO_2, it is important to utilize carbon dioxide emission locally, wherever possible, in order to avoid expensive transportation.

For example, in recent years, conversion of CO_2 to power using on-site molten carbonate fuel cell technology is being pursued by ExxonMobil and Fuel Cell Energy Corp. Carbon dioxide conversion to construction materials by carbonation technology applied to several types of waste materials can also be done near the sites where waste materials are generated. Many direct and indirect utilizations of CO_2 are graphically illustrated in Figure 1.6 [21]. Market size and GHG mitigation potentials for several sectors are illustrated in Table 1.6 [21]. CO_2 utilization strategies are also discussed in a number of literature publications [19,22,23].

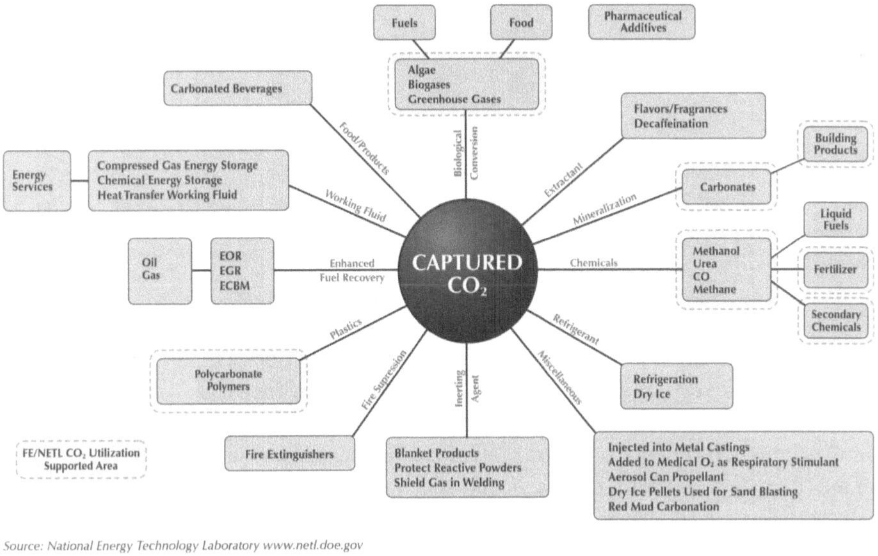

FIGURE 1.6 Schematic illustration of the uses of CO_2. The sizes of bubbles do not correspond to the size of the market or their carbon utilization and storage potentials. (Image courtesy of National Energy Technology Laboratory from CO_2 Utilization Focus Area [21,24].)

TABLE 1.6
Market Size and GHG Mitigation Potential of Selected CCU Sectors [21]

Market Size: $ Billion	2020	2025	2030	GHG Mitigation: Billions of Metric Tons of CO_2	2020	2025	2030
Concrete	60	200	400	Concrete	a	0.7	1.4
Fuels	5	60	250	Fuels	a	a	2.1
Aggregates	4	30	150	Aggregates	a	0.7	3.6
Algae Ag/feed products	3	10	120	Algae Ag/feed products	a	a	1.2
Algae fuels/chemicals	2	4	200	Algae fuels/chemicals	a	a	2
Polymers	1	3	25	Polymers	a	a	a
Commodity chemicals	0	5	12	Commodity chemicals	a	a	a

[a] Less than 0.5 billion tons CO_2.

TABLE 1.7
Industrial Process Utilization of CO_2 as a Raw Material for Synthesis of Organic Compounds in 2006 [8]

Industrial Processes That Utilize CO_2 as Raw Material	World Capacity Per Year [Million Tons]	Amount of Fixed CO_2 [Million Tons]	Balance [%]
	Chemical Synthesis		
Salicylic acid	0.07	0.025	0.04
Urea	143	105	87.61
Cyclic carbonates	0.08	0.04	0.05
Poly (propylene carbonate)	0.07	0.03	0.04
	Fuel Synthesis		
Methanol	20	2	12.25
Synthetic natural gas	-	-	-
Other fuels	-	-	-

While several utilization processes outlined in Figure 1.6 are commercialized, the total CO_2 use is very small compare to CO_2 generation. Since CO_2 molecule is very stable, indirect methods to convert CO_2 to useful products need to be implemented. So far, these methods are not used to their fullest potentials. The use of selective chemical and biological reaction pathways and efficient catalysts and novel techniques are needed to favor its conversion. Current estimations show that the chemical industry could contribute to convert around 1% of the global CO_2 emissions into chemical products [8]. Table 1.7 illustrates the utilization of CO_2 in different chemical conversion processes as reported in 2006.

Tables 1.6 and 1.7 show that the amount of CO_2 utilized today for the production of organic chemicals and fuels (methanol) is small (around 100 million tons) compared to today's global CO_2 emissions of around 30 billion tons. Only around 10% of the global crude oil consumption is used today in the chemical industry. It is often used as liquid fuels such as gasoline, diesel, and heavy oil. In recent years, the numbers shown in Tables 1.6 and 1.7 are improved with the advent of new catalysts and more innovative processes.

The current largest use of CO_2 resides in the synthesis of urea, which is a much used fertilizer. About 110 megatons of CO_2 are used yearly for this chemical synthesis [8,25]. The chemical reaction involved in the production of urea is known as:

$$CO_2 + 2NH_3 \rightarrow CO(NH_2)_2 + H_2O \tag{1.3}$$

CO_2 is likewise used to produce salicylic acid, which is found in pharmaceuticals and cyclic organic carbonates [8,25]. What is more, salicylic acid is produced by the reaction of sodium phenolate with CO_2 to produce sodium salicylate. The formed sodium salicylate is then converted into salicylic acid by the addition of sulfuric acid.

Sodium sulfate is obtained as a by-product while aspirin is produced from salicylic acid. The reaction path involved in the production of salicylic acid and aspirin can be written as:

$$CO_2 + C_6H_5ONa \rightarrow C_6H_5(COONa)OH \quad \textbf{sodium salicylate} \tag{1.4}$$

$$(H^+) \rightarrow C_6H_5(COOH)OH \quad \textbf{salicylic acid} \tag{1.5}$$

$$(CH_3CO)_2O \rightarrow C_6H_5(COOCH_3)COOH \quad \textbf{aspirin} \tag{1.6}$$

Methanol, used as a chemical reagent or as a fuel, is produced by the reaction of carbon dioxide and hydrogen in the presence of a catalyst. Furthermore, methanol can also be dehydrated to form gasoline-like fuels [8,25]. The chemical reaction involved in the production of methanol from CO_2 and H_2 is:

$$CO_2 + 3H_2 \rightarrow CH_3OH + H_2O \tag{1.7}$$

In addition, it has to be noted that supercritical carbon dioxide is a known solvent for promoting difficult chemical reactions [8,25]. Figure 1.7 is an easy way to show the possible range of reactions for CO_2. In these reactions, CO_2 can be used either as the main compound in reactions or as a source of carbon or oxygen [8,25].

In recent years, significant efforts are made to convert CO_2 to useful products by mineral carbonation, novel heterogeneous catalysis, thermochemical, photochemical,

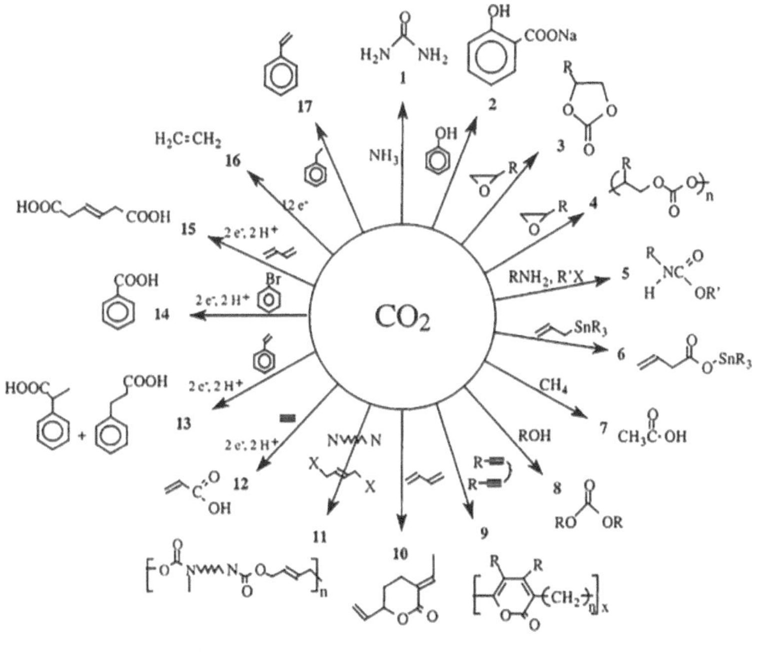

FIGURE 1.7 Illustration of transformations of CO_2 [8].

electrochemical and biological processes, carbonate fuel cell, and plasma-assisted catalysis. Remaining six chapters of the book are devoted to examine applications of various novel technologies for converting carbon dioxide to useful end-products. These approaches are clearly attractive from the economical point of view, and if commercially successful, they can reduce carbon emission in a significant way. While several of these approaches are still in the development stage, they seem to have bright economic potential. Since waste streams are distributed in nature with varying size and composition of carbon, often the effective technology needs to be targeted to the specific waste stream.

1.5 ORGANIZATION OF THE BOOK

In general, treatment strategies for CO_2 work in three parts: capture, concentrate and/or purify if needed, and utilize (directly or through conversion) or store for long term. In the present book, we will address all three steps of treatment strategies. In the past, major efforts have been made for the capture of CO_2 emission from power plants and industries, concentrate the stream, and sequester it underground or under the seawater. Methods were also adopted to add pre- and post-combustion or oxy-combustion treatments to reduce CO_2 emission. The strategies for CO_2 concentration, transport, and storage are discussed in detail in Chapter 2. As mentioned before, while effective, this strategy turned out to be expensive particularly for the treatment of carbon dioxide coming from the power industry. Since currently there is a wide gap between CO_2 emission and CO_2 utilization, CO_2 storage is still important in order to avoid CO_2 emission in the environment.

In recent years, CO_2 utilization strategy (CCUS) has become very important. CO_2 can be utilized either directly or it can be converted to useful chemicals and fuels by different methods of conversion. Various applications for direct utilization of CO_2 streams are described in Chapter 2. Indirect utilization of CO_2 is more complex, and several strategies that are currently being examined are described in Chapters 3–7. As mentioned before, for both direct and indirect utilizations of CO_2, it is important to utilize carbon dioxide emission locally, wherever possible, in order to avoid expensive transportation.

Chapter 3 examines the viability of mineral carbonation process. In recent years, carbon dioxide is converted to carbonates by mineral carbonation of natural and synthetic Ca- and Mg-bearing materials and industrial wastes containing calcium, magnesium, and silicate oxides, some of which can be used to produce components of construction materials. Mineralization involves the reaction of minerals (mostly calcium or magnesium silicates) with CO_2 to give inert carbonates. The reaction to form carbonates itself requires no energy inputs and actually releases heat, although significant energy is typically required to generate the requisite feed minerals. When possible, the use of industrial waste is favored. The current bottleneck, however, for viable mineral carbonation processes on an industrial scale is the reaction rate of carbonation. The chapter also examines the role of mineral carbonation in the productions of synthetic aggregates, cement, and concrete for construction industry. New formulations of materials such as concrete will require testing and property validation before being accepted by users and regulators for the market.

Chapter 4 examines various types of biological treatments to convert carbon dioxide into useful biochemicals, biomaterials, and biofuels. Carbon dioxide can be biologically converted to biofuels by green algae, microalgae, cyanobacteria, chemolithotrophs, and bio-electrochemical systems. The chapter also discusses anaerobic and gas fermentations to convert CO_2 to syngas. Six major pathways for CO_2 fixation and conversion are also discussed. Several factors have expanded the repertoire of biobased products that can be synthesized directly from CO_2, including the large number of CO_2-utilizing microorganisms, genetic modification of microorganisms, and tailoring enzymatic/protein properties through protein engineering. Biological utilization has a large range of potential uses in the development of commercial products, including various biofuels, chemicals, and fertilizers. However, biological utilization rates and scalability remain challenges with significant potentials for the future.

Chapter 5 examines the role of heterogeneous catalysis for thermal and electrochemical conversions of carbon dioxide to useful chemical and fuels. The use of fuel cells as part of CCUS strategy is also discussed. Carbon dioxide can be chemically converted to alcohols, acids, hydrocarbons, polymer precursors, carbon monoxide, and carbon nanotubes. Carbon dioxide can also be converted to liquid fuels, fertilizer, polymers, and secondary chemicals. These conversions require catalysts to overcome kinetic barriers. Because carbon in CO_2 is in its most highly oxidized form, many of the resulting reactions are reductions, either through the addition of hydrogen or electrons. Catalysts are critical not only for making the transformation possible, but also for reducing the energy inputs to (ideally) the minimum amount dictated by the thermodynamics of the transformation, and discovery of appropriate catalysts and development of energy-efficient processes are current bottlenecks. However, combination of C1 and C2 carbon productions from CO_2 followed by the Fischer-Tropsch synthesis provides a fertile avenue for productions of higher hydrocarbons, alcohols, acids, and fuels. The chapter also examines in detail electro-catalysis process to produce a variety of chemicals like CO, methane, methanol, formaldehyde, formic acid, ethylene, ethanol, propanol, etc. New immobilized catalysts on the electrodes are being constantly evaluated. Finally, the chapter also examines the role of high-temperature fuel cells (particularly, direct carbon fuel cell, SOEC, and molten carbonate fuel cell) for conversion of carbon and CO_2 to power, heat, hydrogen, and other valuable chemicals.

Chapter 6 examines the role of solar energy-based thermal, photo-thermal, and photocatalytic conversions of CO_2 to valuable chemicals and fuels. Homogeneous, heterogeneous, and immobilized homogeneous catalysts are considered. The use of renewable fuel like solar energy takes special meaning because it makes CO_2 conversion as a part of green technology. Photo-electro catalysis and novel PV/EC concept are also examined in detail. The latter approach has a strong commercial potential. While many ideas presented in this chapter are not yet successful at large scale, they present great potential for future success.

Finally, Chapter 7 presents the rapidly growing concept of plasma-activated catalysis for CO_2 conversion to many valuable fuels and chemicals. The chapter examines the roles of nonthermal plasma like DBD and warm plasmas like MW and GA and GAP on plasma-activated catalysis for four sets of reactions; CO_2 dissociation to produce CO; CO_2 hydrogenation to produce CO by reverse water gas shift reaction, to produce methane or to produce methanol; and artificial photosynthesis ($CO_2 + H_2O$) to produce syngas, and dry reforming of methane to produce a variety of chemicals and fuels. The effectiveness of plasma-activated catalysis is examined in detail. While this approach is not quite ready for commercialization, it offers many positive features like flexibility, ease of operation at the industrial scale, convenient operating conditions, ease of scaleup, etc. The approach has a bright potential for converting CO_2 to useful products.

A number of studies have attempted to estimate the market for carbon utilization products, and one published study has reported that the future market could be as high as $800 billion by 2030, utilizing 7 billion metric tons of carbon dioxide per year (CO_2 Sciences, Inc., 2016). Carbon dioxide is fundamentally a low-value, low-energy waste gas, which is often available in large quantities in single locations. If treatment strategies outlined above are successful, not only CO_2 emission to the environment will be prevented, but CO_2 can also become a very valuable raw material for chemicals, fuels, and power.

REFERENCES

1. National Academies of Sciences, Engineering, and Medicine 2019. *Gaseous Carbon Waste Streams Utilization: Status and Research Needs.* Washington, DC: The National Academies Press. https://doi.org/10.17226/25232.
2. IPCC 2007. *Climate Change 2007: Mitigation of Climate Change. Contribution of Working Group III to the Fourth Assessment Report of the Intergovernmental Panel on Climate Change* [B. Metz, O.R. Davidson, P.R. Bosch, R. Dave, L.A. Meyer (eds.)]. Cambridge and New York: Cambridge University Press, 863 pp. ISBN: 9781139468640.
3. IPCC 2014. *Climate Change 2014: Mitigation of Climate Change. Contribution of Working Group III to the Fifth Assessment Report of the Intergovernmental Panel on Climate Change* [O. Edenhofer, R. Pichs-Madruga, Y. Sokona, E. Farahani, S. Kadner, K. Seyboth, A. Adler, I. Baum, S. Brunner, P. Eickemeier, B. Kriemann, J. Savolainen, S. Schlömer, C. von Stechow, T. Zwickel, J. C. Minx (eds.)]. Cambridge and New York: Cambridge University Press.
4. Boden, T. A., G. Marland, and R. J. Andres 2017. *Global, Regional, and National Fossil-Fuel CO_2 Emissions.* Oak Ridge, TN: Carbon Dioxide Information Analysis Center, Oak Ridge National Laboratory, U.S. Department of Energy, doi: 10.3334/CDIAC/00001_V2017.
5. "Inventory of U.S. Greenhouse gas emissions and sinks, 1990–2016", EPA report 430-R-18-003. Available at https://www.epa.gov/ghgemissions/inventory-us-greenhouse-gas-emissions-and-sinks. Washington, D.C.: EPA (accessed April 12, 2018).
6. Shah, Y. T. 2021. *Hybrid Power.* New York, NY: CRC Press, Taylor and Francis.
7. Shah, Y. T. 2021. *Hybrid Energy Systems-Strategy for Industrial Decarbonization.* New York, NY: CRC Press, Taylor and Francis.
8. Nizio, M. 2016. Plasma catalytic process for CO_2 methanation. Catalysis. Université Pierre et Marie Curie - Paris VI. English. NNT: 2016PA066607.
9. Samimi, A., and S. Zarinabadi 2012. Reduction of greenhouse gases emission and effect on environment. *Journal of American Science* 8:1011–1015.
10. Chapter 8 energy-related carbon dioxide emissions. U.S. Energy Information Administration/International Energy Outlook 2010 (accessed February 18, 2016).
11. Sullivan, B.P., K. Krist, and H.E. Guard 1993. *Electrochemical and Electrocatalytic Reactions of Carbon Dioxide.* Amsterdam: Elsevier Science Publishers.

12. Fechete, I., and J. Vedrine 2015. Nanoporous materials as new engineered catalysts for the synthesis of green fuels. *Molecules* 20(4):5638.
13. Aresta, M., A. Dibenedetto, and A. Angelini 2014. Catalysis for the valorization of exhaust carbon: From CO_2 to chemicals, materials, and fuels. Technological use of CO_2. *Chemical Reviews* 114(3):1709–1742.
14. Xiaoding, X., and J. A. Moulijn 1996. Mitigation of CO_2 by chemical conversion: Plausible chemical reactions and promising products. *Energy & Fuels* 10(2):305–325.
15. Ma, J., et al. 2009. A short review of catalysis for CO_2 conversion. *Catalysis Today* 148(3–4):221–231.
16. Behr, A. 1985. The synthesis of organic chemicals by catalytic reactions of carbon dioxide. *Bulletin des Sociétés Chimiques Belges* 94(9):671–683.
17. Zhang, G., et al. 2013. A comparison of Ni/SiC and Ni/Al_2O_3 catalyzed total methanation for production of synthetic natural gas. *Applied Catalysis A: General* 462–463:75–81.
18. Zangeneh, F. T., S. Sahebdelfar, and M. T. Ravanchi 2011. Conversion of carbon dioxide to valuable petrochemicals: An approach to clean development mechanism. *Journal of Natural Gas Chemistry* 20(3):219–231.
19. Songolzadeh, M., M. Soleimani, M. T. Ravanchi, and R. Songolzadeh 2014. Carbon dioxide separation from flue gases: A technological review emphasizing reduction in greenhouse gas emissions. *The Scientific World Journal* 2014:34, doi: 10.1155/2014/828131.
20. Psarras, P. C., S. Comello, P. Bains, P. Charoensawadpong, S. Reichelstein, and J. Wilcox 2017. Carbon capture and utilization in the industrial sector. *Environmental Science & Technology* 51(19):11440–11449, doi: 10.1021/acs.est.7b01723.
21. Bobeck, J., J. Peace, and F. M. Ahmad, Carbon utilization— A vital and effective pathway for decarbonization a report a report by *Center for Climate and Energy Solutions.*
22. IEA (International Energy Agency) 2017. *World Energy Outlook, 2017.* Available at https://www.iea.org/ weo2017/ (accessed October 10, 2018).
23. Supekar, S. D., and S. J. Skerlos 2014. Market-driven emissions from recovery of carbon dioxide gas. *Environmental Science & Technology* 48(24):14615–14623, doi: 10.1021/es503485z.
24. Buchanan, M. 2017. Accelerating breakthrough innovation in carbon capture, utilization and storage, *A Report of the Carbon Capture, Utilization and Storage Experts' Workshop*, September 26–28, Houston, TX, Department of Energy, Washington, D.C.
25. Arakawa, H., et al. 2001. Catalysis research of relevance to carbon management: Progress, challenges, and opportunities. *Chemical Reviews* 101(4):953–996.

2 Methods for Carbon Dioxide Capture/Concentrate, Transport/Storage, and Direct Utilization

2.1 INTRODUCTION

Energy has been a need of flourishing civilization, but utilization of conventional energy resources based on fossil fuels is creating environmental problems such as emissions of greenhouse gases, particulate matter, smoke, etc. [1–6]. About 82% of the energy required all over the world is generated from fossil fuels [7] through various modes. This is resulting in the production of carbon dioxide (CO_2) and its release into the environment. The CO_2 is a greenhouse gas (GHG) primarily responsible for global warming. This chapter is aimed for a brief review of various physical and chemical technologies for CO_2 capture, concentration, transport, and sequestration mechanisms. The chapter also examines the direct utilization of CO_2 streams. Different methods of carbon separation and concentration techniques including absorption into liquid, gas-phase separation, adsorption on solid, hybrid processes such as adsorption-membrane systems, cryogenic separation, etc. are briefly discussed. Major purpose of these technologies is to concentrate carbon dioxide streams at a level (generally over 90% concentration) so that it can be economically transported to the sequestration sites or can be directly or indirectly utilized.

Carbon capture and sequestration, known as CCS, is one of the technological steps toward the clean energy generation. The CCS refers to the process of capturing CO_2 at its source, concentrating it, and storing it before its release to the atmosphere. The worldwide efforts on CCS were started in March 1992 at Amsterdam where many scientists and engineers from various countries gathered in the First International Conference and discussed about Carbon Dioxide Removal. It was established that clean energy can be produced by either removal of carbon from the fuel itself or removal from post-combustion exhaust gases [8]. The CCS methodology can reduce or even eliminate the CO_2 emission to the atmosphere, and thereby clean energy can be produced [9,10]. While CCS does not make use of concentrated CO_2 stream, it is a defensive and cost strategy that allows the use of fossil fuels without harming the environment. Ultimately this strategy needs to be changed to CCUS strategy where

DOI: 10.1201/9781003229575-2

TABLE 2.1
Amount of CO_2 in Flue Gases of Power Plants [6]

Method	Concentration of CO_2 (Vol. %)
Coal fired boiler	14
Natural gas fired boiler	8
Natural gas combined cycle	4
Natural gas partial oxidation	40
Coal oxygen combustion	>80

CO_2 stream can be used to make valuable products before making a decision of its storage or sequestration [11].

As shown in Chapter 1, the amounts of CO_2 generation and addition in environment by power plants are very large. Furthermore, as shown in Table 2.1, different fuel usage and different methods of combustion generate different levels of CO_2 in the flue gas. Typically, a coal-fired power plant with a capacity of 1,000 MWe generates approximately 30,000 tons of CO_2 per day [12,13]. The CO_2 released by power plants can be mitigated by CCS techniques, but the cost is quite high [14]. An integrated CCS system will include the three main steps: (a) capturing and separating the CO_2, (b) compression and transportation of the captured CO_2 to the sequestration site, and (c) sequestration of CO_2 in geological reservoirs or the oceans. The main options for sequestration include (a) use of deep saline reservoirs, (b) injection into the deep ocean [15], and (c) sequestration into tar-sand reservoirs. Another better option for sequestration involves the injection of CO_2 into hydrocarbon deposits to enhance oil recovery (EOR), shale gas recovery (EGR), or production of coal-bed methane (ECBM). This method uses CO_2 stream to produce fuels along with its sequestration. This is a part of direct utilization of CO_2 as described later in this chapter. The deep saline formations (100–1,000 GtC) and oceans (1,000 GtC) are having the highest world sink capabilities of CO_2 disposal options [16].

The CCS can be implemented in two ways: (a) pre-combustion CCS process, where carbon is captured during fuel processing itself, before combustion of fuel for generation of energy and (b) post-combustion CCS process, where separation of CO_2 from combustion products, i.e., flue gases, is done after combustion of the fuel. Other option is oxy-combustion where carbon is burned with pure oxygen to generate pure CO_2 stream. This, however, requires a very expensive separation of O_2 from air. Removing CO_2 from the atmosphere by enhancing its uptake in soils and vegetation (e.g., afforestation) or in the ocean (e.g., iron fertilization) is yet another form of sequestration. Pre-combustion, post-combustion, and oxy-combustion are graphically illustrated in Figure 2.1.

Pre-combustion is mainly applied to coal-gasification plants, while post-combustion and oxy-fuel combustion can be applied to both coal- and gas-fired plants. Post-combustion technology is currently the most mature process for CO_2 capture [17]. On the cost side, Gibbins and Chalmers [18] compared the three technologies for both gas- and coal-fired plants. They reported that for coal-fired plants the pre-combustion technology presented the lowest cost per ton of CO_2 avoided,

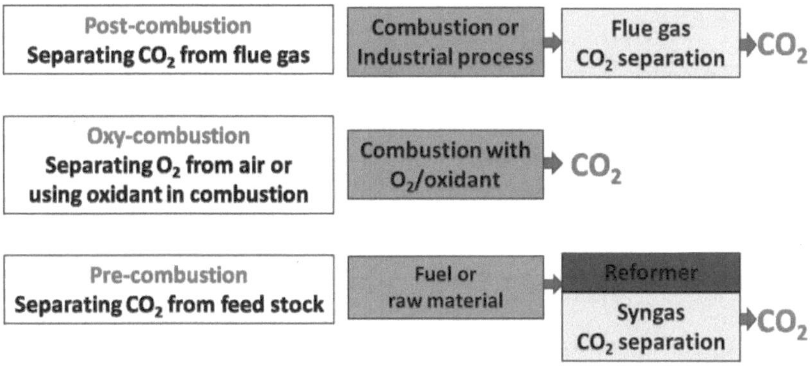

FIGURE 2.1 Three pathways for CO_2 capture. (Image courtesy of Gassnova SF [1].)

while the post-combustion and oxy-fuel technologies are of similar costs. However, for gas-fired plants, the cost per ton of CO_2 avoided for the post-combustion capture was almost 50% lower than the other two capture technologies. Moreover, the post-combustion CO_2 capture is normally the least efficient option, with an energy penalty of about 8% and 6% for the coal- and gas-fired plants, respectively [19].

The main advantage of post-combustion capture is its easy integration capability with the existing power plants, but the partial pressure and concentration of CO_2 are very low in the flue gases. For transportation and storage of CO_2, a minimum concentration should be reached. The required extra energy and extra costs of carbon capture to attain a minimum required concentration are significantly high.

Pre-combustion carbon capture is mostly used in process industries. There are also full-scale CCS plants in some industries which use this method [20]. The amount of CO_2 is much higher in the gas mixture in this process than the conventional flue gas mixture. Due to the higher pressure and lower gas volume, less energy is required in this process compared to post-combustion capture, but still, the energy penalty is high. Pre-combustion is mainly used in integrated gasification combined cycle technology. This technology demands a huge auxiliary system for smooth operation. Therefore, the capital cost of this system is too high compared to other systems.

On the other hand, carbon capture processes without requiring separation are comparatively novel in power generation. There is no full-scale operational plant based on these processes. There are some pilot-scale operation and some subscale demonstration plants under development using oxy-fuel combustion [4,21,22]. The most promising step regarding oxy-fuel combustion is the 50 MWth demonstration power plant built in Texas by Net Power using the concept of the Allam cycle. It ensures near zero emission. This method has some other advantages like reduction in equipment size, compatibility with various kinds of coals, and no need for an onsite chemical plant [21].

The process, however, requires a large amount of high-purity oxygen. Therefore, an energy-intensive ASU is needed for oxygen production. Membrane-based technology for air separation may compete with cryogenic ASU through a higher degree of integration into the power cycle [23]. Due to this ASU and CO_2 compression unit

TABLE 2.2
Efficiency Comparison of Power Generation with Different Carbon Capture Processes [6]

Fuel Type	Process	Net Efficiency	Net Power (MW)
Coal (bituminous)	Without carbon capture	44	758
	Pre-combustion	31.5	676
	Post-combustion	34.8	666
	Oxy-combustion	35.4	532
	Oxy-combustion (Allam cycle)	51	226
Natural gas	Without carbon capture	55.6	776
	Pre-combustion	41.5	690
	Post-combustion	47.4	662
	Oxy-combustion	44.7	440
	Oxy-combustion (Allam cycle)	59	303

used in this process, net power output decreases significantly. Along with these, there are some technical uncertainties that demand more research to understand the full-scale operation. However, since no extra cost is required for CO_2 separation, this process remains a promising one to produce electricity at a lower cost while confirming near zero emission.

A comparison of the thermal efficiency of power plants with different CO_2 capture processes is provided in Table 2.2. The efficiencies shown in the table are based on the lower heating value of the fuel. Bituminous coal is considered for coal-based power plants due to its extensive use in power production [4]. The Selexol process is taken into consideration for pre-combustion carbon capture in an IGCC GE-type gasifier.

When coal is used as a fuel, post-combustion and oxy-combustion carbon captures show an almost similar drop in efficiency. An interesting observation in this comparison is the efficiency of the Allam cycle. The targeted efficiency of the Allam cycle is almost the same as the reference power plant efficiency without capture. If this cycle can be implemented commercially at a larger scale, the overall power generation efficiency will increase while ensuring total carbon capture.

When natural gas is used as fuel, the pre-combustion carbon capture shows a 14% drop in the efficiency from the reference power plant, whereas the post-combustion carbon capture shows an 8% drop. The traditional oxy-combustion process exhibits an efficiency of 44.7%. The Allam cycle shows an extraordinary performance whose efficiency happens to be over 3% points higher than that of the reference combined cycle without CO_2 capture. From the efficiency comparison of Table 2.2, it may be concluded that the Allam cycle is expected to be the leading technologies in the near future for fossil fuel-based power generation. The 50 MWth Allam cycle provides the basis for deployment of large-scale facilities. Currently, 300 MW natural gas-fired plants are under development.

Conventional carbon capture process results in the reduction of efficiency. More fuel is burnt per unit of electricity production due to this inefficiency which leads to more production of CO_2. Also, the processes used for capturing carbon dioxide may affect the environment in different ways other than the direct emission of CO_2. For example, different substances used for separating and capturing CO_2 may have undesired effect on the human body and environment. Using a solid sorbent covered with coating was experimented to reduce the formation of dust from the substance [24]. This could also reduce the capacity of the substance to capture carbon dioxide. Also, stripping of organic solvent from membranes and sorbents is suggested to prevent undesired odor. Before employing carbon capture, it should be ensured that reducing CO_2 is not being achieved at the cost of other environmental impacts.

Life cycle assessment (LCA) of the plants is necessary to properly understand the environmental impacts of the carbon capture methods. Schreiber et al. [25] used the LCA methodology for post-combustion carbon capture using MEA whose impact on the environment and human health was investigated for five power plants. The global warming potential (GWP), human toxicity potential (HTP), acidification potential (AP), photo oxidant formation potential, and eutrophication potential (EP) were considered as impact categories. As expected, GWP was much lower with MEA compared to the power plants without capture whereas HTP was three times higher with MEA plants. Schreiber et al. [25] concluded that upstream and downstream processes such as emissions from fuel and material supply, waste disposal, and wastewater treatment influence the environmental impact measures for power plants with carbon capture. Viebahn et al. [26] revealed about a 40% increase in AP, EP, and HTP when post-combustion carbon capture was implemented in a power plant.

A similar result was found by Veltman et al. [27]. They showed that a power plant with post-combustion capture yields a tenfold increase in toxic impacts on freshwater compared to a plant without capture. Impacts on other categories were negligible. Degradation of MEA resulted in the emission of ammonia, acetaldehyde, and formaldehyde. Cuellar-Franca et al. [28] compared life cycle environmental impacts of carbon capture and storage with carbon capture and utilization. GWP with utilization was much greater than that with storage. The highest reduction of GWP was found for pulverized coal and IGCC plants employing the oxy-fuel capture method as well as combined cycle gas turbine plants equipped with a post-combustion capture technology.

Pehnt et al. [29] showed that a conventional power plant operating on coal with post-combustion carbon capture would result in an increase in the environmental impact in almost all categories except GWP. Solvent degradation and energy penalty due to CO_2 capture process are the main reasons for this increase. Pre-combustion capture showed a decrease in all the environmental impact categories compared to a conventional power plant. They identified oxy-fuel combustion as the most potential process to reduce all the environmental impact categories if co-capture of other pollutants can be achieved.

Nie et al. [30] investigated comparative environmental impacts of post-combustion and oxy-fuel combustion carbon captures. Their analysis showed that almost all environmental impact categories except GWP would increase with post-combustion carbon capture. The same is true for oxy-fuel combustion except GWP, AP, and EP. However, the amount of increase of these impact categories was found to be less in oxy-fuel combustion compared to the post-combustion carbon capture.

2.2 PHYSICAL AND CHEMICAL SEPARATIONS AND CAPTURE/CONCENTRATE TECHNOLOGIES

No matter what process is used, as described above, to reduce carbon emission during combustion process for power generation, it is important to concentrate carbon dioxide stream with low concentration to high concentration before it can be transported, utilized, and/or sequestered. The physical and chemical separation technologies used for this purpose are:

1. Membrane separation process
2. Adsorption-based systems
3. Solvent-based scrubbing process
4. Hydrate-based separation
5. Cryogenic separation process

In this section, we briefly evaluate each of these technologies.

2.2.1 Membrane Separation Process

The membrane separation process contains a specially designed membrane sieve that separates molecules based on their molecular size. Several demonstrations of CO_2 separation have been performed, notably the separation of CO_2 from CH_4, CO_2 from air and CO_2, CO, H_2S, and H_2O from a mixture of gases [31,32]. Use of membranes for removing CO_2 provides versatility, adaptability, environmentally friendly process, easy operation, and requiring less space and light in weight. They are also cost-effective (once developed on a commercial level), produce minimal waste, and can be adapted to a variety of carbon sequestration schemes.

Membranes in the application are polymeric gas permeation membranes (PGPM), facilitated transport membranes (FTM), hollow fiber gas–liquid membrane contactors, inorganic membranes, mixed matrix membranes (MMM), and ceramic membranes. Low manufacturing cost of polymeric membranes is of great interest for industrial applications, but they generally exhibit selectivity about 5–10-fold lower than those of inorganic membranes. The inorganic membranes are useful for CO_2 separation processes at high temperatures due to their robust thermal, chemical, and mechanical stability [32]. Polymer membranes with better plasticization suppression properties are useful for CO_2 separation [31]. Hasebe et al. [33] fabricated high gas permeable separation membranes containing silica nanoparticles, a type of MMM. They reported that gas transport channel formed by the nanoparticles can enhance the gas permeability without significant decrease in gas selectivity and the syntheses of silica nanoparticles are cost-effective. The development of ceramic and metallic membranes [34] and polymeric membranes [35] for membrane diffusion could produce membranes significantly more efficient for CO_2 separation than liquid absorption processes. Brunetti et al. [36] conducted a general review on current CO_2 separation technology using membranes and compared other separation technologies such as adsorption and cryogenic. They pointed out that the performance of a membrane system is strongly affected by the flue gas conditions such as low CO_2

concentration and pressure. A two-stage membrane-based process with boiler air feed as a sweep stream to increase the CO_2 concentration for CO_2 capture was studied and optimized by Mat and Lipscomb [37]. With the use of the facilitated transport membrane, the CO_2 separation is feasible, even for low CO_2 concentration (about 10% in flue gas) and it is possible to achieve more than 90% CO_2 recovery and with a purity in the permeate above 90% CO_2 [38].

A. **Polymeric membranes**

The applications of polymeric membranes for large CO_2 capture processes in the power and industrial sector have not yet matured. Key challenges are related to the low partial pressure of CO_2 and the large scale required for flue gas treatment. For membranes to be cost-effective, further innovations in process design and membrane materials are thus needed. In order to improve permeability and selectivity for CO_2, novel membranes such as facilitated transport membranes (FTMs)—that incorporate a "carrier component" in the structure itself—are attractive at the low CO_2 partial pressure experienced in post-combustion applications.

Polymeric membrane technology has been demonstrated at relatively large scale at the National Carbon Capture Center (NCCC) in the US state of Alabama at a 20 ton per day (tpd) scale. It employs the Membrane Technology and Research, Inc. (MTR) membrane (see Figure 2.2). A similar 6 tpd CO_2 capture demonstration has been executed at the NCCC using the Air Liquide cold membrane approach [39]. Moreover, the performance of FTMs has been evaluated at the Norcem cement factory in Norway employing hollow fiber membrane modules with 18 m^2 of membrane area [40]. The test results indicated 70 mol% CO_2 purity can be easily achieved in a single stage. Polymeric membranes also present good stability—they have been exposed to high concentrations of SO_2 and NOx for a long period of time without significant performance change.

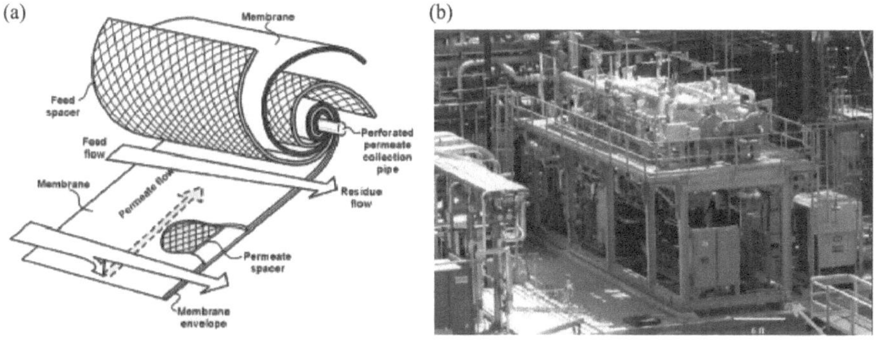

FIGURE 2.2 (a) Spiral-wound membrane module and (b) installed 20 ton per day membrane demonstration facility at the National Carbon Capture Center in the United States. ((a) Image courtesy of National Energy Technology Laboratory; (b) Image courtesy of T. Merkel and B. Freeman, Membrane Technology and Research, Inc. [1].)

B. **Inorganic membrane systems**

Inorganic membranes are mainly targeted for process integration at high temperatures; however, this increases the potential for reaction with gas components at the surface and for diffusion processes inside the material. This may lead to the reduction of membrane lifetime.

A survey of integration possibilities for high-temperature membranes in power generation cycles with CO_2 capture is given by Bredesen et al. [41]. In pre-combustion CO_2 capture schemes, ion-conducting oxygen transport membranes (OTMs) can be used for synthesis gas production either by partial oxidation of natural gas or by gasification of coal or biomass. Air Products and Praxair have developed and demonstrated this membrane technology in units designed to produce up to 100 tons/day of O_2 and 1 MW of net power. Smaller-scale experiments have demonstrated high-purity O_2 production for over 15,000 hours.

The dense metal H_2-selective membrane technology allow production of CO_2-free hydrogen from syngas. Recent efforts are focused on the development and experimental verification of membranes employing thinner palladium layers (i.e., <5 microns). This technology has been used for coal-derived syngas at the University of North Dakota Energy and Environmental Research Center [42], and at the NCCC in Alabama [43] under adverse industrial conditions and for H_2 separation from a syngas-side stream of the Statoil Methanol Plant at Tjeldbergodden, Norway [44,45].

C. **Ceramic membranes**

This membrane type is at an early development stage compared with the membranes described above. Research directions have been to combine metal or oxide electron conductors with proton-conducting oxides to obtain mixed-conducting membrane materials. Other directions have been to develop a single-phase mixed conductor (e.g., $BaCeO_3$-based). More research is needed related to both the transport properties and the stability of the proton-conducting membrane concepts. As alternatives to H_2 separation, high-temperature ceramic-carbonate dual-phase membranes can be applied to separate CO_2 from the other gas species at temperatures above 400°C. Dual-phase membranes may be applied in both post- and pre-combustion schemes [46], but further work is needed to demonstrate their potential.

According to DOE [1], in order to improve membrane technology, two research challenges of particular interest:

1. Transport phenomena in new membrane materials
2. Fabrication and use of the novel membrane systems in effective process design

Next-generation materials such as thermally rearranged polymers (TRs); polymers of intrinsic microporosity (PIM); mixed matrix membranes (MMMs) employing MOFs, zeolites, or other nanoparticles; supported ionic liquid membranes; carbon molecular sieve (CMS) membranes; and FTMs have shown potential for enhanced performance related to selectivity and permeance. However, transport phenomena through these novel materials are less well understood. As pointed out by DOE report, an understanding of transport processes occurring at the interface is very important for designing high-performance membranes. In addition to understanding transport phenomena, the challenge is to use these novel membrane materials in effective membranes and membrane modules. The large-scale fabrication of a membrane consisting of a dense, thin layer of next-generation material on a support has many challenges. Membrane module design should be optimized for using these next-generation, high-flux, thin membranes on optimized support structures [1].

2.2.2 Adsorbent-Based Systems

Adsorption is an attractive technology for a number of reasons. It can be retrofitted to any power plant should the adsorption column be optimized to ensure an acceptable footprint and cost. In addition, it can cover a wide range of temperature and pressure conditions so that low-, medium-, and high-temperature adsorbents can be used and adsorbents for both pre- and post-combustion settings can be designed. Another strength of the adsorption is the potentially minimal environmental footprint vis-à-vis amine-based solvents, which tend to decompose and form toxic and/or corrosive compounds. The use of waste materials as adsorbents could potentially enhance the sustainability of the process.

The adsorbent beds are regenerated, that is, release of adsorbate, by pressure swing, temperature swing, and washing methods [47]. The solid adsorbents are classified into amine-based (such as silica gels, activated carbon, tetraethylene pentaamine), alkali (earth) metal-based adsorbent (such as CaO, MgO/ZrO_2, and MgO/Al_2O_3), and alkali metal carbonate solid sorbents (such as Na_2CO_3 and K_2CO_3, MgO, ZrO_2, SiO_2, Al_2O_3, TiO_2, CaO, and zeolites). These three types of adsorbents have different adsorption environment conditions and CO_2 capture capacity. The amine-based adsorbents operate at −20°C to 75°C in the absence of water vapor and at a pressure of 1 bar with CO_2 capturing capacity of 4.3 mmol/g. The alkali metal-based adsorbents operate at high temperatures between 600°C and 650°C in the absence of water vapor with CO_2 capturing capacity of 1.39 mmol/g. Finally, alkali metal carbonate operates at low temperatures with water vapor and has a CO_2 capturing capacity of 2.49 mmol/g [48]. The carbonate systems are based on the ability of a soluble carbonate to react with CO_2 to form a bi-carbonate, which when heated releases CO_2 and reverts to a carbonate.

Chemically adsorbed materials offer the potential to enable higher selectivity and promotes a higher capacity of adsorbed CO_2 at relatively low pressures compared to physically adsorbed materials. However, they also require more energy for regeneration. The adsorbed CO_2 can be recovered by swinging the pressure (PSA), temperature (TSA), or sometimes by applying vacuum (VSA). PSA is a commercially available technology for CO_2 recovery from power plants that can have efficiency higher than 85% [49,50]. In this process, pressure is used to adsorb and desorb CO_2. In TSA, the adsorbed CO_2 will be released by increasing the system temperature using hot air or steam injection. The regeneration time is normally longer than PSA, but CO_2 purity higher than 95% and recovery higher than 80% can be achieved [51]. Operating cost of a specific TSA process was estimated to be of the order of 80–150 US$/ton CO_2 captured [52]. The adsorption approach relies on advanced materials with appropriate nanoscale features. Advances in design based on the molecular building block approach, coupled with computer simulations and better means to probe materials in situ, have led to the ability to molecularly design new CO_2 capture materials.

In recent years, new porous adsorbent materials like metal-organic frameworks (MOFs), covalent organic frameworks, and several other classes of porous polymer materials are examined [53]. A common feature of these materials is tunable micropores (<2 nm pores) based on modular synthetic routes and the ability to probe structure at molecular levels [54,55]. New inventions include MOFs with amine-appended micropores for capture in wet gas [55]; MOFs with fluorinated pores that achieve selective CO_2 sorption through a combination of thermodynamic and kinetic means [56]; MOFs that use a unique mechanism to bond CO_2 giving large capacity changes over small temperature ranges [57]; and solids capable of physisorbing CO_2 in the presence of water vapor [58,59]. The effective deployment of practical adsorption-based technologies depends on the development of made-to-order adsorbents expressing mutually two necessary requisites: (a) high selectivity/affinity for CO_2 and (b) excellent chemical stability in the presence of impurities in the flue gas, such as H_2O, SOx, and NOx. A significant stride in the field of atomistic computer modeling of porous sorbents has allowed a precise picture of the interaction between the adsorbent and CO_2 at the molecular level which can accurately predict a material's adsorption properties. They suggest the need for the discovery of new adsorbents offering in parallel the requisite chemical stability (wet conditions and tolerance of impurities), high selectivity, high capacity (8–15 wt%), and moderate heat of adsorption in the physisorption range of 35–54 kJ/mol.

2.2.2.1 Chemical Looping Systems

While chemical looping is a part of solid adsorption method, it is often separated. There are two types of chemical looping systems used in practice as shown in Figure 2.3.

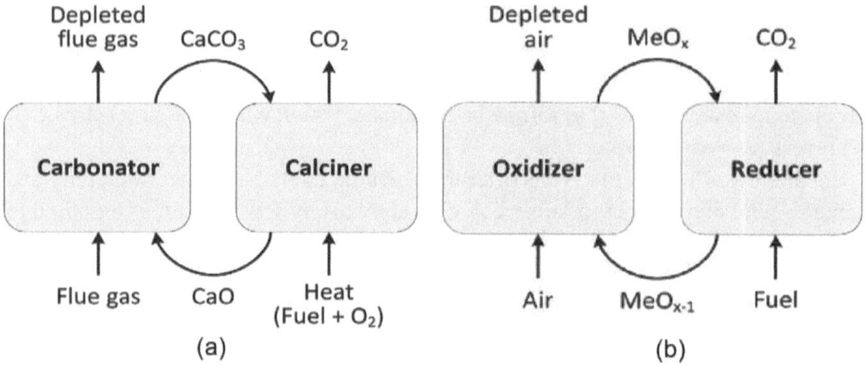

FIGURE 2.3 The two looping systems: (a) the calcium looping system and (b) the chemical looping combustion system. (Image courtesy of Jochen Ströhle, Technische Universität Darmstadt [1].)

As shown in Figure 2.3a, the calcium looping process is based on the chemical absorption of CO_2 by a metal oxide to form a solid carbonate. The carbonate is then regenerated, releasing CO_2 at increased temperature. This cycling process is also called "carbonate looping." Calcium oxide (CaO) is the most promising base material, since limestone is abundant worldwide at a very low cost. (In addition, potassium- and magnesium-based materials and hydrotalcites, zirconates, and silicates have displayed promising developments.) The CaO-based cycle is therefore also called "calcium looping." Calcium looping typically harnesses each reaction direction in a separate fluidized bed reactor, and the calcium compounds are cycled between the two. In the carbonator, CO_2 in a mixed gas stream combines with CaO exothermically, leaving a mostly carbon-free exhaust gas. The solid passes to the calciner, where it is heated to decompose endothermically to generate a pure CO_2 stream. Key to the efficiency of the technology is that the exothermic CO_2 capture step takes place at a temperature at which the liberated heat can be used in an efficient steam cycle to generate electricity and potentially repower existing plants.

The most straightforward variant of calcium looping is based on using an oxy-fired calciner to provide heat for sorbent regeneration, which however requires energy for air separation. The need for oxygen is around half that of a pure oxy-fuel-based power plant. This variation is demonstrated at a semi-industrial scale [60]. Various concepts such as sorbent regeneration by steam hydration, recarbonation, pretreatment of limestone (e.g., doping with HBr), or synthesized sorbent materials have been developed along with novel reactor concepts such as in situ combustion/carbonation and indirect calcination (e.g., through heat pipes) are proposed to increase the long-term cyclic capacity of the fuel and thus reduce the fuel and oxygen consumption in the calciner. There has recently been significant interest in designing calcium looping plants that are flexible in power production. Alongside traditional turn-down and turn-up, medium-term (a few hours') storage of calcined sorbent is being studied. Calcium looping has synergies with industries beyond power generation. It can produce a pure stream of hydrogen alongside a stream of CO_2 during hydrogen manufacture from methane using sorbents. Spent CaO from calcium looping can be used as a raw material for cement production, eliminating a major waste stream while capturing CO_2 emissions from the plant. A similar reuse of sorbents can be achieved in primary steel manufacturing. Other innovative new applications continue to develop.

In chemical looping combustion (Figure 2.3b), a metal oxide is used as an oxygen carrier instead of using pure oxygen directly for the combustion as in the case of oxy-fuel combustion. During the process, the metal oxide is reduced to metal while the fuel is being oxidized to CO_2 and water. The metal is then oxidized in another stage and recycled in the process. Water, the process by-product, can be easily removed by condensation, while pure CO_2 can be obtained without the consumption of energy for separation. Depending on the selected oxygen carrier material and reactor design, the redox chemical looping method can be configured for efficient gasification of carbon-based fuels to syngas—a building block for a wide variety of high-value chemical products—or can be designed for complete carbon-based fuel combustion for power generation applications, i.e., chemical looping combustion. The indirect combustion of the carbon-based fuel with oxygen supplied by the oxygen carrier intermediate from air eliminates the need for an air separation unit to perform oxy-combustion-based carbon capture. The result is a potentially highly efficient process scheme. The DOE milestone plan identifies chemical looping as the technology that has the highest cost-reduction benefit among all the methods, but it is also the one that is farthest from commercialization realization [61].

There are a wide variety of metal oxides that are of low-cost and suitable for this process including Fe_2O_3, NiO, CuO, and Mn_2O_3. The effectiveness of different metal oxides in this process has been studied by various researchers [62–67]. Adánez et al. [65] found that support inert materials can be used to optimize the performance of the metal oxides, but the choice of inert material will depend on the type of metal oxide used. Lyngfelt et al. [68] studied experimentally the feasibility of chemical looping in a boiler with a design of two interconnected fluidized beds. This technology has been reviewed recently by Lyngfelt and Mattisson [69]. Both Lyngfelt and Mattisson [69] and Adanez et al. [70] found that this process is a very promising technology for CO_2 capture. Erlach et al. [71] compared the CO_2 separation of IGCC using pre-combustion with that of chemical looping combustion and found that the net plant efficiency of the latter is 2.8% higher than the former case.

Several large pilot plants, up to the MWth scale, have recently been operated to discern the feasibility of their commercialization for CO_2 capture [72]. One major challenge is to reach full conversion of the fuel in the reducer, particularly when solid fuels are used. Recently, much research has addressed the development of so-called CLOU (chemical looping with oxygen uncoupling) materials, since they have the capability to release molecular oxygen under reducing conditions, which strongly enhances the conversion of the fuel [60]. However, currently available CLOU materials suffer from rather high production costs and only moderate material integrity. New reactor concepts, e.g., with internal or multiple stages, have been proposed for the reducer to improve fuel conversion.

Several overarching challenges for adsorbents and looping technologies include:

1. Design and construct tailor-made materials with desired requisites for specific carbon capture applications.
2. Obtain better understanding of the structure–properties relationship at various levels.
3. Enhance long-term reactivity, recyclability, and robust physical properties of materials within their working cycles.
4. Understand the relationship between material and process integration to produce optimal capture designs for flexible operation.
5. The, oxygen carriers during chemical looping must be morphologically stable over thousands of cycles while maintaining their oxygen capacity and mechanical stability despite particle collision, thermal shock, and gaseous impurities. The relationship between the process design and material design is very important in developing a competitive chemical looping system.

2.2.3 Solvent-Based Scrubbing Process

Solvent-based scrubbing process involves the use of different types of solvents. First-generation technology used MEA, DEA, and related solvents. Second-generation amine scrubbing involves numerous special types of solvents including piperazine and others. Solvent-based scrubbing also includes aqueous ammonia scrubbing, water lean solvents, and multi-phase solvents. The use of these solvents are briefly described below.

2.2.3.1 First Generation of Amine Based Scrubbing

In this process, a liquid sorbent is used to separate the CO_2 from the flue gas. The sorbent can be regenerated through a stripping or regenerative process by heating and/or depressurization. Typical sorbents include monoethanolamine (MEA), diethanolamine (DEA), and potassium carbonate [73]. Among the various aqueous alkanolamines, such as MEA and DEA, Veawab et al. [74] found that MEA is the most efficient one for CO_2 absorption with efficiency over 90%. Subsequently, Aaron et al. [34] conducted a review on various CO_2 capture technologies and concluded that the most promising method for CO_2 capture for CCS is absorption using MEA.

While this process is the most mature method for CO_2 separation [75], it is generally deemed uneconomical as it results in large equipment sizes and high regeneration energy requirements (about 30% of the energy produced) to release the CO_2 from MEA. The regeneration heat energy may be received from the solar heating system. Apart from this, the additives can help to improve the system performance and the design modifications are possible to drop capital costs and increase energy integration [75]. The energy consumption required to regenerate the solvent can be reduced by using ejector technology into post-combustion carbon capture. One other important challenge for the large deployment of this technology for CCS is its potential amine degradation, resulting in solvent loss, equipment corrosion, and generation of volatile degradation compounds [76,77]. Moreover, amine emissions can degrade into nitrosamines and nitramines [78], which are potentially harmful to the human health and the environment.

In order to overcome the limitations of energy-intensive process MEA scrubbing, another technique called reactive hydrothermal liquid-phase densification (rHLPD) is used to solidify monolithic material without using high-temperature kilns. The integration of MEA-based CCS processing and mineral carbonation by using rHLPD technology results in the formation of a mineral (wollastonite $CaSiO_3$), which has a high compressive strength of ~121 MPa. The produced material, similar to Portland cement, can be used as value-added binding material for construction and infrastructure development [79].

2.2.3.2 Second-Generation Amine Scrubbing

Second-generation amines and processes have been developed by more than ten companies and organizations. Two amine scrubbing systems have been commercialized at coal-fired power plants. The MHI solvent KS-1 with an "energy saving process" and the Cansolv solvent DC-103, with lean vapor compression, are operating near design conditions with a reboiler heat duty of 2.3–2.4 GJ/ton CO_2 removed [80], compared with 3.5 GJ/ton [81] for first-generation amine scrubbing.

The thermodynamic efficiency of second-generation amine scrubbing is ~50%. The ideal work requirement to separate CO_2 from coal-fired flue gas and compress it to 150 bar is 110 kwh/ton CO_2. The Cansolv and MHI commercial units are operating with an estimated total electricity burden (including heat) of 220–240 kwh/ton

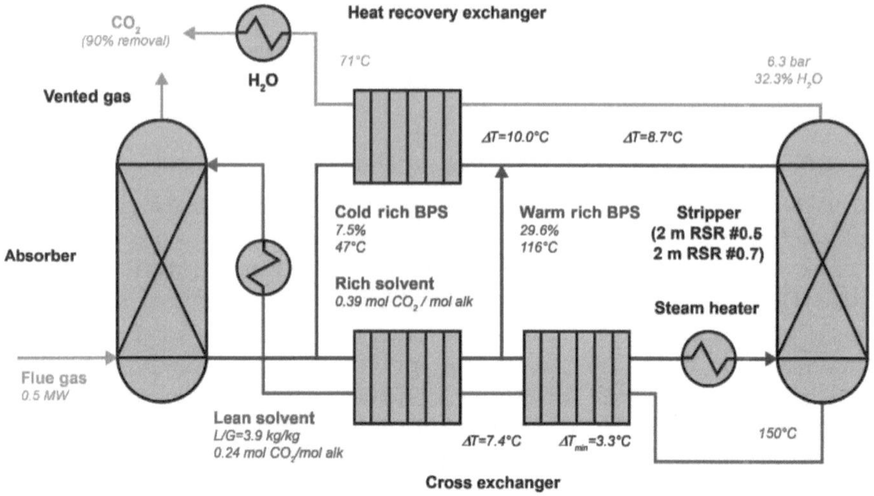

FIGURE 2.4 Example of second-generation solvent process [1].

CO_2 removed. A reversibility analysis of aqueous piperazine (PZ) with an advanced flash stripper (AFS) shows that this typical second-generation system requires 230 kwh/ton (see Figure 2.4). Piperazine has been found to react much faster than MEA, but because it has a larger volatility than MEA, its application in CO_2 absorption is more expensive and is still under development [4]. The irreversibility of this system is distributed among several unit operations, so there are no easy means of improving second-generation amine scrubbing.

In summary, energy improvements have achieved an equivalent electricity requirement of close to 230 kwh/ton CO_2 at >50% thermodynamic efficiency for second-generation amine scrubbing; improvement opportunities are limited by the ideal work requirement. It should be emphasized that, in cases of implementation of CS in energy-intensive industries like cement and steel, solvent regeneration may in some cases make use of excess heat. This shows the significant benefit of heat-driven separation processes. If steam must be derived from a separate boiler, parameters related to primary energy usage or cost would be a more appropriate performance indicator than the equivalent electricity requirement. Remaining challenges to address are oxidation and amine aerosol research to eliminate uncertainty.

2.2.3.3 Aqueous Ammonia Scrubbing of CO_2

The ammonia-based carbon capture technology can be divided into the normal temperature method (15°C–30°C) and the low-temperature method (2°C–10°C). In ammonia-based wet scrubbing of CO_2, the flue gas is passed through aqueous ammonia. The ammonia and its derivatives react with CO_2 via various mechanisms, one of which is the reaction of ammonium bicarbonate. In this mechanism, the lower heat of reaction for amine-based systems results in energy savings. The ammonia-based absorption has a number of other advantages such as the potential for high CO_2 capacity, lack of degradation during absorption/regeneration, tolerance to oxygen in the flue gas, and low cost [74]. Based on the thermodynamic analysis and process simulation, it was found that the equilibrium regeneration energy can be reduced to 1,285 kJ/kg CO_2 and the energy consumption for the NH_3 abatement system is 1,703 kJ/kg CO_2. As this process is operated at room temperature, the additional energy consumption for the cooling of the flue gas and the absorbent can be avoided [82]. The ammonium bicarbonate is used by plants as fertilizer and converts into biomass. The gasification of biomass again would give fuel for energy generation. Therefore, CCS by CO_2 conversion into fertilizer is the most convenient and sustainable process as CO_2 is recycled in the environment and the environment remains carbon neutral.

Post-combustion CO_2 capture (PCC) with solar-assisted chilled-ammonia-based CO_2 capture system in a coal-fired power plant was undertaken for study under different meteorology conditions. It was found from the economic viewpoint that prices of the solar thermal collector and the equipment of the phase change materials (PCM) have clear impacts on the levelized costs of electricity (LCOE) and the cost of CO_2 removed (CCR). The prices of solar thermal collectors vary from location to location resulting in varying cost of PCC system.

2.2.3.4 Water-Lean Solvents

Water-lean solvent systems are solvent processes that retain the chemical selectivity of water-based solvents. The goal is to reduce the energy required to regenerate the solvent by exploiting the lower specific heats of organics compared with that of water. From a time and cost perspective, these third-generation solvents have the potential to use first- and second-generation aqueous amine infrastructure, enabling a potential rapid ascent up the development ladder.

All water-lean solvents use one of the three known CO_2 binding chemistries: carbamate, alkylcarbonate, and azoline-carboxylate (see Figure 2.5). All reported formulations are nonvolatile by design to minimize fugitive emissions and lessen environmental impacts. Water-lean solvents have enthalpies of CO_2 absorption that are comparable to those of aqueous solvents, ranging from −50 to −90 kJ/mol

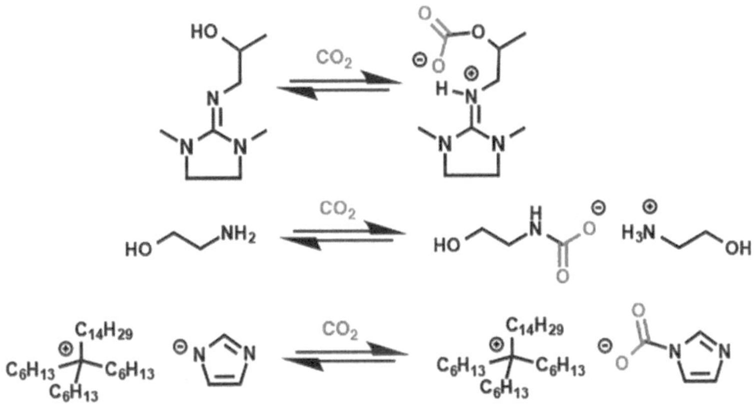

FIGURE 2.5 Binding mechanisms of CO_2. (Image courtesy of Pacific Northwest National Laboratory [1].)

CO_2, indicating similar selectivity and viability for post-combustion CO_2 capture. Water-lean solvents exhibit unique physical and thermodynamic properties—such as physical state (e.g., solid, liquid), contact angle, wettability, viscosity, and volatility, and thermodynamic properties—such as thermal conductance and solvation free energy—that differ substantially from those of aqueous solvents. Some properties may be detrimental to CO_2 capture performance (e.g., higher viscosity and lower thermal conductance), whereas the lower specific heat and higher physical solubility of CO_2 are beneficial. Other unique solvent behaviors include: a higher rate of mass transfer than aqueous solvents, an inverse temperature dependence (a solvent absorbs faster the colder it gets), phase changes (e.g., the ability to concentrate CO_2-containing material to regenerate), and the ability to destabilize the CO_2 carrier (resulting in a reduction in the temperature needed for CO_2 release). In each case, these properties may be exploited to improve process efficiency.

Notable water-lean solvents have reported significant reductions in reboiler duty, from 1.7 to 2.6 GJ/ton CO_2, compared with the US DOE NETL Case 10 baseline (3.5 GJ/ton CO_2) or the performance of second-generation amine scrubbing (2.2–2.4 GJ/ton). The combination of lower reboiler duties and regeneration temperatures below 100°C enable higher plant efficiencies, translating into projections of a 2.1%–7.1% increase in net plant efficiency compared with the NETL Case 10 baseline. The costs of using these solvents can be reduced using second-generation amine infrastructure (e.g., AFSs) and novel configurations designed and optimized for the physical and thermodynamic properties unique to these solvents [1].

Most water-lean solvents have demonstrated an acceptable level of water tolerance, and all solvents report a steady-state loading of water (up to 10 wt%) without a need for exotic water management infrastructure. Water-lean solvents may be less corrosive than aqueous solvents. Some amines and amidines have been shown to act as corrosion inhibitors. Additionally, their lower water content reduces the amount of carbonic acid in solution resulting in less corrosion. The reduced corrosion may enable the use of cheaper alloys of steel in process infrastructure, potentially reducing costs. Water-lean solvents are expected to be roughly as durable against flue gas impurities as their aqueous counterparts, as they both entail the same functional groups, and the amount of the impurities is fixed by the flow rate of flue gas.

2.2.3.5 Multiphase Solvents

Multiphase solvents for CO_2 capture consist of solvent systems capable of forming more than one liquid phase (demixing solvents) or liquid/solid systems (precipitating solvents). Phase change as part of an absorption/desorption cycle opens a range of opportunities for improved performance of the capture process:

- Formation of a high-density CO_2-rich phase so that only a part of the solvent requires regeneration. An extremely concentrated phase with high reactant and CO_2 concentration may enable significantly reduced heat requirements in terms of both sensible heat and steam requirements for CO_2 stripping. The low-density phase can contain practically no CO_2 and is recycled to the absorber, whereas all of the CO_2 is present in the high-density phase.

- Intensified desorption at lower temperatures (<100°C), using low-value heat from waste streams.
- High-pressure desorption due to extreme CO_2 backpressure from the concentrated solvent, enabling a significant reduction in the CO_2 compression cost.
- Precipitation of bound CO_2, or the reactant itself enabling a buildup of bound CO_2 at constant equilibrium pressure, and/or a Le Chatelier shift of equilibria to favor a high loading capacity.

The systems currently under development most commonly perform as one homogeneous liquid phase in the absorber section; whereas a phase change or immiscibility occurs as a result of salt solubility limits, or of the immiscibility of hydrophobic/lipophilic amines or other reactants with limited solubility in the base solvent. Systems with combinations of amines, inorganic salts, organic solvents, and water are being studied for this purpose (see Figure 2.6). There is also a close relationship with and significant potential to use multiphase effects in recently developed nonaqueous/water-lean solvent systems [1].

2.2.3.6 Scientific Challenges

Costs along with corrosion, degradation, volatility, aerosols, and the capture plant footprint are significant issues with first- and second-generation aqueous amine systems for CO_2 capture.

The major research challenge is in designing high-performance solvent systems by mastering the ability to accurately predict the chemical and physical properties of potential liquid absorbents for CO_2 capture based on fundamental molecular understanding. Solvent development needs to be integrated with process development.

Solvent-based CO_2 capture requires large equipment and is capital intensive. The technology is readily available for commercial deployment, yet there is significant

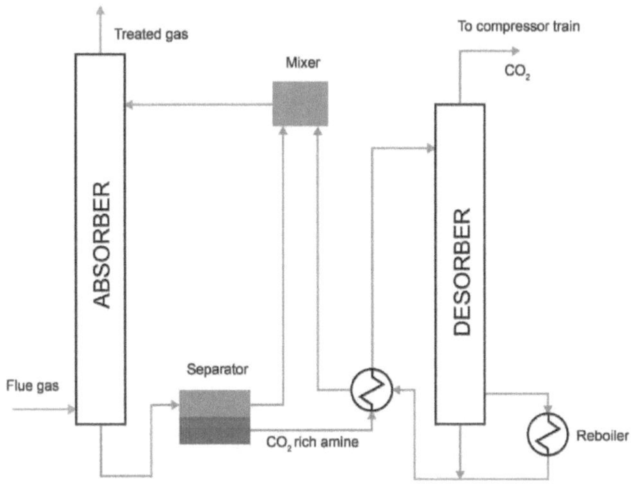

FIGURE 2.6 Conceptual two-phase solvent process flow diagram. (Image courtesy of SINTEF [1].)

potential for capital cost reduction. Capital costs account for up to 50% of total project expenditures. The absorber can represent 50% of total capital costs, and the stripper up to 20%. Reducing the size and/or cost of these units could, therefore, positively impact deployment by significantly reducing costs. Process intensification is an area of chemical engineering focused on developing strategies to dramatically reduce the sizes of existing processes [1].

2.2.4 Hydrate-Based Separation

Hydrate-based CO_2 separation is a new technology by which the exhaust gas containing CO_2 is exposed to water under high pressure forming hydrates. The CO_2 in the exhaust gas is selectively engaged in the cages of hydrate and is separated from other gases. The mechanism is based on the differences in phase equilibrium of CO_2 with other gases, where CO_2 can form hydrates easier than other gases such as N_2 [83]. This technology has the advantage of small energy penalty (6%–8%) [84] and the energy consumption of CO_2 capture via hydrate could be as low as 0.57 kWh/kg-CO_2 [83]. Improving the hydrate formation rate and reducing hydrate pressure can improve the CO_2 capture efficiency [83]. Tetrahydrofuran (THF) is a water-miscible solvent, which can form solid clathrate hydrate structures with water at low temperatures. So the presence of THF facilitates the formation of hydrate and is frequently used as a thermodynamic promoter for hydrate formation. Englezos et al. [85] found that the presence of a small amount of THF substantially reduces the hydrate formation pressure from a flue gas mixture (CO_2/N_2) and offers the possibility to capture CO_2 at medium pressures. Recently, Zhang et al. [86] studied the effects and mechanism of the additive mixture on the hydrate phase equilibrium using the isochoric method and confirmed the effect of THF on hydrate formation. USDOE considers this technology to be the most promising long-term CO_2 separation technology identified today and is currently in the R&D phase [84,87,88].

2.2.5 Cryogenic Separation Process

Cryogenic separation is a CO_2 removal process using distillation at very low temperature and high pressure. In this technique, flue gas is passed through cooling media. The flue gas containing CO_2 is cooled to de-sublimation temperature (−100°C to −135°C), where the solidified CO_2 is separated from other gases and compressed to a high pressure of 100–200 atmospheric pressure. The amount of CO_2 recovered can reach 90%–95% of the flue gas [4]. There are two cryogenic systems: flash separation with internal cooling and separation with distillation column. Since the distillation is accompanied at extremely low temperature and high pressure, it is an energy-intensive process estimated to be (600–660) kWh per ton of CO_2 recovered in the liquid form [4,89]. Numerous patented processes have been developed, and research has mainly been focused on cost optimization [90]. The evaluation of low-temperature processes for producing high-purity, high-pressure CO_2 from oxy-fuel combustion flue gas through simulation and modeling in Aspen HYSYS has also been investigated [91]. The cryogenic CO_2 capture (CCC) process appears to consume 30% or more less energy and money than other major competing carbon capture processes. In addition, the CCC process enjoys several ancillary benefits, including (a) it is a minimally invasive bolt-on technology, (b) it provides highly efficient removal of most pollutants (Hg, SOx, NO_2, HCl, etc.), (c) possible energy storage capacity, and (d) potential water savings. Cryogenic Carbon Capture^{TM} (CCC) is a promising, transformational post-combustion carbon capture technology. The CCC process can reduce CO_2 emissions by over 95% at a cost of less than $45/ton of CO_2 avoided and has a parasitic load of less than 17%. This is about half the energy and cost of currently available technologies and meets the DOE goals for carbon capture systems. Additionally, the CCC process easily retrofits existing plants, recovers water from flue gas, enables energy storage, and robustly handles most impurities in the gas stream.

The CCC External Cooling Loop (CCC ECLTM) process presented by Baxter et al. [92] under DOE supported project (see Figure 2.7) dries and cools flue gas from existing systems to near ambient temperature, provides pressure sufficient to overcome pressure drop, cools it to a temperature slightly above the point where CO_2 forms a solid, condenses CO_2 in a desublimating heat exchanger as it further cools the gas, precipitating CO_2 as a solid, separates the solid from the gas, pressurizes the CO_2, and reheats the CO_2 and the remaining flue gas by cooling the incoming gases. A thermodynamic feature of CO_2 in most flue gases (<16% CO_2 on a dry basis) is that CO_2 will not form a liquid phase at any temperature or pressure. Rather, the CO_2 desublimates in a thermodynamically pure solid phase or remains as a vapor. The CCC process produces CO_2 in a high pressure, liquid phase, and the remaining flue gas at ambient pressure. Both streams are at near ambient temperature. CO_2 capture efficiency depends primarily on the coldest temperature the flue gas obtains. At 1 bar, the process captures 99% of the CO_2 from typical coal flue gases at −207°F (−133°C) and 90% at −179°F (−117°C). Furthermore, the captured CO_2 is typically 99.4%–99.99% pure.

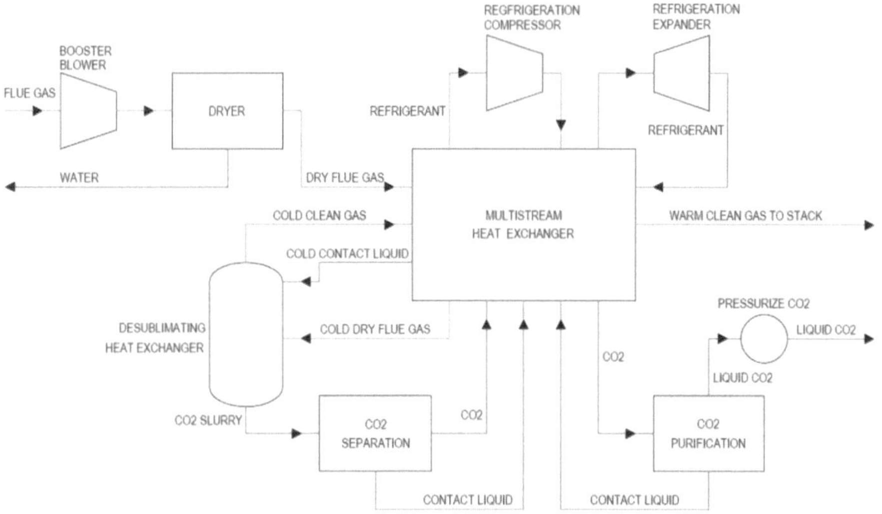

FIGURE 2.7 High-level flow diagram of the CCC ECLTM process [1,92].

These are relatively mild conditions compared with competing oxy-fuel technologies, which require temperatures of about −200°C with a slightly larger volumetric flow rate of gas and requires substantial purification steps after the combustor. Most carbon capture technologies also produce a gaseous CO_2 stream that requires compression to 100–150 bar, whereas CCC produces a liquid CO_2 stream that requires fewer resources to compress, including capital and operating costs and energy. Additionally, most alternative processes exhibit rapidly increasing costs and energy demands as capture increases above 90%. Due to the relatively small difference in temperature as the capture increases (~15°C between 90% and 99% capture), CCC does not exhibit this same rapid increase.

2.3 TRANSPORTATION OF CAPTURED CO_2

Captured CO_2 is required to be transported up to a suitable location. The CO_2 can be transported by pipelines, trucks, and ships. The mode of CO_2 transportation can be selected based on separation and capturing site, and sequestration site. According to IPCC-2005 report [93], at present pipeline transportation is much more mature and viable technology for onshore transport, but shipping of captured CO_2 is economically viable under specific conditions. The transportation cost ranges 1–8 US \$/t CO_2 transported per 250 km pipeline or shipping for mass flow rates of 5 (high end) to 40 (low end) Mt CO_2/year. Pipelines are considered to be the most viable of high volume of CO_2 through long distances [94].

For long lifetime, pipelines are also the most efficient way for CO_2 transport for power plants. For shorter period road and rail tankers are more competitive [95]. The cost of transport varies considerably with regional economic situation. Supercritical is the preferred state for CO_2 transported by pipelines, which implies that the pipelines operative temperature and pressure should be maintained within the CO_2 supercritical envelop, i.e., above 32.11°C and 72.9 atm [96]. The typical range of pressure and temperature for a CO_2 pipeline is between 85 and 150 bar, and between 131°C and 441°C to ensure a stable single-phase flow through the pipeline [97]. The drop in pressure due to the reduction of the hydraulic head along the pipeline is compensated by adding recompression stations. Larger diameter pipelines allow lower flow rates with smaller pressure drop and therefore a reduced number of recompression stations; on the other hand, larger pipelines are more expensive therefore a balancing of costs needs to be considered [97]. Impurities in the CO_2 stream represent a serious issue because their presence can change the boundaries of the pressure and temperature envelope within which a single-phase flow is stable. Furthermore, the presence of water concentration above 50 ppm may lead to the formation of carbonic acid inside the pipeline and cause corrosion problems. Hydrates may also form that may affect the operation of valves and compressors.

Currently, pipelines are mostly used for EOR projects. The oldest is the Canyon Reef Carriers pipeline, a 225 km pipeline built in 1972 for EOR in Texas (USA). The longest is the 800 km Cortez pipeline which is carrying 20 million tons/year of CO_2 from a natural source in Colorado to the oilfields in Denver City, Texas since 1983 [97]. CO_2 pipelines are mostly made of carbon steel and composed of insulated 12 m sections with crack arresters every 350 m and block valves every 16–32 km.

The onshore pipelines are buried in trenches of about 1 m deep. Offshore pipelines in shallow water also need to be deployed in trenches as protection from fishing and mooring activities. Deep-water pipelines generally do not need to be buried unless their diameter is below 400 mm [97,98].

An integrated network, where different sources will merge for their final transport to the storage areas, can reduce the total pipelines length by 25%, but it will require that all sources produce CO_2 stream with the same quality (e.g., pressure, T, water content) before being combined together [99]. When the flow managed through a network of pipelines increases, there is an exponential decrease in the cost of transport; models highlight that the cost for transporting CO_2 along a 1,000 km pipeline is around 8 USD/ton for a mass flow of 25 Mt CO_2/year with a further reduction down to 5 USD/ton if the flow increases to 200 Mt CO_2/year [100]. Further cost saving maybe achieved from the reuse of existing gas pipelines. One of the biggest uncertainties is the effects on the pipelines' integrity of long-term exposure to CO_2 fluxes in terms of corrosion and potential brittle fractures propagation due to the sharp cooling of the pipelines in case of leak of supercritical CO_2 [101].

2.4 CO_2 SEQUESTRATION METHODS

As mentioned in Chapter 1, collecting dilute CO_2 waste stream, concentrating its composition of carbon dioxide, transporting it, and sequestering it is a defensive strategy adopted by many power plants and other sources of dilute CO_2 stream emissions. Even with an increasing use of CO_2 for chemicals, materials, fuels, and power, the carbon capture and sequestration (CCS) has become essential to prevent the generated CO_2 reaching into the atmosphere. Annually about 3 Gt carbon dioxide, which is around one-eighth of current global CO_2 production, needs to be sequestrated [102]. In the United States, the Southeast Regional Carbon Sequestration Partnership has identified more than 900 large stationary sources of CO_2 that contributes 31% of the country's CO_2 stationary source emissions. The CO_2 sequestration can be done in geological formations, deep oceans, saline aquifers, and tar-sands. Generally, CO_2 is stored at depths between 800 and 1,000 m [28,93]. A schematic showing both terrestrial and geological sequestration of carbon dioxide emissions from heavy industry, such as a chemical plant is illustrated in Figure 2.8.

2.4.1 Geological Sequestration of CO_2

The injection and storage of captured CO_2 into the used oil wells, mined coal mines beneath the earth and deep saline aquifers, is termed as geological sequestration. The geologic injection that could also be considered is the use of abandoned, uneconomic coal seams. Geological storage is at present considered to be the most viable option for the storage of the large CO_2 quantities needed to effectively reduce global warming and related climate change [103–106]. It has been shown that CO_2 storage potential can reach 400–10,000 GT for deep saline aquifers compared with only 920 GT for depleted oil and gas fields and 415 GT in unmineable coal seams [107]. The CO_2 injections in geological formations are usually performed for enhanced hydrocarbon recovery in oil and gas reservoirs, and storage and sequestration in saline aquifers.

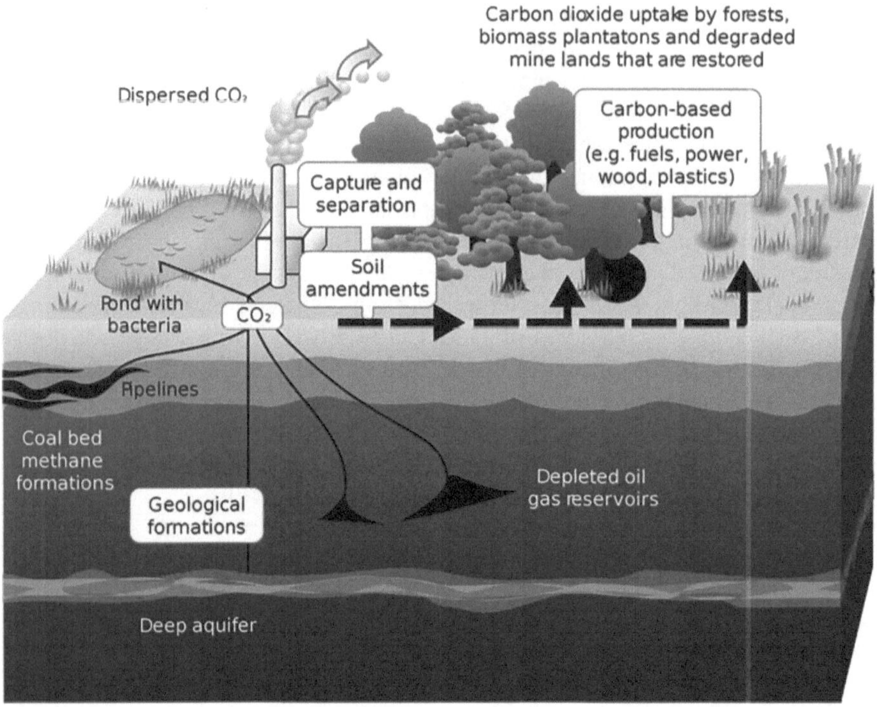

FIGURE 2.8 Schematic of both terrestrial and geological sequestration of carbon dioxide emissions from heavy industry, such as a chemical plant [1].

Once CO_2 is injected into the formation, it diffuses through the pore structure of coal and is physically adsorbed; thus retention on a permanent basis is possible. The chemical reactions between brine ions and CO_2 molecules and consequent reactions with mineral grains are also important processes [108]. The CO_2 geo-storage efficiency in oil wells is strongly affected by the wettability of CO_2-brine-mineral system at storage conditions. Water-wetness decreases with increase in CO_2-wetness, which results in reducing both structural and residual trapping capacities. Use of nanofluid, e.g., silicon dioxide (SiO_2) nanoparticles, renders CO_2-wet calcite to water-wet, which enhances CO_2 geo-storage potentials [109].

The geologic sequestration of CO_2 has a higher expected retention rate and expected residence times are at least thousands of years. The consideration of carbon credits should be made on the retention ability of the geologic reservoir. A typical geological storage site can hold several tens of million tons of CO_2 trapped by different physical and chemical mechanisms [110]. General requirements for geological storage of CO_2 include appropriate porosity, thickness, and permeability of the reservoir rock, a cap rock with good sealing capability, and a stable geological environment [111]. Requirements such as distance from the source of CO_2, effective storage capacity, pathways for potential leakage, and in general economic constraints may limit the feasibility of being a storage site. The amount of CO_2 that leaks into the atmosphere should be considered as the difference of the amount sequestered in the geologic formation versus the actual quantity remaining [112,113]. Coal beds often contain large amounts of methane. The extraction of this methane could represent a value-added process. Currently, Burlington Resources is injecting 70,000 tons of CO_2 per year into a deep coal formation located in the San Juan Basin [113,114]. A similar small-scale project was undertaken by the Alberta Research Council in Canada and reported that by using CO_2 instead of water to flood the bed, there exists a higher potential to recover the methane efficiently and also sequester the CO_2. While this sounds ideal, much further research is needed in this area to understand and optimize the process. Worldwide storage capabilities for CO_2 within deep coal beds are estimated to be up to 150 Gt [13].

Bachu [115] described the criteria and approaches for selecting suitable geological sites for storing CO_2, including the tectonic setting and geology of the basin, its geothermal regime, hydrology of formation waters, hydrocarbon potential, and basin maturity. In addition, economic aspects related to infrastructure and socio-political conditions also affect the site selection. Deep ocean storage is also a feasible option for CO_2 storage although environmental concerns (such as ocean acidification and eutrophication) will likely limit its application. Different geological settings have different criteria of consideration for their reliability as CO_2 storage areas.

2.4.2 Sequestration in Saline Aquifers

Large deep formation of porous rocks is known as saline aquifers. These are basically porous sandstones and limestones, which contain large amount of brine water in their pore space. Disposal of CO_2 from stationary sources (e.g., fossil-fueled power plants) into brackish (saline) aquifers has been suggested as a possible means for reducing emissions of greenhouse gases into the atmosphere. The CO_2 at first compressed at very high pressure, about 95 bar or higher [116], and then injected into the saline aquifers, where aquifer water is replaced by the CO_2 which occupies the porous space.

Deep aquifers at 700–1,000 m below ground level often host high salinity formation brines [117]. Deep saline aquifers can be found in widespread areas both onshore and offshore and are considered to have enormous potential for storage of CO_2. Yang et al. [105] conducted a review on the characteristics of CO_2 sequestration in saline aquifers, including CO_2 phase behavior, CO_2–water–rock interaction, and CO_2 trapping mechanisms that include hydrodynamic, residual, solubility, and mineral trapping [107,118,119]. Over the past two decades, several pilot and commercial projects for CO_2 storage on saline or deep saline aquifers have been launched. Statoil's Sleipner project in the North Sea, as part of a commercial natural gas operation, stores around 1 Mt CO_2/year in a deep saline aquifer hosted in the Utsira Sand formation, about 1,000 m below the seafloor with an available volume for CO_2 storage in the order of $6.6\times10^8 m^3$ [120–122]. Other projects of different scales (i.e., commercial, pilot, and demonstration) are described by Rai et al. [123], Michael et al. [124], and Global CCS Institute [101]. Gorgon and the Latrobe Valley projects in Australia have much larger CO_2 injection capacity (Z4.5 Mt/year). White et al. [125] conducted a comprehensive review on the storage of the captured CO_2 in deep saline aquifers. Myer [106] has reviewed the global status of geological CO_2 storage.

2.4.3 CO_2 Sequestration into Deep Ocean

The direct injection of CO_2 into the ocean can reduce the peak atmospheric CO_2 concentrations and their rate of increase. However, using this method, it is estimated that around 15%–20% of the CO_2 injected into the ocean will leak back into the atmosphere over hundreds of year [113]. Oceans cover more than 70% of Earth's surface and are the biggest natural CO_2 sink. It is estimated that oceans contain about 38,000 Gt of carbon and take up carbon from the atmosphere at a rate of about 1.7 Gt annually. At the same time, oceans produce 50–100 Gt carbon (in the form of phytoplankton) annually, which is greater than the intake by terrestrial vegetation [117]. The carbon inventory in the ocean is enormous at about 50 times greater than that of our atmosphere [126].

At depths greater than 3 km, CO_2 will be liquefied and sunk to the bottom due to its higher density than the surrounding seawater [115,127]. Mathematical models suggest that CO_2 injected in this way could be kept for several hundred years [128]. House et al. [127] further showed that injecting CO_2 into deep-sea sediments at a depth greater than 3 km can provide permanent geological storage of CO_2 even with large geomechanical perturbations. Therefore, deep ocean storage can present a potential sink for large amounts of anthropogenic CO_2. However, this approach is more controversial than other geological storage methods. Injecting large amounts of CO_2 directly into our oceans may affect the seawater chemistry (such as reducing its pH) causing ocean acidification, which may lead to disastrous consequences to the marine ecosystem [129–133]. Although the IPCC has recognized the potential of ocean CO_2 storage, it also noted its local risks and suggest more studies on its feasibility and long-term effect on the marine ecosystem.

2.4.4 Tar-Sand CO_2 Sequestration

In this process, compressed CO_2 at 200 bar and 400°C is injected into the deep sea oil-bitumen sand bed. At the depth of about 600–1,000 m, the CO_2 will exist as a supercritical fluid [134] with specific gravity of somewhere between 0.6 and 0.8. The supercritical CO_2 is buoyant in the saline formation water and will rise until it encounter a seal. Bitumen is soluble in CO_2 and becomes liquid, which can be extracted easily from unmineable bitumen seams.

2.4.5 Opportunities and Challenges of CCS

The carbon sequestration beneath the ocean and saline aquifers has great potential and can save millions of tons of CO_2 emission to the atmospheres. Over the period, the stored carbon again may convert into fuel, which may be explored in future. On the contrary, there are challenges and problems related to the stored carbon. Injection of CO_2 into saline aquifers will give rise to a variety of coupled physical and chemical processes, including pressurization of reservoir fluids, immiscible displacement of an aqueous phase by the CO_2 phase, partial dissolution of CO_2 into the aqueous phase, chemical interactions between aqueous CO_2 and primary aquifer minerals, and changes in effective stress which may alter aquifer permeability and porosity, and may give rise to increase in seismic sensitivity as well.

Nonisothermal effects may arise from phase partitioning, chemical reactions, and compression/decompression effects. Many of the important processes involve nonlinear effects and dependencies on pressure, temperature, and fluid composition. If geo-sequestration of CO_2 is to be employed as a key emission reduction method in the global efforts to mitigate against climate change, simple yet robust screening of the risks of disposal in brine aquifers will be needed [135]. The CCS technologies require water, and huge amount of wastewater is generated; therefore, cost-effective water treatment technologies are also required for site-specific cases [136].

Many countries including the United States, Canada, Brazil, United Kingdom, Germany, Spain, Norway, Sweden, Algeria, Saudi Arabia, India, Australia, China, Indonesia, China, Taiwan, Hong Kong, Japan, South Korea, etc. are working on CCS. Most of the CCS sites are concentrated around the coal fields, oil fields, and fuel-processing plants, where low-carbon fuel is produced along with CCS. The cement industries and power plants are the most significant industrial sectors with high CO_2 discharge. Pilot and demonstration projects are essential to develop carbon capture for this major CO_2 discharging sector. A comprehensive review on worldwide pilot projects for CCS along with technologies used, CO_2 capturing capacity, and sequestration method is given in references [137–139].

One of the challenges of CCS method is risk of leakage after storage of CO_2 into different formations. A number of circumstances, such as leakage through existing or induced faults and fractures, leakage along a spill point, caprock failure or permeability increase and leakage along a well and wellhead failure, are possible for the leakage of CO_2 from the target reservoirs [140]. Injection of captured carbon at depth of sea may contaminate groundwater through leakage. This leads to understanding for design and implementation of appropriate monitoring and control system, both for serving the purpose and assurance of environmental safety [141]. In the storage of supercritical CO_2, if there is any leakage of stored CO_2, then it will rise in saline water until it encounters a seal as the supercritical CO_2 floats in the saline formation water. Therefore, determination of the effectiveness of such seals will be a necessary part for the appraisal of suitable sites for CO_2 storage. The seal integrity of shattered oil and gas wells will be relatively well known, but the deep saline aquifers will be less well-understood. So, considering such formations for secure CO_2 storage will represent significant challenges. The uncertainties in quantifying leakage rates and the expected cost of leakage risk is unlikely to significantly hinder global CCS deployment or the effectiveness of policy for mitigating climate change [142].

While the CCS technologies today are well understood and effective, and can probably provide what is expected, there are some outstanding technical concerns including the development of indigenous and lower-cost CCS technologies, integration and deployment of CCS technologies, regulations and protocols for sequestration site characterization, characterization of sequestration site leakage and mitigation, and technical basis for monitoring, verification, operational protocols, and risk characterization. There are multiple hurdles including cost-effective and viable technologies for implementing CCS technologies [4]. These also include technical readiness and maximizing early investment. The infrastructure to transport CO_2 (e.g., trucks, pipelines) is a key enabler for the commercial deployment of CCS system [143]. Initially, some incentives and Government actions for this infrastructure are needed to build networks sufficient for large-scale commercial CCS deployment. The maturity of various CCS technologies discussed in this chapter is illustrated in Table 2.3 [5].

TABLE 2.3
Status of Various Technologies [5]

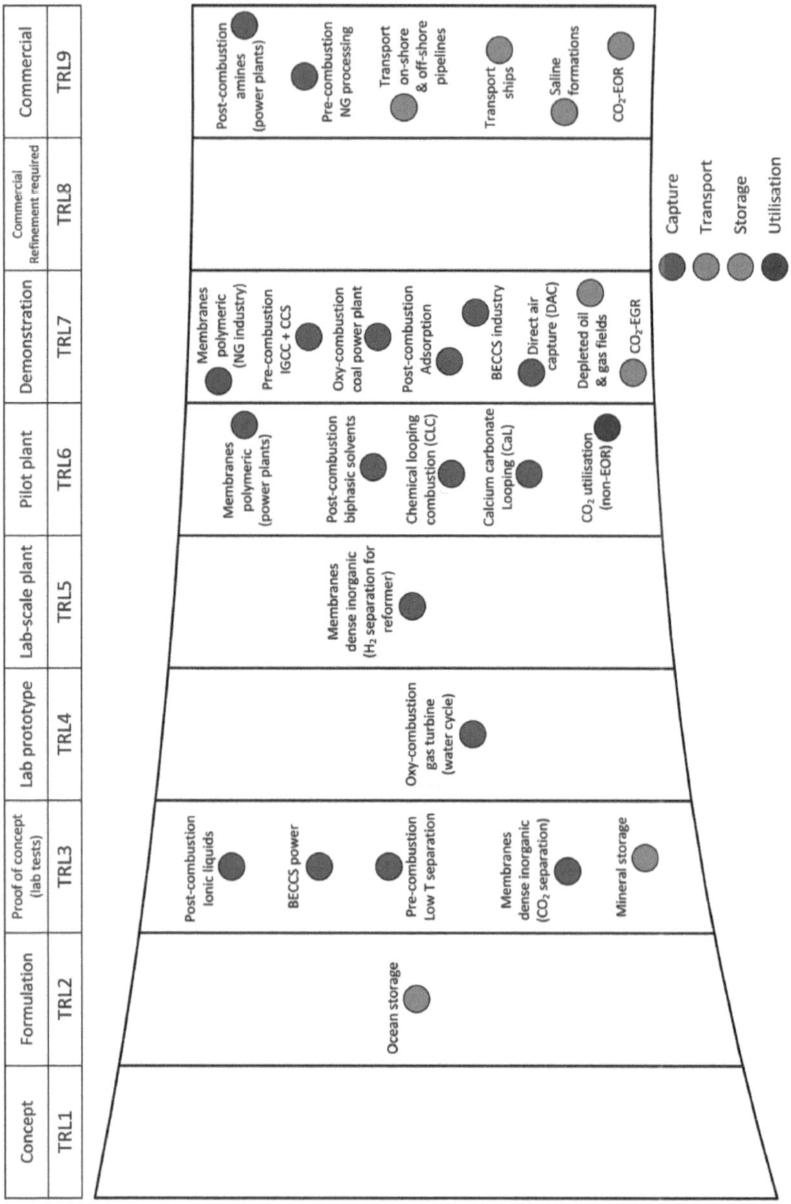

2.5 DIRECT UTILIZATION OF CO_2

While CCS strategy has been essential in order to meet the environmental requirements, in recent years, more efforts are directed toward CCUS strategy which includes utilization of CO_2 for useful purposes. There are two types of utilization: direct and indirect. In direct utilization, CO_2 streams are used for useful purposes, while in indirect utilization, CO_2 is converted to valuable chemicals, materials, fuels, fuel additives, and power. Various technologies used for indirect utilization of CO_2 are described in detail in Chapters 3–7. Here we briefly examine methods of direct utilization of CO_2.

The National Academy Report [144] and DOE report [1] point out that carbon dioxide has been used for decades for various industrial applications such as enhanced oil and gas recovery, as a refrigerant, as an extractive solvent, and as an additive in food and beverage products and for technologically mature processes that do not involve chemical transformations. While only a very small percentage of emitted CO_2 (of the order of 2%–3%) is used for this purpose, they play an important part in overall treatment strategies. Some of the direct use of carbon dioxide for chemical purpose or to improve chemical, food, pharmaceutical, or manufacturing processes can be listed as:

1. Refrigerant; use of dry ice
2. Extractant; flavors/fragrances, decaffeination
3. Carbonated beverages
4. Fire extinguishers
5. Inerting agents; like blanket products, protect carbon powder, and shield gas in welding
6. Miscellaneous; such as injection into metal castings, added to medical O_2 as a respiratory stimulant, aerosol can propellant, dry ice pellets used for sandblasting, and red mud carbonation.

Besides the ones mentioned above, the direct use of CO_2 for enhanced oil and gas recovery is particularly important because it serves multiple purposes. It not only helps extra productions of oil and gas, but the method also sequesters CO_2 underground for long period. DOE considers this to be an important strategy for not only treating CO_2 emission but also for improving the productivity of oil and gas industry. Because of its importance, we briefly examine these direct utilization methods in some details.

While some of the challenges with using CO_2 for enhanced hydrocarbon recovery are unique—such as low permeability and chemo-morphological heterogeneity of the reservoirs—it is now possible to transfer understanding of geologic CO_2 storage and enhanced hydrocarbon recovery from conventional reservoirs to accelerate the use of CO_2 for enhanced hydrocarbon recovery. Further, the development of several in-operando synchrotron and neutron scattering, tomography, and spectroscopy techniques has enabled a fundamental understanding of CO_2-heterogeneous surface interactions at multiple length scales. Specialized laboratory-scale instrumentation, such as high-pressure fluid displacement pumps and custom-designed reactor systems, have been developed to mimic in situ geologic environments. At the field scale, the development of advanced well-diagnostic tools for various geological applications, including the detection of natural gas hydrates, is now transferable for accelerating the use of CO_2 as a working fluid in extreme environments. Novel synthetic materials can also be used to provide insights into these highly heterogenous geologic materials.

2.5.1 Use of CO_2 for EOR

CO_2 has two characteristics that make it a good choice for this purpose: it is *miscible* with crude oil, and it is less expensive than other similarly miscible fluids. When we inject CO_2 into an oil reservoir, it becomes mutually soluble with the residual crude oil. This occurs most readily when the CO_2 density is high (when it is compressed) and when the oil contains a significant volume of "light" (i.e., lower carbon) hydrocarbons (typically a low-density crude oil). Below some minimum pressure, CO_2 and oil will no longer be miscible. As the temperature increases (and the CO_2 density decreases), or as the oil density increases (as the light hydrocarbon fraction decreases), the minimum pressure needed to attain Carbon Dioxide-Enhanced Oil Recovery increases. For this reason, oil field operators must consider the pressure of a depleted oil reservoir when evaluating its suitability for CO_2 enhanced oil recovery. Low-pressured reservoirs may need to be re-pressurized by injecting water. When the injected CO_2 and residual oil are miscible, it enables the CO_2 to displace the oil from the rock pores, pushing it toward a producing well. This is graphically illustrated in Figure 2.9. As CO_2 dissolves in the oil, it swells the oil and reduces its viscosity; which helps to improve the efficiency of the displacement process. Often, CO_2 floods involve the injection of volumes of CO_2 alternated with volumes of water; *water alternating gas* or WAG floods. This approach helps to mitigate the tendency for the lower viscosity CO_2 to finger its way ahead of the displaced oil. Once the injected CO_2 breaks through to the producing well, any gas injected afterward will follow that path, reducing the overall efficiency of the injected fluids to sweep the oil from the reservoir rock.

The US EOR production plot shown in Figure 2.10 illustrates how the use of CO_2 EOR has been a very productive strategy over more than 40 years. Currently, the Denver Unit of Shell oil company produces about 31,500 barrels of oil per day, of which 26,850 is incremental oil attributable to the CO_2 flood. The Wasson Field's Denver Unit CO_2 EOR project has resulted in more than 120 million incremental barrels of oil thru 2008. Furthermore, in CO_2 EOR projects, all of the injected CO_2 either remains sequestered underground or is produced and re-injected in a subsequent project, making the notion of using captured anthropogenic CO_2 for EOR in places far removed from natural sources of CO_2 a likely possibility. Companies have already launched several examples of this approach. For years, ExxonMobil Corp. has sold CO_2 from its La Barge, Wyoming gas processing facility to area oil producers for use in CO_2 EOR projects. The company currently captures four million metric tons of CO_2 per year for this purpose. Another major CO_2 EOR project using industrially sourced CO_2 is located at Weyburn oil field, a Williston basin reservoir just across the U.S. border in Saskatchewan, Canada. EnCana Corp., a Canadian

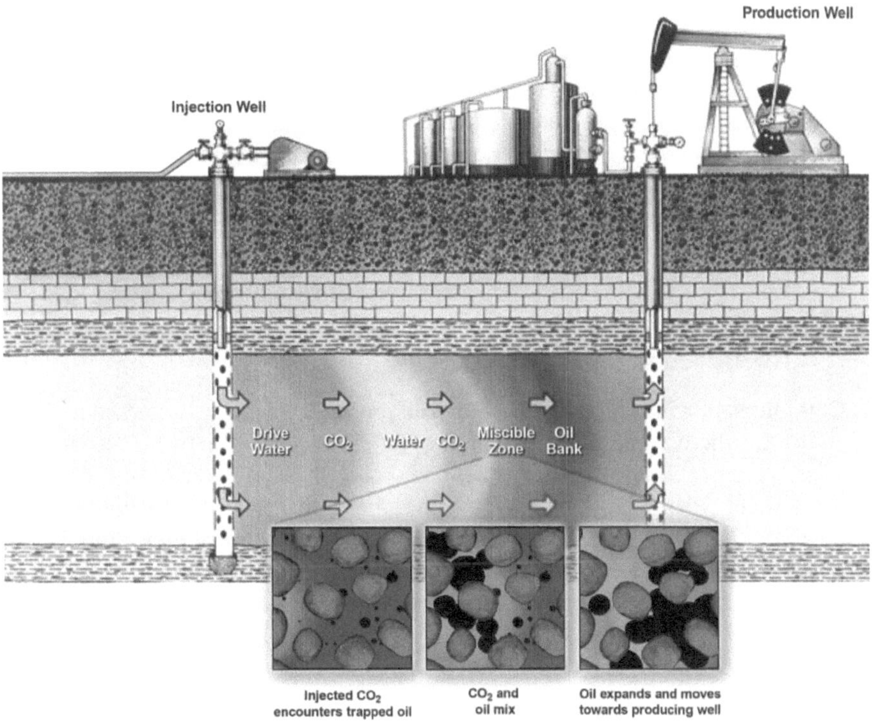

FIGURE 2.9 Enhanced oil recovery using CO_2 and water [145].

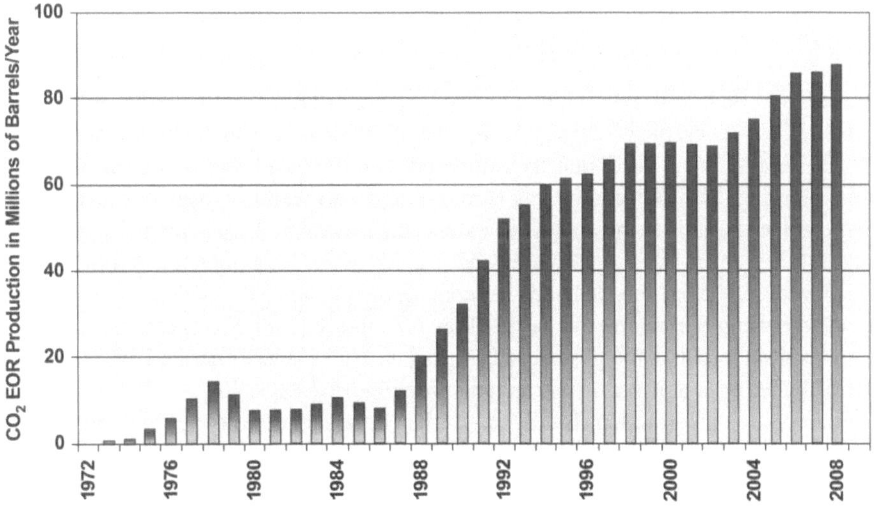

FIGURE 2.10 US CO_2 EOR production [145].

company, injects about 95 million cubic feet (4,935 metric tons) per day of CO_2 into Weyburn, a 55-year-old field, to recover an incremental 130 million barrels of oil via miscible or near-miscible displacement. The CO_2 is sourced from the lignite-fired Dakota Gasification Company synthetic fuels plant in North Dakota, and delivered via a 205-mile pipeline. EnCana estimates that as much as 585 billion cubic feet (30 million metric tons) of CO_2 can be permanently sequestered underground through the project, while boosting the synfuels plant's revenues by about $30 million per year and extending the Weyburn field's life by 20–25 years.

2.5.2 Recent Advancements in Direct Use CO_2 for Enhanced Shale Gas Recovery

As CO_2 interacts with rocks like shale in the presence of groundwater, some have reported that there is significant growth of interesting pore networks leading to increased permeability. Potential chemical and physical interactions between CO_2 and shale when CO_2 is employed as an alternative hydrofracturing fluid are illustrated in Figure 2.11. A fundamental physical and chemical understanding of CO_2–water–solid interactions in extreme environments and in confined regions of variable geometry from the molecular to the field scales is essential for the effective utilization of CO_2 as a working fluid (e.g., fracking fluid) in a wide range of geologic environments. The use of CO_2 in energy recovery is important for unlocking the potential for both conventional and unconventional hydrocarbon resources while improving overall sustainability, including net carbon emission and water requirements. Accelerated innovations in this area are possible through advanced materials discovery to tune materials chemistry in complex environments, the development of well-defined protocols for controlled laboratory- and field-scale measurements, and dynamic optimization of the fate of a fluid containing various components originating in rocks and CO_2 based on dynamic upscaling and downscaling studies.

For high oil extraction efficiency, CO_2 is generally employed; it is supercritical at reservoir conditions and helps recovery through swelling and viscosity reduction [146,147]. However, the efficiency of the overall process is reduced by the buoyancy of CO_2 because of its lower density compared with the host brine in the aquifer, viscous fingering, and CO_2 channeling through high-permeability layers. To overcome these problems, CO_2 is injected as a foam, which reduces the gas mobility in high-permeability zones such as fractures and gravity-override regions that contain little oil. Moreover, the foam is designed to break in the presence of residual oil, enabling contact between the oil and the gas to allow oil capture [148]. Foam is identified as a complex fluid in which the gas phase is segregated in bubbles separated by thin films (called lamellae) stabilized by surfactants (see Figure 2.12).

Foam with a gas fraction (fg) of 0.5–0.6 has been successfully applied; but larger fgs, up to 0.8–0.9, can be reached by using nanoparticles instead of or in conjunction with traditional surfactants [149]. One of the new applications of CO_2 foam as a fluid is in hydraulic fracturing for shale-gas production. Here, the major advantages are the smaller volume of water required to perform the operation and the rheological properties of the foam that allow a better control of fluid mobility in fractures.

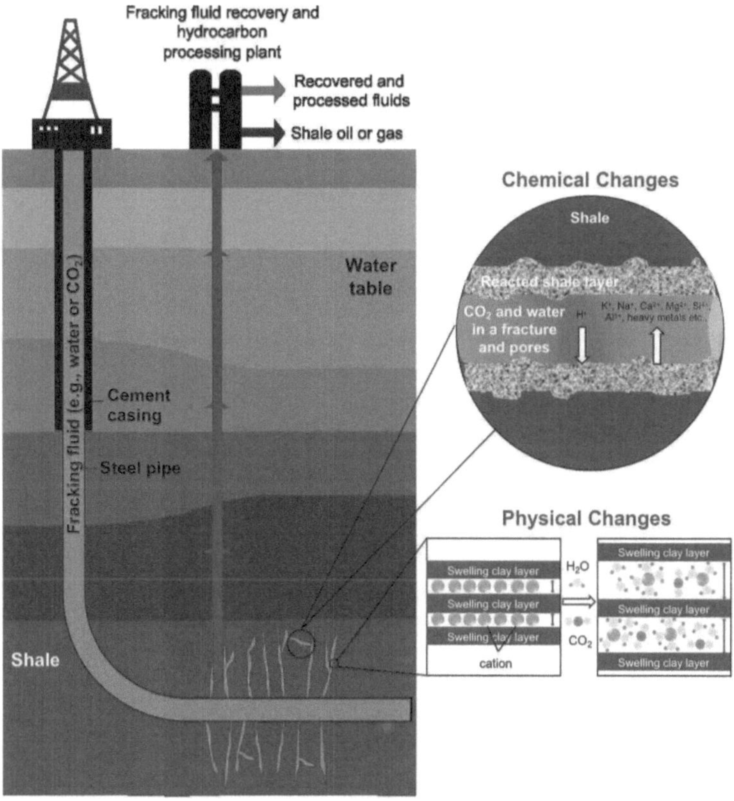

FIGURE 2.11 Potential chemical and physical interactions between CO_2 and shale when CO_2 is employed as an alternative hydrofracturing fluid [1].

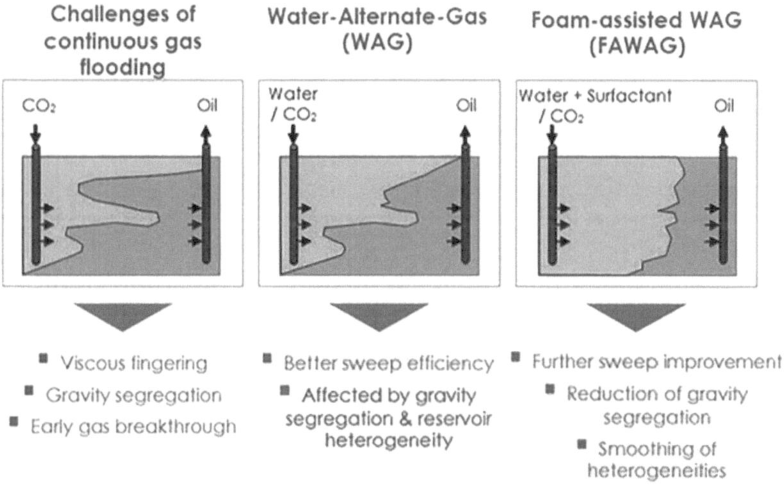

FIGURE 2.12 Schemes for injecting CO_2 foam for EOR applications and its advantages compared with continuous CO_2 flooding. (Image courtesy of EOR alliance [1].)

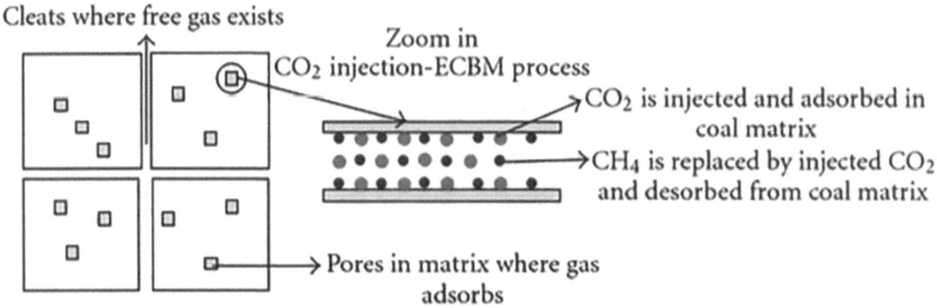

FIGURE 2.13 A schematic representation of CO_2 sequestration-ECBM production [154].

Employing smart nanoparticles to stabilize the foam could significantly reduce the consumption of water, particularly in arid regions, and maximize the use of CO_2. The major challenges of this approach are:

- Understanding the behavior of a CO_2 foam moving through media of variable permeability (nano- and microporous and fractured) at the reservoir conditions of temperature, pressure, and salinity
- Understanding the reactivity between CO_2 foam and the shale to predict carbon storage

2.5.3 ECBM Recovery Using CO_2

The process of ECBM and storage of CO_2 in coal seams involves capturing CO_2 from a flue gas stream, compressing it for transport to an injection site, followed by injection of CO_2 into the coal to enhance methane recovery and/or store CO_2. Methane desorbs from the micropores of the coal matrix when the hydrostatic pressure is reduced, such as from the drilling of a well, and flows through the cleats to a wellbore. The main methods which can induce methane release from coal formations are to reduce the overall pressure, usually by dewatering the formation, generally through pumping; or to reduce the partial pressure of the methane by injecting another inert gas into the formation, such as CO_2, where the methane on the surface gets displaced by the other gas (see Figure 2.13). Dewatering and reservoir pressure depletion is a simple but relatively inefficient process, recovering less than 50% of the gas in place. Lowering the hydrostatic pressure in the coal seam accelerates the desorption process. Once dewatering has taken place and the pressure has been reduced, the released methane can be produced. CBM wells initially primarily produce water; then gas production eventually increases, while water production declines. Some wells do not produce any water and begin producing gas immediately, depending on the nature of the fracture system. Hydraulic fracturing or other completion enhancement methods are used to assist recovery but, even so, because permeability is normally low, many wells at relatively close spacing must be drilled to achieve economic gas production.

CBM production potential is determined by a number of factors that vary from basin to basin, and include fracture permeability, development history, gas migration, coal maturation, coal distribution, geologic structure, well completion options, hydrostatic pressure, and produced water management. In most areas, naturally developed fracture networks are the most sought-after areas for CBM development. Areas, where geologic structures and localized faulting have occurred, tend to induce natural fracturing, which increases the production pathways within the coal seam. When injected, CO_2 moves through the coal seam along its natural fractures (the cleat system), and from there diffuses to the coal micro-pores where it is preferentially adsorbed. In coal, CO_2 has a higher affinity to become adsorbed onto the reservoir rock surfaces than methane that is naturally found within them. Upon injection, the CO_2 displaces methane from some of the adsorption sites. The ratio of CO_2 to methane varies from basin to basin, but has been linked to the maturity of the organic matter in the coal. As much as another 20% could potentially be recovered through the application of CO_2-ECBM [150]. In addition, the fact that some CBM is high in CO_2 content shows that, at least in some instances, CO_2 can safely remain stored in coal for geologically significant time periods [151].

Several ECBM/CO_2 storage field-level tests on the injection and storage of CO_2 in coals, combined with ECBM, were conducted in the San Juan Basin of the United States. One of the longest-operated field pilots was the Allison Unit CO_2-ECBM pilot [152]. During 6 years of operation (1995–2001), approximately 335,000 metric tons of CO_2 were injected into the 900 m deep Fruitland coal seams. The project recovered 45 million cubic meters (1.6 Bcf) of incremental CBM and stored 270,000 metric tons of CO_2. The main conclusions drawn from the project were:

(a) CO_2 injection into coal can significantly improve methane recovery; recovery was improved from 77% (under traditional practices) to 95% (using CO_2 injection) of original gas in place within the central pilot area;

(b) injectivity losses are likely when CO_2 is first introduced into the coal seam; initial CO_2 injectivity was reduced by 60% (with coal permeability reduced by an order of magnitude near the wells); the loss of CO_2 injectivity was modest and a steady rebound in CO_2 injectivity was noted with time as methane was produced;

(c) improvements are required in reservoir simulation models to properly capture the interaction of CO_2 injection, methane release, and the coal reservoir; though existing reservoir simulation models provide a reasonable match of project performance; and

(d) advances in well injectivity technology could unlock the massive CO_2 storage potential of CBM resources in deep coals; particularly if technology is developed to overcome reduced injectivity due to matrix swelling.

The potential benefits of using CO_2/N_2 mixtures to possibly overcome the limitations from swelling associated with injecting pure CO_2 was a primary feature of the Tiffany ECBM pilot [153]. BP (formerly Amoco) began to investigate ECBM techniques in the late 1980s. Building on laboratory and pilot tests, after 9 years of primary CBM production, N_2 injection commenced in January 1998; utilizing ten newly drilled directional N_2 injection wells, and later into two additional converted production wells. The results showed a steep increase in methane production accompanied by the rapid breakthrough of N_2. This breakthrough resulted from a ten-fold increase in the cleat permeability. The results indicate that in cases where the rank and permeability are not adequate for ECBM and storage operations, there may be opportunities to look at pulsing and/or mixing N_2 into the injection stream to improve injectivity during storage and ECBM operations. Various other large- and small-scale testings have shown the workability of this method for methane production and CO_2 sequestration [151].

REFERENCES

1. Buchanan, M. 2017. Accelerating breakthrough innovation in carbon capture, utilization and storage, a Report of the Carbon Capture, Utilization and Storage Experts' Workshop, September 26–28, Houston, TX, Department of Energy, Washington, D.C.
2. Salvi, B. L., and S. Jindal 2019. Recent developments and challenges ahead in carbon capture and sequestration technologies. *SN Applied Sciences* 1:885. https://doi.org/10.1007/s42452-019-0909–2.
3. Sood, A. and S. Vyas 2017. Carbon capture and sequestration – A review. In: *2017 IOP Conference Series: Earth and Environmental Science* 83 012024.
4. Leung, D. Y. C., G. Caramanna, and M. M. Maroto-Valer 2014. An overview of current status of carbon dioxide capture and storage technologies. *Renewable and Sustainable Energy Reviews* 39:426–443. https://doi.org/10.1016/j.rser.2014.07.093.
5. Bui, M., et al. 2018. Carbon capture and storage (CCS): The way forward. *Energy & Environmental Science* 11:1062.
6. Sifat, N. S., and Y. Haseli 2019. A critical review of CO_2 capture technologies and prospects for clean power generation. *Energies* 12:4143. doi: 10.3390/en12214143. www.mdpi.com/journal/energies.
7. World Bank Data 2013. Available at http://data.worldbank.org/indicator/EG.USE.COMM.FO.ZS?end=2013&start=1992 (accessed January 3, 2017).
8. Salvi, B. L., and K. A. Subramanian 2015 Sustainable development of road transportation sector using hydrogen energy system. *Renewable and Sustainable Energy Reviews* 51:1132–1155. https://doi.org/10.1016/j.rser.2015.07.030.
9. Damen, K., M. V. Troost, A. Faaij, and W. Turkenburg 2007. A comparison of electricity and hydrogen production systems with CO_2 capture and storage—part B: Chain analysis of promising CCS options. *Progress in Energy and Combustion Science* 33:580–609.
10. Kama, A. A. L., M. Fodha, and G. Lafforgue 2013. Optimal carbon capture and storage policies. *Environmental Modeling & Assessment* 18:417–426. https://doi.org/10.1007/s10666-012-9354–y.
11. Bauer, N., O. Edenhofer, H. Held, and E. Kriegler 2005 Uncertainty of the role of carbon capture and sequestration within climate change mitigation strategies. *Greenhouse Gas Control Technologies*:931–939.
12. Hitchon, B. 1996 *Aquifer Disposal of Carbon Dioxide*. Sherwood Park: Geoscience Publishing.
13. Pruess, K., and J. Garcia 2002. Multiphase flow dynamics during CO_2 disposal into saline aquifers. *Environmental Geology* 42:282–295.
14. Audus, H. 1997. Greenhouse gas mitigation technology: An overview of the CO_2 capture and sequestration studies and further activities of the IEA greenhouse gas R&D programme. *Energy* 22(2/3):217–221.
15. Das, L. M. 1991. Safety aspects of a hydrogen-fuelled engine system development. *International Journal of Hydrogen Energy* 16(9):619–624.
16. Wang, W., and Y. Cao 2012. A combined thermodynamic and experimental study on chemical-looping ethanol reforming with carbon dioxide capture for hydrogen generation. *International Journal of Energy Research* 1:1. https://doi.org/10.1002/er.2976.

17. Bhown, A. S., and B. C. Freeman 2011. Analysis and status of post-combustion carbon-dioxide capture technologies. *Environmental Science & Technology* 45:8624–8632.
18. Gibbins, J., and H. Chalmers 2008. Carbon capture and storage. *Energy Policy* 36:4317–4322.
19. 2007. International Energy Agency Report. Capturing CO_2. IEA Greenhouse GasR&D Program 2007. ISBN: 978-1-898373-41-4.
20. Jansen, D., M. Gazzani, G. Manzolini, E. van Dijk, and M. Carbo 2015. Pre-combustion CO_2 capture. *International Journal of Greenhouse Gas Control* 40:167–187.
21. Theo, W. L., J. S. Lim, H. Hashim, A. A. Mustaa, and W. S. Ho 2016. Review of pre-combustion capture and ionic liquid in carbon capture and storage. *Applied Energy* 183:1633–1663.
22. Stanger, R., et al. 2015. Oxyfuel combustion for CO_2 capture in power plants. *International Journal of Greenhouse Gas Control* 40:55–125.
23. Pfaff, I., and A. Kather 2009. Comparative thermodynamic analysis and integration issues of CCS steam power plants based on oxy-combustion with cryogenic or membrane based air separation. *Energy Procedia* 1:495–502.
24. Wilberforce, T., A. Baroutaji, B. Soudan, A. H. Al-Alami, and A. G. Olabi 2019. Outlook of carbon capture technology and challenges. *Science of the Total Environment* 657:56–72.
25. Schreiber, A., P. Zapp, and W. Kuckshinrichs 2009. Environmental assessment of German electricity generation from coal-fired power plants with amine-based carbon capture. *International Journal of Life Cycle Assessment* 14:547–559.
26. Viebahn, P., J. Nitsch, M. Fischedick, A. Esken, D. Schüwer, N. Supersberger, U. Zuberbühler, and O. Edenhofer 2007. Comparison of carbon capture and storage with renewable energy technologies regarding structural, economic, ecological aspects in Germany. *International Journal of Greenhouse Gas Control* 1:121–133.
27. Veltman, K., B. Singh, and E. G. Hertwich 2010. Human and environmental impact assessment of postcombustion CO_2 capture focusing on emissions from amine-based scrubbing solvents to air. *Environmental Science & Technology* 44:1496–1502.
28. Cuéllar-Franca, R. M., and A. Azapagic 2015. Carbon capture, storage and utilisation technologies: A critical analysis and comparison of their life cycle environmental impacts. *Journal of CO_2 Utilization* 9:82–102.
29. Pehnt, M., and J. Henkel 2009. Life cycle assessment of carbon dioxide capture and storage from lignite power plants. *International Journal of Greenhouse Gas Control* 3:49–66.
30. Nie, Z., A. Korre, and S. Durucan 2011. Life cycle modelling and comparative assessment of the environmental impacts of oxy-fuel and post-combustion CO_2 capture, transport and injection processes. *Energy Procedia* 4:2510–2517.
31. Adewole, J. K., A. L. Ahmad, S. Ismail, and C. P. Leo 2013. Current challenges in membrane separation of CO_2 from natural gas: A review. *International Journal of Greenhouse Gas Control* 17:46–65. https://doi.org/10.1016/j.ijggc.2013.04.012.
32. Zhang, Y., J. Sunarso, S. Liu, and R. Wang 2013. Current status and development of membranes for CO_2/CH_4 separation: A review. *International Journal of Greenhouse Gas Control* 12:84–107. https://doi.org/10.1016/j.ijggc.2012.10.009.
33. Hasebe, S., S. Aoyama, M. Tanaka, and H. Kawakami 2017. CO_2 separation of polymer membranes containing silica nanoparticles with gas permeable nano-space. *Journal of Membrane Science* 536:148–155. https://doi.org/10.1016/j.memsci.2017.05.005.
34. Aaron, D., and C. Tsouris 2005. Separation of CO_2 from flue gas: A review. *Separation Science and Technology* 40:321–348..
35. Yave, W., A. Car, S. S. Funari, S. P. Nunes, and K. V. Peinemann 2009. CO_2-philic polymer membrane with extremely high separation performance. *Macromolecules* 43:326–333.

36. Brunetti, A., F. Scura, G. Barbieri, and E. Drioli Membrane technologies for CO_2 separation. *Journal of Membrane Science* 359:115–125.
37. Mat, N. C., and G. G. Lipscomb 2017. Membrane process optimization for carbon capture. *International Journal of Greenhouse Gas Control* 62:1–12. https://doi.org/10.1016/j.ijggc.2017.04.002.
38. Hussain, A., and M.-B. Hägg 2010. A feasibility study of CO_2 capture from flue gas by a facilitated transport Membrane. *Journal of Membrane Science* 359:140–148. https://doi.org/10.1016/j.memsci.2009.11.035.
39. Chaubey, T. 2017. CO_2 capture by cold membrane operation with actual power plant flue gas. In: DEFE0013163, *Proceedings of the 2016 NETL CO_2 Capture Technology Meeting*, Pittsburgh, PA, August 8–12, 2016.
40. Hägg, M.-B., A. Lindbråthen, X. He, S. G. Nodeland, and T. Cantero 2017. Pilot demonstration reporting on CO_2 capture from a cement plant using hollow fiber process. *Energy Procedia* 114:6150–6165.
41. Bredesen, R., K. Jordal, and O. Bolland 2004. High-temperature membranes in power generation with CO_2 capture. *Chemical Engineering and Processing* 43:1129–1158.
42. Schwartz, J., D. Makuch, D. J. Way, J. Porter, J., N. Patki, M. Kelley, J. Stanislowski, and S. Tolbert 2015. Advanced hydrogen transport membrane for coal gasification, final report DE- FE0004908, US Department of Energy.
43. Castro-Dominguez, B., I. P. Mardilovich, R. Ma, N. K. Kazantzis, A. G. Dixon, and E. Ma. 2017. Performance of a pilot-scale multitube membrane module under coal-derived syngas for hydrogen production and separation. *Journal of Membrane Science* 523, 515–523.
44. Peters, T. A., R. Bredesen, and H. J. Venvik 2017. Pd-based membranes in hydrogen production: Long time tests and contaminant effects. In Drioli, E., and Barbieri, G., eds., *Membrane Engineering for the Treatment of Gases*. RSC Publishing. ISBN 978-1-78262-875-0.
45. Peters, T. A., P. M. Rørvik, T. O. Sunde, M. Stange, F. Roness, T. R. Reinertsen, J. H. Ræder, Y. Larring, and R. Bredesen 2017. Palladium (Pd) membranes as key enabling technology for pre- combustion CO_2 capture and hydrogen production. *Energy Procedia* 114:37–45.
46. Anantharaman, R., T. A Peters, W. Xing, M. L. Fontaine, and R. Bredesen 2016. Dual-phase high temperature membranes for CO_2 separation—Material development and performance assessment. *Faraday Discussions* 192:251–269.
47. Rubin, E. S., H. Mantripragada, A. Marks, P. Versteeg, and J. Kitchin 2012. The outlook for improved carbon capture technology. *Progress in Energy and Combustion Science* 38:630–671.
48. Li, L., N. Zhao, W. Wei, and Y. Sun 2013. A review of research progress on CO_2 capture, storage, and utilization in Chinese Academy of Sciences. *Fuel* 108:112–130.
49. Takamura, Y., S. Narita, J. Aoki, and S. Uchida 2001. Application of high-PSA process for improvement of CO_2 recovery system. *Canadian Journal of Chemical Engineering* 79:812–816.
50. McKee, B. 2002. Solutions for the 21st Century: Zero emissions technology for fossil fuels. France: IEA, Com. Energy Research and Technology. OECD/IEA.
51. Clausse, M., J. Merel, and F. Meunier 2011. Numerical parametric study on CO_2 capture by indirect thermal swing adsorption. *International Journal of Greenhouse Gas Control* 5:1206–1213.
52. Kulkarni, A. R., and D. S. Sholl 2012. Analysis of equilibrium-based TSA processes for direct capture of CO_2 from air. *Industrial & Engineering Chemistry Research* 51:8631–8645.
53. Slater, A. G., and A. I. Cooper 2015. Function-led design of new porous materials. *Science* 348(6238):aaa8075.

54. Yaghi, O. M., M. O'Keefe, N. W. Ockwig, H. K. Chae, M. Eddaoudi, and J. Kim 2003. Reticular synthesis and the design of new materials. *Nature* 423:705–714.
55. Fracaroli, A. M., H. Furukawa, M. Suzuki, M. Dodd, S. Okajima, F. Gándara, J. A. Reimer, and O. M. Yaghi. 2014. Metal-organic frameworks with precisely designed interior for carbon dioxide capture in the presence of water. *Journal of the American Chemical Society* 136: 8863–8866.
56. Nugent, P., Y. Belmabkhout, S. D. Burd, A. J. Cairns, R. Luebke, K. Forrest, T. Pham, S. Ma, B. Space, L. Wojtas, M. Eddaoudi, and M. J. Zaworotko 2013. Porous materials with optimal adsorption thermodynamics and kinetics for CO_2 separation. *Nature* 495:80–84.
57. McDonald, T. M. 2015. Cooperative insertion of CO_2 in diamine-appended metal-organic frameworks. *Nature* 519:303.
58. Bhatt, P. M., Y. Belmabkhout, A. Cadiau, K. Adil, O. Shekhah, A. Shkurenko, L. J. Barbour, and M. Eddaoudi 2016. A fine-tuned fluorinated MOF addresses the needs for trace CO_2 removal and air capture using physisorption. *Journal of the American Chemical Society* 138(29):9301–9307.
59. Shimizu, G., R. Vaidhyanathan, S. Iremonger, K. Deakin, J.-B. Lin, and K. W. Dawson 2014. Metal organic framework, production and use thereof. Patent WO 2014138878 A1.
60. Abanades, J. C., B. Arias, A. Lyngfelt, T. Mattisson, D. E. Wiley, H. Li, M. T. Ho, E. Mangano, S. Brandani 2015. Emerging CO_2 capture systems. *International Journal of Greenhouse Gas Control* 40:126–166.
61. Fan, L.-S. 2017. *Chemical Looping Partial Oxidation: Gasification, Reforming, and Chemical Syntheses.* Cambridge: Cambridge University Press.
62. Ishida, M., M. Yamamoto, and T. Ohba 2002. Experimental results of chemical looping combustion with $NiO/NiAl_2O_4$ particle circulation at 12001C. *Energy Conversion and Management* 43:1469–1478.
63. Cho, P., T. Mattisson, and A. Lyngfelt 2002. Reactivity of iron oxide with methane in a laboratory fluidized bed–application of chemical-looping combustion. In: *Proceedings of the 7th International Conference on Circulating Fluidised Beds*, Niagara Falls, Ontario, May 5–7, pp. 599–606.
64. Brandvoll, Ø., and O. Bolland 2004. Inherent CO_2 capture using chemical looping combustion in a natural gasfired power cycle. *Journal of Engineering for Gas Turbines and Power* 126:316–321.
65. Adánez, J., L. F. de Diego, F. García-Labiano, P. Gayán, A. Abad, and J. M. Palacios 2004. Selection of oxygen carriers for chemical-looping combustion. *Energy Fuels* 18:371–377.
66. Zafar, Q., T. Mattisson, and B. Gevert 2005. Integrated hydrogen and power production with CO_2 capture using chemical-looping reforming-redox reactivity of particles of CuO, Mn2O3, NiO, and Fe_2O_3 using SiO_2 as a support. *Industrial & Engineering Chemistry Research* 44:3485–3496.
67. Li, F., S. Luo, Z. Sun, X. Bao, and L. S. Fan 2011. Role of metal oxide support in redox reactions of iron oxide for chemical looping applications: Experiments and density functional theory calculations. *Energy & Environmental Science* 4:3661–3667.
68. Lyngfelt, A., B. Leckner, and T. Mattisson 2001. A fluidized-bed combustion process with inherent CO_2 separation; application of chemical-looping combustion. *Chemical Engineering Science* 56:3101–3113.
69. Lyngfelt, A., and T. Mattisson 2011. Materials for chemical-looping combustion. In: Stolten, D., Scherer, V., eds. *Efficient Carbon Capture for Coal Power Plants.* Weinheim: Wiley-VCH Verlag GmbH & Co. KGaA, pp. 475–504.
70. Adánez, J., A. Abad, F. Garcia-Labiano, P. Gayan, and L. de Diego 2012. Progress in chemical- looping combustion and reforming technologies. *Progress in Energy and Combustion Science* 38:215–282.

71. Erlach, B., M. Schmidt, and B. Tsatsaronis 2011. Comparison of carbon capture IGCC with pre-combustion decarbonisation and with chemical-looping combustion. *Energy* 36:3804–3815.
72. Ohlemüller, P., J. P. Busch, M. Reitz, J. Strôhle, and B. Epple 2016. Chemical-looping combustion of hard coal: Autothermal operation of a 1 MWth pilot plant. *Journal of Energy Resources Technology* 138(4):042203.
73. Hendriks, C. 1995. *Energy Conversion: CO_2 Removal from Coal-Fired Power Plant.* Netherlands: Kluwer Academic Publishers.
74. Veawab, A., A. Aroonwilas, and P. Tontiwachwuthiku 2002. CO_2 absorption performance of aqueous alkanolamines in packed columns. *Fuel Chemistry* 47:49–50.
75. Figueroa, J. D., T. Fout, S. Plasynski, H. McIlvried, and R. D. Srivastava 2008. Advances in CO_2 capture technology—The U.S. Department of Energy's Carbon Sequestration Program. *International Journal of Greenhouse Gas Control* 2:9–20. https://doi.org/10.1016/s1750-5836(07)00094–1.
76. Rochelle, G. T. 2012. Thermal degradation of amines for CO_2 capture. *Current Opinion in Chemical Engineering* 1–2:183–190.
77. Fredriksen, S. B., and K. J. Jens 2013. Oxidative degradation of aqueous amine solutions of MEA, AMP, MDEA, Pz: A review. *Energy Procedia* 37:1770–1777.
78. da Silva, C. F., et al. 2013. Intercalation of amines into layered calcium phosphate and their new behavior for copper retention from ethanolic solution. *Open Journal of Synthesis Theory and Applications* 2:1–7.
79. Li, Q., S. Gupta, L. Tang, S. Quinn, V. Atakan, and R. E. Riman 2016. A novel strategy for carbon capture and sequestration by rHLPD processing. *Frontiers in Energy Research* 3, Art. No. 53. https://doi.org/10.3389/fenrg.<jtl>.2015.00053.
80. Rochelle, G. T. 2014. From Lubbock, TX, to Thompsons, TX: Amine scrubbing for commercial CO_2 capture from power plants. In: *Plenary presentation, 12th International Conference on Greenhouse Gas Control Technologies*, GHGT-12, Austin, TX, October 5–9.
81. DOE (Department of Energy)/NETL (National Energy Technology Laboratory). 2007. *Cost and Performance Baseline for Fossil Energy Plants—Volume 1: Bituminous Coal and Natural Gas to Electricity*, DOE-NETL-2007/1281, US Department of Energy, May.
82. Kozak, F., A. Petig, E. Morris, R. Rhudy, and D. Thimsen 2009. Chilled ammonia process for CO_2 capture. *Energy Procedia* 1:1419–1426.
83. Fan, S., Y. Wang, and X. Lang 2011. CO_2 capture in form of clathrate hydrate-problem and practice. In: *Proceedings of the 7th International Conference on Gas Hydrates (ICGH 2011)*, UK, July 17–21.
84. Elwell, L. C., and W. S. Grant 2006. Technology options for capturing CO_2 – Special reports. *Power* 150:60–65.
85. Englezos, P., J. A. Ripmeester, R. Kumar, and P. Linga 2008. Hydrate processes for CO_2 capture and scale up using a new apparatus. In: *Proceedings of the International Conference on Gas Hydrates*, ICGH.
86. Zhang, Y., M. Yang, Y. Song, L. Jiang, Y. Li, and C. Cheng 2014. Hydrate phase equilibrium measurements for (THFþSDSþCO2þN2) aqueous solution systems in porous media. *Fluid Phase Equilibria* 370:12–18.
87. Babu, P., R. Kumar, and P. Linga 2013. Progress on the hydrate based gas separation (HBGS) process for carbon dioxide capture. *AICHE Annual Meeting*, SanFrancisco, 3–8 November 2013.
88. Sun, D., and P. Englezos 2014. Storage of CO_2 in a partially water saturated porous medium at gas hydrate formation conditions. *International Journal of Greenhouse Gas Control* 25:1–8.
89. Gottlicher, G., and R. Pruschek 1997. Comparison of CO_2 removal systems for fossil fuelled power plants. *Energy Conversion and Management* 38:S173–S178.

90. Tuinier, M. J., M. V. S. Annaland, G. J. Kramer, and J. A. M. Kuipers 2010. Cryogenic CO_2 capture using dynamically operated packed beds. *Chemical Engineering Science* 65:114–119.
91. Besong, M., M. M. Maroto-Valer, A. Finn 2013. Study of design parameters affecting the performance of CO_2 purification units in oxy-fuel combustion. *International Journal of Greenhouse Gas Control* 12:441–449.
92. Baxter, L., A. Baxter, C. Bence, D. Frankman, C. Hoeger, A. Sayre, K. Stitt, and S. Chamberlain 2016. Cryogenic carbon capture. In: Krutka, H., Tri-State Generation and Transmission Association, and Zhu, F., eds. , *CO_2 Summitt II: Technologies and Opportunities*. UOP/Honeywell, ECI Symposium Series, Santa Ana Pueblo, New Mexico, April 10–14.
93. Metz, B., O. Davidson, H. de-Coninck, M. Loos, and L. Meyer 2005. *IPCC Special Report on Carbon Dioxide Capture and Storage, Intergovernmental Panel on Climate Change*. Cambridge University Press. http://www.ipcc.ch/pdf/special- reports/srccs, (accessed December 10, 2014).
94. Svensson, R., M. Odenberger, F. Johnsson, and L. Stromberg 2004. Transportation systems for CO_2 – Application to carbon capture and storage. *Energy Conversion and Management* 45:2343–2353.
95. Norisor, M., A. Badea, and C. Dinca 2012. Economical and technical analysis of CO_2 transport ways. *UPB Scientific Bulletin, Series C* 74(1):127–138.
96. Johnsen, K., K. Helle, S. Roneid, and H. Holt 2011. DNV recommended practice: Design and operation of CO_2 pipelines. *Energy Procedia* 4:3032–3039.
97. Forbes, S. M., P. Verma, T. E. Curry, S. J. Friedmann, and S. M. Wade 2008. *Guidelines for Carbondioxide Capture, Transport, and Storage*. World Resources Institute, 144 p.
98. 2011. FEED. CO_2 transport pipeline. Except for public use. Vattennfal Europe. Job n.P10111, 143 p.
99. 2010. International Energy Agency Report. CO_2 pipeline infrastructure: An analysis of global challenges and opportunities. Element Energy Limited. Final Report, 134 p.
100. Chandel, M. K., L. F. Pratson, and E. Williams 2010. Potential economies of scale in CO_2 transport through use of a trunk pipeline. *Energy Conversion and Management* 51:2825–2834.
101. Rabimdran, P., H. Cote, and I. G. Winning 2011. Integrity management approach to reuse of oil and gas pipelines for CO_2 transportation. In: *Proceedings of the 6th Pipeline Technology Conference*, Hannover Messe, Hannover, Germany, April 4–5.
102. Snieder, R., and T. Young 2009. Facing major challenges in carbon capture and sequestration. *GSA Today* 19(11):36–37.
103. Celia, M. A., and J. M. Nordbottena 2009. Practical modeling approaches for geological storage of carbon dioxide. *Ground Water* 47:627–638.
104. Van der Zwaan, B., and K. Smekens 2009. CO_2 capture and storage with leakage in an energy-climate model. *Environmental Modeling & Assessment* 14:135–148.
105. Yang, F., B. J. Bai, D. Z. Tang, S. Dunn-Norman, and D. Wronkiewicz 2010. Characteristics of CO_2 sequestration in saline aquifers. *Petroleum Science* 7:83–92.
106. Myer, L. 2011. Global status of geologic CO_2 storage technology development. United -States carbon sequestration council report July 2011.
107. Rackley, S. A. 2010. *Carbon Capture and Storage*. Burlington: Butterworth-Heinemann, Elsevier.
108. Javadpour, F. 2009. CO_2 injection in geological formations: Determining macroscale coefficients from pore scale processes. *Transport in Porous Media* 79:87–105. https://doi.org/10.1007/s11242-008-9289–6.

109. Al-Anssaria, S., M. Arif, S. Wang, A. Barifcani, M. Lebedev, and S. Iglauer 2017. CO_2 geo-storage capacity enhancement via nanofluid priming. *International Journal of Greenhouse Gas Control* 63:20–25. https://doi.org/10.1016/j.ijggc.2017.04.015.
110. Doughty, C., B. M. Freifeld, and R. C. Trautz 2008. Site characterization for CO_2 geological storage and vice versa: The Frio brine pilot, Texas, USA as a case study. *Environmental Geology* 54:1635–1656.
111. Solomon, S., M. Carpenter, and T. A. Flach 2008. Intermediate storage of carbon dioxide in geological formations: A technical perspective. *International Journal of Greenhouse Gas Control* 2:502–510.
112. Herzog, H. 1999. An introduction to CO_2 separation and capture technologies. Energy Laboratory Working Paper, Massachusetts Institute of Technology, Cambridge. https://sequestration.mit.edu/pdf/introduction_to_capture.pdf. (accessed June 19, 2017).
113. Stewart, C., and M. Hessami 2005. A study of methods of CO_2 capture and sequestration—The sustainability of a photosynthetic bioreactor approach. *Energy Conversion and Management* 46:403–420.
114. Beecy, D., and V. Kuuskraa 2001. Status of U.S. geologic carbon sequestration research and technology. *Environmental Geosciences* 8(3):152–159.
115. Bachu, S. 2000. Sequestration of CO_2 in geological media: Criteria and approach for site selection in response to climate change. *Energy Conversion and Management* 41:953–701.
116. Soong, Y., S. W. Hedges, B. H. Howard, R. M. Dilmore, and D. E. Allen 2014. Effect of contaminants from flue gas on CO_2 sequestration in saline formation. *International Journal of Energy Research* 38:1224–1232. https://doi.org/10.1002/er.3140.
117. Yamasaki, A. 2003. An overview of CO_2 mitigation options for global warming–emphasizing CO_2 sequestration options. *Journal of Chemical Engineering of Japan* 36:361–375.
118. Gunter, W. D. 2000. CO_2 sequestration in deep unmineable coal seams. In: *Conference Proceedings; CAPP/CERI Industry Best Practices Conference*, pp. 1–19.
119. DoE US 1999. Carbon sequestration research and development; DOE/SC/FE-1. Washington DC: U.S. Department of Energy.
120. Arts, R., P. Zweigel, and A. Lothe 2000. Reservoir geology of the Utsira Sand in the Southern Viking Graben area–A site for potential CO_2 storage. In: *62nd EAGE Meeting*, paper B-20, Glasgow.
121. Korboel, R., and A. Kaddour 1995. Sleipner Vest CO_2 disposal—Injection of removed CO_2 into the Utsira formation. *Energy Conversion and Management* 36:509–512.
122. Kongsjorden, H., O. Karstad, and T. A. Torp 1997. Saline aquifer storage of carbon dioxide in the Sleipner Project. *Waste Management* 17:303–308.
123. Rai, V., N. C. Chung, M. C. Thurber, and D. G. Victor 2008. PESD carbon storage project database. Working paper#76, The program on energy and sustainable development, Stanford University, Available at S SRN: http://ssrn.com/abstract=1400118; http://dx.doi.org/10.2139/ssrn.1400118.
124. Michael, K., et al. 2010. Geological storage of CO_2 in saline aquifers – A review of the experience from existing storage operations. *International Journal of Greenhouse Gas Control* 4:659–667.
125. White, C. M., B. R. Strazisar, E. J. Granite, J. S. Hoffman, and H. W. Pennline 2003. Separation and capture of CO_2 from large stationary sources and sequestration in geological formations — Coalbeds and deep saline aquifers. *Journal of the Air & Waste Management Association* 53:645–715.
126. International Energy Agency Report. Improvements in power generation with post-combustion capture of CO_2. IEA Greenhouse Gas R&D Programmes, PH4/33.

127. House, K. Z., D. P. Schrag, C. F. Harvey, and K. S. Lackner 2006. Permanent carbon dioxide storage in deep-sea sediments. *Proceedings of the National Academy of Sciences of the USA* 103:12291–12295.
128. Adam, E. E., and K. Caldeira 2008. Ocean storage of CO_2. *Elements* 4:319–324.
129. Seibel, B. A., and P. J. Walsh 2001. Potential impacts of CO_2 injection on deep-sea biota. *Science* 294:319–320.
130. Hall-Spencer, J. M., et al. 2008. Volcanic carbon dioxide vents show ecosystem effects of ocean acidification. *Nature* 45:96–99.
131. Rodolfo-Metalpa, R., et al. 2011. Coral and mollusc resistance to ocean acidification adversely affected by warming. *Nature Climate Change*. http://dx.doi.org/10.1038/NCLIMATE1200.
132. Espa, S., G. Caramanna, and V. Bouche 2010. Field study and laboratory experiments of bubble plumes in shallow seas as analogues of sub-seabed CO_2 leakages. *Applied Geochemistry* 25:696–704.
133. Caramanna, G., Y. Wei, M. M. Maroto-Valer, P. Nathanail, and M. Steven 2013. Laboratory experiments and field study for the detection and monitoring of potential seepage from CO_2 storage sites. *Applied Geochemistry* 30:105–113.
134. Holloway, S. 2007. Carbon dioxide capture and geologic storage. *Philosophical Transactions of the Royal Society A* 365:1095–1107.
135. Mathias, S. A., P. E. Hardisty, M. R. Trudell, and R. W. Zimmerman 2009. Approximate solutions for pressure buildup during CO_2 injection in brine aquifers. *Transport in Porous Media* 79:265–284. https://doi.org/10.1007/s11242-008-9316-7.
136. Castle, J. W., J. R. Wagner, J. H. Rodgers Jr, and G. R. Hill 2011. Water in carbon capture and sequestration: Challenges and opportunities. In: *28th Annual International Pittsburgh Coal Conference 2011*, PCC, 3, pp. 1681–1692.
137. http://www.globalccsinstitute.com/projects/pilot-and-demonstration-projects. (accessed March 4, 2018).
138. http://www.eaton.com/Eaton/OurCompany/SuccessStories/Energy/Skyonic Corporation/index.htm. (accessed March 14, 2018).
139. http://sequestration.mit.edu/tools/projects/kemper.html. (accessed March 14, 2018).
140. Koornneef, J., A. Ramírez, W. Turkenburg, and A. Faaij 2012. The environmental impact and risk assessment of CO_2 capture, transport and storage—An evaluation of the knowledge base. *Progress in Energy and Combustion Science* 38:62–86. https://doi.org/10.1016/j.pecs.2011.05.002.
141. Newmark, R. L., S. J. Friedmann, and S. A. Carroll 2010. Water challenges for geologic carbon capture and sequestration. *Environmental Management* 45(4):651–661.
142. Deng, H., J. M. Bielicki, M., J. P. Fitts, and C. A. Peters 2017. Leakage risks of geologic CO_2 storage and the impacts on the global energy system and climate change mitigation. *Climatic Change* 144:151–163.
143. d'Amore, F., and F. Bezzo 2017. Economic optimisation of European supply chains for CO_2 capture, transport and sequestration. *International Journal of Greenhouse Gas Control* 65:99–116. https://doi.org/10.1016/j.ijggc.2017.08.015.
144. National Academies of Sciences, Engineering, and Medicine 2019. *Gaseous Carbon Waste Streams Utilization: Status and Research Needs*. Washington, DC: The National Academies Press. https://doi.org/10.17226/25232.
145. 2010. Carbon dioxide enhanced oil recovery – Untapped domestic energy supply and long term carbon storage solution, a report by NETL, Department of Energy, Washington, D. C.
146. Orr, F.M. and Taber J.J. 1984. Use of Carbon dioxide in enhanced oil recovery. *Science* 224(4649):563–569. doi:10.1126/science.224.4649.563.
147. Lake, L.W. 1989. *Enhanced Oil Recovery*. Englewood Cliffs: Prentice-Hall Inc., 550 p.

148. Rossen, W. 1996. Foams in enhanced oil recovery. In *Foams: Theory, Measurements, and Applications*, R. K. Prudhomme and S. Khan, eds., New York: Marcel Dekker.
149. Prigiobbe, V. A., A. W. Worthen, K. P. Johnston, C. Huh, and S. L. Bryant. 2016. Transport of nanoparticle-stabilized CO_2-foam in porour media. *Transport in Porous Media* 111(1):265–285.
150. http://www.ieaghg.org/docs/general_publications/8.pdf.
151. Godec, M., G. Koperna, and J. Gale 2014. CO_2-ECBM: A review of its status and global potential. *Energy Procedia* 63:5858–5869.
152. Advanced Resources International, Inc. 2005. The Allison Unit CO_2–ECBM Pilot: A reservoir modelling study, prepared for U.S. Department of Energy Project Number: DE-FC26-0NT40924, February, 2003; and Reeves, S. R., and A. Oudinot. The Allison Unit CO_2-ECBM Pilot – A reservoir and economic analysis, *2005 International Coalbed Methane Symposium*, Paper 0523, Tuscaloosa, Alabama, May 16–20.
153. Advanced Resources International, Inc. 2004. The Tiffany Unit N2 – ECBM Pilot: A reservoir modeling study, prepared for U.S. Department of Energy Project Number: DE-FC26-0NT40924.
154. He, Q., S. D. Mohaghegh, and V. Gholami 2013. A field study on simulation of CO_2 injection and ECBM production and prediction of CO_2 storage capacity in unmineable coal seam. *Journal of Petroleum Engineering* 2013. https://doi.org/10.1155/2013/803706.

3 Carbon Capture by Mineral Carbonation and Production of Construction Materials

3.1 INTRODUCTION

Mineralization converts carbon to a permanent solid form that is thermodynamically favored over CO_2 gas (Figure 3.1). Indeed, more than 90% of carbon on earth currently exists in this state, and ultimately the long-term geological fate of anthropogenic carbon will be in mineral form [1–6]. The past decade has seen incredible progress in advancing the understanding of, and technology for, capturing and converting CO_2 into solid mineral-based materials. Most prior research has focused on the investigation of in situ mineralization of CO_2 in geologic formations with magnesium- and calcium-bearing minerals [7,8]. Ex situ mineral carbonation has also been developed to chemically and physically accelerate mineral carbonation rates [9–13].

As shown in Figure 3.2, the most abundant materials that can fix gaseous CO_2 into solid carbonates are silicate rocks along with basinal brines and seawater. In addition, a wide range of industrial wastes, such as steel slag and fly ash, contain alkaline metals. The use of these industrial wastes in ex situ carbon mineralization is interesting because they often coexist with industrially emitted CO_2. Thus, the energy-intensive transportation of feedstock can be eliminated.

The ex situ carbonation of silicate minerals and alkaline industrial wastes can lead to co-production of valuable by-products (e.g., high-surface-area silica, iron oxide, and extracted rare earth elements) and the chemical and physical properties of

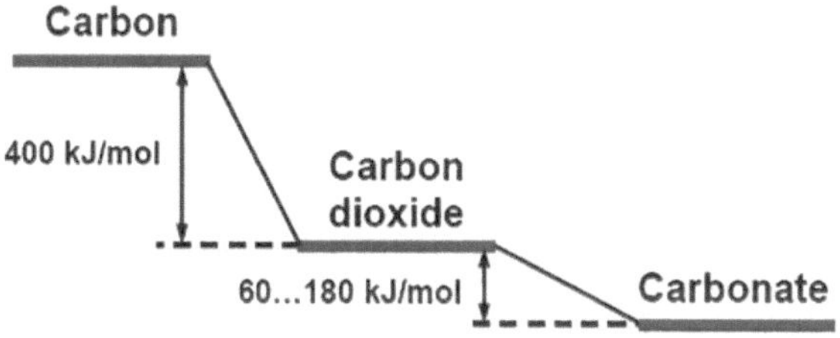

FIGURE 3.1 Thermodynamics of carbon mineralization. (Image courtesy of Ah-Hyung Park, Columbia University [2].)

DOI: 10.1201/9781003229575-3

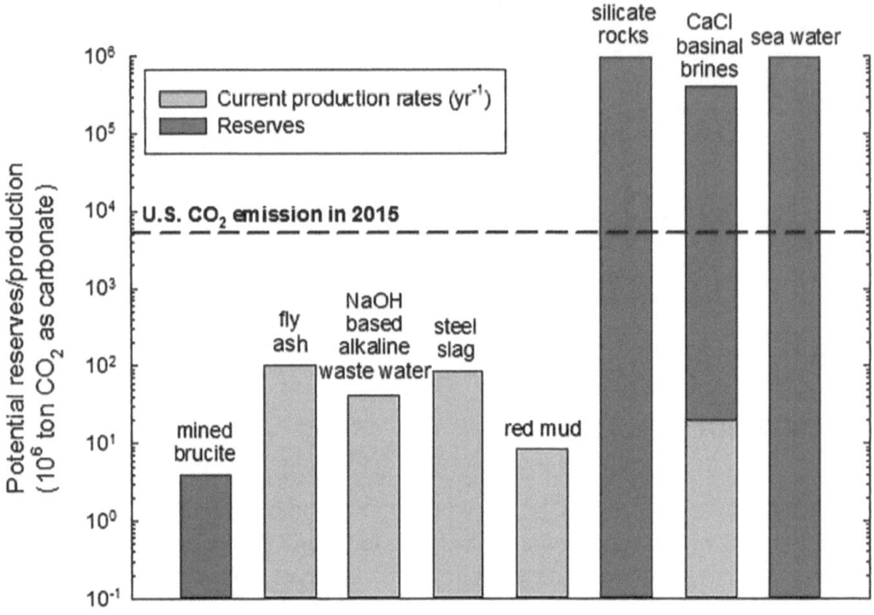

FIGURE 3.2 Various feedstocks for carbon mineralization. (Image prepared by Ah-Hyung Park, Columbia University using data from H. Xie et al. 2016. *Environ. Earth Sci.* 75 and The Statistics Portal [2].)

solid carbonates can be tuned during the carbonation process to match various applications. For example, carbonates and other by-products can be used as construction materials, which generally have high environmental footprints. A built environment that uses CO_2 from fossil fuel combustion on a scale commensurable with historical anthropogenic emissions could potentially achieve the net negative emissions. It can also displace energy-intensive and environmentally unfriendly raw material feedstocks by incorporating industrial waste streams into product fabrication.

A recent market assessment determined that carbonate mineral-based construction materials could reach annual revenues of $1 trillion and the potential to consume 3–6 GT of global CO_2 emissions by 2030 [14]. To achieve this, many different industrial feedstocks and building products are envisioned that share common production pathways and processes. In nature, carbon mineralization is generally considered a slow phenomenon, but in industrial products, it has been observed to proceed at rates orders of magnitude faster than predicted by conventional models.

We currently lack a fundamental understanding of this dynamic and highly heterogeneous reaction system for effective mineral carbonation, but there is consensus that the key to transformative advance lies in optimizing surface reactivity and mass fluxes across mineral–liquid–gas interfaces. To realize the potential of both in situ and ex situ carbon mineralization for carbon storage as well as carbon utilization, we must be able to model and control reaction pathways and rates to accurately estimate carbon storage and utilization potentials and to meet product performance criteria.

3.2 THERMOCHEMISTRY AND POSSIBLE DRIVERS FOR MINERALIZATION

Carbon mineralization (see Figure 3.3) encompasses a complex set of reactions by which CO_2 reacts with a Ca, Mg, and/or Fe oxide bearing phase (MO = metal oxide) leading to the formation of the corresponding solid carbonate phase (MCO_3). A specific feature of this carbon utilization pathway is that the overall reaction is exothermic (e.g., the enthalpy of reaction of CaO to $CaCO_3$ is 90 kJ/mol).

Solid carbonate products can be tailored for physical and chemical properties and applications by controlling feedstock, reaction pathways, and system and heat integration. A gas–solid process at a high temperature can be exploited for enhancing the carbonation kinetics of Ca oxide phases and presents the advantage of not requiring wastewater separation and processing, which presents additional costs and environmental burdens. At ambient conditions, carbonation proceeds via a gas–liquid–solid pathway. Carbon dioxide and reacting metal oxides dissolve in the liquid phase from which carbonates precipitate. Mine waste processing and cement curing are examples of processes employing low liquid-to-solid ratios (wet or thin-film routes). For other applications in which the goal is to maximize CO_2 storage or produce high-value products with greater purities, ex situ slurry phase processes are preferred.

Carbonation is a natural process where CO_2 reacts with different minerals forming solid precipitates leading to the weathering of the rocks. The reactions are spontaneous and exothermic and can be exemplified as (3.1) and (3.2) where calcium and magnesium oxides are considered to react with CO_2.

$$CaO + CO_2 = CaCO_3 \ +179\,\text{kJ/mol} \tag{3.1}$$

$$MgO + CO_2 = MgCO_3 + 118\,\text{kJ/mol} \tag{3.2}$$

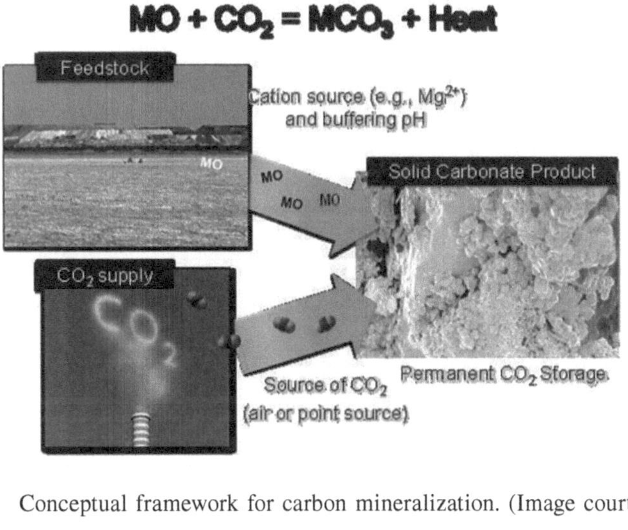

FIGURE 3.3 Conceptual framework for carbon mineralization. (Image courtesy of Greg Dipple [2].)

The most reactive compounds for CO_2 mineralization are oxides of divalent metals, Ca and Mg, and their availability in nature is mainly in the form of silicates, such as olivine ((Mg, Fe)$_2$SiO$_4$) orthopyroxene ($Mg_2Si_2O_6$–$Fe_2Si_2O_6$), clinopyroxene ($CaMgSi_2O_6$–$CaFeSi_2O_6$), and serpentine ((Mg, Fe)$_3$Si$_2$O$_5$(OH)$_4$), the latter originated by the hydration of olivine. When CO_2 dissolves in water, it reacts with these silicates forming corresponding carbonates, where CO_2 is fixed in a mineral form. Mantle peridotite and basalts deposits, enriched in Mg, Fe, and Ca silicates, are the main targets for in situ CO_2 mineralization projects. An example carbonation (CO_2 mineralization) reaction involves the interaction of CO_2 with portlandite to produce calcite and water: $Ca(OH)_2 + CO_2 \leftrightarrow CaCO_3 + H_2O$. The reaction is thermodynamically favored ($\Delta G = -74.61$ kJ/mol) and exothermic ($\Delta H = -68$ kJ/mol) and proceeds spontaneously under ambient conditions at relative humidity greater than 25%.

The use of mineral carbonation process in the past was used largely for CO_2 sequestration. In recent years due to cost limitations, more efforts are directed toward the use of this process for developing more construction materials from a variety of raw materials including industrial wastes. The chapter addresses various ways carbonation process can be used for either CO_2 sequestration or for producing useful products like cements, aggregates, synthetic concrete, and construction materials of varying properties. Mineral carbonation is attractive because of abundance of raw materials like Si, O, Al, Fe, Ca, alkalis of K, Na, and Mg in earth's crust. Furthermore, these elements are well dispersed geographically, economical to extract and according to Biernacki et al. [15] available in quantities that would last more than 100,000 years. Besides the natural availability of these elements, construction and demolition wastes can be used for road-based materials and some concrete aggregates relieving space in landfills [16].

Besides its use for CO_2 sequestration, mineral carbonation offers an attractive route to CO_2 utilization because (a) solid carbonates, the main products of mineral carbonation reactions, are already used in construction materials markets; (b) the chemistry involved in making carbonates based on calcium (Ca) and magnesium (Mg) is well known; (c) carbonation can consume large amounts of CO_2 by chemically binding it into stable, long-lived mineral carbonates; and (d) the reaction of CO_2 with alkaline solids is thermodynamically favored, thereby needing little, if any, extrinsic energy. This chapter assesses the current state of research in mineral carbonation, highlights pathways to convert CO_2 to carbonates consisting of calcite ($CaCO_3$), magnesite ($MgCO_3$), or mixtures of the two, and identifies market and environmental considerations relevant to the commercialization of mineral carbonation technologies. Both Ca and Mg carbonate can be used for aggregates and cement materials needed for construction materials.

Mineral carbonation products may have the potential to utilize up to 1 gigaton (Gt) of CO_2 annually [14] if they were to replace existing products. The large production volumes offer substantial economies of scale resulting in costs on the order of $50 per ton for concrete [17], $100 per ton for OPC, and $10 per ton for aggregates [18]. As pointed out in the national academy report [1], construction industry is, however, very much distributed and localized with 5,500 ready-mix concrete plants, hundreds of post-concrete plants, and nearly 100 OPC plants in United States alone [19]. This distributed nature means that the utilization of CO_2 in construction industry will require infrastructure for distributed CO_2 access and transportation of manufactured products.

While carbonation is, in general, a slow process, it can be accelerated through novel technologies which include various pretreatments and use of additives, temperature, and pressure. The raw materials used to form Ca or Mg carbonates are desalination brines, industrial wastes, portlandite, ordinary Portland cement, and low rank Ca silicates which are mixed with low concentration (30%) or high concentration (50%) CO_2 waste stream under high temperature and pressure to produce Ca and Mg carbonates. Carbonates, commonly in a particulate or granular form, are predominately used in the production of cement and concrete, though they also find use in a variety of industrial applications, including in paper and food production. The market for construction material is enormous; more than 50 billion tons of mineral aggregates and concretes are produced annually. This demand is likely to further increase due to emphasis on so-called "green buildings, bridges and roads".

Depending on their end use, carbonates may be formed as nominally pure compounds or as mixtures with silicates, aluminates, and/or ferritic compositions. Natural carbonates are available from mined geological sources such as limestone quarries. In mineral carbonation, synthetic carbonates are made by contacting alkaline solids, often in an aqueous suspension, with carbon dioxide [3,20]. The carbonation of mineral oxides is thermodynamically favorable and can be accomplished at close to ambient temperature. The mineral carbonation can be carried out "in situ" or "ex situ". Here we examine both options. The "in situ" carbonation is often used for CO_2 sequestration.

3.3 METHODS OF CARBONATION

Mineral carbonations can be carried out in a number of different ways. Various methods are graphically illustrated in Figure 3.4. Methods are generally broken down as gas-solid reactions or aqueous phase reactions. In each case, there are direct and indirect routes. As mentioned earlier, it can be carried out "in situ" or "ex-situ". Many routes require pretreatments of solids to accelerate the process. Here we briefly examine all of these options.

3.3.1 In Situ Carbonation

Mantle peridotite and basalts deposits, enriched in Mg, Fe, and Ca silicates, are the main targets for in situ CO_2 mineralization sequestration projects.

3.3.1.1 Peridotites

Peridotite is a component of ophiolites which are complex geological sequences representing the emplacement on the land of sections of oceanic crust. The world's largest ophiolitic outcrop is the Samail Ophiolite in the Sultanate of Oman extending for about 350 km with a width around 40 km and an average thickness of 5 km; about 30 vol% is composed of mantle peridotite. The mineralogical composition of this peridotite is 74% olivine (partially serpentinized), 24% orthopyroxene, 2% spinel ($MgAl_2O_4$), and traces of clinopyroxene. The Samail Ophiolite is characterized by

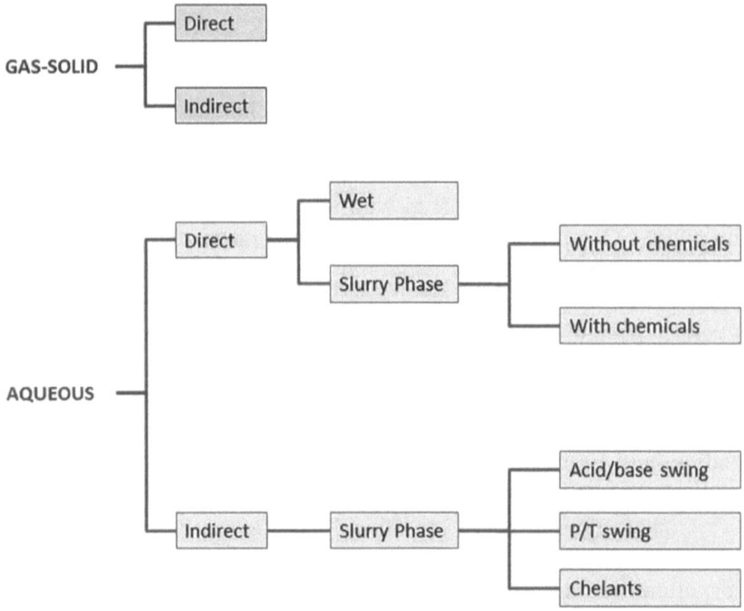

FIGURE 3.4 Main process routes for carbon mineralization [2].

the presence of an internal network of fractures hosting aquifers of variable volumes and chemical compositions, where several mineralized springs emit alkaline waters enriched in carbonates; the origin of those waters is linked to the natural carbonation process acting within the peridotite.

Surface water flows through fractures originating a shallow aquifer open to the atmospheric CO_2 and O_2 fluxes; the water reacts with the peridotite and the pre-existing carbonate rocks in an open system becoming enriched in Mg^{2+} and HCO^3. This water may infiltrate in the deeper regional aquifer which is isolated from the atmospheric fluxes. The chemical reactions with the peridotite will trigger the formation of serpentine, brucite, magnesite, and dolomite; Ca^{2+} and OH will accumulate in the water leading to a strong pH increase up to a value of 12. When these waters emerge at the surface in the alkaline springs the sudden intake of CO_2 from the atmosphere will precipitate Ca-carbonates; the mixing with the shallow aquifer will further precipitate Ca-carbonates and brucite. The formation of carbonates, mostly in the form of terraced travertine around the springs, consumes some OH^- decreasing the pH to values of 8–9. The total volume of carbonate in the Samail Ophiolite is $5.5 \times 10^7 m^3$ with an average age of 26,000 years indicating that about 4×10^7 kg CO_2 per year are consumed by the precipitation of carbonates. This natural process requires long times for the reactions to develop, in the order of magnitude of 50 years for the shallow-water aquifers and up to 5,600 years for the deep reservoirs. Artificial enhancement of carbonation can be achieved by injecting fluids with a higher concentration of CO_2 and increasing the temperature. For example, when injecting CO_2 at 90°C with 100 bar $p_{CO2,}$ about 0.63 kg of CO_2 can be permanently stored as carbonates for each kg of peridotite. A typical in situ mineralization project in the Samail Ophiolite could include the drilling of the peridotite, hydrofracturing of the hosting volume, injection of heated fluids to increase the temperature at 185°C, which is the optimum temperature for olivine carbonation rates, followed by injection of pure CO_2 at 25°C. The exothermal reaction (producing 760 kJ/kg) and the geothermal gradient (up to 20°C km^{-1}) will both contribute to the reduction in the energy needed for heating the fluids. The resulting enhancement of the carbonation rate following this process is considered to be one million times faster than the natural process pace [3].

3.3.1.2 Basalts

The largest presence of basalts is on the oceanic crust. Large outcrops of basalts are also present on the continental crust. Basalts can have a good degree of secondary permeability due to the formation of altered and brecciated horizons or networks of fractures during or after their deposition. The resulting pore space may be filled by circulating water originating from aquifers within the hosting rocks at different depths and mineral concentrations. These aquifers are often enriched in ions including Ca^{2+} and Mg^{2+}, which can react with the injected CO_2 precipitating carbonates and releasing H^+ as in reaction (3.3):

$$\left(Ca^{2+}, Mg^{2+}\right) + CO_2 + H_2O = \left(Ca, Mg\right)CO_3 + 2H^+ \tag{3.3}$$

The reaction rate is controlled by the concentration of H^+ and will not proceed further until these ions are neutralized by He reaction with the hosting rock. Considering olivine and Ca-plagioclase basalts the neutralization process follows reactions (3.4) and (3.5).

$$Mg_2SiO_4 + 4H = 2Mg + SiO_2 \tag{3.4}$$

$$CaAl_2Si_2O_8 + 8H^+ = Ca^{2+} + 2Al^{3+} + 2SiO_2\left(aq\right) + 4H_2O \tag{3.5}$$

The availability of reactive Mg, Al, and Ca silicates is therefore the controlling factor for the development of in situ carbon dioxide mineralization [3].

Following the injection of CO_2 (either as supercritical fluids or as aqueous solution), the dissolution of some minerals and the precipitation of others, mostly carbonates, may change the porosity of the reservoir; carbonate deposition during the first stages of the injection may have adverse effects on the storage potential due to the reduction in available pore space which is progressively filled by minerals, and thus clogging the surrounding of the injection well. Mineral deposition in a more advanced phase of the injection and during the post-injection phase instead is considered an advantage enhancing the trapping potential of the hosting structure.

Injecting CO_2 within the basalts of the ocean seafloor would benefit from a further series of trapping mechanisms in addition to the geochemical transformation of CO_2 in carbonates. The deep-water environment, below 2,700 m, and the cold temperature, below 2°C, will make the injected CO_2 denser than the surrounding seawater, allowing it to sink with a gravitation-trapping mechanism; the same environmental parameters are also favorable to the formation of CO_2 hydrates, where the CO_2 molecule is "encaged" within a lattice of ice strongly reducing its solubility in water in the case of seepage. Lastly, the thick sedimentary cover of the seafloor will form a low-permeability layer further reducing the probability of leakage. Feasibility studies are carried out to examine the potential of basalts, both onshore and offshore, for CO_2 storage [21].

3.3.2 Ex Situ Carbonation

Most processes under consideration for mineral carbonation focus on metal oxide (such as calcium and magnesium) bearing materials, whose corresponding carbonates are not soluble. Moreover, since waste materials rich in calcium oxide are conveniently located close to the CO_2 emission source, they have also been targeted as MC feedstock. We briefly examine the processes and methods developed for both rocks and waste resources for CO_2 sequestration.

Since oxides and hydroxides of Ca and Mg are not abundant, silicate rocks containing the desired Mg and Ca have been targeted for mineral carbonation. Table 3.1 summarizes the main minerals available and their performance in terms of mass ratio of ore necessary to carbonate the unit mass of CO_2 (R_{real}) and reaction efficiency (E_{CO2}). Serpentine, olivine, and to less extent wollastonite because of its lower abundance, are preferred based on performance and availability. The sequestration of CO_2 in carbonates can be achieved through various process routes, such as: (a) direct carbonation (DC) is the simplest approach, where a Ca-/Mg-rich solid is carbonated in a single process step. DC can be further divided into gas–solid carbonation and direct aqueous mineral carbonation. The direct aqueous mineral carbonation route with

TABLE 3.1
Mineral Chemistry, Carbonation Potential, and Reactivity

Rock	Mineral	Mg	Ca	Fe^{2+}	R_{real}	E_{CO2} (%)
Serpentine	Antigorite	24.6	0.1	2.4	2.1	92
Serpentine	Lizardite	20.7	0.3	1.5	2.5	40
Olivine	Fayalite	0.3	0.6	44.3	2.8	66
Olivine	Forsterite	27.9	0.1	6.1	1.8	81
Feldspar	Anorthite	4.8	10.3	3.1	4.4	9
Pyroxene	Augite	6.9	15.6	9.6	2.7	33
Basalt		4.3	6.7	6.7	4.9	15
Oxide	Magnetite	0.3	0.6	21.9	5.5	8
Ultramafic	Talc	15.7	2.2	9.2	2.8	15
Ultramafic	Wollastonite	0.3	31.6	0.5	2.8	82

(Carbonation test conditions: 80% 37 mm feed; 1 hour; 185 1C; P_{CO2} = 150 atm; 15% solids; 0.64 M NaHCO, 1 M NaCl). R_{real} = mass ratio of ore necessary to carbonate unit mass of CO_2; E_{CO2}% = reaction efficiency, % stoichiometric conversion of Ca, Fe^{2+}, and Mg cations in silicate feed to carbonate [3,22].

the aid of pretreatments (DCP) is considered as the state of the art and is typically selected to compare other technologies. (b) Indirect carbonation in which reactive Mg/Ca oxide or hydroxide first extracted from the feedstock and then in the second step leached cations are reacted with CO_2 to form the desired carbonate.

3.3.2.1 Direct Carbonation

The direct route is best suited to treating pure and concentrated CO_2 streams, which are generally obtained in the chemical and fertilizer production plants; nonferrous metal, iron, and steel production facilities; and cement plants.

The most straightforward process route is the direct gas–solid carbonation, and it was first studied by Lackner and co-workers [23]. Various reactions depending on the feedstock are possible. As an example, the direct gas–solid reaction of olivine is given:

$$Mg_2SiO_4(s) + 2CO_2(g) = 2MgCO_3(s) + SiO_2(s) \tag{3.6}$$

High CO_2 pressures (100–150 bar) are necessary in order to obtain reasonable reaction rates. Direct aqueous carbonation involves carbonic acid route process in which CO_2 reacts at high pressure (100–159 bar) in an aqueous suspension with olivine or serpentine [24,25]. Firstly, CO_2 dissolves in water and dissociates to bicarbonate and H^+ resulting in a pH of about 5.0–5.5 at high CO_2 pressure:

$$CO_2(g) + H_2O(l) = H_2CO_3(aq) = H^+(aq) + HCO_3(aq) \tag{3.7}$$

Mg^{2+} is then liberated from the mineral matrix by H^+:

$$Mg_2SiO_4(s) + 4H^+(aq) = 2Mg^{2+}(aq) + SiO_2(s) + 2H_2O(l) \tag{3.8}$$

Finally, Mg^{2+} reacts with bicarbonate and precipitates as magnesite:

$$Mg^{2+}(aq) + HCO^3(aq) = MgCO_3(s) + H^+(aq) \tag{3.9}$$

The direct carbonation is often carried out with mechanical, chemical, or thermal pretreatments to enhance the carbonation reaction rates and efficiencies through surface area increase. Two major processes have been developed: high-energy mechanical grinding and chemical leaching, although other methods such as thermal- and mechano-chemical-pretreatments have also been reported. The mechanical grinding approach aims at disintegrating particles or increasing the surface area. Particle size reduction takes place to reduce the particle size to 300 mm, which may be necessary to liberate valuable mineral grains. Crushing is normally performed on dry materials using compression equipment such as jaw or cone crushers. Instead, grinding is accomplished by abrasion and impact of the ore by the free motion of unconnected grinding media such as rods, balls, or pebbles [26].

The US National Energy Technology Laboratory (NETL) developed a direct carbonation process involving grinding of magnesium (or calcium) silicates at 150°C–200°C, 100–150 bar, where 0.64 M $NaHCO_3$ and 1 M NaCl were added to the solutions [27]. $NaHCO_3$ was used to turn to slightly alkaline pH the solution in order to facilitate carbonate precipitation. Olivine carbonation proceeded to over 80% in 6 hours. Wollastonite was found to be the most reactive, reaching over 70% in 1 hour, and unlike the magnesium minerals, the wollastonite reaction proceeded rapidly in distilled water [27]. The carbonation of olivine and wollastonite was controlled by the surface area consistent with the shrinking-core model, in which the particle surface reacts to release magnesium into solution, leaving a shrinking core. The higher wollastonite efficiency was related to the much higher precipitation rate for $CaCO_3$ compared to $MgCO_3$, which is four orders of magnitude lower than those of $CaCO_3$ [28].

While various pretreatment options such as ultrasonic treatment and wet grinding in caustic solution have been tested, they did not result in a higher reactivity [29]. The major problem with many other pretreatment options is the high energy input required [30]. Extensive studies on the mechanical activation of silicates were performed at NETL [31,32] and were reviewed later by Huijgen and Comans [30] and Sipila et al. [33]. The major conclusions made were that high-energy attrition grinding of silicates resulted in a higher conversion rates but consumed excessive amount of energy.

Direct carbonation of serpentine requires thermal pretreatment to remove hydroxyl groups, resulting in the chemical transformation to pseudo-forsterite. Serpentine requires heating treatment above 630°C to remove chemically bound water from the lattice [34].

$$Mg_3Si_2O_5(OH)_4 - (MgO)_3(SiO_2)_2 + 2H_2O \tag{3.10}$$

The NETL findings indicate that the reaction rate for serpentine was slow if water (OH groups) was not removed. Thermally treated serpentine at 630°C for 2 hours reached 65% CO_2 storage capacity. Similar results were obtained with high energy attrition grinding, but with a substantial associated energy penalty [27]. Partial dehydroxylation with heat integration (63% decrease in energy requirement for thermal-activation) has led to an overall mineral carbonation process estimated cost of $ 70 per tCO_2 captured [35,36], compared to $210 per tCO_2 captured in the NETL process [27].

Brent and co-workers [37] explored a better use of the system heat in order to avoid the drawbacks of serpentine thermal-chemical activation. The process is being exploited by Orica, a large Australian company. The energy savings were obtained in Shell's direct pure and flue gas mineralization technology. Since flue gas with 10% vol CO_2 has a much lower solubility than pure CO_2 under pressure, leaching of cations in the presence of CO_2 took place at a much slower rate. To avoid this, the flue gas was brought into contact with the mineral slurry prior to the precipitation stage in a separate slurry mill at ambient temperature [38]. The slurry mill achieved both a huge reduction of particle size and the formation of carbonate intermediates other than bicarbonate, for instance, hydromagnesite. Shell's thermal activation, which can reduce energy requirement up to 63%, was employed in this process [39].

Shell has developed an aqueous slurry-based mineralization technology suitable for both serpentine and olivine mineral rocks. The process comprises pretreatment, leaching, and precipitation steps, where activation of serpentine is achieved by both mechanical and thermal means [38]. The overall process resembles that developed by NETL, but operates under less stringent process conditions. The slurry from the leaching step is pressurized (up to 45 bar) and heated up to 110°C–140°C in the precipitation step. Here, precipitation of dissolved $Mg(HCO_3)_2$ takes place as well as transformation of hydromagnesite into magnesite.

A different approach has been developed by Calera (see Figure 3.5), which owns a demonstration plant at the gas-fired Moss Landing power plant (USA). The Moss Landing plant has demonstrated to capture flue gas CO_2 from a 10 MW power generator at 90% efficiency for about 2 years [40,41], This process [42] claims to reach a conversion of 98% by using large amounts of NaOH and/or electricity [43]. In addition to capturing and mineralizing CO_2, the SkyMine process also claims the possibility to clean SO_x and NO_2 from the flue gas, and remove heavy metals, such as mercury. A joint venture, namely, Skyonic Corporation, which includes BP and ConocoPhillips, constructed a commercial CO_2 capture plant to remove 83,000 tCO_2 per year from a cement plant (130,000 including the reduced emissions in producing backing soda). The strength of the process is represented by the possibility to produce valuable carbon-negative products (e.g., hydrochloric acid and sodium bicarbonate) using low-cost chemical inputs in a low energy requirement capture mineralization plant [44].

Organic acids and their anions may affect mineral weathering rates by three possible mechanisms: (a) changing the dissolution rate far from equilibrium either decreasing solution pH or forming complexes with cations at the mineral surface, which provides a new parallel reaction mechanism for the detachment of material from the mineral surface; (b) ability to make aqueous complexes with aqueous metals that would otherwise inhibit rates; and (c) changing the ions speciation in solution, which affects the dissolution rate of minerals [45,46]. Far from equilibrium, the dissolution rates of most silicate minerals increase exponentially with increasing

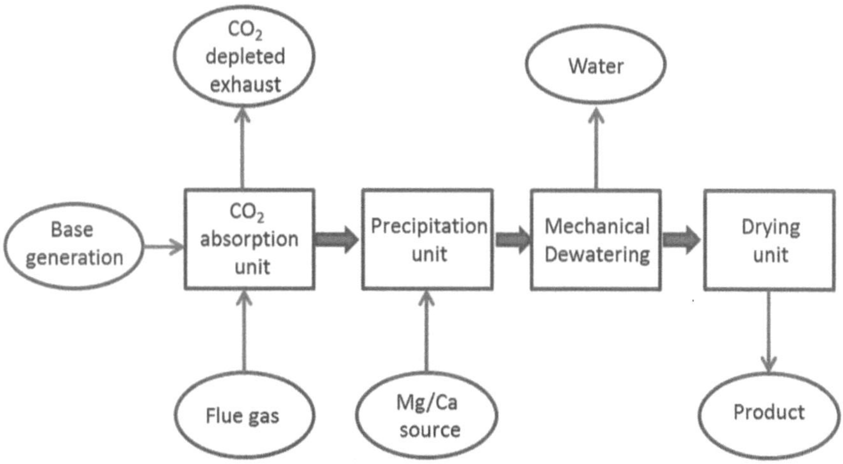

FIGURE 3.5 Calera process diagram [3].

hydrogen ion concentration (low pH) in the solution. The pH effect can be explained by the fact that sorption of protons on an oxide surface polarizes the metal–oxygen bonds, weakening the bonding with the underlying lattice [45]. The contrasting effects of organic acids on steady-state forsterite dissolution rates with increasing pH were related to their aqueous speciation, since these organic species are in the form of neutral species at acidic pH, but as negatively charged aqueous species under mild acidic and neutral conditions [47].

3.3.2.2 Indirect Carbonation

Indirect mineral carbonation refers to processes that take place in more than one stage. Indirect carbonation typically involves the extraction of reactive components (Mg^{2+}, Ca^{2+}) from minerals, using acids or other solvents, followed by the reaction of the extracted components with CO_2 in either the gaseous or aqueous phase. Pure carbonates can be produced using indirect methods, due to the removal of impurities in previous carbonate precipitation stages [48]. A wide number of strong acids and bases such as HCl, H_2SO_4, and HNO_3 have been employed for the dissolution of silicate rocks [49].

In a typical capture process, CO_2 is first absorbed by chemicals (e.g., NH_3) and then desorbed (to recover the sorbent) and compressed for transportation, where stripping and compression consumes about 70% of the total CCS energy consumption. Since CO_2 captured as sodium carbonate/bicarbonate is directly used in the proposed mineral carbonation, there is no need for desorption and compression of CO_2. This process as other pH swing processes is also able to separate three different products: silica, magnesite, and iron oxide [10,48]. This process could also be integrated with the chilled ammonia CO_2 capture process, which has been demonstrated to capture more than 90% of CO_2 (from 3% to 15% CO_2 in flue gas) [50] and an estimated energy penalty of 477 kW h per tCO_2 [51].

The main drawback of the aqueous pH swing ammonium-based process (see Figure 3.6) is the large amount of water that needs to be separated from the salts during the regeneration step. Wang and Maroto-Valer [52] indicated that when a

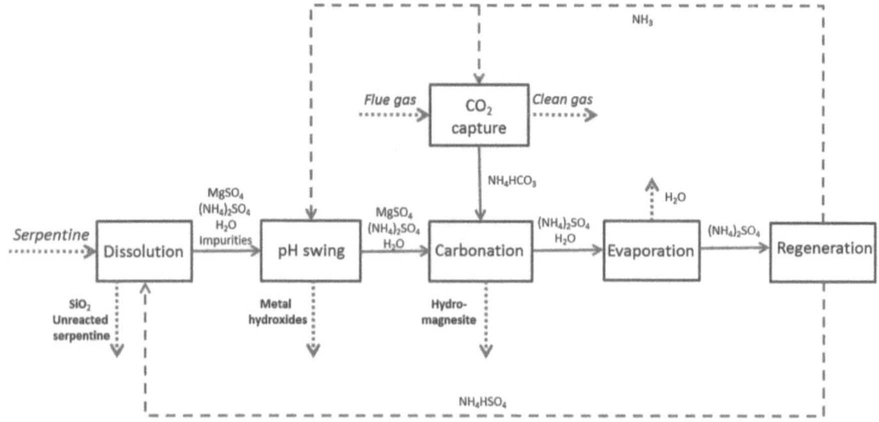

FIGURE 3.6 pH-swing CO_2 mineral carbonation process with recyclable ammonium salts [3].

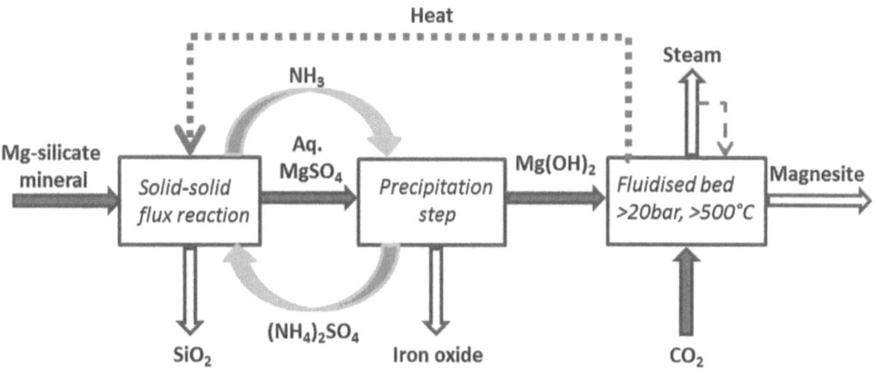

FIGURE 3.7 Åbo Akademi mineral carbonation process [3].

solid to liquid (S/L) ratio or 50 g/L was used, 50–56 tH_2O were required to sequester 1 tCO_2. Since water evaporation is a high-energy penalty process, they attempted to reduce the water usage in the system [52]. When the S/L ratio increased to 300 g/L, 16 tH_2O were required to sequester 1 tCO_2. However, since the CO_2 fixation efficiency decreased to 46.6%, a larger amount of reactants (serpentine and salts) were required. Moreover, the amount of water to be evaporated is still too high and alternative separation methods need to be investigated in order to make this process economically feasible.

A two-step process that also uses ammonium salts has been recently developed (see Figure 3.7) by Fagerlund et al. [53,54].

3.3.3 Comparison of MC Options

In situ MC has a great potential in terms of the volume of CO_2 which could be permanently fixed within the hosting rocks as solid carbonates thus reducing the risk of potential seepage from the storage site. There is a large availability of minerals, which can react in situ with the injected CO_2, both onshore and offshore and often close to anthropogenic sources of CO_2. In situ MC can also be beneficial for the worldwide development of storage projects. Abundant onshore and offshore basalts and peridotites are available for in situ low-temperature mineralization. The largest layered onshore basalt formations are located in India (provinces of Deccan Traps), United States (Columbia River basalts), Russia (Siberian Traps), and UAE/Oman [55]. The current limits of in situ carbonation are due to the slow pace of the process and the need for artificial ways of enhancement of the chemical reactions which require a large amount of energy. Identifying specific sites where natural characteristics such as geothermal gradients are favorable to the carbonation process may reduce the associated costs.

Ex situ MC presents intrinsic materials handling issues, due to the large mineral requirements and associated reaction products, which result in a large process scale (larger than actual power plant materials handling), and it seems to be only employable to existing small-medium emitters. MC may be suitable to large emitters if the new plants are designed with the required infrastructures. Since for small-medium emitters, geologic sequestration may not be an economically viable option, and there are no commercialized processes that specifically address this technology gap, MC may target this market. Large ultramafic rock deposits within a 100–200 km radius of power/industrial plants emitting over 1 Mt per year CO_2 are available in South Africa, China, Russia, Kazakhstan [56], New South Wales in Australia [36], United States, and Europe [27]. However, not all these resources are easily accessible. Mg-bearing silicates such as serpentine and olivine represent the most suitable mineral resources, while other Mg-silicates and Ca-silicates are less attractive due to their low Mg content and/or low availability.

Moreover, a number of large-scale industrial wastes can be considered as feedstock for CO_2 mineralization. Regardless of several benefits, such as avoiding costs for mining and transport, the current CO_2 mineralization technologies developed for wastes still cannot compete with geological storage in terms of potential quantity and cost of sequestrated CO_2. The ones that appear to carbonate easily under mild conditions (contain free lime, do not require additional grinding, bind CO_2 effectively even from dilute flue gases, etc.) and have a high carbon sequestration capacity (APC residues, APMW, OS FA) are only available locally or in too small quantities to make a global impact. However, especially in countries that lack geological storage, these options should be considered (for instance OS FA could capture 10%–12% of CO_2 emitted from OS-based heat and power sector [22].

Overall, the processes that are attracting major attention (see Figure 3.8) and that seem to be viable at this point have in common the potential production of sellable products, the co-removal of different pollutants from the flue gas, and process integration essential to lower the costs. The conceptual integration of high temperature and pressure industrial mineral carbonation facility into a developing mine site has been recently demonstrated to be feasible at an operating cost of B$83 per tCO_2 [57].

Direct gas-solid processes, which require temperatures up to 500°C and fine grinding of minerals (5–35 mm), achieve low capture efficiency and are not viable on the industrial scale at the current scale of development. On the contrary, it is well documented in the literature that the presence of water considerably enhances the

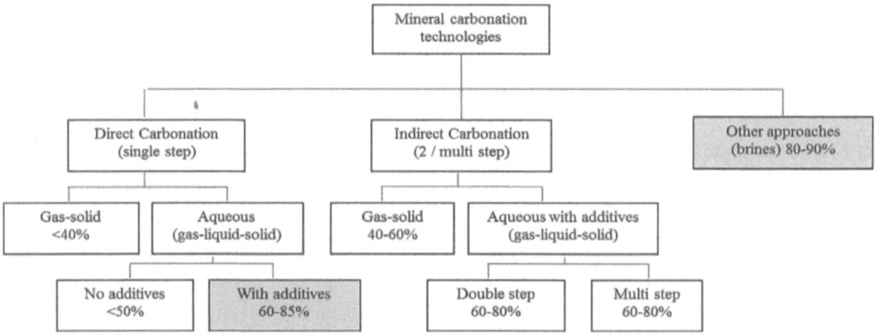

FIGURE 3.8 Mineral carbonation process routes. In dark the most promising technologies at the current state of research and development [3].

reaction rate in the carbonation process. Feedstock pretreatment by fine grinding, thermal activation, and chemicals in direct aqueous carbonation processes shows significant improvements in CO_2 capture efficiency (up to 85% with pure CO_2 stream) [27]. Meanwhile, the regeneration and recyclability of additives (NaOH, $NaHCO_3$) still need to be addressed. The NETL-modified processes proposed by Brent and Shell make use of the low-grade heat from power plants and from the serpentine thermal-activation to decrease the overall energy consumption. However, no public data are available to estimate the potential deployment and costs associated with these processes.

Aqueous mineral carbonation can be carried out via direct and indirect pathways (see Figure 3.4). Direct carbonation refers to processes whereby the extraction of mineral ions from the feed material and the carbonation reaction occurs simultaneously as a single step. At elevated temperatures and pressures, conversion is rapid, whereas at ambient conditions, such as in mine waste piles, direct conversion proceeds at a slower pace but on a scale of millions of tons, considering the large amounts of material that can take part in the reaction under in situ conditions. On the other hand, carbon mineralization via the indirect route involves two or more stages. It requires the extraction of mineral ions from the feed material (dissolution) as a preliminary procedure prior to magnesium carbonate precipitation. By separating the dissolution and precipitation steps, carbonation can be completed under milder conditions because of the independent optimization of each step. This leads to an enhancement of the overall carbonate conversion rate while also generating a product with higher purity and hence potential for utilization. Acceleration of the dissolution step can be achieved through energy-consuming pretreatments (e.g., heat activation), chemical additives (e.g., Mg-, Ca-, and Si-targeting ligands [9–13]), and carbonic anhydrase (biocatalyst [58]). In the case of chemically and biochemically enhanced carbon mineralization, the recyclability of the chemical additives would be one of the critical factors for economic feasibility.

Multistep aqueous indirect processes in the presence of additives are also able to reach high carbonation efficiency using mild process conditions and short residence time as a result of faster reaction kinetics in the presence of additives. However, the energy-intensive chemical regeneration step is slowing the development of this group of technologies. Also, the use of catalytic enzymes such as carbonic anhydrase is unlikely to be effective due to their instability and very high costs. At the current state of the art, indirect routes seem to be still too expensive to be competitive as CCS technology for large deployment.

3.4 RAW MATERIALS AND ASSOCIATED TECHNOLOGIES

As discussed earlier, the largest carbon mineralization potential lies with natural silicate minerals. There are, however, many industrial processes producing reactive materials and CO_2 gas streams at a scale that an indirect approach can be applied to both concentrated and diluted CO_2 streams, including untreated flue gases at both small and large scales [59,60]. The industrial environments and feedstocks are highly variable, but all share alkaline reaction conditions and similar reaction pathways. In this section, the discussions focus on mineral and industrial wastes that are readily available for CO_2 utilization and storage. The chemistry and process technology applied for wastes can also be used to provide insights into in situ carbon mineralization occurring in geologic formations into which CO_2 is injected to be stored.

Magnesium-rich ores from mining can be used as feedstocks for carbon mineralization, including dunite, harzburgite, and serpentinite, composed primarily of silicate minerals (e.g., olivine, serpentine, pyroxene). These ores liberate magnesium in the form of cations in aqueous media, which subsequently react with CO_2 to form the carbonate solid product. The reactivity of silicate minerals is enhanced through pretreatment, either in the form of grinding, as in mine waste, or in the form of thermal activation to elevated temperatures (some as high as 700°C) [32]. The substantial, established reserves of these ores, and the possibility of producing products of commercial value following reaction with CO_2, make this utilization technology appealing. The technology, however, requires some pretreatments of ore which should be done with a minimum consumption of energy. The dissolution of the silicate minerals results in a silica-rich passivation layer that limits the overall conversion. Thus, an understanding of the reaction kinetics and mechanisms associated with the transition from the surface reaction kinetic limited reaction regime to the mass transfer limited regime is important to optimize the extent of the carbon mineralization for various minerals.

Some of the drawbacks of mineral carbonation of primary earth minerals could be avoided by using solid wastes generated from large-scale industrial processes such as coal or oil shale-fired power plant, solid-waste incinerator, cement plant, steel and paper industry, and many others as a feedstock [29,61]. This approach has a number of advantages:

(a) these materials are often associated with CO_2 point source emissions;

(b) they tend to be chemically less stable than geologically derived minerals [39] and thus require a lower degree of pretreatment and less energy-intensive operating conditions to enhance carbonation yields [39];

(c) waste materials could supply a readily available source of calcium or magnesium mineral matter (preferably in the form of CaO or $Ca(OH)_2$) without the need for mining; they are typically fine-grained with high reactive surface areas (CKD, CBD, and AODS);

(d) hazardous waste can be reclassified through pH-neutralization and mineral transformation (MSWI, APC, asbestos tailings, RM, OS FA); and finally,

(e) the end product of the sequestration step may be amendable for reuse in products such as road base or other construction material [62] as well as pure and precipitated Ca or Mg carbonates [61,63]. On the other hand, the amount of industrial waste materials available is relatively limited and rather unpredictable due to developments in technology (changes in availability and chemical composition) and legislation issues. Some of the waste materials for mineral carbonation are illustrated in Figure 3.2. In recent times, the research has focused on assessing and maximizing the storage of CO_2 by optimizing the operating conditions including pressure, temperature, liquid-to-solid ratio, gas humidity, gas flow rate, liquid flow rate, particle size, and solid pretreatment [62,64,65].

Generally, the waste carbonation reaction could occur in four routes: (a) conversion inside the solid particle, (b) $CaCO_3$ crystallization on the surface, (c) $CaCO_3$

precipitation in bulk solution, and (d) attachment on solid solution. According to Huntzinger et al. [66], Huijgen et al. [39,67], and Uibu and Kuusik [68], the main mechanisms affecting the rate and extent of carbonation are transportation-controlled mechanisms such as CO_2 and Ca^{2+}-ions diffusion to/from reaction sites, boundary layer effects (diffusion across precipitate coatings on particle surface, dissolution of $Ca(OH)_2$ at the particle surface), and pore blockage/precipitate coating. Typically, the classical shrinking core-type model has been used for describing heterogeneous solid–fluid reactions for determination of the rate-limiting mechanism [66,69–71].

Asbestos, copper, nickel, platinum deposits, diamondiferous kimberlite pipes, and podiform chromite deposits hosted by dunite, serpentinite, and gabbronorite produce tailings suitable for mineral carbonation [59,72]. Also, the bauxite residue (red mud) from alumina processing is a suitable feedstock for CO_2 sequestration [73,74]. In addition to CO_2 storage, the carbonation treatment also improves the properties of wastes, especially in the case of asbestos tailings and red mud, enabling safer landfilling or reuse [2]. Production of 1 t of asbestos (4 Mt globally) generates ca. 20 t of tailings [59]. The tailings from chrysotile processing are often associated with residual asbestos and, are therefore classified as hazardous wastes. Carbonation of asbestos tailings could be useful in several ways, as they contain up to 40% MgO, the mining and size reduction are already done and the asbestiform nature of the mineral is destroyed. Thus, both the remediation of hazardous waste and the sequestration of CO_2 could be simultaneously achieved [27].

Alkaline mine waste and tailings generated by the mining of nickel, chrome, platinum group metals, diamond, chrysotile, copper, and gold are also reactive with CO_2 for carbon mineralization. The use of tailings in accelerated applications is geographically limited to mine sites situated within a few hundred kilometers of CO_2 point sources. The most studied tailings are those mainly composed of ultramafic rocks, because of their high storage potential (around 0.40 t CO_2/t). Experiments with serpentinite-based residues showed promising carbon mineralization rates from both concentrated and pure CO_2 streams under aqueous conditions. Combination with metals extraction and value-added products increases the potential application of such routes [75]. Documented rates of CO_2 uptake at mine sites in Canada, the United States, Australia, and Norway range from 0.3 to 6 kg CO_2/m^2 per year, which is 2–5 orders of magnitude greater than natural weathering rates for Earth's major river catchments.

As high-grade sulfide deposits are almost depleted and laterites require more complex processing than sulfide ores, the nickel industry has focused on low-grade sulfide resources, often hosted in ultramafic rocks [59]. Valorizing these ultramafic tailings could make marginal nickel projects economically feasible. Teir et al. [76,77] extracted magnesium from serpentinite. The Mg extracts were carbonated (carbonate conversion 94%) in a multistage process with a carbonate conversion of 94% and producing individual precipitates of silica, iron oxide, and hydromagnesite of 93%–99% purity [77].

As mentioned above, red mud (RM) is the caustic waste material of bauxite ore processing for alumina extraction. Producing 1 t of alumina generates 1.0–1.5 t of highly alkaline RM (70 Mt annually) [74]. Three largest components of RM are Fe_2O_3 (30%–60%), Al_2O_3 (10%–20%), and SiO_2 (3%–50%) with less than 10%

concentrations of Na_2O, CaO, and TiO_2 each [78]. Mineral carbonation of RM reduces its toxicity and leaching behavior in terms of long-term storage in addition to CO_2 sequestration [74]. Carbonated RM can also be used for various applications such as fertilizers, brick and tile industry, plastics industry, wastewater treatment, and cement production [73]. RM is generally carbonated via a direct process route at ambient temperatures and pressures [73,74,78] and a sequestration capacity of 0.04–0.05 tCO_2 per t RM has been reported. At Kwinana in Western Australia, Alcoa operates a residue carbonation plant, where gaseous CO_2 from a nearby ammonia plant is contacted with RM, reducing the pH of the slurry to a less hazardous level and capturing in the process 0.030–0.035 tCO_2 per t of RM [40].

Industrial wastes such as APC wastes and ashes from solid fuel combustion often contain a considerable amount of free lime. For aqueous carbonation processes, irreversible hydration of calcium oxide is followed by simultaneous dissolution of $Ca(OH)_2$ and dissociation of aqueous CO_2 precedes the carbonation reaction. As the Ca^{2+}-ions are converted to $CaCO_3$ and precipitated out, more $Ca(OH)_2$ dissolves to equalize the Ca^{2+} concentration. APC residues are formed in the process of the flue gas treatment and typically contain a mixture of fly ash, unburned carbon, and unreacted lime. Due to the lime content (typically pH 4–12), and high concentration of heavy metals (Zn, Pb, Cd, Cr, Cu, Hg, Ni), soluble salts and chlorinated compounds, APC residues are classified as a hazardous waste [59,79]. High percentage of readily active calcium hydroxides makes the carbonation of APC residues potentially suitable for CO_2 sequestration [79]. Also, the APC carbonation products present a pH value that meets the regulatory limits (pH of 9.5) [80]. Incineration of MSW generates solid residues, bottom ash and air pollution control (APC) residues, and atmospheric CO_2 emissions [59,81]. As Ca and Mg contents of MSWI BA (nonhazardous waste) are typically too low for significant CO_2 sequestration, the mineral carbonation technique is mainly applied to achieve a chemically stable structure with improved leaching behavior [79] for different applications (e.g., secondary building material in road sub-bases, wind and noise barriers, etc.) [59,81].

Coal-fired power plants generate annually 12,000 Mt CO_2 and 600 Mt fly ash (FA) [59,82]. About 30% of coal FA is utilized for construction materials [82]. Coal FA is a fine powder (particle size typically 10–15 mm), whose composition varies depending on the mineral content of fuel. Generally, it consists of an amorphous aluminosilicate glass matrix ($Si_xAl_yO_z$) and recrystallized minerals, including quartz (SiO_2), cristobalite (SiO_2), and mullite ($3Al_2O_32SiO_2$) [82]. FA from oil-shale combustion has been investigated as a potential sorbent for mineral carbonation. The maximum CO_2 sequestration potential of FA from Ca-rich lignite type coal or oil shale ashes [83] can be as high as 43%–49%. Studies have mainly been focused on the direct aqueous carbonation route under mild process conditions with either water [83–85] or brine [86] as the reaction medium or by natural weathering over a longer period of time [85]. A pilot-scale mineral carbonation process was developed and tested by reacting coal FA with flue gases in a fluidized bed reactor at a 2,120 MW coal-fired power plant in Point of Rocks, USA.

Alkaline industrial solid residues are highly reactive with CO_2 because of their formation at high temperature and high surface areas from comminution. Their colocation with large industrial CO_2 emitters reduces transportation requirements.

Economic co-benefits include the immobilization of potential contaminants and extraction of high-value commodities such as rare earth elements. Direct, indirect, dry, wet, and aqueous processing routes have been tested, giving a broad portfolio of application opportunities. Similar process pathways exist for the treatment of waste concrete, coal fly ash, cement kiln dust, paper mill waste, municipal solid waste incineration residues, and even asbestos wastes [87]. Cellulose pulp production for paper manufacture results in the formation of several types of alkaline paper mill wastes (APMW), which typically contain 45%–82% [88] free CaO and are therefore suitable sorbents for mineral carbonation. The pulp mills also generate CO_2, which could be used to carbonate the APMW. Produced $CaCO_3$ could be utilized in the pulp and paper industry or sold as a value-added by-product [59,88].

Steelmaking processes generate significant amounts of CO_2 (0.28–1 t-CO_2 per t-steel [89]), accounting for 6%–7% of global CO_2 emissions [90]. Also, globally, these processes generate about 315–420 Mt per year [61]. Iron and steel slags consist mainly of Ca-, Mg-, Al-silicates and oxides in numerous combinations [91]. Their annual total CO_2 emissions are estimated to be up to 171 Mt of CO_2 [92], representing about 0.6% of global CO_2 emissions from fuel combustion [59]. In general steel-making slags require grinding as carbonation pretreatment [59], but the cost of mining and transportation to CO_2 emission sites can usually be avoided. Mineral carbonation of steel slag is in most cases carried out in a water slurry phase (L/S 4 1 w/w) at ambient [89,93] or elevated pressure and temperature [39,67,94].

Quaghebeur et al. [5] examined mineral carbonation opportunities for the recycling of steel slags and other alkaline residues that are currently landfilled. The Carbstone process was initially developed to transform non-hydraulic steel slags [stainless steel (SS) slag and basic oxygen furnace (BOF) slags] in high-quality construction materials. The process makes use of accelerated mineral carbonation by treating different types of steel slags with CO_2 at elevated pressure (up to 2 MPa) and temperatures (20°C–140°C). For SS slags, raising the temperature from 20°C to 140°C had a positive effect on the CO_2 uptake, strength development, and the environmental properties (i.e., leaching of Cr and Mo) of the carbonated slag compacts. For BOF slags, raising the temperature was not beneficial for the carbonation process. Elevated CO_2 pressure and CO_2 concentration of the feed gas had a positive effect on the CO_2 uptake and strength development for both types of steel slags. In addition, the compaction force had a positive effect on the strength development. The carbonates that are produced *in situ* during the carbonation reaction act as a binder, cementing the slag particles together. The carbonated compacts (Carbstones) have technical properties that are equivalent to conventional concrete products. An additional advantage is that the carbonated materials sequester 100–150 g CO_2/kg slag. The technology was developed on lab scale by the optimization of process parameters with regard to compressive strength development, CO_2 uptake, and environmental properties of the carbonated construction materials. The Carbstone technology was validated using (semi-)industrial equipment and process conditions.

Steel-making by-products, including basic oxygen furnace and electric arc furnace slag, are derived from the reaction between flux materials (e.g., limestone and dolomite) and the silica/silicate impurities in iron ore. The major constituents of slag include CO_2-reactive phases such as calcium-containing silicates, CaO (i.e., free lime), and MgO. In terms of abundance and reaction rates, slag is one of the most promising feedstocks for carbon mineralization. In addition, its high CaO content hinders direct use in bound building materials, since it causes a decrease in durability due to hydration and swelling when exposed to moisture. Carbon dioxide treatment and conversion to carbonates circumvent these issues, thereby increasing the utilization potential of the slag. Several types of process routes have been tested, including a gas–solid route to be applied directly on slag as it is poured out of the furnace [95].

Low-rank calcium silicates ($CaO/SiO_2 < 2$, molar ratio) are well known to carbonate, especially in the presence of water and elevated CO_2 concentrations. This reaction can be used to carbonate a range of otherwise slightly hydraulic low-calcium silicates including rankinite (Ca_3SiO_7) and wollastonite (Ca_3SiO_5). The carbonation of low-rank silicates in such a manner yields intermixed calcium carbonate and amorphous silica, which serve as effective binding agents [96]. However, to realize favorable reaction kinetics, carbonation reactions in these systems need to be carried out using concentrated CO_2; otherwise, the kinetics are typically too slow for practical exploitation.

While low-rank silicates can be carbonated, the constituents of anhydrous OPC ($CaO/SiO_2 \geq 2$, molar ratio) are much more resistant to carbonation. This is because, in the presence of water, there exists a competition between hydration and carbonation, with hydration being the preferred pathway [97]. Carbonation of hydrated OPC and traditional concrete proceeds very slowly under ambient conditions due to the hindered diffusion of CO_2 [98]. As a result, it can take years for just the first few centimeters of a concrete to carbonate. Despite this slow rate, the tremendous volume of concrete emplaced worldwide can serve, over time, as a considerable sink for CO_2 [99–101]. Furthermore, crushed cementitious construction and demolition wastes in the form of fine particulates could achieve substantial CO_2 uptake.

Cement kiln dust (CKD) is a fine by-product of Portland cement and lime high-temperature rotary kiln production [82]. The cement industry generates 0.15–0.20 t of CKD per ton of cement (world output 2.8 Gt) [59,82] and 5% of global CO_2 emissions [102]. Typical CKD contains 38%–48% CaO and 1.5%–2.1% MgO [59,66,82], but a significant amount of CKD is already carbonated (CKD contains 46%–57% $CaCO_3$ [66]). Cement bypass dusts (CBD), which are removed after kiln firing, have much lower carbonate content than CKD, and therefore much higher potential to capture CO_2 (0.5 t-CO_2 per t-CBD) [103]. Waste cement is a by-product obtained from the aggregate recycling process, where waste concrete is pulverized and classified to separate the aggregate from the waste cement. According to Bobicki et al. [59], waste cement has a potential to store up to 61 Mt CO_2 considering the annual waste concrete production of 1,100 Mt from EU, USA and China together [59]. However, the majority of waste cement is currently already reused in construction applications [82]. Teramura et al. [104] used a CO_2-activated hardening process to produce building materials, where waste cement was mixed with water (50% H_2O) before molding it into bricks, curing with CO_2 and drying overnight. Kashef-Haghighi and Ghoshal [105] achieved a carbonation efficiency of 18% and an E_{CO2} of 8.9% by curing fresh concrete blocks in a flow-through reactor (20% CO_2 in N_2,

20 1C and 60 minutes). Katsuyama et al. [106] and Iizuka et al. [107] produced high-purity $CaCO_3$ using an indirect aqueous carbonation route for the extraction of Ca^{2+} from cement waste by pressurized CO_2 (30 bar) and subsequent carbonation at reduced pressures (1 bar).

The carbonation of hydrated lime (portlandite, $Ca(OH)_2$, often called slaked lime) has been practiced over millennia [108]. Hydrated lime mortars take up atmospheric CO_2 at ambient conditions over long periods of exposure, resulting in the formation of calcium carbonate. This process, while far too slow to be practical as a route for CO_2 utilization (similar to the carbonation of hydrated OPC and traditional concrete under ambient conditions), can be greatly accelerated under suitable conditions of temperature, pressure, CO_2 concentration, and relative humidity. Similar to atmospheric carbonation, accelerated carbonation of portlandite also results in the formation of a monophasic $CaCO_3$ product (often calcite) with robust cementation properties. The substantial tendency of portlandite to carbonate allows for accelerated carbonation to be accomplished even using relatively dilute streams of CO_2 ($\geq$5% CO_2)—for example, using flue gas emitted by coal, natural gas, or cement plants—as long as suitable conditions of relative humidity and temperature can be maintained. Furthermore, portlandite offers among the highest CO_2 uptake per unit mass (0.59 g CO_2 per gram of portlandite), and its fast reaction kinetics allow near-complete carbonation to be achieved over the course of hours [20].

During carbonation of cementitious materials, a sequence of individual steps occur: (a) CO_2 diffusion in air and (b) permeation through the solid is followed by (c) solvation of CO_2(g) to CO_2(aq), (d) hydration of CO_2(aq) to H_2CO_3, (e) ionization of H_2CO_3 to H^+, HCO^3 and $CO_3{}^2$, (f) dissolution of cementitious phases (Ca_3SiO_5, Ca_2SiO_4) releasing Ca^{2+} and $SiO_4{}^2$ ions, (g) nucleation of $CaCO_3$ and calcium–silicate–hydrate gel, (h) precipitation of solid phases, and (i) secondary carbonation by converting calcium–silicate–hydrate gel ultimately to silicate hydrate gel and $CaCO_3$ [87,109]. The extent and rate of carbonation depend mainly on the diffusivity and reactivity of CO_2, which in turn depend on the binder type and hydration degree as well as pore type and process conditions (CO_2 partial pressure, relative humidity, temperature, and pressure) [109].

When CO_2 in a vapor state is injected into brines at ambient temperature and pressure, hydrated carbonate products form spontaneously, as long as sufficient pH buffering is provided. This approach results in favor of the formation of nesquehonite ($Mg(HCO_3)(OH){\cdot}2H_2O$) at the expense of other reaction products [110]. The approach has been used, for example, to produce nesquehonite-based products that yield comparatively low compressive strengths, on the order of 8 MPa [111]. The value of alkaline brines for carbon mineralization is in the substantial quantities of calcium and magnesium (and occasionally strontium) that can trap low-pressure impure CO_2 as soluble bicarbonates or be precipitated as solid carbonate minerals. Brines have a significant capacity to store CO_2 and also a relevant potential to produce value-added products, such as fillers and pigments for paper, polymers, and coatings. Carbonation of oil/gas brines also leads to co-precipitation of toxic heavy metals (e.g., Pb, Ni, Zn), which reduces the environmental impact of brines discharged into surface or subsurface waters.

As shown above, industrial (alkaline) wastes such as fly ash, slags, mine tailings, cement kiln dust, and air pollution control residues constitute the by-products of coal combustion, metal processing and mining, OPC production, and waste incineration, respectively. Such residues span a diversity of compositions as a function of the (a) parent coal composition, (b) ore-refining process, (c) ore composition, (d) waste make-up, and/or (e) combustion processes. Each of these industrial wastes features intrinsic alkalinity (e.g., they are mixtures of Ca and Mg with Al, Si, and alkalis) that can be neutralized by reaction with CO_2. However, the effectiveness of carbonation of these waste depends on the level of alkalinity. As an example, fly ashes can be broadly characterized as being Ca rich (Class C) or Ca poor (Class F) following standardized classifications. The extent of carbonation in such systems is critically linked to their mobile (readily reactive) Ca content. As a result, Ca-poor fly ashes offer little if any carbonation and strength gain compared to their Ca-rich counterparts. In fly ashes and other alkaline wastes the Ca and Mg present is not all mobile; as a result, such wastes often display carbonation extents that are substantially inferior to those inferred from their simple oxide (CaO or MgO) compositions [112].

3.5 CHALLENGES AND PERSPECTIVES FOR CARBONATION PROCESSING

There are several challenges that need to overcome to make the best use of MC for CO_2 sequestration and utilization. Some of these are outlined below.

3.5.1 Required High Levels of pH in the Solution

The success of carbonate processing requires alkaline precursors and high pH levels in the solution at all times. Alkaline precursors are progressively neutralized by contact with CO_2. High pH levels promote carbonate precipitation. These requirements are difficult because the production of the alkaline precursors and reactants typically requires high-temperature activation. For example, the production of portlandite ($Ca(OH)_2$) requires the thermal desorption of CO_2 from limestone (i.e., at $T \approx 800°C$, $p = 1$ bar) to produce lime (CaO), which is subsequently hydrated to form portlandite. The portlandite thus formed, following contact with CO_2, carbonates to re-form calcite. This high-temperature process is, at best, CO_2 neutral when viewed from the perspective of the mineralized CO_2. The development of truly CO_2-neutral or CO_2-negative pathways requires the development of new scalable low-temperature or hydrothermal routes for producing alkaline precursors, such as from waste streams containing Ca and Mg. Furthermore, the dissolution of CO_2 in water produces carbonic acid in equilibrium with CO_2 at pH level of 6.35. $CaCO_3$ solubility in water decreases with increase in pH. In order for $CaCO_3$ to sustain precipitation, there is a need for alkaline buffering. Therefore, increasing the pH is the key means of accelerating carbonation rates. This requires the development of chemical additives that can enhance carbonation rates by affecting reaction controls. Enhancing growth rates, facilitating surface dehydration, and surface complexation that promotes subsequent adsorption and heterogeneous growth can improve carbonation rates.

3.5.2 Availability and Suitability of CO_2 Streams

For most alkaline reactants, kinetics is favored by increasing CO_2 concentration [113–115]. While significant industrial sources of concentrated CO_2 streams are available, the vast majority of waste CO_2 streams are dilute ($\leq$25% CO_2 v/v). Thus, for most alkaline feedstocks there is a need for either low-cost CO_2 capture and concentrate systems or the ability to handle large volumes of gas streams containing components other than CO_2. The distance between dilute CO_2 stream availability and major construction markets (cities) is a major issue. This may necessitate (a) the co-location of manufacturing facilities for carbonated materials alongside CO_2 emissions sites, which may also offer waste heat that favors the advancement of chemical reactions, or (b) transporting CO_2 to production and consumption centers. This strategy may impose additional costs related to CO_2 capture and/or transport when CO_2 distribution pipeline network is not available. Due to low profit margin in construction industry, it is important to consume low-concentration CO_2 stream near its site of production. Unlike high-concentration CO_2 stream, low-concentration CO_2 stream can be obtained at low or no cost or at presumably the equivalent of a landfill tipping fee.

3.5.3 Issues Related to Feedstock, Precursors, and Products

MC requires some precursors that are cheap and abundant for large-scale operation. This may be problematic if the process becomes too large. OPC and coal combustion wastes are produced at industrial scales worldwide. As such, global production of these materials is currently on the order of 4.1 billion and 1 billion tons on an annual basis for OPC and coal combustion wastes, respectively, with pricing on the order of \$100 per ton, and \$10–\$100 per ton (depending on location and the material's suitability for use in traditional concrete), respectively. Reactants appropriate for carbonation, including portlandite and wollastonite ($CaSiO_3$, a low-rank silicate), are currently produced at a global level approaching 350 million tons [116] and 0.8 million tons, respectively. In principle, OPC, hydrated lime, and wollastonite can all be produced using similar facilities (such as typical OPC plants), as long as suitable raw materials are available in the local vicinity. As such, based on volume scaling and the availability of raw materials, it might be possible to produce portlandite and wollastonite at cost parity to OPC, and with considerably reduced CO_2 intensity due to the reduced synthesis temperatures needed [117] and their lower molar content of CaO per unit mass. While coal combustion wastes are currently readily available, this may change with a transition away from coal-fired electricity generation. It is unlikely that the production of carbonated aggregates will be produced at cost parity compared to quarrying. Therefore, in the absence of suitable incentives, carbonated binders may offer a more market-viable proposition than carbonated aggregates.

Shipping of construction materials, whether as constituents or as preformed elements, is expensive on account of their substantial bulk. Thus, their transport is cost effective only over a few hundred miles, if by road. While rail or over-water transport may be more cost effective in some cases, it typically requires shipment of much larger quantities of product. These logistical limitations suggest that carbonation technologies may be most effectively implemented at multiple discrete sites rather than at large centralized operations, reducing the economies of scale.

3.5.4 Feedstock Effects on Physical and Chemical Barriers for Carbonation

National academy report [1] points out various physical and chemical barriers to carbonation created by the selection of raw materials. For fresh OPC-based concrete, reaction rates are slow at ambient conditions; carbon dioxide concentration in the slurry phase is limited, and carbonation is impeded in the presence of moisture. For mature OPC-based concrete, under ambient conditions and in the mass transfer control regime, it takes years for carbonation to occur. For low-ranked calcium silicates, a high concentration of CO_2 is needed to achieve sufficiently fast reaction kinetics. For hydrated lime and in dilute CO_2, favorable reaction conditions require the maintenance of slightly above ambient temperature conditions. Finally, for industrial wastes like fly ash, slags, etc., the heterogeneity of wastes results in broad variations in reaction kinetics, with their CO_2 uptake being substantially lower than that estimated from their bulk oxide composition.

3.5.5 Construction Codes and Standards for Products

In general, it is desirable that a new product achieves performance equivalence, or ideally performance benefits, vis-à-vis a product that it is intended to displace. While this argument is logical for the vast majority of applications in construction, its application in the context of OPC-based products may differ. This is because the construction sector has gained empirical confidence in the use of OPC and traditional concrete as construction materials, and construction industry standards are notoriously slow to change. Furthermore, construction standards and codes are often jurisdictional (e.g., city by city or state by state). This results in fragmented compliance and acceptance standards that may inhibit or delay the market entry and adoption of new products. A further complication is that construction standards are often prescriptive rather than performance based; as such, they often define the compositions of materials that can be used. Pending external forcing through legislation, preferential government purchasing programs, or the imposition of CO_2 taxes and penalties, this is a substantial challenge that has to be overcome before the use of carbonated materials can become widespread within the construction sector. Among other elements, this requires moving to a system of harmonized performance-based standards, which will accelerate the adoption and acceptance of new materials for construction applications and new tools that can model and predict long-term performance of new materials.

3.5.6 MC Cost and Commercial Viability

One of the major challenges for CCS including MC projects is the cost. Recently, an estimated transport and storage cost of B$17 per tCO_2, which is about double the cost associated to geological storage in sedimentary basins, has been associated to in situ

MC in basaltic rocks [118]. Therefore, the total cost of in situ MC will be in the range of 72–129 per tCO_2 (considering a CO_2 capture cost of \$55–112 per tCO_2), which is by far larger than the recent European carbon market CO_2 price of B\$7 per tCO_2. It should be noted that geological storage costs do not take into account potential long-term monitoring costs due to the un-reactivity of dry CO_2 in sedimentary rocks. Furthermore, the in situ MC option drastically reduces potential leakages [118].

Main operating costs for MC are dependent on plant size, pretreatment (grinding feedstock and thermal-treatment), operating conditions (mixing, high temperature/pressure), additives (extraction of reactive species), and separation/disposal of the reaction products. Unfortunately, in the absence of availability of commercial data, most estimates are based on laboratory or pilot-scale operations. Based on NETL work, sequestration costs for a direct process with pretreatments are estimated to be 50, 90, and 210 \$ per tCO_2 for olivine, wollastonite, and serpentine, respectively. NETL process estimates an overall mineral carbonation process cost of A\$ 70 per tCO_2 avoided. NETL also estimates that the direct use of thermal heat instead of electrical energy, coupled to partial dehydroxylation with heat integration can lead to a 63% decrease in energy requirement for thermal activation [35,36,118,119].

The cost of direct aqueous carbonation of concrete waste and steel slag is estimated to be in the range of US\$8–104 per t-CO_2, depending on the operation conditions [120]. Using the indirect aqueous carbonation route with the production of value-adding products such as high-purity PCC or hydromagnesite from EAF slags or serpentine would require chemicals (HCl, HNO_3, CH_3COOH, NaOH) for a cost of \$600–4,500 per t-CO_2 if not regenerated. An analysis performed by IEA GHG (2000) found this approach economically unattractive. A similar conclusion was reached by Teir et al. [76] who used HCl/HNO_3 for the dissolution of the feedstock and NaOH in the precipitation step.

Carbonation technologies, which produce building materials or aggregates, still need to be demonstrated at a scale sufficient to prove their commercial viability on a large scale [121]. From the technologies examined here, it appears that the processes that have advanced at the demonstration phase are those that use alkalinity generated by electrolysis of brines, saltwater, or alkaline wastes as feedstock. Table 3.2 summarizes the primary benefits from the use of CO_2 in terms of CO_2 avoided the energy required to obtain the carbonation products and market values. Although these "advanced" carbonation processes are viewed as an attractive concept, for CO_2 uptake, they are not as competitive as large-scale capture and geologic storage and they should mainly focus on serving the construction market.

3.5.7 Factors Affecting Carbonation Dynamics

A number of factors affect carbonation dynamics and CO_2 uptake of various mineral carbonation approaches considered here. First, carbonation reactions, while often unaffected by flue gas contaminants such as particulate matter and acid gases (e.g., SO_x and NO_x), are sensitive to the presence and state (liquid or vapor) of moisture, which may be present or liberated over the course of carbonation. This is because water, although appropriate for accelerating carbonation reactions on mineral surfaces, when condensed within pores retards the diffusive transport of CO_2 (and hence

TABLE 3.2
Potential Uses of Carbonation Products from Some MC Processes [121]

Carbonation Process	Amount of CO_2 Utilized	Value of By-Products ($ Per tCO_2)	Energy Penalty for By-Product Process (%)	CO_2 Emissions Avoided	Products Market and Size	Market Size (Billion $ Per Year)
Skyonic	Cl_2: 14 Mt per year; Na_2CO_3: 20 Mt per year; H_2: 836 Mt per year	Na_2CO_3: B300 $ per t, H_2: B10 $ per t, Cl_2: 240 $ per t	20	2.9 t per tCO_2; captured	Solvay process (Na_2CO_3 or $CaCO_3$) (3.4–9 billion $ per year)	3.4–9
Calera	Sand and aggregate market: 1,500 Mt per year; cement: 24 Mt per year	Aggregate: 7 $ per t, cement: 100 $ per t	8–28	0.5 t per tCO_2 Captured	$CaCO_3$ for cement, aggregates 31 billion % per year	21
Alcoa	2–23 Mt per year	10–300 $ per t	n.a.	n.a.	n.a. About 500 billion$ per year	B500

carbonation kinetics) toward the reactants. Therefore, it is important to control liquid water saturation levels within microstructures, such as when carbonation is carried out in precast components. Second, the solubility of CO_2 decreases with increasing temperature, reducing the concentration of CO_2 in the liquid and promoting precipitation of calcite by decreasing its solubility. Hence, the temperatures over which carbonation reactions are carried out need to be considered and controlled. Third, as is typical for fluid-solid reactions, the surface area of the reactants (particle size) affects reaction kinetics. Although fine particles can accelerate kinetics, their production is energy intensive due to the need for grinding. In addition, their use may complicate slurry processing, due to their tendency to agglomerate, and result in refined microstructures that show retarded liquid and vapor transport.

All of the approaches described above result in the formation of calcium carbonate polymorphs (aragonite, vaterite, calcite [122,123]) and in certain cases hydrous silica [124], or hydrated calcium silicates, as the reaction products that glue the composite together, thereby imparting strength [109]. These products are stable under ambient conditions of temperature, pressure, and CO_2 concentration over geological time scales, as evidenced by the widespread deposits of limestone in nature.

Assuming a need for an alternative to OPC at today's production levels, the amount of carbon dioxide that can be utilized—by mineral carbonation—can be established by considering the ability of diverse alkaline solids (e.g., fly ash, blast furnace slag, portlandite, and low-rank silicates) to react with carbon dioxide, thereby producing alkaline carbonates. A range of carbon dioxide uptake can be established by considering fly ashes and portlandite. Fly ashes feature a terminal carbon dioxide uptake of

≤0.05 g CO_2 per gram of fly ash, and portlandite features a CO_2 terminal uptake of 0.54 g CO_2 per gram of portlandite over time scales of tens of hours. These uptakes suggest global carbon dioxide utilization levels ranging between 0.2 billion tons for fly ash [1] and 2.2 billion tons for portlandite, if sufficient quantities of these materials were available. Since carbon dioxide mineralization resulting in cementation is expected to be achieved using a diversity of alkaline precursors, with a range of carbon dioxide uptake capacities, a rough estimate of approximately one billion tons of carbon dioxide on an annual basis is reasonable.

The introduction of carbon dioxide into the "cementing formulation" can be plausibly accomplished by the following two pathways: (a) injecting carbon dioxide into the fresh concrete over a short period or (b) exposing preformed components of structural components to vapor-phase carbon dioxide, in dilute or concentrated form on the order of hours, within reactors. Expectedly, these pathways result in different levels of carbon dioxide uptake. For example, carbon dioxide injection into fresh concrete results in uptake that is limited by the overall rate of carbon dioxide reaction with calcium and magnesium, which in turn may be limited by the solubility of carbon dioxide in alkaline aqueous solution (i.e., ≤0.01 g CO_2 per gram of cementitious components). On the other hand, the exposure of preformed components results in carbon dioxide uptake that is limited by the nature of the reactant used and potentially the geometry of the body, which may result in carbon dioxide transport limitations such that the carbon dioxide uptake typically ranges between 0.05 and 0.50 g CO_2 per gram of cementitious components.

3.5.8 Additional Impeding Factors

National academy report [1] points out that carbonation is commonly impeded by one of the three factors: (a) competitive reactions, (b) the formation of surface-passivating films, and (c) the presence of water. The first factor is often a function of the tendency of solids to hydrate versus carbonate. The second factor refers to dense films of calcite that form on carbonating surfaces and hinder further contact between the reactant solid and CO_2. The third factor relates to the fact that water favors carbonation when adsorbed on surfaces [125] but inhibits carbonation when condensed within the pores of a micro- or mesoporous solid [126,127]. Besides these factors, there are barriers for MC depending on raw materials. For hydrated lime in dilute CO_2 condition, an above-ambient temperature is required. High-concentration CO_2 stream is required for low-rank Ca silicates. The heterogeneity of industrial waste requires broad variations in reaction kinetics, with their CO_2 uptake being substantially lower than that estimated from their bulk oxide composition. Slow reaction, mass transfer limitation, and impedance by water are key barriers to MC of fresh and mature OPCs.

While challenging, addressing these timely topics is now within grasp as a wide range of advanced characterization tools that can provide the insights and data needed (such as real-time particle size, shape, and crystallinity analysis; Raman spectrometry; x-ray diffractometry; x-ray tomography; and x-ray absorption, among others) have recently been developed and are becoming accessible to the wider carbon mineralization community. Likewise, molecular modeling techniques and computational tools to express them have become increasingly accurate and reliable, and experimental techniques to generate high-resolution time- and scale-dependent data have arrived and can be readily exploited, with the right resources.

3.6 MINERAL CARBONATION PRODUCTS AND THEIR UTILIZATION

We break down products and their utilization into three parts. While the major use of MC is for CO_2 sequestration and development of construction materials, MC process can also be used to develop other valuable products and recover metals. Here we examine the role of MC in these three areas separately. The nature of process and products obtained by MC is graphically illustrated in Figure 3.9.

A. **Perspectives on MC use in Construction industry**

Three major products of mineral carbonations are aggregates, OPC, and concrete, all used for various construction projects. These constitute the second largest material flow behind water in the global market. Each year, nearly 30 billion tons of concrete [128] are generated globally from a production base of 4.1 billion tons of OPC [116] and nearly 52 billion tons of mineral aggregates which are used in concrete, barriers and road-based applications [129]. Natural and synthetic sources of construction materials can be potentially replaced by aggregates and binding agents produced from mineral carbonation.

Concrete is a mixture of Portland cement with sand and gravel aggregate. Cement production exceeds four billion tons per year with a carbon footprint of about 800 kg CO_2 per ton of cement, making it the largest contributor in the industrial chemical sector. It contributes 5%–7% of global anthropogenic CO_2 production and thus has the potential for reducing

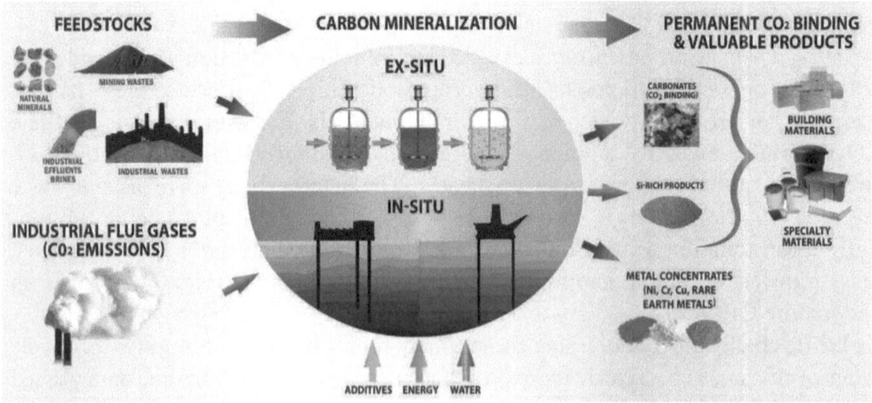

FIGURE 3.9 Scheme of carbon mineralization and of the range of its products. (Image courtesy of Florent Bourgeois, Laboratoire de Génie Chimique; Au-Hung Park and Xiaozhou Sean Zhou, Columbia University [2].)

anthropogenic CO_2 emissions significantly (estimated at 1–2 billion tons per year). Through carbon mineralization, CO_2 emissions from cement manufacturing could be used for the production of concrete. The CO_2 can be mineralized using a variety of feedstocks producing carbonated aggregates. These aggregates can be used to manufacture concrete or concrete-based building materials, hence replacing gravel aggregates.

Direct and indirect carbon mineralization of various natural and industrial alkaline materials offers the opportunity to produce a range of tailored carbonation products and by-products as construction and building materials [13,130]. The range of products including those with high values can be expanded by process modification or post-processing. Applications include structural materials (e.g., cements, concrete, and mortars), nonstructural materials (e.g., for road-base, erosion, sea, and flood protection barriers), and calcium- and magnesium-based carbonates that may be used for different applications (e.g., as additives for protective coatings such as paints and polymers). specialty materials, metal concentrates, etc. A scheme of carbon mineralization and of the range of its products, which depends on feedstock, operating route, and reaction conditions, is illustrated in Figure 3.9. In construction applications, carbonate solids can provide cementitious binding of the components of the building product, or structural support (as an aggregate or filler).

The carbonation of alkaline industrial waste of different types can also be used for other purposes. Carbonation can reduce the toxicity of the waste, make them nonhazardous, immobilize or extract heavy metals, stabilize chemically metastable mineral phases, or even store energy or provide carbon to microalgae. Carbonation of waste materials also eliminates the need for waste to discard in the landfill. The building sector presents a very broad and significant carbon footprint. Along with the generation of products with improved technical properties, carbonation makes a significant impact on the reduction of CO_2 emission.

A variety of technical approaches can be taken for carbon dioxide utilization in cement and concrete industry (e.g., [131,132]). The reaction of calcium silicates, both within traditional Portland cements and novel non-hydraulic cements, with CO_2 to form calcium carbonate leads to hardening and solidification of the treated binder. Conventional moist curing can be complemented or replaced by CO_2 curing of traditional precast concrete masonry products. Ready-mix concrete using CO_2 as an admixture improves compressive strength. In addition, aggregates formed through carbon mineralization of alkaline waste can displace natural aggregate input into concrete and hence further reduce CO_2 emissions. For this latter application, carbon mineralization is employed in a cold-bonding granulation process, during manufacturing and/or curing under a CO_2-rich atmosphere, to produce **synthetic aggregates** from waste residues of incineration fly ash and steelmaking slag.

Carbon mineralization can also be used to manufacture blocks or compacts containing finely milled steel slags that, by CO_2 curing at elevated temperature and pressure, achieve high mechanical performance without the use of cementitious binders and also provide a permanent storage of CO_2 [5]. Adequate compression strengths can also be obtained by CO_2 curing under ambient conditions and thereby reducing the slag content. Leaching properties of incineration bottom ash can also be changed by carbon mineralization such that it can be used as aggregates or filler materials. This also has an environmental impact. In short, CO_2 emissions from cement manufacturing can also be used in the curing of concrete. The introduction of CO_2 into the curing process promotes carbon mineralization reactions. The mineralized CO_2 is incorporated into the concrete matrix, enhancing strength. The strength benefits available through carbon mineralization present an opportunity to reduce the amount of cement used for an equivalent strength-grade concrete. Reduction in the carbon footprint of concrete through mineralization of CO_2 has the promise to transform the built environment without sacrificing the quality of construction materials.

The supply of natural materials required for construction industries is diminishing as construction projects globally expands. Construction material has three components; aggregates, cement (binding agent), and concrete. The final product concrete is used for numerous construction projects like buildings, roads, bridges, pavements, etc. Mineral aggregates, which range in size from micrometers to centimeters, are granular materials and they constitute roughly 60%–80% of a typical concrete, with the rest consisting of a binding phase, or matrix. Currently, the vast majority of aggregates used globally are natural materials including sand, gravel, and crushed rock, and other virgin materials mined from quarries, gravel pits, sea beds, and riverbeds. Near urban areas where many new constructions take place, these natural sources are depleting and they are also not available through new mining activities which are restricted due to their environmental impacts.

Secondary and manufactured aggregates [133] which include recycled concrete, by-products from industrial processes (e.g., blast furnace slag), and mineral aggregates that are synthetically manufactured (like geosynthetic aggregate and carbonate-cemented aggregate) for use in concrete [134] currently contributes only a small portion of aggregate need. However, due to the factors mentioned above, the emphasis on synthetic aggregates has been globally increasing. In Europe, 10% of aggregates are produced from recycled and synthetic materials. This trend is further accelerated due to limited and expensive landfill capacity [135]. In future, MC is expected to represent a promising avenue for aggregate production, as long as the synthetic products are economical and meet relevant quality and performance standards.

Cement manufacturing plants are close to limestone quarries which are generally away from an urban environment. Both limestone and cement are heavy, low-cost materials, and the cost of shipping over distances greater than about 250 miles, unless carried out over water, is often prohibitive. Synthetic cementing systems based on mineral carbonation, does not harden until all the mixed ingredients make contact with CO_2, thereby offering greater control over the workability environment. This is significant as the ease of controlling the rheology of such systems makes them potentially better suited for advanced (e.g., additive) manufacturing of structural components which feature superior strength-to-weight ratio, optimized topology, and complex geometries that could not be fabricated using existing casting or molding-based techniques. Process flexibility provided by synthetic cement from MC is not possible with cement from natural sources.

Cement manufacturing process emits CO_2. Most strong and reliable ordinary Portland cement (OPC) is produced from limestone, silica, clay, and iron compounds, along with other additives. Its production and use have a large carbon footprint, generating approximately 0.7 tons of CO_2 emissions per ton of cement produced and consumed globally. These CO_2 emissions stem largely from the thermal decomposition of limestone to lime (primarily $CaCO_3$ to CaO) and from fuel burning to provide the heat required to drive the formation of clinker phases in the kiln. Although, improvement in the kiln thermal efficiency and combination of fly ash and slags with OPC in the binder section have reduced CO_2 emission in the concrete production [136], these reductions have been more than offset by the global increase in cement consumption over the same period.

The mixture of aggregates (primarily sand and stone), cement, water, and chemical additives creates the final product concrete. In this mixture, cement is the binding agent. The mixture only takes few hours before it hardens. Concrete, also a heavy, low-cost product, is often produced closer to the site of consumption. Concrete is most often used as ready-mix concrete or in the form of precast components. In the United States, ready-mix concrete, precast concrete, and concrete masonry comprise about 55%, 25%, and 15% of the overall concrete market, respectively. The differences in these market segments are broadly operational. Concrete is commonly classified by its compressive strength following 28 days of aging, which determines the load-bearing capacity of a given concrete formulation and the types of structural or nonstructural components it may be used for. Building technology using concrete is thus heavily regimented and somewhat inflexible. Synthetic concrete obtained from MC can provide the needed flexibility to the construction industry

B. **Other utilization pathways of MC**

As illustrated in Figure 3.9, MC produces a number of products that can be utilized to make the overall process economically more viable. This utilization depends on the need and the nature of the process employed. Since the market generally requires high purity products, processes that generate multiple separated products are preferred. For these reasons, indirect processes which may be more suitable for controlling the morphology and particle size of precipitated products for high-value applications are generally preferred compared to direct mineralization technologies [137,138].

Applications of carbonate products can be divided into low and high volumes and low and high qualities (or tech) based on their uses. For the MC products to be commercially used, there are particle size, distribution and low level of contaminants specifications and quality criteria that must be met. The feedstock from iron/steel works which can easily lead to iron oxide during MC can be suitably used for construction and filler applications. Among the low-tech, high-volume applications, the use of MC products as liming agents to buffer the acidity of soils are promising but require the MC products to be free from potential pollutants that might derive from particular flue gas or mineral wastes converted into carbonates. Another suitable application is land reclamation from the sea in coastal areas and mine reclamation using silica, magnesium, and calcium carbonates [139].

High-end applications usually require stringent specifications. Monodisperse nanoparticles uniform in size, shape, and composition have a wide number of applications in industry, such as catalysts, chromatography, ceramics, pigments, pharmacy, photographic emulsions, etc. [140]. MC can be used to produce silica in the amorphous phase and with particles smaller than about 30 mm, which could serve as a pozzolanic cement replacement material or as a filler [141]. High purity SiO_2 (>98.5%) from MC, can be used for deoxidizer in steel making, circuit boards, ceramic matrix composites, and semiconductors [103,139]. Similar purity is expected for ceramics applications, while slightly lower purity would be required for use a refractory material (95% SiO_2) and iron and steel making (90% SiO_2). Such purity requirements will most likely need additional post-processing. Fine powder of amorphous silica from MC can be a high-quality reactive cement additive [139].

Calcium carbonate is extensively used as a novel functional material in several fields such as plastics, rubber, paint, printing ink, weaving, toothpaste, make-up, and foodstuffs. Calcium carbonate is a product in MC processes that use inorganic wastes or calcium silicates, such as wollastonite. Nano-sized, high-performance, low-cost fillers from calcium carbonates in the form of ground calcium carbonate (GCC) and precipitated calcium carbonate (PCC) can be developed by MC [139].

Calcium carbonate precipitate in six different forms, namely amorphous calcium carbonate (ACC), hexahydrate calcium carbonate (HCC), monohydrate calcium carbonate (MCC) and the polymorphs calcite, aragonite and vaterite, which have the trigonal, orthorhombic, and hexagonal crystal system, respectively [142] each with specific applications and requirements. For the PCC applications, several physical and chemical properties, such as particle size average and distribution, morphology, specific surface area, polymorph or the chemical purity are very important in determining the potential market [143]. The different polymorphs of $CaCO_3$ can have different functions as additives. For example, dispersion can be increased if cubic $CaCO_3$ is added as an addition in paint; acicular or rod-like $CaCO_3$ has a reinforcing effect on rubber and plastics; and spherical $CaCO_3$ has a significant impact on the brightness and transparency of ink [144]. $CaCO_3$ with different polymorphs, morphologies, and grain sizes can be obtained by controlling the initial concentration of the reagents, stirring speed, pH, type and amount of additives, and other reaction conditions. For example, changing carbonation time and after aging, different $CaCO_3$ polymorphs can be generated [145], Changes in operating variables or addition of ethyl trimethyl ammonium bromide cationic surfactant (2%) can change particle sizes and morphologies of precipitated $CaCO_3$ [140].

Carbonation efficiency and modification of reaction products properties can also be accomplished with the use of enzymes such as carbonic anhydrase (CA). The carbonation capacity of alumina immobilized CA was found to be 25% lower compared to that obtained in the presence of free CA [146]. Favre et al. [147] reported that at higher pH, calcite and vaterite were observed while at lower pH, only calcite was favored. CA can be prepared in a number of different ways and encapsulated in MOF. The nature of MOF is determined by the $CaCO_3$ morphology [148].

Steel slag can be utilized as PCC if calcium is selectively extracted prior to carbonation to fulfill the requirements of purity and crystal shape. Zevenhoven and co-workers selectively extracted calcium from the slag with an aqueous solution of ammonium salt (NH_4NO_3, CH_3COONH_4, or NH_4Cl) producing PCC from the steel slag derived calcium-rich solution with properties comparable to the PCC produced by conventional methods. This method, however, deemed expensive in order to get high dissolution efficiency and presented less brightness compared to traditional PCC resulting in a decrease of its market value. Despite this, the separation of iron oxide before the carbonation stage can enhance the quality of PCC produced by this method [137,138,149]. Recently, an innovative synthesis of the goethite–calcite nanocomposite involving three sequential precipitation steps has been proposed [150].

As shown in the rest of the book, other methods have been used to convert CO_2 into chemicals and fuels. Compared to the utilization of MC products as construction or filling materials, which could, in theory, absorb Gt of CO_2, industrial utilization of CO_2 as solvent and reactant amounts to only 0.5 wt% (128 Mt per year). The main advantage of other methods are that they allow another methods for the productions of chemicals and fuels while removing CO_2 from the environment.

Finally, an ex situ indirect process route can produce **high-purity mineral carbonates**, such as precipitated calcium carbonates or precipitated magnesium carbonates, which have even greater value as a chemical feedstock for paper and paint production, displacing resource extraction routes for conventional materials [13,151,152]. Hydrated Mg-carbonate minerals produced through carbon mineralization of brines and alkaline solids have high solubility, which facilitates their use as a **carbon storage and transport medium**. Microalgae requires carbon source to generate biofuels. In the absence of a point source of CO_2, carbon may be supplied to the algae through the dissolution of carbonate minerals. Thus, an innovative carbon mineralization process consists of capturing CO_2 into hydrated minerals to be provided to microalgae as a supplement has been used. A similar novel process may exploit the large enthalpy of hydrated Mg-carbonate minerals' dissolution–precipitation cycles to **store energy** or serve as a pathway to capture CO_2 from dilute or dirty sources and convert it to a high-purity stream for other utilization processes. Table 3.3 outlines an order of magnitude of the current and potential future CO_2 consumption. This table

TABLE 3.3
Current and Potential Future of CO_2 Consumption

CO_2 Uses	Existing (Future) CO_2 Demand (Mt Per Hour)
Enhanced oil recovery	30–300 (300)
Urea	5–30 (30)
Food and beverage	B17 (35)
Water treatment	1–5 (5)
Other	1–2 (6)
Enhanced coal bed methane recovery	30–300
CO_2 concrete curing (MC)	30–300
Algae cultivation	4,300
Mineralization (MC)	4,300
Red mud stabilization (MC)	5–30
Baking soda (MC)	1
Liquid fuels (methanol, formic acid)	4,600

Source: Modified from Ref. [40].

shows that EOR and urea yield boosting are technologies already in use. MC technologies, algae cultivation, and potentially ECBM could utilize flue gas directly and therefore would not require a conventional capture plant to deliver a concentrated CO_2 stream [40]. A semi-quantitative ranking process identified mineralization technologies (mineral carbonation and concrete curing), EOR, EGR, and ECBM and algae cultivation having the greatest potential to accelerate alternative forms of CCS.

C. **Enhanced metal recovery by MC**

An additional desirable outcome for carbon mineralization is to engineer processing pathways that simultaneously extract metal value from such feedstocks while binding CO_2 to the feedstock material itself. The underlying concept uses CO_2 from flue gases to extract metal phases from such feedstocks. The process could be referred to as **enhanced metal recovery** (EMR), in that it has some similarity to enhanced oil recovery (EOR). This concept adds economic value to carbon mineralization associated with both metal concentrate and carbonate production, the latter being tied to the production of building materials. For example, new production pathways using CO_2 as a reactive agent to process nickel laterites would produce both nickel and carbonate building materials. Similar approaches could extract rare earth metals from existing stockpiles of mining waste. EMR, applied to primary mineral resources, could be envisioned ex situ and/or in situ. The carbonated products could then be used for producing building materials, or they could be left in the ground. An alternative deployment would use carbonate solid precipitates to bind and immobilize deleterious metals and materials (e.g., asbestos) within mine waste and the use of EMR for ex situ carbon mineralization using silicate minerals and alkaline industrial wastes that contain valuable metals and rare earth elements [2].

Feedstocks for carbon mineralization—such as silicate ores, mine tailings, and industrial residues like steel slag—typically contain many metals of value, such as iron, nickel, vanadium, chromium, platinum, and rare earths. An innovative application of carbon mineralization referred to as enhanced metal recovery (EMR) exploits this process to store CO_2 as mineral carbonates and to recover valuable metals. The reaction of the alkaline feedstocks with dissolved CO_2, an acidifying agent, releases elements such as magnesium, calcium, or lithium, as well as other valuable elements (e.g., rare earth elements) into solution [2].

As illustrated in Figure 3.10, three potential EMR routes are envisioned:

1. In situ, in which a CO_2-saturated solution is injected into geologic formations.
2. Ex situ EMR as a heap leaching process, by which a CO_2-saturated solution is fed to heaps of industrial residues or mine tailings.
3. Ex situ EMR in a dedicated processing plant, employing either direct or indirect carbon mineralization reactions. Reaction conditions can be engineered depending on the types of feedstock and metals that are to be extracted.

In all EMR routes, CO_2 is bound to the carbonate product, which is either valorized, for example, as a construction material (for the ex situ cases) or bound in the ground (in situ route). The metal-rich solution is subsequently processed by hydrometallurgy to recover the metals and metalloids. In EMR, the carbon mineralization costs may be partially offset by the value of the recovered metals [2].

3.7 DEMONSTRATION AND COMMERCIAL PROJECTS AND THEIR CHALLENGES

So far, mineral carbonation has been implemented only in a few demonstration plants: the first is the Calera process, in the gas-fired Moss Landing plant (USA). The plant showed the technical capacity to capture CO_2 (30 kt per year) from a 10 MW power generator at 90% efficiency, with an associated energy penalty of 10%–40%. However, the potential impact on water balances and hydrology from extraction and reinjection of brines and the conclusion that the tested brines (technically unsuitable), seawater (too costly), and alkaline wastes (limited availability) render this process unsuitable for operations at a significant scale [89].

Another brine-based process (Skyonic) approached the commercialization stage. Skyonic is currently retrofitting Capitol's cement mill (San Antonio, USA) owned by Capitol Aggregates. This process directly processes flue gas and produces hydrochloric acid, bleach, chlorine, and hydrogen. The scale of products generated compared to current markets being too small to hinder its development at a larger scale.

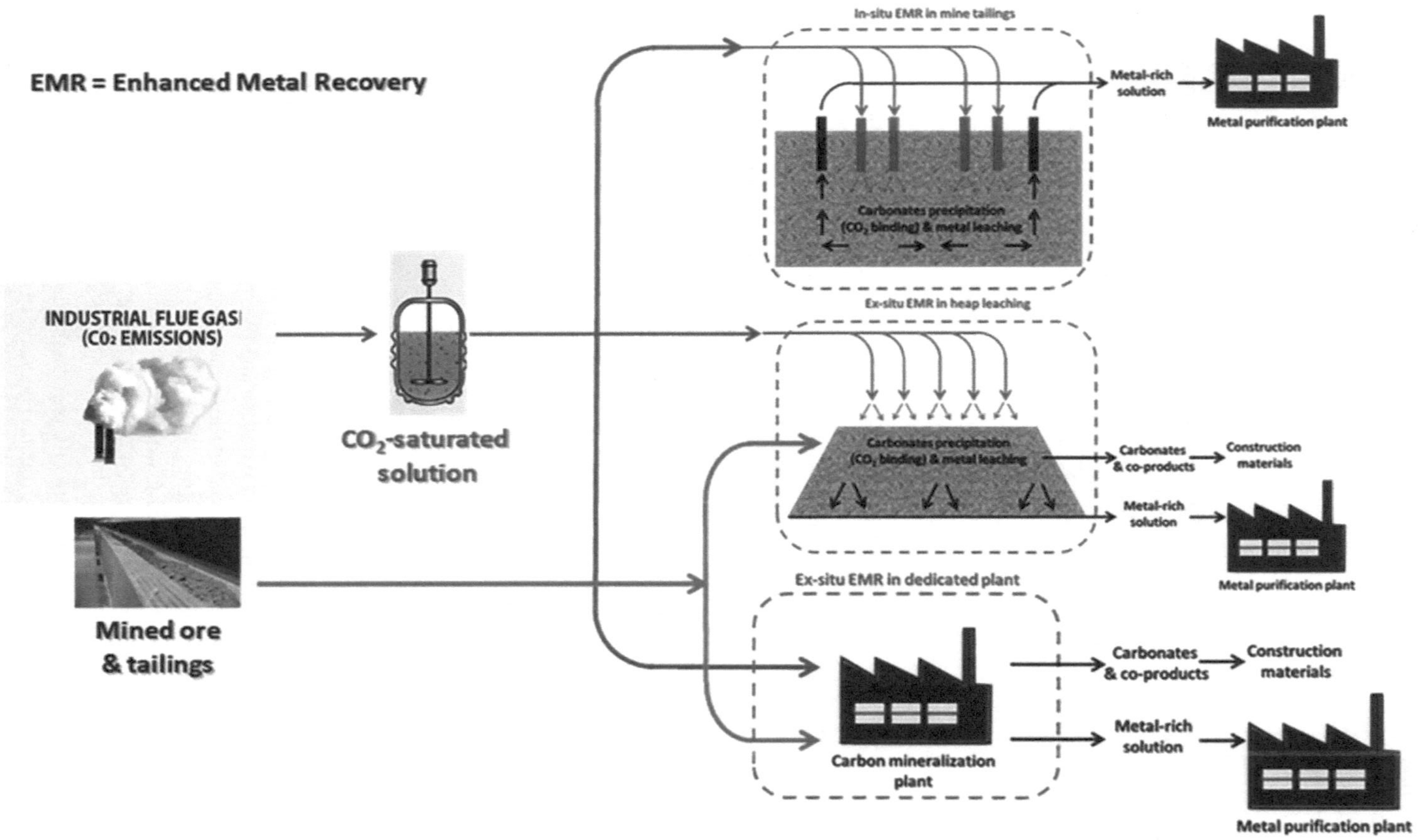

FIGURE 3.10 Enhanced metal recovery and CO_2 utilization and storage [2].

Only few projects based on inorganic wastes have moved to the commercial or even small-scale demonstration phase. For example, a pilot-scale mineral carbonation process that uses coal FA has been installed at a 2,120 MW coal-fired power plant to reduce CO_2, SO_2, and Hg emissions in Point of Rocks, USA [153]. Also, accelerated carbonation has been applied for the commercial production of aggregates from APC residues [154] and in a residue carbonation plant for red mud stabilization at Kwinana in Western Australia [40].

Carbonation of red mud has been run by Alcoa since 2007 locking 70 ktCO_2 per year generated in a nearby ammonia plant [91]. However, B30t red mud per tCO_2 is used, which is about ten times the typical rate of serpentine rock usage. Also, this technology requires a concentrated and preferably high-pressure source of CO_2 (85% pure) and needs to be located in reasonable proximity to an alumina refinery [155–157].

The CarbFix demonstration project (in situ MC), where 5% CO_2 in water has been injected in porous basalts near the continental margins, has recently shown that it is feasible to sequester more than 80% of CO_2 injected in less than 1 year at 20°C–50°C [118]. This mineral trapping pathway avoids one of the major drawbacks associated with geological storage in sedimentary basins, since CO_2 dissolved in water is not buoyant and also offers a storage potential one order of magnitude higher than the potential CO_2 emissions from burning all fossil fuel resources [118].

Besides the ones mentioned above, according to the national academy report [1], there are substantial efforts underway to produce construction materials via mineral carbonation. Most are being undertaken by startup companies and precommercial entities, which fall into two groupings: (a) those producing binding agents in concrete (examples include Solidia Technologies, Carbon Upcycling UCLA, CarbonCure, and Carbstone Innovation) and (b) those producing synthetic aggregates (examples include Carbon8 Systems and Blue Planet). Brief descriptions of reactants, products, and processes used in these plants are described in Table 3.4.

3.8 FUTURE REQUIREMENTS FOR MINERAL CARBONATION

Despite the large resources available for CO_2 sequestration and the clear advantages over geological storage, the costs of both in situ and ex situ MCs are currently too high for a large deployment of the technology, and new systems are being investigated to overcome the technical issues, process energy economics, chemical reaction rates, materials handling and market acceptance issues. Some of the issues that need to be addressed are as follows:

1. Representative raw materials comparison; research to develop new material formulations with novel properties and to advance the use of additive manufacturing to construct components with superior strength-to-weight ratio, optimized topology, and more complex geometries compared to what can be made with existing construction methods
2. Processes performance data; technical criteria require for fulfillment of scale and engineering requirements of construction materials
3. MC integration with point source; process design research to integrate mineral carbonation processes with existing carbon dioxide capture technologies

TABLE 3.4
Startup and Pre-Commercial Companies for MC

Company	Process, Reactants and Products
• Solidia Technologies	• Low-rank silicates as the reactant for carbonate mineralization • CO_2 uptake of up to 0.38 g CO_2 per gram of reactant • Well suited for ex situ production, similar to precast concrete. • Products include intermixed $CaCO_3$ and SiO_2 (hydrous silica) • Robust cementing behavior of product
• Carbon Upcycling, UCLA	• Portlandite and industrial wastes as the primary reactants for carbonate mineralization • CO_2 uptake on the order of 0.59 g CO_2 per gram of portlandite • Approach is suited for co-located ex situ production, such as, alongside a fossil-fuel power plant or an OPC production plant, which serve as a source of solid reactants, waste CO_2, and/or heat
• Carbon Cure	• Injects pure CO_2 into ready-mix concrete formed during initial mixing with the CO_2 • Uptake of approximately ≤0.01 g CO_2 per gram of reactants • The reaction products are formed in situ, then blended with the binder, thereby forming carbonate compounds • Increased and somewhat accelerated strength gain as compared with traditional concrete • Approach is currently being implemented across numerous ready-mix concrete plants in the United States
• Carbstone Innovation	• Using slags produced from iron and steel manufacturing as the alkaline substrate for mineral carbonation • Possible CO_2 uptake on the order of 0.10 g CO_2 per gram of slag • Approach is best suited for ex situ production alongside a source of both slags and waste CO_2 • Produce a diversity of preformed product
• Carbon 8 systems	• Using heterogeneous air pollution control (APC) residues as the alkaline reagent for the production of carbonate aggregates – CO_2 uptake around 0.12 g CO_2 per gram of solid reactants • Approach has achieved commercial operations in the United Kingdom based on its ability to encapsulate and isolate APC residues in a carbonate matrix
• Blue Bonnet	• Using alkaline rock and industrial wastes as the primary reactants for carbonate mineralization • Process is thought to be compatible with the use of dilute CO_2 waste streams and is therefore well suited for co-location alongside fossil-fuel power plants or OPC production plants that could serve as a source of solid reactants and/or waste CO_2

National Academy Report [1] points out that the scalability and market viability of these approaches are affected by a diversity of factors, including (a) the purity and the availability of CO_2, (b) the availability of low-cost alkaline reagents and/or facilities for their manufacture at scale, (c) the low-cost, commoditized nature of the existing analogous products, OPC and concrete, (d) restrictive building codes and standards wherein compliance is often a function of the material composition (e.g., OPC-based chemistries) rather than their engineering performance, and (e) the net amount of CO_2 utilization that can be achieved.

4. Complete information on cost/energy balance for thermal activation; research to develop energy- and carbon dioxide–efficient pathways and processes for producing alkaline solids that can be readily carbonated and do not require high-temperature activation
5. Sufficient knowledge of indirect carbonation fundamentals; more research is needed for controlling relative rates of carbonation and hydration; research to develop additives for enhanced carbon dioxide solubility or structure-directing agents that accelerate particle growth
6. Sufficient knowledge of carbonation fundamentals using flue gas
7. Lack of assessed reactor technology options and cost studies
8. Process scale and materials handling issue; Infrastructure for physical handling of CO_2
9. Environmental impact of large mining operations
10. Research to develop physical and instrumental assessment methods, improved modeling, performance-based criteria for product properties and develop new analytical tools for studying carbonation reactions in dense and viscous suspensions

Ex situ mineral carbonation with inorganic wastes could be part of an integrated approach to carbon sequestration, which combines remediation of hazardous wastes such as asbestos tailings and use of readily available fine industrial wastes such as EAF and cement-kiln dusts to meet CO_2 emission goals. On the other hand, in situ carbonation may be viable for large-scale emitters with proper financial incentives. The use of MC for the construction industry is the best long-term strategy.

Carbon mineralization presents great potential as a permanent CO_2 storage treatment and for obtaining products and materials that may be used for other applications. However, to achieve high reaction rates, material pretreatment, enhanced operating conditions, and the use of precaptured CO_2 (or reagents and additives) have generally been reported to be necessary, thus leading to processes being too energy-intensive and costly.

With regard to in situ carbonation of mine tailings, the rates and extent of carbon mineralization are limited by the availability of CO_2. Thus, enhancing the delivery of CO_2 to reaction sites is a priority for maximizing the benefits of this approach. Enhanced delivery of CO_2 is expected to push these systems to the limits of cation availability, but the ability to deliver CO_2 is limited by mass transfer rates within alkaline mineral–fluid–gas systems. For some mines, this limit may be on par with the greenhouse gas production associated with mine operations, opening up the possibility that some mine operations could become greenhouse gas neutral. As CO_2 delivery limits are overcome, the next limit to carbonation rates will be imposed by the rate of metal extraction from the mine waste feedstock. Activation of feedstock can dramatically improve reaction performance, but surface control of these reactions needs to be reduced with a better understanding of prevailing reaction mechanisms.

Matching and guaranteeing the required performance over time is a very important factor for the application of MC in the building sector. This requires an extensive standard testing methods to ensure that the new materials are safe to use for a range of infrastructure or building applications. These tests are expensive and time consuming, requiring months or years of analysis. Uncertainty in long-term performance of MC-created concrete is an issue. This requires a thorough understanding of binding characteristics and its long-term stability of chemical bonding created by carbon mineralization.

The current state of the art of carbon mineralization use in construction is limited to precast building materials and nonreinforced applications. The implementation of carbon mineralization may have a big breakthrough if the technology is adapted to replace the cement binder in ready-mix concrete and in reinforced (concrete) building applications. This can only happen if mechanical behavior and environment quality of carbonated products are further improved; which in terms require a better fundamental understanding of surface reactivity and mass fluxes in cement under different environment. This lack of a fundamental understanding hinders the large-scale adoption of CO_2 in concrete production. The reliable use of industrial feedstock such as slag or mine waste as aggregate requires better understanding of their physical and chemical interactions on various microscale processes such as altering reaction chemistry, obstructing mass transport, deregulating dissolution rates, crystallizing unexpected impurity phases, or immobilizing important reactive components such as CO_2. The use of MC as a carbon storage and transport medium (e.g., for micro-algae cultivation) requires a well-constrained process that includes thermodynamic conditions for precipitation and dissolution and kinetics of precipitation of hydrated minerals within a brine and their dissolution in the presence of algae.

The MC process involves a heterogeneous system with solid-liquid and gas reactions mostly occurring at the interfaces. The relevant thermodynamic properties and reaction rates along with governing mass transfer limitations are very complex and dependent upon the operating conditions. Feedstock composition and mineralogy play an important role on CO_2 reactivity and this could also be time and sample dependent in the given environment. Pretreatments also have important effects on the reaction kinetics with feedstock. Mineralogy is a key parameter for CO_2 feedstock reactivity and affects the release in solution of major components and trace contaminants such as metals and metalloids, thus impacting product properties and valorization potential. Even trace amounts of chemical constituents and mineral phases can profoundly alter reaction pathways and rates. The challenges of carbonation research thus primarily come from the complexity of the mineral system, which contains not only various cations (Ca/Mg/Si) but also compounds with different states of carbonation and hydration [132]. Thermodynamics and prevailing mass transfer and kinetic forces dictate the overall behavior of the carbonation process and that need to be understood at microscale over a period of time [3,158].

Multiscale geochemical thermodynamic models that cover leaching processes, interfacial water transportation, and mineral carbonation at microscale, mesoscale, and macroscale need to be systematically addressed. High-resolution experimental systems based on spectrum analysis and calorimetric analysis need to be developed for thermodynamic data characterization and thermodynamic model validation [159]. Multiple conditions (e.g., pH, high temperature, high pressure, gas species) should be analyzed in the experimental system. The building of a thermodynamics database with standard data format is also necessary. Recent development of more robust force fields such as CLAYFF [160] have enabled us to develop an advanced understanding of diffusive transport, sorption, and ion exchange behaviors at solid–fluid interfaces. Emerging research frontiers in the area of carbon mineralization include developing convergence in predictive behaviors using advanced computational tools, and synchrotron and neutron scattering and spectroscopic measurements. Available online tools can help in this regard. More effort is needed to find or develop additives for enhancing carbon mineralization kinetics that can be regenerated energy-efficiently, so as not to hamper the whole process in terms of costs and overall environmental impacts.

MC involves both inorganic and biological processes. Hence, biological routes to carbon mineralization should also be considered, using active biological agents or bio-derived substances (e.g., enzymes such as carbonic anhydrase, and other organic macromolecules). Biomimetic routes might be developed to mimic the local chemistry and mass transport mechanisms that organisms use in building their well-structured and complex tissues. This could lead not only to novel routes for accelerated carbon mineralization, but also the production of novel bio-inspired materials. An integrated technology capable of obtaining a product with suitable properties for utilization and also significant CO_2 uptake, employing flue gas (or other diluted CO_2 sources such as syngas and biogas from anaerobic biodegradation processes with its contaminants) should be developed. It should be carried out with ultramafic feedstocks and ores or other types of industrial residues such as steel slag and should be optimized to maximize carbonate yields within timeframes associated with CO_2 capture in aqueous phase.

Carbon mineralization as a pathway to utilization presents multiple potential assets: it allows CO_2 to be stored in a permanent solid form, improves the properties of alkaline waste materials and/or construction materials, and provides products for different applications and even for transport and temporary storage of carbon and/or energy. Carbon mineralization systems and their thermodynamic and kinetic multiphase behavior are not, however, sufficiently understood to allow the design to be optimized for energy-efficient applications. The tailoring of carbon mineralization processes for different types of feedstocks aimed at achieving specific products/applications may be achieved only upon a deep understanding of carbonation systems and the reactions that occur within.

Being able to predict the dynamic speciation of major and trace elements during treatment and the product properties of solid feedstock/CO_2 systems in aqueous or gas systems is key to identifying and controlling carbon mineralization reaction pathways and rates. Current models cannot reproduce the most basic processes observed in experiments. To overcome these constraints, fundamental research is necessary to derive insight into the interaction among solid, liquid, and gas phases. Controlling bulk physical and chemical properties during the production process by real-time monitoring of the gas and liquid permeability of the compact and carbon mineralization product properties is absolutely necessary to optimize the overall reaction and obtain a product responsive to specific performance criteria. The achievement of this objective would result in a large capacity to generate construction and building materials based on the use of gas streams rich in CO_2, which would hence be permanent.

REFERENCES

1. National Academies of Sciences, Engineering, and Medicine 2019. *Gaseous Carbon Waste Streams Utilization: Status and Research Needs*. Washington, DC: The National Academies Press, https://doi.org/10.17226/25232.
2. Buchanan, M. 2017. Accelerating breakthrough innovation in carbon capture, utilization and storage, *A Report of the Carbon Capture, Utilization and Storage Experts' Workshop*, September 26–28, Houston, TX, Department of Energy, Washington, D.C.

3. Sanna, A., M. Uibu, G. Caramanna, R. Kuusikb, and M. M. Maroto-Valer 2014. A review of mineral carbonation technologies to sequester CO_2. *Chemical Society Reviews* 43:8049–8080 (an open access paper).
4. Ho, H.-J., A. Iizuka, E. Shibata, H. Tomita, K. Takano, and T. Endo 2020. CO_2 utilization via direct aqueous carbonation of synthesized concrete fines under atmospheric pressure. *ACS Omega* 5(26):15877–15890.
5. Quaghebeur, M., P. Nielsen, L. Horckmans, and D. Van Mechelen 2015. Accelerated carbonation of steel slag compacts: Development of high-strength construction materials, *Proceedings of the Fifth International Conference on Accelerated Carbonation for Environmental and Material Engineering* (ACEME 2015, *Front. Energy Res.*, 17 December 2015, https://doi.org/10.3389/fenrg.2015.00052.)
6. Skocek, J., M. Zajac, M. Ben Haha 2020. Carbon capture and utilization by mineralization of cement pastes derived from recycled concrete. *Scientific Reports* 10:5614, https://doi.org/10.1038/s41598-020-62503-z www.nature.com/scientificreports.
7. Gadikota, G., A.-H. A. Park, P. Kelemen, and J. Matter 2014. Chemical and morphological changes during olivine carbonation for CO_2 storage in the presence of NaCl and $NaHCO_3$. *Physical Chemistry Chemical Physics* 16:4679–4693.
8. Gadikota, G., E. J. Swanson, H. Zhao, and A.-H. A. Park 2014. Experimental design and data analyses for accurate estimation of reaction kinetics and conversion for carbon mineralization. *I&EC Research* 53(16):6664–6676.
9. Pan, S.-Y., T.-C. Liang, A.-H.A. Park, and P.-C. Chiang 2018. An overview: Reaction mechanisms and modelling of CO_2 utilization via mineralization. *Aerosol and Air Quality Research* 18:829–848.
10. Park, A.-H.A., and L.-S. Fan 2004. CO_2 mineral sequestration: Physically activated aqueous carbonation of serpentine and pH swing process. *Chemical Engineering Science* 59:5241–5247.
11. Park, A.-H A., R. Jadhav, and L.-S. Fan 2003. CO_2 mineral sequestration: Chemically enhanced aqueous carbonation of serpentine. *CJChE* 81(3–4):885–890.
12. Swanson, E. J., K. Fricker, M. Sun, and A.-H.A. Park 2014. Directed precipitation of hydrated and anhydrous magnesium carbonates for carbon storage. *Physical Chemistry Chemical Physics* 16(42):23440–23450.
13. Zhao, H., Y. Park, D. H. Lee, Y. N. Jang, K. S. Lackner, and A.-H.A. Park 2013. Tuning dissolution kinetics of wollastonite via chelating agents for CO_2 sequestration with integrated synthesis of precipitated calcium carbonates. *Physical Chemistry Chemical Physics* 15(36):15185–15192.
14. The Global CO_2 Initiative 2016. Carbon dioxide utilization (CO2U): ICEF roadmap 1.0.
15. Biernacki, J. J., et al. 2017. Cements in the 21st century: Challenges, perspectives, and opportunities. *Journal of the American Ceramic Society* 100(7):2746–2773.
16. EPA (U.S. Environmental Protection Agency) 2016. *Advancing Sustainable Materials Management: 2016 Recycling Economic Information (REI) Report.* Available at https://www.epa.gov/smm/recycling-economic- information-rei-report (accessed October 10, 2018).
17. Villere, P. 2015. *2015 Industry Data Survey. The Concrete Producer.* Available at https://www.concreteconstruction.net/producers/2015-industry-data-survey_o (accessed September 11, 2018).
18. USGS 2018. *Mineral Commodity Summaries 2018.* Reston, VA: U.S. Geological Survey, 200 p, doi: 10.3133/70194932.
19. Portland Cement Association 2013. *U.S. Portland Cement Industry: Plant Information Summary.* Skokie, IL: Portland Cement Association.
20. Vance, K., G. Falzone, I. Pignatelli, M. Bauchy, M. Balonis, and G. Sant 2015. Direct carbonation of $Ca(OH)_2$ using liquid and supercritical CO_2: Implications for carbon-neutral cementation. *Industrial & Engineering Chemistry Research* 54(36): 8908–8918.

21. P. Falkowski, R. J. Scholes, E. Boyle, J. Canadell, D. Canfield, J. Elser, N. Gruber, K. Hibbard, P. HoÅNgberg, S. Linder, F. T. MacKenzie, B. Moore, T. Pedersen, Y. Rosenthal, S. Seitzinger, V. Smetacek, and W. Steffen 2000. The global carbon cycle: a test of our knowledge of earth as a system. *Science* 290:291–296.
22. Uibu, M., M. Uus, and R. Kuusik 2009. CO_2 mineral sequestration in oil-shale wastes from Estonian power production. *Journal of Environmental Management* 90:1253–1260.
23. Lackner, K. S., D. P. Butt, and C. H. Wendt 1997. Progress on binding CO_2 in mineral substrates. *Energy Conversion and Management* 38:S259–S264.
24. O'Connor, W. K., D. N. Nielsen, S. J. Gerdemann, G. E. Rush, R. P. Waltera, and P. C. Turner 2001. Research status on the seqestration of carbon dioxide by direct aqueous mineral carbonation. *18th Annual International Pittsburgh Coal Conference*, Newcastle, Australia.
25. O'Connor, W. K., D. C. Dahlin, D. N. Nilsen, R. P. Walters, and P. C. Turner 2001. Carbon dioxide sequestration by direct aqueous mineral carbonation. *Proceedings of the 25th International Technical Conference on Coal Utilization & Fuel Systems*, Clear Water, FL.
26. Haug, T. A. 2010. Dissolution and carbonation of mechanically activated olivine: Investigating CO_2 sequestration possibilities, PhD thesis, Norwegian University of Science and Technology, Trondheim, ISBN 978-82-471-1961-7, http://www. Diva portal.org/smash/get/diva2:303780/FULLTEXT01.pdf.
27. Gerdemann, S. J., W. K. O'Connor, D. C. Dahlin, L. R. Penner, and H. Rush 2007. "Ex situ mineral carbonation". *Environmental Science & Technology* 41:2587–2593.
28. Pokrovsky, O. S., and J. Schott 1999. Dolomite surface speciation and reactivity in aquatic systems. *Geochimica et Cosmochimica Acta* 63:881–897.
29. Wee, J.-H. 2013. A review on carbon dioxide capture and storage technology using coal fly ash. *Applied Energy*, 106:143–151.
30. Huijgen, W. J. J., and R. N. J. Comans 2005. ECN-C-05-022, Energy Research Centre of The Netherlands, Petten, The Netherlands.
31. O'Connor, W. K., D. C. Dahlin, G. E. Rush, S. J. Gerdemann, L. R. Penner, and R. P. Nielsen 2004. Ex-Situ and In-Situ Mineral Carbonation as a Means to Sequester Carbon Dioxide, Albany Research Center, DOE/ARC-TR-04-002, Albany, OR.
32. Gerdemann, S. J., D. C. Dahlin, W. K. O'Connor 2002. Factors Affecting Ex-Situ Aqueous Mineral Carbonation Using Calcium and Magnesium Silicate Minerals, *6th International Conference on Green House Gas Control Technologies*, Kyoto, Japan.
33. SipilaÅN, J., S. Teir, and R. Zevenhoven 2008. Carbon dioxide sequestration by mineral carbonation: Literature review update 2005–2007, Abo Akademi University report VT 2008-1.
34. Chizmeshya, et al. 2002. First principles studies of mineral carbonation reaction processes in serpentine minerals, *2nd Annual Conference on Carbon Capture & Sequestration*, Pittsburgh.
35. Boerrigter, H. 2008. A process for preparing an activated mineral. Patent number: WO2008142025A2.
36. Balucan, R. D., B. Z. Dlugogorski, E. M. Kennedy, I. V. Belovac, and G. E. Murch 2013. *International Journal of Greenhouse Gas Control* 17:225–239.
37. Brent G. F. 2009. Integrated chemical process. Patent number: 20090305378A1.
38. Verduyn, M., H. Geerlings, G. van Mossel, and S. Vijayakumari 2011. Review of the Various CO_2 Mineralisation Product Forms. *Energy Procedia* 4:2885–2892.
39. Huijgen, W. J. J. and R. N. J. Comans 2006. Carbonation of steel slag for CO2 sequestration: leaching of products and reaction mechanisms. *Environmental Science & Technology* 40:2790–2796.
40. Parsons Brinckerhoff and Global CCS Institute, 2011. Accelerating the uptake of CCS: Industrial use of captured carbon dioxide, 279, http://www.globalccsinstitute.com/publications/acceleratinguptake-ccs-industrial-use-captured-carbon-dioxide.

41. Calera 2013. Calera process, http://www.calera.com/index.php/technology/the_science/.
42. Andersen, S. O., D. Zaelke, O. Young, H. Ahmadzai, F. Anderson, M. Atkinson, E. Carson, R. J. Carson, S. Christensen, J. S. J. Van Deventer, S. Hanford, V. Hoenig, A. Miller, M. Molina, L. Price, V. Ramanathan, H. Tope, J. Wilkinson, and M. Yamabe 2011. Scientific synthesis of calera carbon sequestration and carbonaceous by-product applications, Consensus Findings of the Scientific Synthesis Team, http://www.igsd.org/climate/documents/Synthesis_of_Calera_Technology_Jan2011.pd.
43. Jones, J. D. 2010. US Pat. 7,727,374, Skyonic Corporation.
44. Skyonic 2013, http://skyonic.com/wp-content/uploads/2010/ 02/Skyonic-Groundbreaking-Release-September-30–2013.pdf.
45. Drever, J. I., and L. L. Stillings 1997. The role of organic acids in mineral weathering. *Colloids and Surfaces A* 120:167–181.
46. Olsen, A. A., and J. D. Rimstidt 2008. Oxalate-promoted forsterite dissolution at low pH. *Geochimica et Cosmochimica Acta* 72:1758–1766.
47. Declercq, J., O. Bosc, and E. H. Oelkers 2013. Do organic ligands affect forsterite dissolution rates?. *Applied Geochemistry* 39:69–77.
48. Eloneva, S., A. Said, C.-J. Fogelholm, and R. Zevenhoven 2012. Preliminary assessment of a method utilizing carbon dioxide and steelmaking slags to produce precipitated calcium carbonate. *Applied Energy* 90:329–334.
49. Lin, P. C., C. W. Huang, C. T. Hsiao, and H. Teng 2008. Magnesium hydroxide extracted from a magnesium-rich mineral for CO_2 sequestration in a gas-solid system. *Environmental Science & Technology* 42:2748–2752.
50. Gal, E. 2006. Ultra cleaning of combustion gas including the removal of CO_2, WO/2006/022885.
51. Kothandaraman, A. 2010. PhD thesis, Massachusetts Institute of Technology, http://sequestration.mit.edu/pdf/Anush a_Kothandaraman_thesis_June2010.pdf.
52. Wang, X., and M. M. Maroto-Valer 2013. Optimization of carbon dioxide capture and storage with mineralisation using recyclable ammonium salts. *Energy* 51:431–438.
53. Fagerlund, J., E. Nduangu, and R. Zevenhoven 2011. *Energy Procedia* 4:4993–5000.
54. Fagerlund, J., E. Nduangu, I. Roma˜o, and R. Zevenhoven 2012. *Energy* 41:184–191.
55. Broecker, W. S. 2008. *Elements* 4:4295–4297.
56. Bodénan, F., F. Bourgeois, C. Petiot, T. Augé, B. Bonfils, C. Julcour, F. Guyot, A. Boukary, J. Tremosa, A. Lassin, and P. Chiquet. 2013. *4th International Conference on Accelerated Carbonation for Environmental and Materials Engineering*, Leuven, Belgium, April 10–12.
57. Hitch, M., and G. M. Dipple 2012. Economic feasibility and sensitivity analysis of integrating industrial-scale mineral carbonation into mining operations. *Minerals Engineering* 39:268–275.
58. Patel, T., A.-H. A. Park, and S. Banta 2013. Periplasmic expression of carbonic anhydrase in *Escherichia coli*: A new biocatalyst for CO_2 hydration. *Biotechnology and Bioengineering* 110:1865–1873.
59. Bobicki, E. R., Q. Liu, Z. Xu, and H. Zeng 2012. Carbon capture and storage using alkaline industrial wastes. *Progress in Energy and Combustion Science* 38:302–320.
60. Power, I. M., A. L. Harrison, G. M. Dipple, S. A. Wilson, P. B. Kelemen, M. Hitch, and G. Southam 2013. Carbon mineralization: From natural analogues to engineered systems. *Reviews in Mineralogy and Geochemistry* 77:305–360.
61. Eloneva, S., S. Teir, J. Salminen, C.-J. Fogelholm, and R. Zevenhoven 2008. Steel converter slag as a raw material for precipitation of pure calcium carbonate. *Industrial & Engineering Chemistry Research* 47:7104–7111.
62. Huntzinger, D. N., J. S. Gierke, L. L. Sutter, S. K. Kawatrad, and T. C. Eisele 2009. Mineral carbonation for carbon sequestration in cement kiln dust from waste piles. *Journal of Hazardous Material* 168:31–37.

63. Velts, O., M. Uibu, J. Kallas, and R. Kuusik 2011. Waste oil shale ash as a novel source of calcium for precipitated calcium carbonate: Carbonation mechanism, modeling, and product characterization. *Journal of Hazardous Material* 195:139–146.
64. Costa, G., R. Baciocchi, A. Polettini, R. Pomi, C. D. Hills, and P. J. Carey 2007. Current status and perspectives of accelerated carbonation processes on municipal waste combustion residues. *Environmental Monitoring and Assessment* 135:55–75.
65. Chang, E. E., C.-H. Chen, Y.-H. Chen, S.-Y. Pan, and P.-C. Chiang 2011. Performance evaluation for carbonation of steel-making slags in a slurry reactor. *Journal of Hazardous Material* 186:558–564.
66. Huntzinger, D. N., J. S. Gierke, K. Kawatra, T. C. Eisele, and L. L. Sutter 2009. Carbon dioxide sequestration in cement kiln dust through mineral carbonation. *Environmental Science & Technology* 43:1986–1992.
67. Huijgen, W. J. J., G.-J. Witkamp, and R. N. J. Comans 2005. Mineral CO_2 sequestration by steel slag carbonation. *Environmental Science & Technology* 39:9676–9682.
68. Uibu, M., and R. Kuusik 2009. Mineral trapping of CO2 via oil shale ash aqueous carbonation: process rate controlling mechanism and developments of continuous mode reactor system. *Oil Shale* 26:40–58.
69. Lee, D. K. 2004. An apparent kinetic model for the carbonation of calcium oxide by carbon dioxide. *Chemical Engineering Journal* 100:71–77.
70. Shih, S.-M., C. u.-S. Ho, Y.-S. Song, and J.-P. Lin 1999. Kinetics of the reaction of $Ca(OH)_2$ with CO_2 at low temperature. *Industrial & Engineering Chemistry Research* 38:1316–1322.
71. Lekakh, S. N., C. H. Rawlins, D. G. C. Robertson, V. L. Richards, and K. D. Peaslee 2008. Kinetics of Aqueous Leaching and Carbonization of Steelmaking Slag. *Metallurgical and Materials Transactions B* 39:125–134.
72. Wilson, S. A., G. M. Dipple, I. M. Power, J. M. Thom, R. G. Anderson, M. Raudsepp, J. E. Gabites, and G. Southam 2009. Carbon dioxide fixation within mine wastes of ultramafic-hosted ore deposits: Examples from the Clinton Creek and Cassiar Chrysotile deposits, Canada. *Economic Geology* 104:95–112.
73. Bonenfant, D., L. Kharoune, S. b. Sauve, R. Hausler, P. Niquette, M. Mimeault, and M. Kharoune 2008. CO_2 sequestration by aqueous red mud carbonation at ambient pressure and temperature. *Industrial & Engineering Chemistry Research* 47: 7617–7662.
74. Yadav, V. S., M. Prasad, J. Khan, S. S. Amritphale, M. Singhand, and C. B. Raju 2010. Sequestration of CO2 using red mud. *Journal of Hazardous Material* 176:1044–1050.
75. Pasquier, L. C., G. Mercier, J. F. Blais, E. Cecchi, and S. Kentish 2016. Technical and economic evaluation of a mineral carbonation process using southern Québec mining wastes for CO_2 sequestration of raw flue gas with by-product recovery. *International Journal of Greenhouse Gas Control* 50:147–157.
76. Teir, S., S. Eloneva, C.-J. Fogelholm, and R. Zevenhoven 2009. Fixation of carbon dioxide by producing hydromagnesite from serpentinite. *Applied Energy* 86:214–218.
77. Teir, S., R. Kuusik, C.-J. Fogelholm, and R. Zevenhoven 2007. Production of magnesium carbonates from serpentinite for long-term storage of CO_2. *International Journal of Mineral Processing* 85:1–15.
78. Sahu, R. C., R. K. Patel, and B. C. Ray 2010. *Journal of Hazardous Material* 179:29–34.
79. FernaÅLndez Bertos, M., X. Li, S. J. R. Simons, C. D. Hills, and P. J. Carey 2004. *Green Chemistry* 6:428–436.
80. Ecke, H. 2003. Sequestration of metals in carbonated municipal solid waste incineration (MSWI) fly ash. *Waste Management* 23:631–640.
81. Rendek, E., G. Ducom, and P. Germain 2006. Carbon dioxide sequestration in municipal solid waste incinerator (MSWI) bottom ash. *Journal of Hazardous Material* B128:73–79.

82. Sanna, A., M. Dri, M. R. Hall, and M. Maroto-Valer 2012. Waste materials for carbon capture and storage by mineralisation (CCSM) – A UK perspective. *Applied Energy* 99:545–554.
83. Back, M., M. Kuehn, H. Stanjek, and S. Peiffer 2008. Reactivity of Alkaline Lignite Fly Ashes Towards CO_2 in Water. *Environmental Science & Technology* 42:4520–4526.
84. Montes-Hernandez, G., R. Perez-Lopez, F. Renard, J. M. Nieto, and L. Charlet 2009. Mineral sequestration of CO(2) by aqueous carbonation of coal combustion fly-ash. *Journal of Hazardous Material* 161:1347–1354.
85. Muriithi, G. N., L. F. Petrik, O. Fatoba, W. M. Gitari, F. J. Doucet, J. Nel, S. M. Nyale, and P. E. Chuks 2013. Comparison of CO_2 capture by ex-situ accelerated carbonation and in in-situ naturally weathered coal fly ash. *Journal of Environmental Management* 127:212–220.
86. Nyambura, M. G., G. W. Mugera, P. L. Felicia, and N. P. Gathura 2011. Carbonation of brine impacted fractionated coal fly ash: implications for CO_2 sequestration. *Journal of Environmental Management* 92:655–664.
87. Pan, S. Y., E. E. Chang, P. Chiang 2012. CO_2 capture by accelerated carbonation of alkaline wastes: A review on its principles and applications. *Aerosol and Air Quality Research* 12:770–791.
88. Perez-Lopez, R., G. Montes-Hernandez, J. M. Nieto, F. Renard, and L. Charlet 2008. Carbonation of alkaline paper mill waste to reduce CO_2 greenhouse gas emissions into the atmosphere. *Applied Geochemistry* 23:2292–2300.
89. Bonenfant, D., L. Kharoune, S. Sauve, R. Hausler, P. Niquette, M. Mimeault, and M. Kharoune 2008. CO_2 Sequestration Potential of Steel Slags at Ambient Pressure and Temperature. *Industrial & Engineering Chemistry Research* 47:7610–7616.
90. Doucet, F. J. 2009. Effective CO_2 - specific sequestration capacity of steel slags and variability in their leaching behaviour in view of industrial mineral carbonation. *Minerals Engineering* 23:262–269.
91. Teir, S., S. Eloneva, C.-J. Fogelholm, and R. Zevenhoven 2007. Dissolution of steelmaking slags in acetic acid for precipitated calcium carbonate production. *Energy* 32:528–539.
92. Eloneva, S., S. Teir, J. Salminen, C.-J. Fogelholm, and R. Zevenhoven 2008. *Energy* 33:1461–1467.
93. Chang, E. E., A.-C. Chiu, S.-Y. Pan, Y.-H. Chen, C.-S. Tan, and P.-C. Chiang 2013. Carbonation of basic oxygen furnace slag with metalworking wastewater in a slurry reactor. *International Journal of Greenhouse Gas Control* 12:382–389.
94. Chang, E. E., S.-Y. Pan, Y.-H. Chen, C.-S. Tan, and P.-C. Chiang 2012. Accelerated carbonation of steelmaking slags in a high-gravity rotating packed bed. *Journal of Hazardous Material* 227–228:97–106.
95. Santos, R. M., D. Ling, A. Savaramini, M. Guo, J. Elsen, F. Larachi, G. Beaudoin, B. Blanpain, and T. Van Gerven 2012. Stabilization of basic oxygen furnace slag by hot-stage carbonation treatment. *Journal of Chemical & Engineering* 203:239–250.
96. Bukowski, J. M., and R. L. Berger 1979. Reactivity and strength development of CO_2 activated non-hydraulic calcium silicates. *Cement and Concrete Research* 9(1):57–68.
97. Young, J. F., R. L. Berger, and J. Breese 1974. Accelerated curing of compacted calcium silicate mortars on exposure to CO_2. *Journal of the American Ceramic Society* 57(9):394–397.
98. Kashef-Haghighi, S., and S. Ghoshal 2013. Physico–chemical processes limiting CO_2 uptake in concrete during accelerated carbonation curing. *Industrial & Engineering Chemistry Research* 52(16):5529–5537.

99. Brady, P. V., J. L. Krumhansl, and H. W. Papenguth 1996. Surface complexation clues to dolomite growth. *Geochimica et Cosmochimica Acta* 60(4):727–731.
100. Galan, I., C. Andrade, P. Mora, and M. A. Sanjuan 2010. Sequestration of CO_2 by concrete carbonation. *Environmental Science & Technology* 44(8):3181–3186.
101. Possan, E., W. A. Thomaz, G. A. Aleandri, E. F. Felix, and A. C. P. dos Santos 2017. CO_2 uptake potential due to concrete carbonation: A case study. *Case Studies in Construction Materials* 6:147–161.
102. Huntzinger, D. N., and T. D. Eatmon. 2009. A life-cycle assessment of Portland cement manufacturing: Comparing the traditional process with alternative technologies. *Journal of Cleaner Production* 17:668–675.
103. Sanna, A. 2013. *Fossil Fuels: Sources, Environmental Concerns and Waste Management Practices* [R. Kumar (ed.)]. New York: NovaScience Pub. Inc., ch. 6, pp. 199–208.
104. Teramura, S., N. Isu, and K. Inagaki 2000. New building material from waste concrete by carbonation. *Journal of Materials in Civil Engineering* 12:288–293.
105. Kashef-Haghighi, S., and S. Ghoshal 2009. *Industrial & Engineering Chemistry Research* CO_2 sequestration in concrete through accelerated carbonation curing in a flow-through reactor. 49:1143–1149.
106. Katsuyama, Y., A. Yamasaki, A. Iizuka, M. Fujii, K. Kumagai, and Y. Yanagisawa 2005. Development of a process for producing high-purity calcium carbonate ($CaCO_3$) from waste cement using pressurized CO_2. *Environmental Progress* 24:162–170.
107. Iizuka, A., M. Fujii, A. Yamasaki, and Y. Yanagisawa 2004. A new CO_2 sequestration process via carbonation of waste cement. *Industrial & Engineering Chemistry Research* 43:7880–7887.
108. Carran, D., J. Hughes, A. Leslie, and C. Kennedy 2012. A short history of the use of lime as a building material beyond Europe and North America. *International Journal of Architectural Heritage* 6(2):117–146.
109. Fernandez Bertos, M., S. J. R. Simons, C. D. Hills, and P. J. Carey 2004. A review of accelerated carbonation technology in the treatment of cement-based materials and sequestration of CO_2 *Journal of Hazardous Material* B112, 193–205.
110. Mignardi, S., C. De Vito, V. Ferrini, and R. F. Martin 2011. The efficiency of CO_2 sequestration via carbonate mineralization with simulated wastewaters of high salinity. *Journal of Hazardous Materials* 191(1):49–55.
111. Glasser, F. P., G. Jauffret, J. Morrison, J.-L. Galvez-Martos, N. Patterson, and M. S.-E. Imbabi 2016. Sequestering CO_2 by mineralization into useful nesquehonite-based products. *Frontiers in Energy Research* 4(3), doi:10.3389/fenrg.2016.00003.
112. Monkman, S., and Y. Shao 2006. Assessing the carbonation behavior of cementitious materials. *Journal of Materials in Civil Engineering* 18(6):768–776.
113. Cui, H., W. Tang, W. Liu, Z. Dong, and F. Xing. 2015. Experimental study on effects of CO_2 concentrations on concrete carbonation and diffusion mechanisms. *Construction and Building Materials* 93:522–527.
114. Sun, J., M. F. Bertos, and S. J. Simons. 2008. Kinetic study of accelerated carbonation of municipal solid waste incinerator air pollution control residues for sequestration of flue gas CO2. *Energy & Environmental Science* 1(3):370–377.
115. IEA report on GHG emissions from major industrial sources-2000, International energy Agency, Paris, France.
116. USGS (U.S. Geological Survey) 2017. Mineral commodity summaries 2017: U.S. Geological Survey:202 p, https://doi.org/10.3133/70180197.
117. Zulumyan, N., A. Mirgorodski, A. Isahakyan, H. Beglaryan, A. Gabrielyan, and A. Terzyan 2015. A low-temperature method of the β-wollastonite synthesis. *Journal of Thermal Analysis and Calorimetry* 122(1):97–104.
118. Gislason, S. R., and E. H. Oelkers 2014. Carbon storage in basalt. *Science* 344:373–374.

119. Fedoroc˘kova, A., M. Hreus, P. Raschman, and G. Suc˘ik 2012. *Minerals Engineering* 32:1–4.
120. Stolaroff, J. K., G. V. Lowry, and D. W. Keith 2005. *Energy Conversion and Management* 46:687–699.
121. CSLF 2012. Annual Meeting Documents Book, CSLF-T-2012-10, 4 September 2012, Perth, Australia.
122. Chang, R., D. Choi, M. H. Kim, and Y. Park 2017. Tuning crystal polymorphisms and structural investigation of precipitated calcium carbonates for CO_2 mineralization. *ACS Sustainable Chemistry & Engineering* 5(2):1659–1667.
123. Chang, R., S. Kim, S. Lee, S. Choi, M. Kim, and Y. Park 2017. Calcium carbonate precipitation for CO_2 storage and utilization: A review of the carbonate crystallization and polymorphism. *Frontiers in Energy Research* 5(17), doi: 10.3389/fenrg.2017.00017.
124. Ashraf, W., and J. Olek 2016. Carbonation behavior of hydraulic and non-hydraulic calcium silicates: Potentialof utilizing low-lime calcium silicates in cement-based materials. *Journal of Materials Science* 51(13):6173–6191.
125. Beruto, D. T., and R. Botter 2000. Liquid-like H_2O adsorption layers to catalyze the $Ca(OH)_2/CO_2$ solid–gas reaction and to form a non-protective solid product layer at 20ÅãC. *Journal of the European Ceramic Society* 20(4):497–503.
126. Burkan Isgor, O., and A. G. Razaqpur 2004. Finite element modeling of coupled heat transfer, moisture transport and carbonation processes in concrete structures. *Cement and Concrete Composites* 26(1):57–73.
127. Goracci, G., M. Monasterio, H. Jansson, and S. Cerveny 2017. Dynamics of nano-confined water in Portland cement: Comparison with synthetic C-S-H gel and other silicate materials. *Scientific Reports* 7(1):8258.
128. World Business Council for Sustainable Development 2009. *The Cement Sustainability Initiative.* Available at https://www.wbcsd.org/Sector-Projects/Cement-Sustainability-Initiative (accessed October 10, 2018).
129. Tepordei, V. 1997. Natural aggregates—Foundation of America's future. U.S. Geological Survey Fact Sheet 144-97, USGS, Washington, D.C.
130. Gadikota, G., K. Fricker, S.-H. Jang, and A.-H. A. Park 2015. Carbonation of silicate minerals and industrial wastes and their potential use as sustainable construction materials. *Advances in CO_2 Capture, Sequestration, and Conversion* [He et al., (eds.)], ACS Symposium Series, Washington, DC: American Chemical Society. ISBN: 9780841230880.
131. Jang, J. G., G. M. Kim, H. J. Kim, and H. K. Lee 2016. Review on recent advances in CO_2 utilization and sequestration technologies in cement-based materials. *Construction and Building Materials* 127:762–773, doi:10.1016/j.conbuildmat.2016.10.017.
132. Zhang, D., Z. Ghouleh, and Y. Shao 2017. Review on carbonation curing of cement-based materials. *Journal of CO_2 Utilization* 21:119–131.
133. Cresswell, D. 2007. MIRO characterisation of mineral wastes, resources and processing technologies—Integrated waste management for the production of construction material. WRT 177/WR0115.
134. Colangelo, F., and R. Cioffi 2013. Use of cement kiln dust, blast furnace slag and marble sludge in the manufacture of sustainable artificial aggregates by means of cold bonding pelletization. *Materials* 6(8):3139–3159.
135. ECO-SERVE Network 2004. Baseline report for the aggregate and concrete industries in Europe, cluster3: Aggregate and concrete production.
136. Snellings, R. 2016. Assessing, understanding and unlocking supplementary cementitious materials. *RILEM Technical Letters* 1:50–55.
137. Teir, S., J. Kettle, A. Harlin, J. Sarlin 2009. *Proceedings of the 3rd International Conference on Accelerated Carbonation for Environmental and Materials Engineering*, ACEME10, Turku, Finland, Nov. 29–Dec. 1, pp. 63–74.

138. Teir, S., S. Eloneva, and R. Zevenhoven 2005. Production of precipitated calcium carbonate from calcium silicates and carbon dioxide. *Energy Conversion and Management* 46:2954–2979.
139. A. Sanna, M. R. Hall, and M. M. Maroto-Valer 2012. Post-processing pathways in carbon capture and storage by mineral carbonation (CCSM) towards the introduction of carbon neutral materials. *Energy & Environmental Science* 5:7781–7796.
140. El-Sheikh, S. M., S. El-Sherbiny, A. Barhoum, and Y. Deng 2013. Effects of cationic surfactant during the precipitation of calcium carbonate nano-particles on their size, morphology, and other characteristics. *Colloids and Surfaces A* 422:44–49.
141. Gronchi, P., T. De Marco, and L. Cassar 2004. US6,716,408 B1.
142. Lindeboom, R. E. F., I. Ferrer, J. Weijma, and J. B. van Lier 2013. *Water Research* Silicate minerals for CO_2 scavenging in autogenerative high pressure digestion. 47:3742–3751.
143. Tai C. Y., and F. B. Chen 1998. Polymorphism of $CaCO_3$, precipitated in a constant-composition environment. *AIChE Journal*:1790–1798.
144 Li, G., Z. Li, and H. Ma. 2013. Comprehensive use of dolomite-talc ore to prepare talc, nano-MgO and lightweight $CaCO_3$ using an acid leaching method. *Applied Clay Science* 86:145–152. doi:10.1016/j.clay.2013.09.015.
145. Li, G., Z. Li, and H. Ma 2013. *International Journal of Mineral Processing* 123:25–31.
146. Wanjari, S., C. Prabhu, N. Labhsetwar, and S. Rayalu 2013. *Journal of Molecular Catalysis B: Enzymatic* 93:15–22.
147. Favre, N., M. L. Christ, and A. C. Pierre 2009. Biocatalytic capture of CO_2 with carbonic anhydrase and its transformation to solid carbonate. *Journal of Molecular Catalysis B: Enzymatic* 60:163–170.
148. Sahoo, P. C., Y. N. Jang, and S. W. Lee 2013. Enhanced biomimetic CO_2 sequestration and $CaCO_3$ crystallization using complex encapsulated metal organic framework. *Journal of Crystal Growth* 373:96–101.
149. Faatz, M., F. GroÅNhn, and G. Wegner 2004. *Advanced Materials* 16:996–1000.
150. Montes-Hernandez, G., F. Renard, R. Chiriac, N. Findling, J. Ghanbaja, and F. Toche 2013. Sequential precipitation of a new goethite–calcite nanocomposite and its possible application in the removal of toxic ions from polluted water. *Chemical Engineering Journal* 214:139–148.
151. Fricker, K., and A.-H. A. Park 2013. Effect of H_2O on $Mg(OH)_2$ carbonation pathways for combined CO_2 capture and storage. *Chemical Engineering Science* 100:332–341.
152. Fricker, K., and A.-H. A. Park 2014. Investigation of different carbonate phases and their formation kinetics during $Mg(OH)_2$ slurry carbonation. *Industrial & Engineering Chemistry Research* 53(47):18170–18179.
153. Reddy, K. J., S. John, H. Weber, M. D. Argyle, P. Bhattacharyya, D. T. Taylor, M. Christensen, T. Foulke, and P. Fahlsing 2011. Simultaneous Capture and Mineralization of Coal Combustion Flue Gas Carbon Dioxide (CO_2). *Energy Procedia* 4:1574–1583.
154. Gunning, P. J., C. D. Hills, and P. J. Carey 2013. *Proceedings of the 4th International Conference on Accelerated Carbonation and Materials Engineering*, Leuven, Belgium, pp. 185–192.
155. Jones, G., G. Joshi, M. Clark, and D. McConchie 2006. Carbon capture and the aluminium industry: preliminary studies. *Environmental Chemistry* 3:297–303.
156. Jones, B. E. H., and R. J. Haynes 2011. Bauxite processing residue: A critical review of its formation, properties, storage, and revegetation. *Critical Reviews in Environmental Science and Technology* 41:271–315.
157. Cooling, D. J. 2007. *Paste 2007—Proceedings of the Tenth International Seminar on Paste and Thickened Tailings*, Australian Center for Geomechanics, eds. A. B. Fourie, and R. J. Jewell, Perth, Australia, pp. 3–15.

158. Wang, T., et al. 2017. Accelerated mineral carbonation curing of cement paste for CO_2 sequestration and enhanced properties of blended calcium silicate. *Chemical Engineering Journal* 323:320–329.
159. Longo, R. C., et al. 2015. Carbonation of wollastonite(001) competing hydration: Microscopic insights from ion spectroscopy and density functional theory. *ACS Applied Materials & Interfaces* 7(8):4706.
160. Cygan, R. T., J.-J. Liang, and A. G. Kalinichev 2004. Molecular models of hydroxide, oxyhydroxide, and clay phases and the development of a general force field. *The Journal of Physical Chemistry B* 108(4):1255–1266.

4 Biological Conversion of Carbon Dioxide

4.1 INTRODUCTION

As mentioned earlier, conversion of CO_2 to useful chemicals and fuels is a win-win strategy. In this regard, biological CO_2 reduction or conversion process can be beneficial for developing carbon-neutral technologies. The integration of CO_2-emitting industrial technologies with CO_2-converting biological systems can be helpful in achieving sustainable value-added products with no or minimal loss of energy and materials that are assuring for improved economics. The CO_2-converting bioprocesses can be directly integrated with the processes emitting a high amount of CO_2. This symbiotic integration can make the whole process carbon neutral [1–3].

In recent years, an energy-neutral and biorefinery approach [4,5] has been used to explore the potential of numerous biological CO_2 utilization methods to decrease the atmospheric CO_2. Biological conversion tends to be slow, less efficient, and costly compared to thermal or catalytic conversion processes. However, its effectiveness toward selectivity and product variations and carbon-neutral footprint and energy neutrality has made it a very viable and important approach for CO_2 utilization. This chapter examines various biological CO_2 utilization methods that can be implemented to generate bio-based products and elaborate on the biology of different sequestration and conversion methods.

There are fundamentally two different approaches for biological conversion examined in the literature. They are (a) photosynthesis-based and (b) nonphotosynthesis-based. The photosynthesis methods include photosynthesis using algae (including green and microalgae) and cyanobacteria, as well as biophotosynthesis and hybrid bioelectrophotocatalysis using inorganic catalysts and microbes. Nonphotosynthetic methods include (a) fermentation (both anaerobic and gas), (b) bioelectrocatalysis, and (c) microbial synthesis using catalysts and enzymes with different types of enabling cycles. Here, we examine each of these techniques in some details with possible products they can generate. Some methods can be hybrid in its nature, and significant efforts are made to improve both photosynthetic and microbial syntheses using genetic manipulations.

Nonphotosynthetic biological systems possess a number of potential advantages over photosynthetic systems. These include a wide variety of organisms, a larger range of potential target chemicals, and the ability to avoid the inefficiency of photosynthesis. Aerobic systems also have the advantage of high productivity, capacity for continuous cultivation, and compatibility with artificial photosynthesis. Some nonphotosynthetic organisms can take advantage of low-cost, low-emission electrons from renewable energy sources, and many can be cultivated with cheap and

DOI: 10.1201/9781003229575-4

ubiquitous gaseous carbon feedstocks comprised of hydrogen, CO, and CO_2, such as industrial waste gas, biogas, or syngas. Microbial organisms are also easier to modify by synthetic biology and genetic engineering than several strains of algae and cyanobacteria.

Both photosynthetic and nonphotosynthetic approaches, however, suffer from poor solvation of CO_2 and H_2 in water, a lack of genetic tools and metabolic understanding, a limited number of strains that have been explored, and limited techniques for downstream processing (DSP) of products.

In general, nonphotosynthetic applications are at an earlier maturity level than photosynthetic ones, and support is warranted on these applications to advance understanding and the maturity of these systems.

4.2 PHOTOSYNTHETIC CO_2 REDUCTION BY ALGAE

CO_2 can be captured and converted to various chemicals fuels, fuel additives, and other value-added products using light energy. In nature, photosynthesis is a natural process for converting CO_2 and H_2O into oxygen by biological transformation. The use of solar energy to cultivate CO_2 reduction by photothermal catalysis, photocatalysis, photoelectrocatalysis, and PV-supported electrocatalysis to form various valuable products like CO, CH_4, methanol, formic acid, and so on is covered in great details in Chapter 6. Here, we strictly focus on CO_2 uptake and reduction by photosynthesis using algae and cyanobacteria. Most of the photosynthetic biological systems follow oxygenic or oxygenic photosynthesis routes [6].

Algae cultivation for CO_2 conversion for biofuel production provides numerous advantages and disadvantages, as shown in Table 4.1. If properly cultivated, however, algae (both green and microalgae) and photosynthetic cyanobacteria can be very valuable sources for biofuel production.

4.2.1 Role of Algae

The term algae broadly refer to any photosynthetic prokaryotic microorganism (cyanobacteria) or eukaryotic microorganism (microalgae) that can be cultivated. Both these types provide a fertile medium for CO_2 uptake and conversion. Algal biomass provides high yields, with more than 30- to 50-fold improvement in oil yield in comparison with common agricultural crops [7]. They also provide better life-cycle analysis compared to nonphotosynthetic methods [8]. Algae can be viewed as self-replicating machines that convert sunlight and CO_2 into value-added products. Products of algae cultivation include a wide range of biofuels, dietary protein and food additives, commodities, and specialized chemicals [2,3].

Several studies have shown that CO_2 mitigation with algae is a sustainable process with simultaneous generation of high-calorific products, such as biodiesel, pigments, fatty acids, etc. [9,10]. Figure 4.1 illustrates the products obtained from CO_2 conversion by photosynthesis using microalgae [11]. The potential utilization of algae is mainly attributed to its wide distribution, high biomass production, capability to adjust in adverse conditions, swift carbon uptake and utilization, and capability to

TABLE 4.1
Advantages and Disadvantages of Algae Based Biofuel Systems

Advantages

1. Algae grows at higher efficiency levels than other biofuel crops.
2. Algae is more productive than other forms of biomass.
3. Algae is a renewable resource.
4. Algae produces viable hydrocarbons for numerous products/high-value feedstock as a replacement of terrestrial biomass/produces low-carbon to high-carbon sources namely carbohydrates, lipids, and proteins.
5. Algae is something that almost anyone can grow/easily cultured, readily and rapidly bio-engineered.
6. Algae can grow in virtually all waters/easy adaptability at various climate conditions/can grow almost anywhere such as ponds, wastelands, etc.
7. Algae biofuels work with our current distribution system.
8. Algae growth helps us to curb our greenhouse gas emissions/sustainable, renewable and environment-friendly.
9. Algae are nontoxic and nonedible.
10. Algae can be used as restoration of contaminated areas and ponds with quick bioremediation and degradable sources and conserve fossil fuel, high CO_2 capture, energy conversion via synthetic photosynthesis.

Disadvantages

1. Algae have the same concerns of monoculture that the agriculture industry experiences.
2. Algae growth may create quality variations during the refinement process.
3. Algae biofuel does not always meet its energy efficiency targets.
4. Algae growth creates regional sustainability problems/indefinite availability of sustainable algae resources for large scale biofuel production.
5. Algae might grow quickly, but it still needs time to produce viable oils.
6. Algae biofuels come with higher production costs/regionally constrained market structures for biofuels/economics need to be improved.
7. Algae growth requires high levels of fertilizer to maximize production.
8. Algae require significant water resources to produce oil for collection.
9. Algae contamination occurs more often with large-scale production methods/harmful algae blooms in global waters/unclear utilization of type of waste materials/may cause neurotoxic problems.
10. Algae utilize arable land/land requirement can be high.
11. Limited expertise for industrial scale biofuel operation/lack of origin and source certification for global monitoring and control of algae fuel.

generate value-added products. Both micro- and macroalgae provide a natural cycle to metabolize inorganic carbon by a photoautotrophic mechanism using carbonic anhydrase enzyme (CAE) [12] resulting in the conversion of CO_2 into valuable products with liberation of oxygen. The generated $NADH_2$ from the electron transport chain combines with the RuBisCo (ribulose-1,5-bisphosphate carboxylase/oxygenase, provided by CAE) and helps in carbohydrate generation from CO_2 and provides the reducing power in the Calvin cycle for glucose synthesis [13]. Calvin cycle is described in detail in Section 4.6.

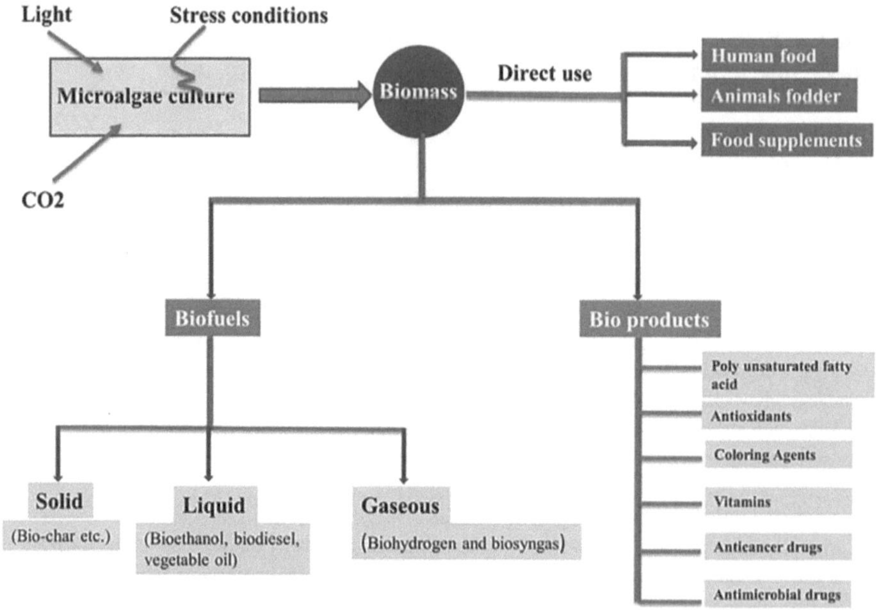

FIGURE 4.1 Microalgae convert atmospheric CO_2 to carbohydrates, lipids, and other valuable bioproducts by using light. Microalgae biomass is a rich source for biofuels and bioactive compounds [11].

4.2.2 Green Algae

Green algae, a common term for eukaryotic single-cellular photosynthetic organisms deriving from several phyla, are a highly diverse group of algae representing many thousands of identified species. Multiple strains have been validated in biomass applications, and their selection is commonly determined by culture conditions. While nature of strain, pH, temperature, flue gas toxicity, and so on affect the cultivation of algae, a continuous supply of CO_2 is also very important. If a continuous supply of CO_2 is not available, onsite compressed CO_2 storage and delivery may be required. The majority of biomass cultivation efforts using green algae have focused on biofuel production. Biodiesel, or fatty acid methyl ester (FAME), production from green algae is now considered as part of a mature industry. While biodiesel can be used as drop-in diesel replacement in many diesel engines, its hygroscopicity, cloud point, and fouling properties have limited widespread adoption. After biomass harvesting, lipids are extracted by one of several methods to prepare isolated neutral lipids, or triacylglyceride (TAG), or total lipids containing both polar and neutral lipids. Methods have been developed to produce biodiesels from total lipid extracts [14,15]. Since in recent years, the use of biodiesel provides smaller fuel market and it is mostly used as an additive for ultra-low-sulfur petroleum diesel to provide added lubricity properties [16], several co-products have been identified that can help make biofuel production more economically feasible.

As pointed out by national academy report [2], the economy of biofuel production from green algae depends significantly on co-products and other downstream operations such as hydrotreatment, pyrolysis, gasification, and so on. Hydrotreatment of lipid extracts derived from algae biomass conversion to produce diesel and gasoline components is a very important part of biofuel industry. Only few companies like Neste Oil and Chevron [17] have the capital infrastructure required to carry out hydrotreatment on a commercial scale. Furthermore, new technology may be required for conversion of algal lipids containing significant portions of polar lipids and hydrophobic pigments along with triglycerides into renewable fuels by hydrotreatment facilities [18]. Hydrogen required for hydrotreatment facilities can be derived from renewable sources like solar, wind, or nuclear source by electrolysis of water. Besides hydrotreatment, liquefaction, pyrolysis, and gasification offer routes to produce a variety of fuels, including methane, ethanol, and fuel oils [10]. Many of these can also be transformed to more complex chemicals by reforming process [19].

In order to make algal process more competitive, generation of co-products is very important. The waste product from unused biomass created during biofuel production is primarily composed of protein and carbohydrates, which can be consumed by anaerobic digestion (AD) to produce biogas, a low-value end product. Or, they can be converted into a variety of more valuable products via combined photosynthetic, fermentative, and chemical methods. Algal carbohydrates can be diverted to fermentative production of ethanol or other commodity chemicals, such as succinic acid [20] from metabolically engineered yeast, while the proteins may be applied to adhesive manufacturing [21].

Co-products also include dietary protein, polyunsaturated fatty acids (PUFAs), and pigments. Algae have long been viewed as an attractive source of dietary protein, both as animal feeds and for human consumption [22]. Green algae contain between 40% and 70% of their dry weight as protein, and the amino acid profile of most algae compares favorably with common food proteins. Protein productivity from algae has been estimated at up to 50 times that of soybeans per acre of land. This suggests that green algae has the potential to supplement or replace crops for animal feed with vastly reduced demands for arable land. Many green algae naturally produce PUFAs that are valuable for humans and animals as food additives. In particular, the omega-3 varieties, eicosapentaenoic acid and docosahexaenoic acid, are valuable for the supplementation of many farmed animals, particularly carnivorous fish such as salmon and tuna [23,24]. In addition, omega-3 fatty acids demonstrate a consumer market as nutraceutical products that is currently dominated by fish oils. Algae are well known for their ability to produce a variety of pigments and could provide colorants as additives to foods, drinks, cosmetics, and a host of other products in order to increase the appeal of the product to the consumer [25]. The use of algae can thus replace the use of fossil fuel for these purposes.

4.2.3 Microalgae for CO_2 Fixation and Biofuel/Co-Products Generation

As shown in Chapter 1, CO_2 sources are varied in their quantity and concentrations. It is also known that a higher concentration (generally greater than 5%) of CO_2 adversely impacts the growth and production rate in most microalgae [12]. Freshwater microalgae *Chlorella* HA-1 exhibited a maximum growth rate when it was evaluated at 5% and 10% of CO_2. However, the growth was inhibited with further increases in CO_2 concentration up to 20%. Thus, the study indicated that *Chlorella* HA-1 can be used to treat industrial flue gas with a maximum CO_2 concentration of 10% [26]. On the other hand, *Chlorococcum littorale* culture isolated from a saline water pond was able to show an almost similar growth rate from 5% to 60% of CO_2, which is a much higher CO_2 concentration than in flue gas (20%). When the CO_2 concentration was boosted to 70%, a decrease in growth rate was identified. As this strain exhibits high growth at high CO_2 concentrations, it can be readily applied for quite a variety of industrial flues gases [27]. Thus, in order to accommodate various concentrations of industrial sources of CO_2 streams, significant research is being carried out to isolate a suitable algal strain for direct and efficient conversion of industrial CO_2 into products such as agar and alginate, etc. As shown in Table 4.2, Kondaveeti et al. [1] have identified CO_2 conversion capabilities of various strains in algae. A selection of particular strain of algae depends not only on its tolerance of CO_2, temperature, and toxic compounds, but also for attaining high growth rates and maximum cell densities [28,29]. As shown in the table, not only there is a significant variation in CO_2 uptake in various strains, but there is also a significant variation in CO_2 uptake in a given strain.

TABLE 4.2
Effect of Algae Type on CO_2 Fixation or Conversion Rate (Data Taken from Refs. [1,30])

Algae	CO_2 Fixation Rate (mg/L/day)	CO_2 Conversion Rate (MicroMole/g of CO_2/h)
Dunaliella tertiolecta	272.4, 313	272–313
Chlorella vulgaris	251.64, 624, 865	7–3,352
Spirulina platensis	318.61, 413	
Botryococcus braunii	496.98, 1,100	1,100
Chlorella sp.UK001	31.8	
Synechocystis aquatilis	1,500	
Chlorella sp.		125–1,246
Nannochloris sp.		57–569
Synechocystis		142
Microcystic sp.		46–493
Synechocystis aquatilis		1,500
Euglena gracilis		36–362
Phaeodactylum tricornutum		27–267
Spirulina sp.		37–373
Scenedesmus sp.		44–436
Anabaena sp.		137–1,373

Growth rate [31] is also dependent on temperature and pH. There are several strains of micro-algae that exhibit a high growth rate in extreme pH and temperature [32]. The most industrially valuable algal product is lipids, which can be used as a source for biofuel productions (e.g., biodiesel) [33]. In order to get an optimum lipid production, it is important to identify right strain and optimize operating conditions of light, pH, and temperature. pH is a very important variable during algae growth; an unoptimized pH can affect the distribution of carbon species and carbon availability, and extreme pH can directly affect the metabolic activity of algae [34,35].

Besides CO_2 concentration, pH, and temperature, the growth rate of algae strains also depends on other impurities like SOx and NOx in flue gases. The dosing of flue gases containing SOx and NOx to algae ponds has shown to increase biomass yields by threefold and high energy generation [36,37]. Recent studies have shown that direct submission of flue gases into algal ponds exhibited a 30% increase in biomass yields over using the equivalent concentration of CO_2 [30,38]. The low levels of NOx and SOx present in the waste stream ultimately can be metabolized by most strains of algae and, as such, require considerations primarily for transport, heat exchange, and time of use [39]. This competence of CO_2 encapsulation by algae can vary according to the algae physiology, pond chemistry, and temperature. About 80%–99% of CO_2 capture can be obtained under optimized conditions with gas residence times as short as 2 seconds [40,41].

While SOx and NOx in flue gas accelerate algae growth, other heavy metal impurities, like arsenic, cadmium, mercury, and selenium [31], pose challenges for biomass cultivation and restrict the use of products for biofuel applications than for the production of protein for animal feed. Algae pond cultivation also needs large land requirement. For a 200 MW/h power plant, a 3,600 acres of algae pond would be required to capture 80% of the plant's CO_2 emissions during the daylight hours, assuming a productivity rate of 20 g dry biomass/m^2/day. Therefore, locating the algae ponds next to CO_2 point suppliers provides numerous potential low-cost and energy-saving advantages [42] with a minimal carbon footprint. Several pilot-scale facilities have implemented this strategy, whereby flue gas is utilized at the site to provide CO_2 to algal cultures [43–45].

The term "microalgae" is generally used for both prokaryotic blue green algae (cyanobacteria) and eukaryotic microalgae including green algae, red algae, and diatoms. Microalgae are considered as bio-factories for the capture of CO_2 and simultaneous production of renewable biofuels, food, animal, and aquaculture feed products and other value-added products such as cosmetics, nutraceuticals, pharmaceuticals, bio-fertilizers, and bioactive substances [46,47]. Microalgae are capable of acquiring inorganic carbon from very low concentration CO_2 streams and possess CO_2 concentrating schemes for efficient photosynthesis [48]. They are also capable of working in extreme environments with simple yet versatile nutrients requirements. Microalgae do not require arable land and are capable of surviving in salineal-kaline water, land, and wastewater [45,49]. Furthermore, microalgae can tolerate notorious waste gasses such as CO_2 and NO_x, SO_x from flue gas, inorganic and organic carbon, N, P, and other pollutants from agricultural, industrial, and sewage wastewater sources, and provide opportunities to transform these waste gases into bioenergy and valuable products with least impact to environment [9,50]. The simple cellular structures and rapid growth of microalgae allows them to have CO_2 fixation efficiency as higher as 10- to 50-folds than terrestrial plants [51,52]. Microalgae can grow under flexible environments such as high concentrations of Ci in the presence of pure or industrial CO_2 streams or with soluble carbonate [53,54]. Microalgae produce lipids, proteins, sugars, and pigments [54]. However, microalgae are most economical when they are used to provide an integrated biorefinery setup where every valuable component like fatty acids for nutraceuticals or biofuels, and proteins for food/feed applications are extracted, processed, and valorized [55,56].

For all biotransformations of CO_2 into products, energy is necessary to chemically reduce the carbon. For this, photoautotrophic microorganisms use sunlight. Photoautotrophic microorganisms such as microalgae use photosynthesis to convert CO_2 into biomass with sunlight. They include a diverse range of organisms found in all aquatic environments, including the oceans, freshwater lakes, and highly alkaline ponds. They typically grow as free-floating cells but can also form dense microbial mats. They take CO_2 and nutrients such as ammonia and phosphate from the water. The cells are rich in proteins that can be used as feed. They can also be induced to accumulate triacylglyceride oils. Biomass can be converted to products via approaches like hydrothermal liquefaction, extraction of oils, or complete biorefinery concepts.

Mass cultivation of the microbes can be achieved in large outdoor growth systems by providing sufficient nutrients, which usually limit growth in the natural environment. The growth systems can be natural lagoons, shallow constructed raceway ponds, or enclosed photobioreactors. Diffusion of CO_2 from the atmosphere is not sufficient to maintain dense and productive cultures. To avoid CO_2 limitation and maximize productivity and solar efficiency, CO_2 is provided to the cultures, typically by sparging with a CO_2-enriched gas, for example, flue gas.

Large-scale cultivation of photoautotrophic microorganisms has key advantages over terrestrial plant crops. They can produce considerably more biomass per unit land area, can utilize land not suitable for traditional agriculture, can make use of brackish and saline water, and can fully utilize nutrients, avoiding cumulative effects from runoff. However, there are challenges in cultivating photoautotrophic microbes cost-effectively. In outdoor environments, productivity can be compromised by fluctuating temperature, light, and oxygen-bearing microorganisms.

4.2.4 Biorefinery Concept of Microalgal Biomass

Just like fossil-fuel refinery, an integrated biorefinery refers to the conversion of biomass into multiple commercially valuable products and fuels [57]. Figure 4.2 depicts a simplistic microalgal-based biorefinery system. While the microalgal biomass consists of lipids (7%–23%), proteins (6%–71%), and carbohydrates (5%–64%), depending upon the microalgal specie and culture conditions [22,58,59], a biorefinery produces valuable products like biogas, biodiesel, biohydrogen, bioethanol, biobutanol, and various other valuable products. An integrated biorefinery also adopts a system approach in which upstream and DSP are integrated with central microalgal biorefinery. Here, we briefly assess the performance of a biorefinery.

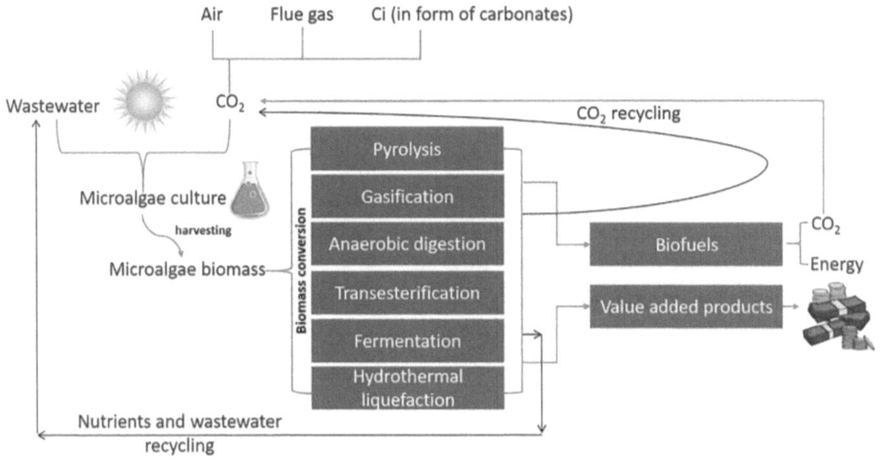

FIGURE 4.2 A simplistic representation of microalgal-based biorefinery system [60].

4.2.4.1 Biodiesel

Microalgae are known to accumulate remarkable amount of lipid. As reviewed by Mata et al. [59], the lipid content of common microalgae such as *Chlorella*, *Dunaliella*, *Isochrysis*, *Nannochloris*, *Nannochloropsis*, *Neochloris*, *Phaeodactylum*, *Porphyridium*, and *Schizochytrium* varies between 20% and 50% of cell dry weight, that can be augmented to higher levels by manipulating environmental and other growth factors, process optimization, and genetic modifications of the production strain. Nitrogen starvation and salinity stress are known to induce an increase in TAG (triacylglycerol) accumulation and relative content of oleic acid in most of the microalgal species [61]. The fatty acid composition of most of the microalgae is dominated by C14–C18 fatty acids, yet the relative composition varies from species to species [62].

Also, the role of HCO_3^- in inducing TAG accumulation has been widely illustrated recently [63–65]. The lipids can be converted into FAMEs via transesterification for biodiesel production. The major by-product glycerol also finds enormous industrial application opportunities. Furthermore, the residual de-oiled microalgal biomass (DMB) can be used for animal feed.

4.2.4.2 Biogas

Microalgal biomass can be efficiently used for the production of biogas, including methane, hydrogen, and biohythane (combination of methane and 5%–25% hydrogen gas) [66]. The resistance of cell wall to enzyme hydrolysis is one of the prime bottlenecks in the AD process. The overall economic feasibility of the process depends on the factors affecting AD, microalgal strain, biomass pretreatment, and culture methods [67]. Later, to make the system economically viable and environmentally sustainable, a closed-loop production scheme is being adopted wherein AD effluents are recycled and used as an input in the first step of AD. Jankowska et al. [67] have presented a detailed review of microalgae's cultivation, harvesting, and pretreatment for AD of biogas production.

4.2.4.3 Bioethanol

The carbohydrate part (mainly glucose, starch, cellulose, and hemicellulose) of the microalgal dry biomass can be used for transforming into bioethanol via fermentation. Although microalgae accumulate relatively low quantities of sugars, the absence of lignin from microalgal structure makes them advantageous over other feedstock such as corn, sugarcane, and lignocellulosic biomass [68,69]. *Isochrysis galbana*, *Porphyridium cruentum*, *Spirogyra* sp., *Nannochloropsis oculate*, and *Chlorella* sp. are mainly exploited microalgae for the production of carbohydrates [70].

4.2.4.4 Biobutanol

The green residual after microalgae oil extraction can be utilized for the production of biobutanol. The higher energy density of biobutanol and its molecular similarity to gasoline makes it more suitable than biomethanol or bioethanol as biofuel. Aside from being a biofuel, it can also be used as a solvent for industrial purposes [71]. Despite having notable significance, a limited number of studies have reported laboratory stage work on the fermentation of microalgae biomass to butanol [72]. Microalgal strains with high starch and convertible sugars concentrations would be ideal for biobutanol production research. *Tetraselmis subcordiformis*, *Chlorella vulgaris*, *Chlorella reinhardtii*, and *Scenedesmus obliquus* could be among the potential candidates [71].

4.2.4.5 Value-Added Products

Apart from the biodiesel and fuels, algae are well known for their usage in the food industry. The commercial products of algal powders are sold in the health and food industries. Algal products, such as unsaturated fatty acids and polysaccharides from microalgae and seaweed, provide a dietary fiber and health benefits. The carotenoids that present in algae serve as antioxidants and vitamin A. In the textile industry, several algal products (pigments) are used as coloring agents. In aquatic fields, algae are used as feedstock in fish, shrimp, and other seafood that are grown in enclosed environments. Carotenoids, such as astaxanthin, beta-carotene, bixin, and fucoxanthin, are used in the cosmetic and pharmaceutical industries [60].

In the context of biorefinery approach, intracellular compounds and metabolites have gained immense importance owing to their high monetary value. Microalgal pigments, chlorophylls a and b, lutein, astaxanthin, β-carotene, phycobilins, C-phycocyanin, have found wide application in dyes, cosmetics, food and feed additives, nutraceuticals and pharmaceuticals, as natural colors, bioactive components, anti-oxidants, nutritive and neuro-protective agents [73]. The commercial products of algal powders are sold in the health and food industries. Algal products, such as unsaturated fatty acids and polysaccharides from microalgae and seaweed, provide a dietary fiber and health benefits. The carotenoids that present in algae serve as antioxidants and vitamin A. In the textile industry, several algal products (pigments) are used as coloring agents. In aquatic fields, algae are used as feedstock in fish, shrimp, and other seafood that are grown in enclosed environments. Carotenoids, such as astaxanthin, beta-carotene, bixin, and fucoxanthin, are used in the cosmetic and pharmaceutical industries [60]. Microalgae are also exploited as rich source of amino acids (leucine, asparagine, glutamine, cysteine, arginine, aspartate, alanine, glycine, lysine, and valine); carbohydrates (β1–3-glucan, amylose, starch, cellulose, and alginates); and vitamins and minerals (vitamin B1, B2, B6, B12, C, and E; biotin, folic acid, magnesium, calcium, phosphate, iodine) that are widely used in food additives, health supplements, and medicine. Microalgae, such as *Nannochloropsis*, *Tetraselmis*, and *Isochrysis*, are used for extraction of long-chain fatty acids popularly known as the omega fatty acids such as DHA (docosahexaenoic acid) and EPA (eicosapentaenoic acid), and have lately gained prime attention as essential for human brain development and health. Other than these, microalgae are also used for the production of extracellular polymeric substances (EPSs), which have many industrial applications and polyhydroxyalkanoates (PHAs). PHAs can be used for manufacturing bioplastics that are very sought after because of their biodegradability [74].

4.2.4.6 System Approach to Biorefinery

The utilization of microalgal biomass for the production of bioproducts within a biorefinery framework is still very expensive [60,74]. In order to make this biorefinery approach economically viable, both upstream processing (USP) and DSP need to be efficiently simplified and integrated with biorefinery. The efficiency of the USP is determined by microalgal strain selection, nutrient supply (CO_2, N, and P), and culture conditions (temperature, light intensity) [75], whereas the constraints at the DSP level are mainly characterized by harvesting, cell disruption, and extraction methods. In DSP, harvesting accounts for 20%–40% of the total production costs and for a multi-product biorefinery, the cost increases to 50%–60% [74].

A. **Upstream Processing**

Screening microalgae strains can be time-consuming and expensive. It is now facilitated by 96-well microplate swivel system (M96SS), and it allows examination of up to 768 microalgal samples at the same time, possible [60,76]. Microalgal production strains can be improved by induced acclimation through manipulation of various environmental stresses [60,77] and showed that mixed diverse community of microalgae, dominated by *Desmodesmus* spp., could be adapted over a time of many months to survive in 100% flue gas from an unfiltered coal-fired power plant containing 11% CO_2. Carbohydrate and starch accumulation in *Chlorella* sp. AE10 was improved by a two-staged process, wherein the CO_2 concentration, light intensity, and nitrogen concentration were changed drastically and cells were diluted at the onset of the second stage resulting in a 42% increase in carbohydrate accumulation [78]. Besides stress manipulation and acclimatization, desirable traits of the microalgal strains can be effectively improved by genetic and metabolic engineering/synthetic biology. Tools such as CRISPER, TAL, and TALEN are used to alter the gene expression unlike gene modification. Synthetic biology engages the use of "biobricks" to create artificial regulatory pathways that can control a desired cellular trait by modifying the metabolism. Interchangeable units such as promoters, ribosome-binding sites (RBS), terminators, *trans*-elements, and regulatory molecules serve as the biobricks [60,79]. Recently, Yang et al. [80] genetically engineered the Calvin cycle of *Chlorella vulgaris* enhancing its photosynthetic capacity by ~1.2-fold. Kuo et al. [81] screened an alkali-tolerant *Chlorella* sp. AT1 mutant strain by NTG (N-methyl-N′-nitro-N-nitrosoguanidine) mutagenesis that survived well 10% CO_2 for prospective CO_2 sequestration.

Large-scale microalgal cultivation and nutrient supply pose huge economic burden. In this regard, biofilm-based attached cultivation is emphasized rather than aqua-suspend methods that have massive water requirement, low biomass productivity, energy intensive and cannot be easily scaled up [60,82]. Microalgal production using wastewater from industrial, agricultural, and sewage sources is a promising way to reduce the ecological footprints substantially [60,83]. Digestates, effluents from biogas production units and AD (containing concentrated nutrients including nitrogen in the form of ammonia, potassium, phosphorous, sulfur, and recalcitrant organic substances), are also used in microalgal cultivation systems. This subject is recently reviewed by Koutra et al. [84].

B. **Downstream Processing**

The main DSP unit operations are harvesting, cell disruption, and extraction. Centrifugation is the most efficient (>95% efficiency) method for harvesting microalgae. However, it is not suitable for large-scale systems due to its cost. Flocculation is a low-cost alternative. Cationic chemical flocculants and polymeric flocculants are generally used [60], but can negatively affect the toxicity of the biomass and output water [46]. Zhou et al. [85] reported a novel fungi-assisted bioflocculation technique, in which filamentous fungal spores were added to the algal culture under optimized conditions and the pellets were formed after 2 days that can be harvested by simple filtration. Attached culture can also make harvesting simple [86]. Conventional disruption methods like bead beating, homogenizers, heating, applying high pressure, and chemicals or enzymes for lysis are costly and pose risk of loss of desired multiproducts in biorefinery concept. Physical disruption by pulsed electric field (PEF) is a promising alternative technology as it is a low-shear technology that operates on low temperature and can aid the extraction of hydrophobic constituents of the biomass [60,74]. In extraction technologies, ionic liquids (ILs) appear to be promising as they are advantageous over conventional solvents. ILs are organic salts that are nonvolatile at room temperature. Also, they can be used for extraction of hydrophilic proteins. Imidazolium-based ILs have been successfully used for cell disruption for lipid extraction from microalgal biomass [87].

4.2.5 Challenges and Benefits in Using Algae

The generation of biofuels and associated co-products using algae faces several challenges and benefits. These challenges and benefits are briefly described below:

1. On an industrial scale, algae are typically growing in photobioreactors, open ponds, or raceway ponds [88]. The use of open ponds can be simpler in architecture and operation. However, these are limited with a large surface area necessity, elevated cultivation cost, high chance of contamination, and low productivity. Photobioreactors with closed systems have been proposed to overcome the above limitations [89]. A detailed review study by Kumar et al. [32] provided great aspects in the design of photobioreactors and challenges noticed in industrial algae growth.
2. Flue gas compositions are made up of different types of gases along with CO_2, which is viable for the growth of few algae. However, high temperatures, fluctuating flue gas composition, and the presence of sulfur and nitrogen can be toxic to some algal strains. Therefore, the selection of suitable algal processes and strains is needed.
3. Light intensity is very important for algae growth [90]. Insufficient light can lead to poor growth, and an increase in light intensity can also show adverse effects on the growth, known as photo inhibition [91].
4. The mixing of cell cultures requires an extensive amount of energy input and constitutes a major portion of operational costs [92]. Stirring of cultures is important to reduce the mass transfer limitations occurring in photobioreactors, which also favorably helps to eliminate oxygen accumulation.
5. In general, the CO_2 uptake/capture with algae is calculated based on carbon concentration in influent and effluent [93]. This, however, falsely assumes that whole CO_2 uptake is carried out by algae and is utilized for its growth. While the carbon content in biomass can deliver a more precise value on the quantity of CO_2-consumption by algae, it also assumes that culture media only has inorganic carbon such as CO_2 and bicarbonate. The presence of organic carbon source complicates the CO_2 uptake. Tang et al. [94] analyzed the rate of CO_2 biofixation for different algae species that were grown in modified Blue-Green 11 (BG11) medium. They calculated the carbon capture rate by varying the CO_2 concentration from 0.03% to 50%. The carbon content in biomass was found to be around 50%, whereas the CO_2 fixation was in the range of 0.105–0.288 g/L day.
6. Photosynthesis is inherently inefficient, as only 3%–6% of total solar radiation energy is captured [95]. In addition, a limitation of algae is the efficiency of the carbon dioxide conversion that limits the annual flux [96]. In order to provide maximum light exposure through large volume and surface area, algal cultivation becomes land and water intensive. Theoretically, to capture all CO_2 from a 10 kiloton/day, power plant would require 25–37 acres of cultivation [2].
7. Biomass cultivation presents capital and operations costs that do not have close parallels in existing large-scale industries.
8. A key benefit to biological systems is their inherent flexibility in terms of feedstocks and environments. There are opportunities to utilize nonarable land and saline water or wastewater for algae cultivation, thereby minimizing competition for natural resources. It is well known that microalgae biofuel production systems exhibit high net energy balance, high water

efficiency, and less land requirement in comparison with first-generation biofuels (plant biomass) [97].

9. Biological systems can tolerate low CO_2 concentrations and impurities in the carbon sources common to industrial power generation. These variables have a considerable impact on capital and operations costs for biological conversion technologies.
10. It is important to account for resource use and environmental impact when comparing algae cultivation to conventional agriculture or other activities. Algae cultivation offers significant productivity improvements as compared to contemporary agriculture. Multiple studies have demonstrated up to 30-fold productivity improvements in oil production [2,98] and 50-fold improvements in protein production from algae when compared to soybean, canola, or corn [99] per acre of land. Although not fully evaluated, protein derived from algal biomass has been proposed for both animal and human consumption, and some companies, for example, Qualitas, already produce specialized algae to enrich animal feeds. The competitive landscape for algal cultivation would shift these commercial applications to prove competitive with current agricultural products such as soy or corn. This issue is further described below.

4.2.6 Factors Affecting Commercialization of Microalgal Technologies

Compared to conventional methods, the microalgae industry is concerned about the two fundamental hurdles, which can be identified as low productivity and high costs [100]. While biofuel and co-products improve the economics of biorefinery, utilization of defatted or DMB for other processes also further helps to develop sustainable biorefineries. DMB can be used as a substrate for other biofuel generation, such as biomethane, biohydrogen, and bioethanol [100]. Thermochemical liquefaction, pyrolysis, and gasification are nonbiological processes in which DMB can be used as substrate to produce fuels and chemicals [100,101]. The biomass of microalgae can be used as inoculum for new processes that reduce the maintenance of parent cultures. This helps in sustainable operation of microalgal bioreactors [101]. DMB is rich in protein and essential amino acids that could be utilized as livestock and aquaculture feeds or as a dietary supplement. DMB has also been tested to replace fishmeal from wild fish caught in an ocean environment. Thus, it can mitigate ocean resource depletion [102]. In wastewater treatment, DMB can be used as an active biosorbent to remove dyes/color and heavy metals [102]. During the extraction of lipids from algal biomass, the biomass used to rupture, which has a higher surface area than the raw biomass. This helps for the efficient functioning of DMB as an active biosorbent [103].

Algae have high potential to store CO_2 as biomass, which is due to high photosynthetic efficiencies and high biomass yields [1–3,60]. Microalgal cultivation was designated as a negative emission technology (NET), which can reduce the impacts of ocean acidification and anthropogenic climate change [104]. Since the scope of microalgae is versatile and possesses the expansive possibility of microalgae cultivation with a self-sustainability approach, it is designated as a blue-bioeconomy.

In consideration of oil extracted from plant biomass versus algal biomass, palm oil can generate a maximum of 4–5 versus 30 ton/ha/yr, thus yielding a fivefold higher product in comparison with a plant crop during similar posttreatment of biomass [97]. In terms of area, algae are known to generate about 15–300 times more oil for biodiesel [105]. In contrast to crop plants' single-harvesting process, algae allow multiple harvesting cycles in the same duration with high yield and greater light and CO_2 capture efficiencies.

Different oil concentrations are noted with various algae species, which is due to differences in the amount of protein/carbohydrates/fats. For instance, *S. maxima* has about 60%–70% w/w proteins, *Porphyidium cruentum* generates around 57% proteins, and *Scenedesmus dimorphus* generates up to 40% w/w of lipids [1,106]. Generations of biomass by pure microalgae for different supply of CO_2 concentrations can also vary significantly. For example, *Chlorella vulgaris* for CO_2 supply in the range of 0.03%–6% generates 0.21 g/L/day, while *Spirulina platensis* at a CO_2 supply of 10% generates 2.91 g/L/day of biomass. On the other hand, *Chlorella* sp.-KR-1 generates 0.6–0.7 g/L/day biomass at CO_2 supply between 10% and 70% [1]. In few species of algae, such as *Botryococcus braunii* and *Chlorella protothecoides*, terpenoid hydrocarbons and lipids are noted, which can be further converted into short-chain hydrocarbons, similar to that of crude oil [107]. On the basis of prevalence of lipids or sugars, algae can be a wonderful substrate for biodiesel or bioalcohol, respectively [108]. Therefore, algae can be a great resource as a biofuel substrate under certain conditions. Several studies have employed different strategies for increasing lipid concentration of algae, such as temperature, CO_2 concentration, light strength/intensity, nutrient starvation, metal, and salinity stress. It is well known that adequate light intensity helps in overproduction of microalgae lipids. It is also recognized that higher lipid concentration is noted at maximum photosynthetic efficiency, which occurs at the light saturation point [109]. Hence, the extreme light intensity can cause photoinhibition, leading to damage of algal photosystems and, thus, reducing lipid accumulation.

In the case of temperature, the optimal value for higher lipid generation achieved does vary from species to species. *C. vulgaris*, a microalgal species, achieved a maximum lipid concentration at 25°C. Microalgae species *S. obliquus* had a lipid concentration of 18% and 40% of dry weight when the temperature was at 20°C and 27.5°C, respectively [110]. This does not mean that an increase in temperature increases total lipid content for all microalgal species. A recent review by Zhu et al. [109] provides detailed strategies for lipid enhancement.

With an increase in algae biodiesel production, several studies had endeavored to calculate the cost of oilgae (algae oil) from large farms. Studies by Benemann and Oswald [111] reported that the cost of oilgae would range from US$39 to US$69 per barrel from 4 km^2 open ponds either by using pure CO_2 or flue gas from coal power stations with productivity in a range of 30–60 $g/m^2/day$ with a 50% algal lipid yield [111]. In recent studies, it is noted that the production of oil from algae would cost around US$66–153 per barrel [112]. Companies generating oilgae are not optimistic in terms of production cost. Seambiotic Ltd. calculated that the cost of dried algae would be US$0.34/kg with a productivity and total lipid content of 20 $g/m^2/day$ and 8%–40%, respectively. Presuming the standard yield of lipid to be 24% without any additional cost, this would be equal to US$1.42/kg for lipid extractions, which is equal to US$209/barrel. With an increase in the yield of lipids to 40%, the lipid extraction would cost around US$0.85/kg, which is equal to US$126/barrel, which points to the requirement of an increase in lipid concentration or decreasing oil prices [111,112]. Accounting for inflation in crude oil prices in 2017, the oil prices seem to be stabilizing around US$51/barrel (https://www.eia.gov/todayinenergy/prices.php, 2017). These estimates suggest that oilgae diesel production needs to be further improved to be economically viable and to compete with conventional fuels [111,112].

4.3 REDUCTION OF CO_2 USING PHOTOSYNTHETIC CYANOBACTERIA

Most types of photosynthetic bacteria derive energy from ATP, which helps in the conversion of CO_2 to biomass and other products [113]. In comparison with eukaryotic plants and algae, photosynthetic bacteria have simpler photosystems due to the presence of pigments on their cytoplasm rather than specialized organelles [114]. Photosynthetic bacteria are classified into five phyla: cyanobacteria, proteobacteria, chlorobi, choloflexi, and firmicutes [114]. Cyanobacteria are capable of fixing atmospheric nitrogen and carbon. Similar to algae, they are distinct and broadly distributed and exist as biofilms or as suspended planktonic cells. Cyanobacteria are found to be a key player by accounting for 20%–30% of the Earth's photosynthetic activity.

Cyanobacteria are being engineered to directly convert solar energy, carbon dioxide, and water to biofuels, and other products. Cyanobacteria-based approaches possess advantages over traditional biological production systems based on plants, green algae, or heterotrophic organisms. For example, cyanobacteria are more amenable to genetic manipulation than are algae and can, therefore, be adapted for the production of a wider range of products. The photosynthetic efficiency of cyanobacteria is two to four times higher than that of plants [115], and their cultivation does not compete with food crops for land usage. Cyanobacteria also do not have the sugar requirements of heterotrophic organisms such as *Escherichia coli* and yeast, although they have a reduced growth rate and fewer available synthetic biology tools compared to these hosts.

There is a plethora of cyanobacterial species, but only a few have been adapted to chemical production. The three predominant strains utilized for chemical production are *Synechococcus elongatus* PCC 7942 (7942), *Synechocystis* sp. PCC 6803 (6803), and *Synechococcus* sp. PCC 7002 (7002). These strains all have sequenced genomes, established culturing methods, and basic metabolic engineering tools [2,116,117], yet each strain presents its own unique advantages and challenges.

Cyanobacteria have been engineered to produce a wide range of fuels, fuel precursors, and commodity chemicals. According to national academy report [2], fuels and fuel precursors that can be produced with cyanobacteria include ethanol, isobutanol, n-butanol, fatty acids, heptadecane, limonene, bisabolene, 2,3-butanediol, 1,3-propanediol, ethylene, glycogen, lactate, isoprene, squalene, and farnesene. Cyanobacteria are also known to produce alkanes and alkenes, which have a desirable property for industrial application [118]. Acyl-acyl carrier protein reductase and aldehyde decarbonylases are the key enzymes noted for efficient conversion of fatty acids of metabolic intermediates to alkanes and alkenes [119]. Iso-butanol has applications as a drop-in biofuel that could be integrated into current energy infrastructure. Fatty acids can be used in the synthesis of biodiesels. Heptadecane can be readily used in fuel production for the synthesis of biodiesel. Limonene has applications as both a biofuel and a solvent. Bisabolene has applications as a biodiesel candidate. 2,3-Butanediol is a commodity chemical used to make synthetic polymers [120] and can readily be converted to methyl ethyl ketone, a fuel additive, and solvent [121]. 1,3-Propanediol has a variety of uses including in polymers, paints, solvents, and antifreeze. Ethylene can be used to generate polyethylene, a widely used polymer, and in its gaseous state can be used to speed the ripening of produce. Glycogen is a common form of energy storage and a potential carbon source for chemical production. It can also be converted into ethanol for use as a biofuel. Biological synthesis of lactic acid for biodegradable polymers has been established in 6803. 3-Hydroxypropionic acid (3-HP), 3-hydroxybuterate (3-HB), and 4-hydroxybuterate (4-HB) all have applications for the synthesis of polymers and plastics used in daily life. Isoprene is commonly used for the production of synthetic rubbers. Squalene is widely used in the food, personal care, and medical industries but its commercial production is unreliable and nonideal. Farnesene has been used as a precursor for high-performance polymers and as a jet fuel candidate. Since productivities and titers are too low, more genetic manipulations and metabolic engineering is needed for commercialization of the technology.

Although few cyanobacterial species are capable of generating ethanol, its production rate needs to increase to make a viable commercial process. Numerous studies [1,122] made an attempt to generate higher alcohols by genetically modified bacteria. These studies suggest that the usage of genetically modified bacteria can be a viable advancement to increase product formation by employing CO_2 as an electron source rather than organics. Cyanobacterial species, such as *Anabaena*, *Aphanocapsa*, *Calothrix*, *Microcystis*, *Nostoc*, and *Oscillatoria*, are equipped for the generation of hydrogen by using a photosynthetic autotrophic mechanism [1,123]. Filamentous cyanobacteria generate H_2 as a by-product during nitrogen fixation under N_2-limiting conditions. Additionally, cyanobacteria can generate H_2 by reversible activity of the hydrogenases enzyme. In general, these species are equipped to use bidirectional or reversible hydrogenase enzymes to oxidize oxygen to generate hydrogen. The enhancement in hydrogen generation can be achieved by blocking the pathways that use the hydrogenases for a reduction reaction [124]. Additionally, due to similarity to algae metabolism, cyanobacteria are well known for their effluent treatment of wastewaters that are rich in nitrogen and phosphorus contaminants.

4.4 FACTORS AND ISSUES AFFECTING PHOTOSYNTHESIS

National Academy Report [2] points out that there are a number of issues that need to be addressed in the use of microalgae and cyanobacteria for photosynthesis. These issues relate to (a) cost, (b) benefits, (c) environmental impacts on their cultivations, (d) land and water need, (e) social acceptability, (f) restrictions in the use of genetically modified organisms, (g) the degree of CO_2 solvation in water, (h) nutrients requirements, (i) downstream burdens, and (j) availability and suitability of CO_2 waste streams. These issues indicate that photosynthesis is only a viable approach when these issues are properly addressed.

Algae are cultivated either in oblong "racecourse" design open ponds or in closed photobioreactor (PBR) with different scales, materials, and designs [2,125] in order to optimize solar radiation upon the biomass culture. The oblong open pond design is least expensive, and in this design, a high rate of water movement can be achieved with paddlewheels or air lift pumps. The simplest and most common PBR design uses a "hanging bag" approach, where polyethylene tubes are suspended vertically with air and CO_2 flow providing agitation and carbon addition [126]. The selection between open pond and closed PBR depends on many factors [127] including capital and operating costs and value of the product. Closed PBR being more expensive is chosen when high-value products with high purity are generated. Open ponds are favored for the production of biofuels given the need to minimize production costs.

There are strict U.S. regulations on what and how much genetically modified organisms can be used due to concerns regarding their containment and spread. The cost and efficiency of CO_2 dissolution in aqueous solution is also a major issue. Simple sparging does not always work. Often this is replaced by amine-based CO_2 concentrator, followed by thermal stripping. Another approach uses carbonate salts to deprotonate solvated CO_2 into soluble bicarbonate. Nutrient fertilizers such as phosphorus and nitrogen also raise concerns that algae cultivation could contribute to eutrophication (excessive algal growth due to the influx of nutrients) in freshwater and coastal zones. The need for excessive amount of freshwater and large land requirements is also a major issue for open pond algae cultivation. Algae are, however, naturally abundant in saltwater, brackish water, and a number of extreme environments. A single power plant may require dozens of hectares of biomass cultivation. However, since algae cultivation does not require arable land, it will not compete with agriculture and can valorize regions with marginal or saline soils.

4.5 HYBRID BIOLOGICAL PROCESSES FOR CONVERSION OF CO_2

There are three ways biological process can be combined with photosynthesis or electrochemical process: biophotosynthesis or biomethanation in the presence of solar energy, bioelectrochemical process, or biophotoelectrocatalytic process. Here, we briefly examine these three hybrid processes.

4.5.1 Biophotosynthesis/Microbial Biomethanation in the Presence of Solar Energy

In this method, methanogens, "methane-producing" microbes, convert CO_2 and H_2O into CH_4 via biophotosynthesis as shown in Figure 4.3a. These microbes use exergonic process of reaction between H_2 and CO_2 as a source of energy that is ultimately stored in the molecules such as ATP. In principle, this CO_2 metabolism reaction does not require solar energy. However, the generation of dihydrogen or other reducing agents using solar energy is detected during methane formation by this method. As shown in Figure 4.3a, a key enzyme for the reduction of carbon dioxide is formylmethanofuran dehydrogenase (fmd) [128]. Here, CO_2 receives a hydride equivalent (H^+ and $2e^-$) from a sulfur-ligated Mo^{4+} or W^{4+} active site of fmd to produce formate (Figure 4.3a). The electrons for the hydride equivalent are carried from the dihydrogen-oxidizing Fe–Ni active site of a nearby hydrogenase to the CO_2-reducing active site via a series of iron–sulfur clusters in the fmd protein.

The formate product then reacts with a primary amine, methanofuran (R^1NH_2 in Figure 4.3a) which as shown in the figure goes through further set of reactions. The final CH_4 evolution step requires that the methyl group will first be transferred from nitrogen to sulfur in the form of the thioether compound R^4SMe (methyl coenzyme M). The thioether and a thiol R^5SH (coenzyme B) then go through further set of reactions to complete the catalytic cycle. Methanogens of the "archaea" class evolved in this process to utilize transition metals (e.g., Fe, Ni, Zn, and Mo) are oxygen-sensitive, as are the hydrogenase enzymes that link the H_2 oxidation and CO_2 reduction reactions [129].

In biomethanation in the presence of solar energy, with the help of CH_4-producing microorganisms as biocatalysts, CH_4 and H_2O are thus produced during the anaerobic reduction of CO_2 by H_2. This reaction occurs at 20°C–70°C and 1–10 bar. The microorganisms are typically contained in a liquid fermentation broth within stirred tank or fixed bed reactors [130]. In order to make this process more successful, several issues need to be resolved which include: (a) engineering bioreactors that can make this link function efficiently, (b) fate of deceased archaea cells, (c) issues pertaining to growth media, (d) mass transfer limitations caused by low solubility and transport of H_2 into water and slow transport of CO_2 in water [130–132].

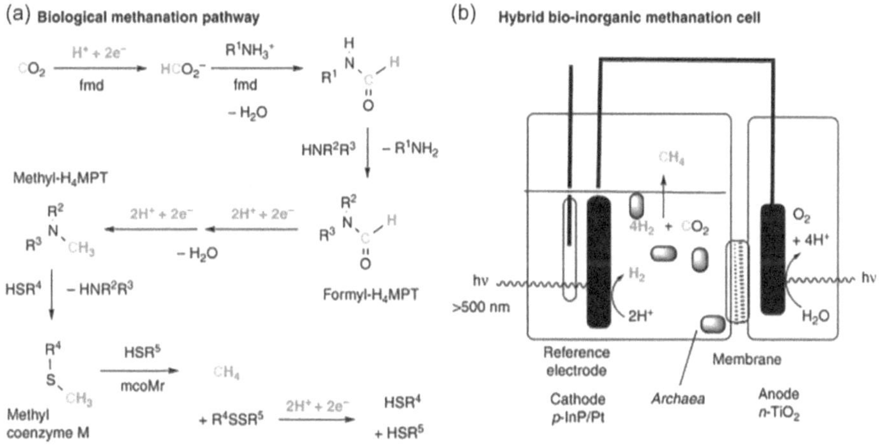

FIGURE 4.3 Biomethanation reaction systems. (a) A simplified scheme describing the methanation of CO_2 catalyzed by archaea. (Reproduced from Ref. [133] (b) A hybrid system for photomethanation utilizing CO_2-metabolizing archaea in the cathode compartment as adapted from Refs. [129,134].

4.5.2 Biophotoelectrocatalysis

In this hybrid method, the inorganic materials produce molecular H_2, which is fed to the microorganisms to fuel the CO_2 reduction reaction. This is carried out in a two-compartment electrolytic cell in which archaea is placed in the cathode compartment. The cell is illustrated by Figure 4.3b [135]. The system can be made completely solar-powered by having a platinum-coated, photoactive *p*-InP cathode to feed the archaea with H_2 and a photoactive n-TiO_2 anode in the second compartment to provide the electrons for H_2 production from water oxidation. This solar-powered cell requires an anion exchange membrane between the compartments to minimize pH changes. The observed Faradaic efficiency of this system was higher (up to 74%) when blue light was filtered out from the cathode compartment, due to the sensitivity of the microbes to these wavelengths. Cathodes supporting immobilized hydrogenases, combined with anodes functionalized with photosystem II, have also been successfully used to form a solar cell capable of generating H_2 [55]. Such a cathode might also be suitable for housing active methanogens, thereby enhancing the ability of archaea to convert CO_2 into CH_4. The conversion of CO_2 to CH_4 can also be carried out by methanogens with the use of electrons directly from electrodes. This will eliminate the need for the electrolysis of water [136].

The hybrid biological approaches for converting CO_2 to CH_4 illustrated in Figure 4.3b are appealing because it can be used for effective and selective conversion of CO_2 to CH_4 even with impure CO_2 streams. High tolerance by these microbes for common impurities like hydrogen sulfide, nitrogen oxides, ammonia, and particulates as well as partial tolerance for oxygen and ethanol [137], which are commonly encountered in flue gas and raw biogas, makes the process very attractive. Similar level of tolerance by inorganic components which play a pivotal role in hybrid bio-inorganic photo-electrocatalytic systems for flue gas and biogas impurities is as yet not known and needs to be determined [132,135,136,138,139].

The hybrid system reported to have a solar-to-chemical efficiency of 10% and an electrical-to-chemical efficiency of 52%. This assumes efficiencies of 20% for solar-to-electrical conversion at the photovoltaic panel, 70% for electrical-to-hydrogen conversion, and 86% for the conversion of CO_2 to CH_4. In order to make this process commercially viable, various issues dealing with the scale-up of this reactor needs to be addressed. For both processes, safety measures for large quantities of genetically modified organisms needed for large-scale process that needs to be addressed to get public acceptance. Literature [130,137] summarizes recently installed methanation projects.

4.5.3 Microbial Electrosynthesis

Microbial electrosynthesis (MES) synthesizes value-added chemicals from CO_2 via electrogenic fermenting cathode microbes (biocathode) and using this cathode electrode as a sole electron source (Figure 4.4) [140]. MES can act as energy storage by converting CO_2 to chemical production such as hydrogen, alcohols, and C_1–C_6 medium chain fatty acids.

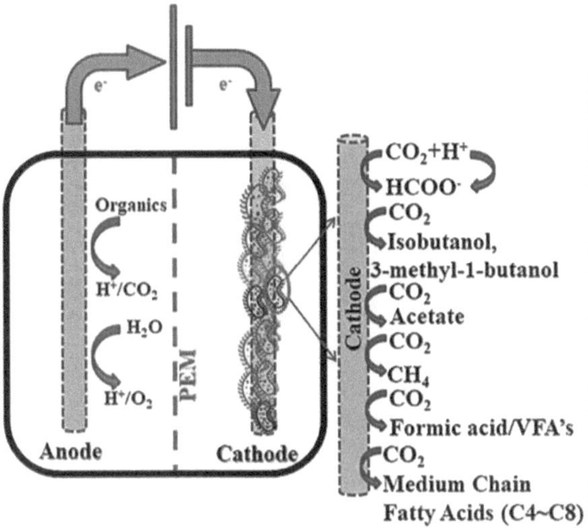

CO_2 reduction to product	Standard Potential (mV)
Methane	-240 ($8e^-$)
Acetate	-280 ($8e^-$)
Butanol	-300 ($24e^-$)
Ethanol	-310 ($12e^-$)
PHB	-310 ($18e^-$)
Butyrate/Butanol	-370 ($4e^-$)
Acetate/Ethanol	-390($4e^-$)
Formate	-410($2e^-$)
Fumarate/Succinate	30($2e^-$)

FIGURE 4.4 Schematic representation of MES system toward [1].

MEC is a modification of MFC which is used for the bioelectrochemical conversion of organics, such as acetate, to generate CO_2, protons, and electrons at the anode. In the MFC, the generated electrons at anode move toward the cathode through an external circuit. The protons are diffused toward the cathode through a separator, typically a proton exchange membrane (PEM, e.g., Nafion-117) or cation exchange membrane (CEM). On the cathode, the diffused protons and electrons are combined together to form water by using O_2 as an electron acceptor (cathodic reduction reaction). When an external electron acceptor, such as O_2, is absent, the diffused protons are converted to H_2 with the help of minimal external voltage (0.2 V), and this mode is now termed as microbial electrosynthesis, MEC [141]. MEC operations are more economic over MFC due to the ecologically beneficial production of valuable products, such as H_2. By the same process, the microbes at biocathodes can shuttle the e^- from the electrode for the reduction of CO_2 to value-added products. This can be achieved by using an anaerobic biotic anode with an electrogenic biocatalyst using organics or with an abiotic anode with an electrolysis process by applying additional potential. The bioprocess for value-added product formation was carried out by numerous metabolic pathways, such as the Calvin Benson cycle or the WLP [1,141]. These cycles are described later in Section 4.6.

MES is capable of generating basic value-added products, such as H_2, acetate, and oxobutyrate [142]. These products can be directly used as fuels or can be used further to generate high-carbon fuels or bioplastics. Initially, Nevin et al. [143] stated reduction of CO_2 to acetate in small concentrations by operating MES at −0.6 V by using an acetogenic *Sporomusaovata* ascathodic biocatalyst [143] and solar energy as an external power source. This study also generated smaller amounts of oxobutyrate (Table 4.3). Nevin et al. [144] achieved higher amount of acetate using mixed cultures.

TABLE 4.3
Bioelectrochemical Generation of Solvents and biofuels from CO_2 under Various Operational and Nutritional Conditions [1]

Reactor	Substrate	Applied Energy (Voltage or Current)	Dominant Catalyst	Products (Microbes or Enzymes)
H-type double chamber (batch)	$CaCO_3$	+0.5 to −0.5V (vs. Ag/AgCl)	*Clostridium sporogenes* BE01	Butanol, ethanol, fatty acids, hydrogen
Double chamber fuel cell (batch)	Butyraldehyde TRIS-HCl buffer +	−0.6 V (vs. Ag/AgCI	Alcohol dehydrogenase enzymes	Butanol
H-type double chamber (batch)	P2 electron carriers in medium + glucose	−0.7 V (vs. Ag/AgCI)	*Clostridium beijerinckii* IB4	Acetone, butanol, ethnol production and butanol as primary product
Two compartment cell (batch)	CAB medium with electron carrier in buffer	−2.5 V poised at cathode	*Clostridium acetobutylicum* ATCC 4259	Butanol and acetone
H-type double chamber(continuous mode)	CO_2 injection + DSMZ medium	−0.6 V (vs. Ag/AgCI)	*Sporomusa*, *Gebacter*, *Clostridium*, *Morella*	Acetate, formate, butyrate, propanol, ethanol and 2-oxobutyrate
H-type double chamber (batch)	CO_2 (no organics in media	−0.8 V (vs. SHE)	*Clostridium* species + Carboxydotrophic mixed culture	Ethanol, butanol, acetate, butyrate
H-type double chamber (batch)	Modified P2 medium SMM medium	1.32 V as applied voltage	*C. pasteurianum*	Butanol and by-product as solvents and acids
Double chamber (batch)	CO_2	−0.55 V (vs. NHE)	N/A	CH_4

Zhang et al. [145] showed that with the use of *S. ovata*, the production rates were 2.3-fold higher using a nickel nanoparticle-coated cathode over a plain graphite cathode (1.13 mM/day). Mohanakrishna et al. [146] reported an acetate generation rate of 4.1 g/L by using enriched single-chamber MES with VITO-activated carbon electrodes. Batlle-Vilanova et al. [147] operated continuous-mode MES for CO_2 reduction and noted an acetate production rate of 0.98 mM.L/day. Marshall et al. [148] reported the highest acetate generation (17.25 mM/day) from CO_2 by using MES with a carbon bed cathode and enriched cultures from methanogenic bacteria and H_2. MES operation with a simultaneous production and extraction mechanism for acetate from CO_2 were operated by Gildemyn et al. [149]. In this study, the maximum concentration of acetate formation was noted as 13.5 g/L, which is found to be the highest to date, whereas the highest production rate of 0.78 g/L/h was registered by LaBelle and May [150]. All these studies pointed out that during long-term operation, the oxidation of acetate to methane is one of the major limitations. The reported variations in acetate production might be due to differences in applied voltage, operational environment, reactor configuration, electrode type, biofilm activity, and type of bioelectrocatalyst at the cathode. Some of the bioelectrochemical generation of solvents and biofuels from CO_2 under various operational and nutritional conditions are illustrated in Table 4.3.

Bajracharya et al. [151] showed that mixed culture gave higher acetate because it has higher H_2 generating potential. The study also showed an increase in CO_2 reduction and production of methane. The increase in H_2 can also be obtained by maintaining higher negative cathode potentials for the reduction of protons to H_2. The leakage of H_2 can be prevented by using high gas absorption electrodes (e.g., platinum single crystal electrode) [1,152].

In MES, acetate is one of the primary products from the reduction of CO_2 [1,153], however with lower yield with no commercial value. A number of studies [1,141] used different starting acids to produce different bioalcohols like ethanol, propanol, butanol, acetone, etc., under different operating conditions of MES. Ethanol and butanol are the major alcohols noted at the applied voltage of 2 V (Cathode potential, −0.9 V).

Several other studies have used different mediators [154] and different operating conditions to optimize the performance of MES. Vassilev et al. [155] operated a dual-cathode MES system for concurrent acetogenesis and solventogenesis followed by carbon chain elongation. This type of system can assist in operational cost reduction with a simultaneous increase in productivity due to the utilization of off-gas [1,156,157]. Most of the product extraction studies of MES include integrated membrane separation with an anion and cation exchange membrane or sequential operation. However, the operation of such processes can be expensive as they often need to be changed because of fouling during longer operation. More work is needed to increase the production rate and decrease operational costs for product extraction and selectivity of product formation.

Research on generation of methane from CO_2 using bioelectrochemical techniques has increased significantly. It has been noted that the optimized electrode potential (direct electron donor) for direct conversion of CO_2 to methane was found to be around −0.6 to −0.7 V vs. Ag/AgCl [1,158]. The standard theoretical potential for the reduction of CO_2 to CH_4 is −0.44 V vs. Ag/AgCl. The required energy for this cathodic reaction can be achieved from the bioanode (organic oxidation) process. However, in practical operation, these can be greatly limited with over-potentials. These over-potentials can be reduced by implementing biocathode systems. Villano et al. [159] demonstrated a double-chamber MEC for methane generation with anode potential of −0.397 V vs. Ag/AgCl by using acetate as an energy source. Simultaneously, a cathode was constantly supplemented along with CO_2 and N_2 for pH correction and carbonate supply. Yasin et al. [160] operated an enriched activated sludge system to generate methane from CO_2. In this system, in the presence of CO_2 and H_2, they noticed a 70-fold increase in methane generation after 72 hours of operation. The MES cells have been proven to be economical and sustainable with CO_2 as a substrate for product formation, such as methanol, ethanol, and formate [161].

Bioelectrochemical systems have the potential to be more productive than biological systems, especially photosynthesis-based systems. Currently, plant solar conversion efficiency of photonic energy caps at 3%–4% at peak growth, whereas photosynthetic microbes grown in optimized bioreactors can reach 5%–7% efficiency [162]. By contrast, photovoltaic devices, which can utilize a greater extent of incident solar energy, have 14%–18% efficiency. Coupling photovoltaic devices with biological CO_2 fixation can result in a more efficient production platform overall.

There are several methods of integrating electrocatalysis into microbial production; the differences between these methods lie in how the electrons are transferred into the biological system [163]. Current research is focused on two main approaches: directly transferring electrons from electrodes to microorganisms and indirectly transferring electrons via electron donors. The direct transfer of electrons from electrodes into microorganisms requires linking intercellular reductions with extracellular electrons. In this type of transfer, electrotrophs typically transfer electrons out of the cell via conduit cytochromes and their accompanying transporters. To achieve bioelectrochemical production, these cellular mechanisms need to be reversed to drive electrons into the cell. This reversion would allow electricity to push biosynthetic pathways toward producing high-value chemicals and fuels. Such approaches have been demonstrated in the production of succinate and acetate. More details on the productions of succinate and acetate are given in NAE report [2], Ross et al. [164] and LaBelle and May [150].

Electrons can be indirectly transferred from electrodes to microorganisms via electrochemically synthesized electron donors such as H_2, formate, ammonia, sulfide, or iron [165]. In this type of transfer, the low redox potentials of H_2 and formate allow a favorable thermodynamic reduction of CO_2 into organic compounds. Ammonia, sulfide, and iron are less thermodynamically favorable because they have higher redox potential. They also require an electron acceptor, such as oxygen, which adds complexity to the electrocatalysis setup and makes these three electron donors less favorable. Of the two more favorable electron donors, formate is a prime target for bioelectrochemical production. H_2 has a low solubility, a low mass transfer rate in cells, and significant safety issues due to combustion concerns inside of pressurized reactors. Formate, by contrast, has high solubility and is readily converted into carbon and energy by cells. However, formate adds complexity to product recovery and can also accumulate and degrade at the anode, reducing yield. To be a viable electron donor, formate needs to be consumed at a rate equal to its production.

Formate-facilitated electron transfer has been demonstrated in the production of alcohols and pyruvate. More details on the productions of alcohols and pyruvate are given in NAE report [2], Lee et al. [166–168], and Tashiro et al. [169].

It is unclear whether the direct or indirect approach will prove to be more scalable for commercialization; each method has inherent challenges that need to be overcome to become industrially viable. The integration of electrocatalysis and microbial production is a relatively new concept which will require more work.

4.6 PATHWAYS FOR CO_2 FIXATION AND CONVERSION

Carbon dioxide is chemically stable and unreactive and must be reduced to enable its incorporation into biological molecules. Autotrophic microorganisms are able to utilize carbon dioxide as their sole carbon source, and a variety of pathways are known to activate and incorporate it into biomolecules essential for growth and replication. Recently, carbon dioxide fixation pathways have received interest for biotechnological applications, since this could provide biological routes for *de novo* generation of fuels and small organic molecules [1].

As shown in Figure 4.5, adaptation of CO_2 to value-added products in the microbial kingdom has been found with six different pathways, such as Calvin Benson Bassham cycle, reductive tricarboxylic acid cycle (reductive/reverse TCA cycle), WLP, 3-hydroxypropionate bicycle, dicarboxylate 4-hydroxybutyrate cycle, and

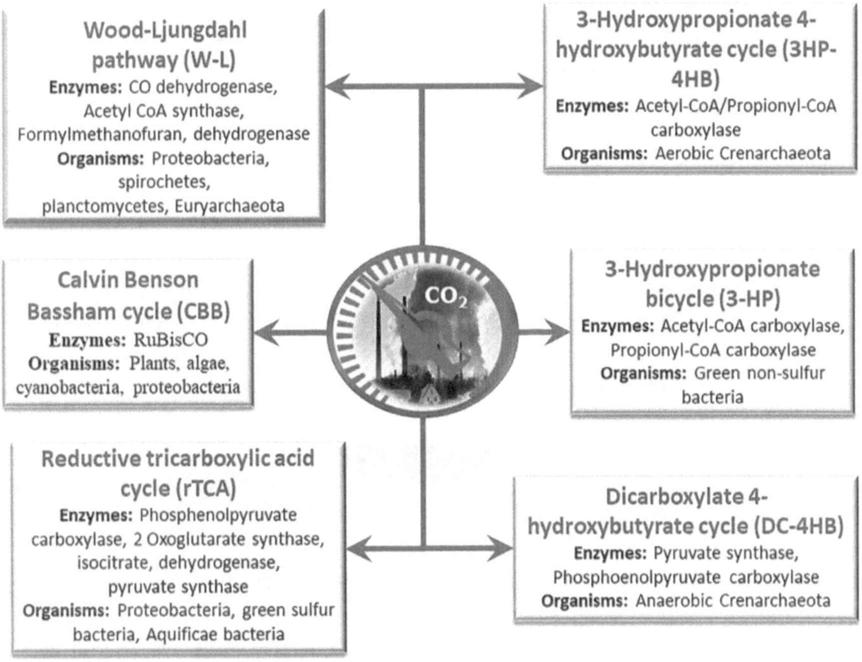

FIGURE 4.5 Various biological pathways for CO_2 metabolism in microbes for carbon fixation [1].

3-hydroxypropionate 4-hydroxybutyrate cycle. These pathways are not found in every species, but specific pathways are found in specific organisms. Among these, the first three routes were found to be dominant for CO_2 conversion.

4.6.1 Calvin Cycle for the Phototrophic Reduction of CO_2

The **Calvin cycle**, or **Calvin–Benson–Bassham (CBB) cycle**, is a series of biochemical redox reactions. Photosynthesis are the chemical reactions that convert carbon dioxide and other compounds into glucose. These reactions occur in the stroma, the fluid-filled area of a chloroplast outside the thylakoid membranes. These reactions take the products (ATP and NADPH) of light-dependent reactions and perform further chemical processes on them. The Calvin cycle uses the reducing powers of ATP and NADPH from the light-dependent reactions to produce sugars for the plant to use. These substrates are used in a series of reduction-oxidation reactions to produce sugars in a step-wise process. There is no direct reaction that converts CO_2 to a sugar because all of the energy would be lost to heat.

In general, three key enzymes and three vital steps of the CO_2 process are noticed in the Calvin cycle [1,4,170–172]. The steps are known as CO_2 fixation, CO_2 reduction, and regeneration of CO_2. In the first step, the 1,5-bisphosphate ribulose bis-phosphate carboxylase (RuBisCo) catalyzes the reaction between CO_2 and 1,5-bisphosphate ribulose, and this process generates 3-phosphoglycerate [1,4,170–172]. Later, 3-phosphoglycerate is further converted into 1,3-diphosphoglycerate by accepting inorganic phosphate (Pi) from ATP and further reduced to 3-phosphate glyceraldehyde with the help of the phosphoglyceraldehyde dehydrogenase enzyme. In the final step, phosphate glyceraldehyde is further transformed to 5-phosphate ribulose via a series of enzymatic reactions following further with its activation by phosphoribulokinase in ATP-dependent condensation to synthesize 1,5-bisphosphate ribulose for CO_2 collection for further cycles [1,4,171,172]. One-sixth of 3-phosphate glyceraldehyde is transformed as sugars, fatty acids, amino acids, etc.

In the dark, plants instead release sucrose into the phloem from their starch reserves to provide energy for the plant. The Calvin cycle thus happens when light is available independent of the kind of photosynthesis (C_3 carbon fixation, C_4 carbon fixation, and Crassulacean Acid Metabolism (CAM)); CAM plants store malic acid in their vacuoles every night and release it by day to make this process work. These reactions are closely coupled to the thylakoid electron transport chain as the energy required to reduce the carbon dioxide is provided by NADPH during the light-dependent reactions. The process of photorespiration, also known as C2 cycle, is also coupled to the Calvin cycle, as it results from an alternative reaction of the RuBisCO enzyme, and its final by-product is another glyceraldehyde-3-P. The Calvin cycle is graphically illustrated in Figure 4.6.

The Calvin–Benson–Bassham (CBB) cycle is a key biological pathway for converting atmospheric CO_2 to organic matter. It is of great significance to the global carbon cycle and crop production, and widely distributed in most autotrophic organisms including plants, algae, cyanobacteria, as well as other photo- and chemoautotrophic bacteria [1,4,170,171]. Metabolic pathways in photoautotrophic organisms related to product formation is illustrated in Figure 4.7. Apart from the CBB cycle, five other carbon fixation pathways have been discovered in nature (which are discussed in this

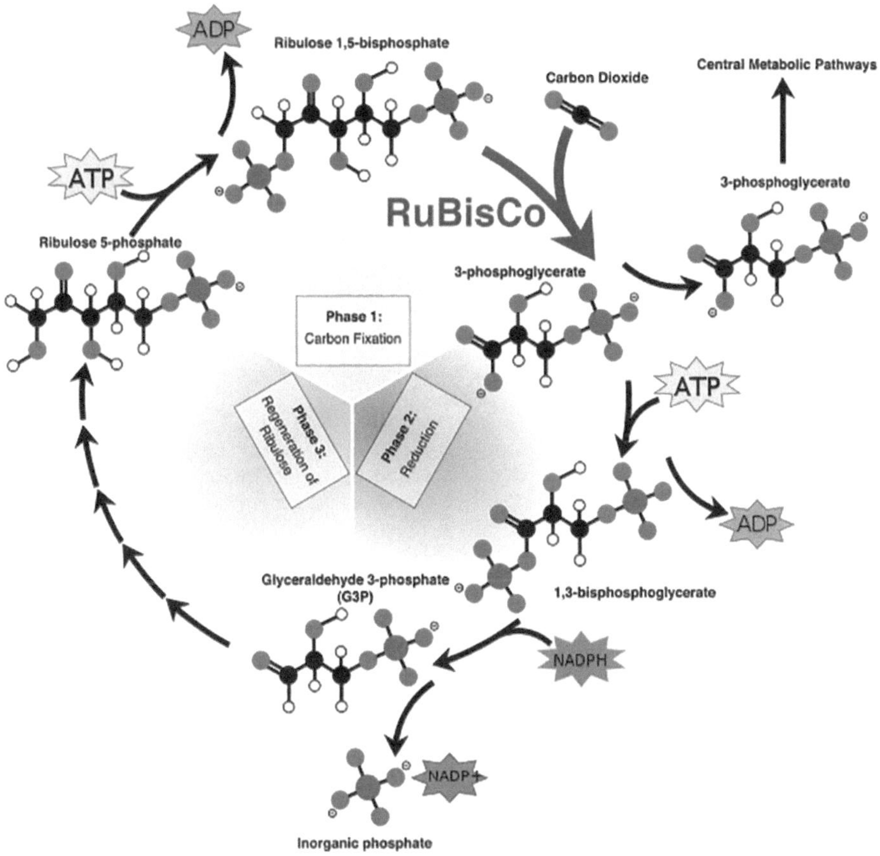

FIGURE 4.6 Overview of the Calvin cycle and carbon fixation [170].

section), among which the reductive acetyl-CoA pathway has the highest CO_2 fixation efficiency under anaerobic conditions, whereby 2 mol CO_2 are fixed into 1 mol of acetyl-coA using 1 mol ATP and 4 mol NAD(P)H [1,4,170,171]. CBB is more energy-intensive, requiring 9 mol ATP and 6 mol NAD(P)H for the fixation of 3 mol CO_2, but is not sensitive to oxygen and is widely distributed in higher plants, algae, and cyanobacteria, which makes improving its efficiency a highly promising prospect [1,4,170,171].

Most research and engineering of the CBB cycle have focused on improving the reaction efficiency of carboxylation by the enzyme RuBisCO, which can be classified into four groups. There are a number of strategies to improve the CO_2 fixation efficiency, including adaptive evolution of RuBisCO catalytic subunits and the promoter of the CBB operon [173], co-expression of auxiliary pathways, and heterologous introduction of highly catalytic RuBisCO. In addition, it was verified that the introduction of a CO_2-concentrating mechanism (CCM) could improve the efficiency of CO_2 fixation [174]. However, there are very few successful cases, probably due to limited improvement of RuBisCO by adaptive evolution, and the long experimental cycles needed for research in plants. *Ralstonia eutropha* H16

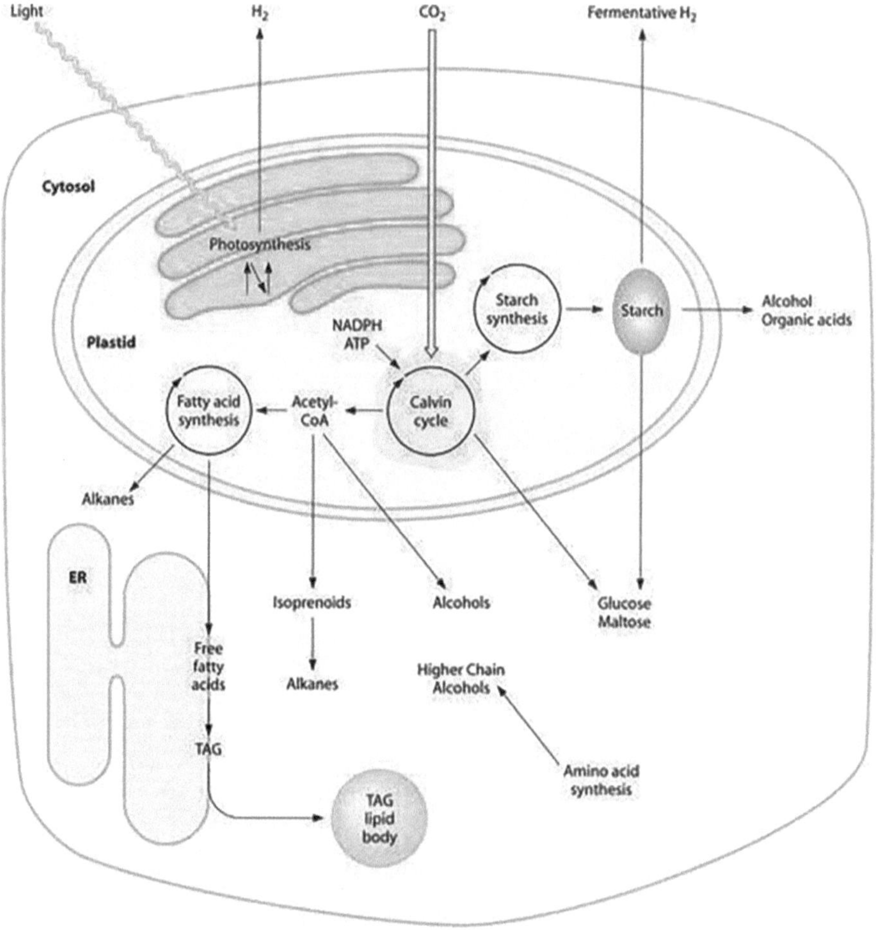

FIGURE 4.7 Metabolic pathways in photoautotrophic organisms related to product formation [3].

(*Cupriavidus necator*) is a Gram-negative facultatively chemoautotrophic bacterium, which utilizes the CBB cycle for carbon fixation. Due to its much shorter generation time compared with plants, *R. eutropha* is a potential platform for optimization of the CBB cycle. In addition, this bacterium has been successfully metabolically engineered to produce various chemicals, such as ethanol [175], isobutanol [176], fatty acids, hydrogen [177], and alkanes [178], which suggest that *R. eutropha* H16 has great potential for development of various biotechnological applications using CO_2 sequestration.

4.6.2 Reverse TCA (Tricarboxylic Acid Cycle) Cycle

The **reverse Krebs cycle** (also known as the **reverse TCA cycle**) is a sequence of chemical reactions that are used by some bacteria to produce carbon compounds from carbon dioxide and water by the use of energy-rich reducing agents as electron donors [179,180]. The reaction is the citric acid cycle run in reverse: where the Krebs cycle takes complex carbon molecules in the form of sugars and oxidizes them to CO_2 and water, the reverse cycle takes CO_2 and water to make carbon compounds. This process is used by some bacteria to synthesize carbon compounds, sometimes using hydrogen, sulfide, or thiosulfate as electron donors. This process can be seen as an alternative to the fixation of inorganic carbon in the reductive pentose phosphate cycle, which occurs in a wide variety of microbes and higher organisms. In contrast to the oxidative citric acid cycle, the reverse or reductive cycle has a few key differences. One of the main differences is the conversion on succinate to 2-oxoglutarate. In the oxidative reaction, this step is coupled to the reduction of NADH. However, the oxidation of 2-oxoglutarate to succinate is so energetically favorable that NADH lacks the reductive power to drive the reverse reaction. In the rTCA cycle, this reaction has to use a reduced low potential ferredoxin. It has been found that some nonconsecutive steps of the cycle can be catalyzed by minerals through photochemistry, while entire two- and three-step sequences can be promoted by metal ion such as iron (as reducing agents) under acidic conditions. However, the conditions are extremely harsh and require 1 M hydrochloric or 1 M sulfuric acid and strong heating at 80°C–140°C. The r-TCA is illustrated in Figure 4.8.

The reductive TCA cycle was first identified in anaerobic *Chloorbium limicola*, a green photoautotrophic sulfur-reducing bacteria. It was also identified

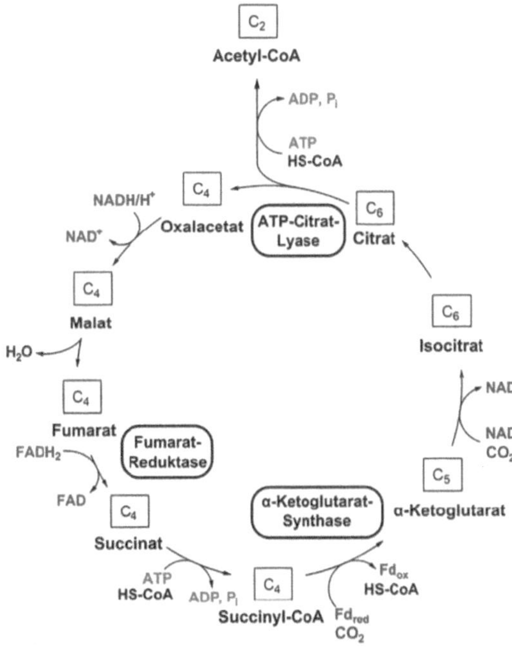

FIGURE 4.8 Overview of r-TCA cycle [180].

in another thermophilic bacterium that uses hydrogen and sulfur as an energy source. The reverse TCA cycle contains four steps of carboxylation, in which, initially, succinyl-CoA is reductively carboxylated with CO_2 by α-Ketoglutarate synthase/2-oxoglutarate synthase (ΔG_0, +19 kJ/mol; first step) to generate α-Ketoglutarate/2-oxoglutarate by utilizing two equivalents of reduced ferrodoxin. Next, α-ketoglutarate/2-oxoglutarate along with CO_2 is converted into isocitrate (ΔG_0, +8 kJ/mol; second step) by using isocitrate dehydrogenase and NADPH. Further, isocitrate is converted into citrate by a cleavage mechanism, followed by oxaloacetate and acetate-CoA by ATP citrate lyase. The terminal compounds are carboxylated with CO_2, which requires a pyruvate synthase enzyme (ΔG_0, +19 kJ/mol). The synthesized pyruvate is activated by pyruvate kinase to yield phosphoenolpyruvate (PEP), which is further converted to bicarbonate by a carboxylation mechanism (ΔG_0, −24 kJ/mol). As an outcome, oxaloacetate is generated and finally converted to succinyl-CoA by a series of enzymes [4].

In the biological rTCA cycle, CO_2 fixation is operated by the two enzyme cofactors: thiamine pyrophosphate (TPP) assists the conversion of acetyl-CoA to pyruvate and succinyl-CoA to α-ketoglutarate [179,181], whereas biotin mediates the formations of oxaloacetate and oxalosuccinate from pyruvate and α-ketoglutarate, respectively [179,182]. These two co-factors have been linked to life's evolution [183]. The replacement of heteroatoms in their ring structures with others (e.g., O or N vs. S) does not inactivate, or in some cases even improves, their functional properties [179,184]. Various heterocyclic compounds with structural features resembling the two have been synthesized under simulated primitive environmental conditions [179,185].

4.6.3 Reverse Acetyl-CoA Cycle

The **Wood–Ljungdahl pathway** is a set of biochemical reactions used by some bacteria and archaea called acetogens and methanogens, respectively. It is also known as the reductive acetyl-coenzyme A (acetyl-CoA) pathway. This pathway enables these organisms to use hydrogen as an electron donor, and carbon dioxide as an electron acceptor and as a building block for biosynthesis. In this pathway, carbon dioxide is reduced to carbon monoxide and formic acid or directly into a formyl group, the formyl group is reduced to a methyl group and then combined with the carbon monoxide and Coenzyme A to produce acetyl-CoA. Two specific enzymes participate on the carbon monoxide side of the pathway: CO dehydrogenase and acetyl-CoA synthase. The former catalyzes the reduction of the CO_2, and the latter combines the resulting CO with a methyl group to give acetyl-CoA. Some anaerobic bacteria and archaea use the Wood–Ljungdahl pathway in reverse to break down acetate. For example, some methanogens breakdown acetate to a methyl group and carbon monoxide, and then reduce the methyl group to methane while oxidizing the carbon monoxide to carbon dioxide. Sulfate-reducing bacteria, meanwhile, oxidize acetate completely to CO_2 and H_2 coupled with the reduction of sulfate to sulfide. When operating in the reverse direction, the acetyl-CoA synthase is sometimes called acetyl-CoA decarbonylase [186,187]. The pathway occurs in both bacteria (e.g. acetogens) and archaea (e.g. methanogens). Unlike the Reverse Krebs cycle

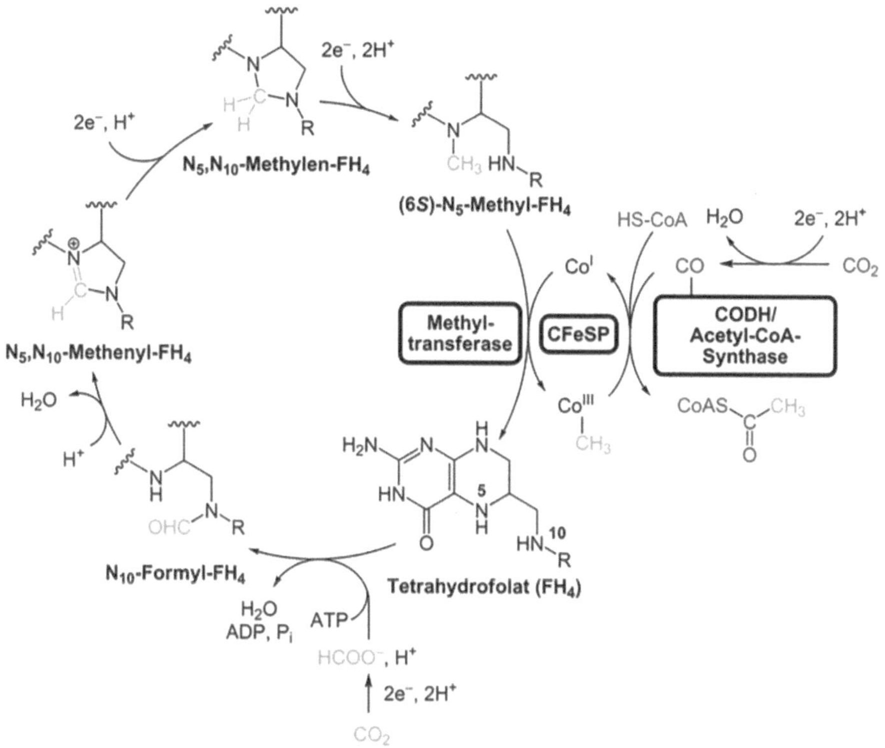

FIGURE 4.9 Overview of W-L pathway [187].

and the Calvin cycle, this process is not cyclic. The overview of W-L pathway is illustrated in Figure 4.9.

The reverse acetyl-CoA pathway is predominantly found in strict acetogenic (Eubacteria) and methanogenic (Euryarchaeota) bacteria. This route was suggested by Ljungdahl and Wood [188], and it was named after them. In this route, initially, CO_2 is converted to formate by using NADH-dependent formate dehydrogenase ($F_{ate}DH$) (ΔG_0 +22 kJ/mol). Later, formate is apprehended by tetrahydrofolate and reduced into a methyl group by generating methyl-H4 folate. Methyltransferase transfers the methyl group of methyl-H4 folate to the cobalt center of hetero dimeric corrinoid iron sulfur protein, thus generating methylated corrinoid protein. Further, CO dehydrogenase (CODH) functions to convert CO_2 into CO via acetyl-CoA synthase accepting the methyl group from CH_3-Co^{III} and converts CO, CoASH, and methyl groups to acetyl-CoA [1,13,146]. As noted in the above enzymatic reaction, the autotrophic mechanism is by either CO_2 or bicarbonate as the carbon source. This mainly depends on the specific enzyme in a specific CO_2 fixation/conversion reaction. The attribute difference is noted in carbon source, and a fast-reliable conversion of CO_2 to bicarbonate is required.

The Wood–Ljungdahl (WL) pathway is one of the most important metabolisms for energy generation and carbon fixation [189]. Although its overall scheme is conserved in Archaea and Bacteria, only the carbonyl branch (CBWL) shares homology, whereas the archaeal and bacterial methyl branches (MBWL) involve different Cl-carriers, cofactors, electron transporters, and enzymes [190]. Other than carbon fixation, the WL pathway can act in reverse to produce reducing power from the oxidation of organic compounds during organo-heterotrophic growth [186,191]. When using the WL pathway for energy generation and carbon fixation, most bacteria produce acetate as an end product (acetogens), whereas most archaea produce methane (CO_2-reducing methanogens). The WL pathway has, therefore, been traditionally linked to methanogenesis in the Archaea. Until recently, all known methanogens were known to fall into two classes (Classes I and II), both belonging to the Euryarchaeota. Irrespective of the type of methanogenesis performed (CO_2-reducing, acetoclastic, and methylotrophic), the representatives of these two classes have been consistently found to share a common set of enzymes for methanogenesis [186,187]. Detail roles of enzymes and two classes of methanogens are described in an excellent article by Borrel et al. [186].

4.6.4 3-Hydroxypropionate Bicycle

The **3-hydroxypropionate bicycle**, also known as the **3-hydroxypropionate pathway**, is a process that allows some bacteria to generate 3-hydroxypropionate utilizing carbon dioxide. In this pathway, CO_2 is fixed (i.e. incorporated) by the action of two enzymes, acetyl-CoA carboxylase, and propionyl-CoA carboxylase. These enzymes generate malonyl-CoA and (*S*)-methylmalonyl-CoA, respectively. Malonyl-CoA, in a series of reactions, is further split into acetyl-CoA and glyoxylate. Glyoxylate is incorporated into beta-methylmalyl-CoA which is then split again through a series of reactions to release pyruvate as well as acetate, which is used to replenish the cycle. This pathway has been demonstrated in *Chloroflexus*, a nonsulfur photosynthetic bacterium; however, other studies suggest that 3-hydroxypropionate bicycle is utilized by several chemotrophic archaea [192].

4.6.5 3HP-4HB and DC-4HB Cycles

The most recently discovered of these are found exclusively in extremely thermophilic archaea as follows: the 3-hydroxypropionate/4-hydroxybutyrate (3HP/4HB) carbon fixation cycle, which operates in members of the crenarchaeal order Sulfolobales [193–195], and the dicarboxylate/4-hydroxybutyrate (DC/4HB) cycle, which is used by anaerobic members of the orders Thermoproteales and Desulfurococcales [193,195]. In both cycles, two carbon dioxide molecules are added to acetyl-CoA (C2) to produce succinyl-CoA (C4), which is subsequently rearranged to acetoacetyl-CoA and cleaved into two molecules of acetyl-CoA. These pathways differ primarily in regard to their tolerance to oxygen and the co-factors used for reducing equivalents as follows: NAD(P)H for the 3HP/4HB cycle and ferredoxin/NAD(P)H for the DC/4HB cycle [193,196,197]. The two archaeal pathways also differ in how they link the CO_2-fixation cycle to central metabolism. In the DC/4HB pathway, pyruvate is synthesized directly from acetyl-CoA using pyruvate synthase. In the 3HP/4HB pathway, another half-turn is required to make succinyl-CoA, which is then oxidized via succinate to pyruvate [193,194].

Metallosphaera sedula is an extremely thermoacidophilic archaeon that grows heterotrophically on peptides and chemolithoautotrophically on hydrogen, sulfur, or reduced metals as energy sources. During autotrophic growth, carbon dioxide is incorporated into cellular carbon via the 3-hydroxypropionate/4-hydroxybutyra te cycle (3HP/4HB). To date, all of the steps in the pathway have been connected to enzymes encoded in specific genes, except for the one responsible for ligation of coenzyme A (CoA) to 4HB. The capacity of two enzymes Msed 0406 and Msed 0394 to use 4HB as a substrate may have arisen from simple modifications to acyl-adenylate-forming enzymes [193]. More details on the role of *Meallosphaera sedula* (optimum growth temperature 73°C, pH = 2) on 3HP/4HB cycle is given in an excellent article by Hawkins et al. [193].

There are 13 enzymes proposed to catalyze the 16 reactions in the 3HP/4HB pathway. The first three enzymes convert acetyl-CoA (C2) to 3HP (C3) via an ATP-dependent carboxylation step. Next, 3HP is converted and reduced to propionyl-CoA, carboxylated a second time, and rearranged to make succinyl-CoA (C4). Succinyl-CoA is reduced to 4HB, which is converted to two molecules of acetyl-CoA in the final reactions of the cycle. Flux analysis and labeling studies have confirmed the operation of this pathway in *M. sedula* [193,195].

All of the enzymes that comprise the first portion of the cycle up to the formation of 4HB have been identified and characterized biochemically in their native or recombinant form, mostly from the extremely thermoacidophilic archaeon *M. sedula* (T = 70°C, pH 2.0) [193,195]. The enzymes involved in the conversion of 4HB to two molecules of acetyl-CoA have not been characterized to the same extent. Activities corresponding to 4-hydroxybutyryl-CoA dehydratase and acetoacetyl-CoA β-ketothiolase have been detected in cell extracts [193,195], although neither enzyme has been purified in its native form or recombinantly produced. Identification of candidates for both of these enzymes has been made based on genome annotation and transcriptomic analysis of autotrophic growth compared with heterotrophy [193,197]. Although neither of the candidate genes for these enzymes has so far been confirmed biochemically, their identity is not in dispute because of strong homology to known versions in less thermophilic organisms. The corresponding gene products in *M. sedula* are Msed_1321 for the 4HB-CoA dehydratase and Msed_0656 for the acetoacetyl-CoA β-ketothiolase. Loder et al. [198] carried out reaction kinetic analysis of 3-hydroxypropionate/4-hydroxybutyrate CO_2-fixation cycle in extremely thermoacidophilic archaea. The 3-hydroxypropionate/4-hydroxybutyrate (3HP/4HB) cycle fixes CO_2 in extremely thermoacidophilic archaea and holds promise for metabolic engineering because of its thermostability and potentially rapid pathway kinetics. A reaction kinetics model was developed to examine the biological and biotechnological attributes of the 3HP/4HB cycle as it operates in *M. sedula*. The model correctly predicted previously observed features of the cycle. The model was then used to assess metabolic engineering strategies for incorporating CO_2 into chemical intermediates and products of biotechnological importance: acetyl-CoA, succinate, and 3-hydroxyproprionate.

4.7 ANAEROBIC AND GAS FERMENTATIONS OF CO_2 FOR VALUE ADDITION

4.7.1 Anaerobic Fermentation of CO_2

Methane is a source of fuel for electricity or heat or a chemical that can be converted to other chemicals or fuels like methanol, ethanol, butanol, biodiesel, etc. In general, methane is generated in two ways: through biotic and abiotic pathways. Abiotic pathways include the catalytic conversion or thermal split of kerogen [199]. Biogenic methane generation is abundantly noticed from AD and from dairy farms. In fact, 20% of global natural gas is being produced from methanogens [200]. Landfill gas is generated by anaerobic digestion of biomass waste within the landfill.

Methanogenic microbes are obligate anaerobes, which are under the Euryarchaeota phyla belonging to the Archaea domain. These obligate anaerobes have the capability of sequestering the CO_2 under optimized specific conditions [1,201]. In gas fermentation, CO_2 sequestration is carried out by hydrogenotrophic methanogens with the help of gamma- and zeta-type carbonic anhydrase enzymes. The CO_2 can be digested with different forms of chemicals, such as bicarbonate, carboxylic acids, alcohols, and biogas with their respective enzymes. However, acetyl-CoA is the central mediator molecule in several microbial processes for product synthesis. Methanogenic and acetogenic bacteria are frequently found in a pair when CO_2 is the reduction medium [1,13]. In conventional AD, the difference in operational conditions affects the behavior and fate of both acetogenic and hydrogenotrophic methanotrophs. Under optimized conditions, the acetogenic bacterium transforms the acetic acid to H_2 and CO_2, which is followed by the conversion to methane via acetate (Eq. 4.1) or the indirect conversion of H_2 and CO_2 (Eq. 4.2) by hydrogenotrophic methanogenic bacterial cells [1,202].

$$CH_3COOH \rightarrow CO_2 + CH_4 \tag{4.1}$$

$$CO_2 + 4H_2 \rightarrow CH_4 + 2H_2O \tag{4.2}$$

For efficient and stable performance in the transformation of CO_2 and H_2 to CH_4 by hydrogenotrophic methanogens, a continuous supplement of molecular hydrogen is required. Based on several studies, the key parameters that control hydrogenotrophic methanogenic activity are identified as pressure, redox conditions, temperature, headspace gas composition, hydraulic retention time, and pH [1,203]. Scherer et al. [204] analyzed the degradation of organics in municipal gray water under thermophilic and mesophilic environments using a lab-scale continuous mode of operation for biogas generation. These processes are also constantly occurring in the landfills with biological waste. The nature of substrate is also important. As indicated by Shah [205], a mixture of substrates often gives better performance of hydrogen and methane generations than a single substrate. The most probable number (MPN) technique illustrated that hydrogenotrophic methanogens (H_2-CO_2 utilizers) were in a range of 10^8–10^{10}/g total solids (TS), which seems to dominate the acetogenic methanotrophs by a factor of 10–10,000, due to short hydraulic retention time (HRT). Similarly,

Ahring et al. [206] noted that hydrogenotrophic methanogens exhibited higher specific methanogenic activity at 65°C. Although these studies prove that the presence of hydrogenotrophic methanogens can enhance methane formation, it is unclear whether enhanced methane formation is by direct or indirect conversion under practical conditions. Microbiology associated with anerobic digestion to produce methane or hydrogen is discussed in great detail in my previous book [205].

Conrad and Klose [207] demonstrated that washed excised roots of rice (*Oryza sativa*) produced H_2, CH_4, acetate, propionate, and butyrate when incubated under anoxic conditions. Acetate production was most pronounced followed by propionate, butyrate, and CH_4, respectively. During this process, hydrogen partial pressures were always high enough to allow exergonic methanogenesis. Radioactive bicarbonate/CO_2 was incorporated into CH_4, acetate, and propionate. The specific radio activities of the products indicated that CH_4 was exclusively produced from H_2/CO_2. The contribution of CO_2 to the production of acetate and propionate was 32%–39% and 42%–61%, respectively. A substantial fraction of propionate was apparently reductively formed from acetate and/or CO_2. The study thus demonstrated an intensive anaerobic dark metabolism of CO_2 on washed rice roots with reduction of CO_2 contributing significantly to the production of acetate, propionate, and CH_4. The CO_2 reduction seemed to be driven by decay and fermentation of root material.

4.7.2 Gas Fermentation of CO_2

Anerobic digestion of biomass can occur with food or waste materials. In order to avoid controversy of using food materials for fuel in anaerobic fermentation, gas fermentation looks to be a viable alternative, in which gaseous substrates (CO_2) are converted to biochemicals. Biomass can be converted to value-added products in a number of different ways including anerobic digestion mentioned above. Among various alternatives, the use of the biomass in its entirety as a feedstock is a key advantage inherent to gas fermentation compared to sugar and cellulosic fermentation to produce low-carbon fuels. Biomass can be gasified to a mixture of carbon monoxide (CO), carbon dioxide (CO_2), hydrogen (H_2), and nitrogen (N_2), also called synthesis gas or syngas. Conversion of biomass to syngas allows for utilization of nearly all the available carbon contained within the biomass, including the otherwise inaccessible lignin fraction, and bypasses the expense and inefficiencies of biomass pretreatment. Also, many industrial processes generate large amounts of carbon gases, which are left unused. CO_2-rich waste gases can be a suitable substrate for gas fermentation. Several microbes have the ability to utilize CO_2 as a carbon source. Among these several microbes, acetogenic bacterium is mostly being studied and shows a capability for industrial applications. These bacterial species are equipped to convert CO_2 and CO by making use of the Wood–Ljungdahl pathway (WLP). Different acetogenic bacteria can naturally generate ethanol or butanol and some other goods of industrial interest, such as 2-oxobutyrate, hexanol, PHA, vitamin B12, polymers, and their precursors for other products.

It is important to note that once biogas is produced, there are two options for conversion to useful products. Traditionally this has been achieved using the Fischer-Tropsch process (FTP), but the technology has some drawbacks and is very capital intensive. The ability to fix gaseous, inorganic carbon into organic material (autotrophy) is also a prerequisite for life, and routes exist in various forms across all domains of life [208]. Eukaryotes (the most common example being photosynthesis in plants), archaea, and bacteria can all fix carbon by reducing CO_2 and/or CO. Anaerobic gas-fermenting bacteria, specifically acetogens, have advantages in low-carbon fuel/chemical production.

It is important to compare FTP against gas fermentation. FTP employs high temperature (150°C–350°C), elevated pressures (30 bar), and heterogeneous catalysts such as cobalt, ruthenium, and iron [209]. In comparison, gas fermentation takes place at 37°C and atmospheric pressure, which presents significant energy and cost savings relative to FTP. FTP, unlike gas fermentation, also requires a fixed H_2:CO ratio of ideally ~2:1 [209]. However, syngas derived from biomass has typically a lower H_2:CO ratio [210,211], often requiring an extra step of water–gas shift reaction [212] at the expense of CO to adjust the H_2:CO for FTP. Although chemical processes are generally considered faster than biological approaches, the latter allow near complete conversion efficiencies due to the irreversible nature of biological reactions [213,214]. Furthermore, the high enzymatic specificities of biological conversions also result in higher product selectivity with the formation of fewer by-products. Crucially, the biocatalysts are also less susceptible to poisoning by sulfur, chlorine, and tars than the inorganic catalysts [210,215], which reduce the gas pre-treatment costs.

With these capabilities (using gaseous carbon, flexible gas compositions, and tolerance to more contaminants over FTP), gas-fermenting microorganisms can make use of a diverse pool of substrates (Figure 4.10). Gas fermentation processes can utilize gasified organic matter of any sort (e.g., municipal solid waste, industrial waste, biomass, and agricultural waste residues) or industrial waste gases (e.g., from steel making, oil refining, or the ferroalloy industry). In this way, gas fermentation can act as a vital bridge in the effort to create sustainable value from waste and enable the perpetual capture of greenhouse carbon in valuable materials. Gas fermentation, therefore, can increase the cyclical carbon emission and fixation in fossil fuel-consuming and carbon-emitting industries. Carbon containing off-gases produced in steel mills, for example, can be sequestered and converted into microbial biomass, fuels, and chemicals. Recycling carbon in this manner can decrease the need for tapping into fossil fuel reserve.

To fix the relatively oxidized carbon contained in these various syngas sources, acetogens (and other gas-fermenting microorganisms) require reducing equivalents in the form of electrons (such as NAD(P)H or reduced ferredoxin) to reduce the carbon to the central building block acetyl-CoA and further to reduced products such as alcohols. CO and H_2 present in syngas themselves can provide these reducing equivalents (see Figure 4.11) by oxidation to CO_2 and water (protons), respectively. Reducing equivalents can also be derived from sources other than the syngas sources discussed above.

H_2 and CO can be produced by electrolysis of water and CO_2, both of which can be supplied power by surplus renewable or nuclear energy. So far, the highest gas fermentation ethanol yields and selectivity have been demonstrated with CO-rich feedstocks [216], providing additional incentive to develop CO_2 electrolysis technology. Technologies to convert CO_2 to CO are, however, at a pre-commercial stage.

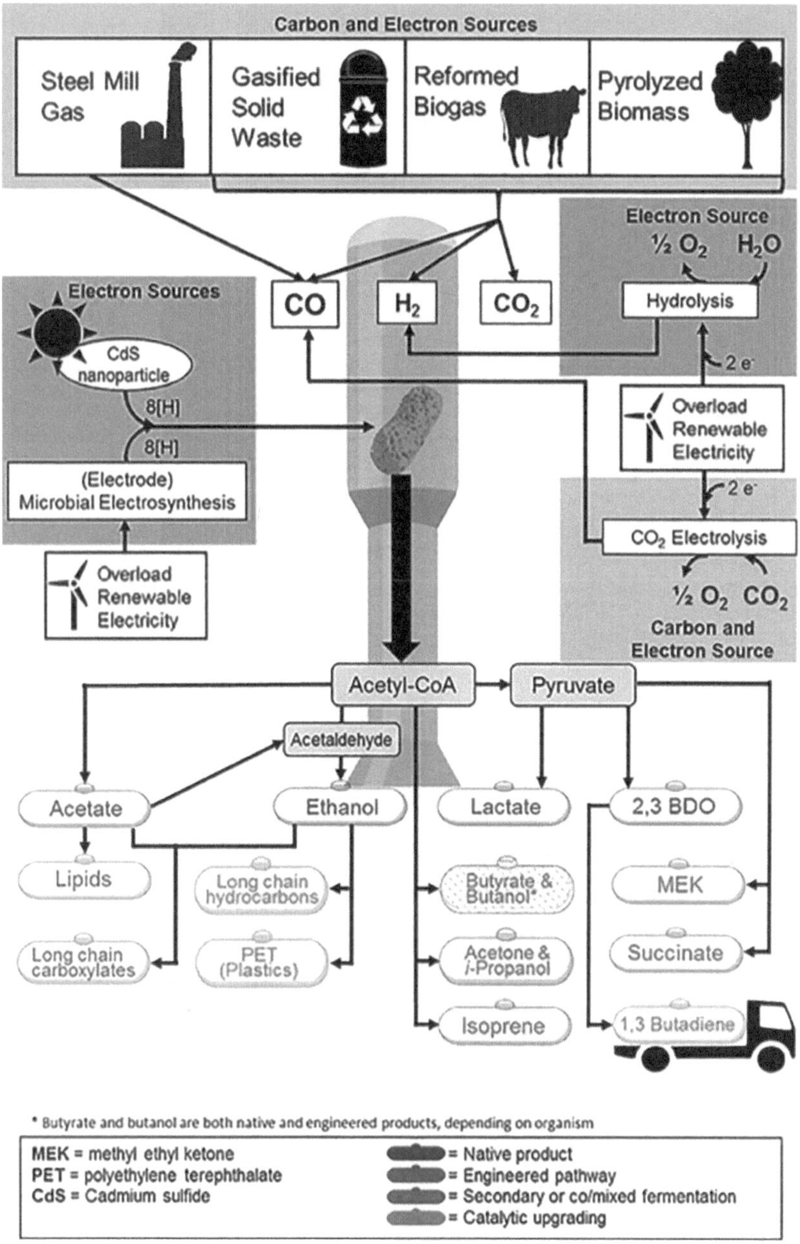

FIGURE 4.10 Overview of feedstock and product options for gas fermentation. Feedstocks to the gas fermentation platform are highlighted in light blue (carbon and electron sources) and green (electron sources). Feedstocks shown are at various stages of commercial deployment. Synthesis of all products shown has been demonstrated including (a) native products (blue text), (b) synthetic products produced through genetic modification (red text), (c) products generated through secondary fermentation of co/mixed cultures (purple text), and (d) products achieved through additional catalytic upgrading (orange text). Acronyms: 2,3-BDO, 2,3-butanediol; MEK, methyl ethyl ketone [210].

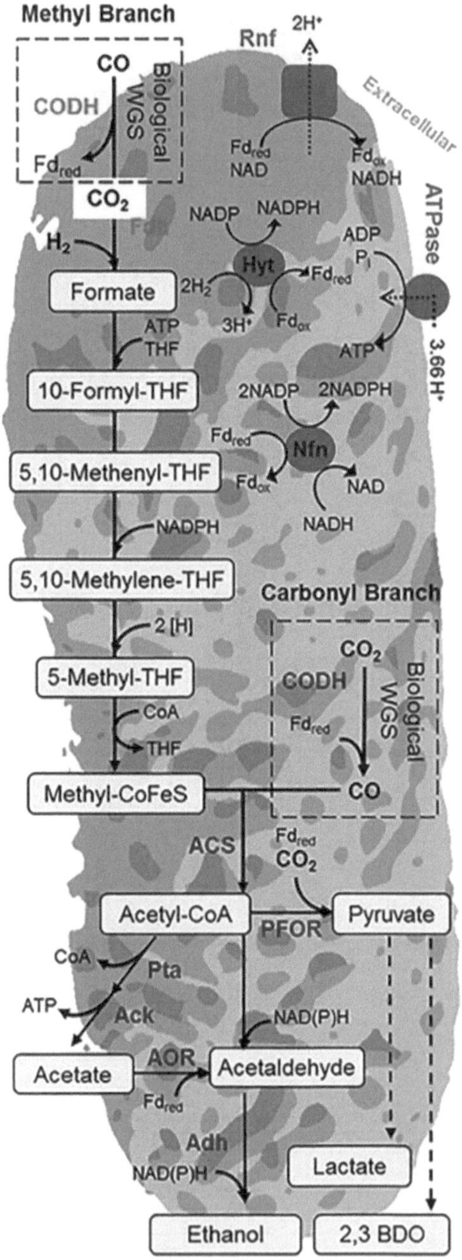

FIGURE 4.11 Overview of the Wood–Ljungdahl pathway (WLP) and energy-conserving mechanisms of acetogen *C. autoethanogenum*. The WLP is central to the gas fermentation platform for carbon fixation. Noteworthy enzymes are labeled in blue. The enzymes involved in energy conservation are shown in purple. Acronyms: 2,3-BDO, 2,3-butanediol; AOR, aldehyde:ferredoxin oxidoreductase; ACS, acetyl-CoA synthase; CODH, carbon monoxide dehydrogenase; Nfn, transhydrogenase; PFOR, pyruvate:ferredoxin oxidoreductase; Rnf, Rhodocbacter nitrogen fixation; THF, tetrahydrofolate; WGS, water-gas shift reaction [210].

As shown in the earlier section, MES can also use electrical current to reduce CO_2 to multi-carbon products [210,217].

Sakimoto et al. [218] described an innovative method for direct electron input to the Wood-Ljungdahl pathway of an acetogen by photosensitizing the microbes. When grown with cadmium nitrate and cysteine, *M. thermoacetica* was able to use biologically generated cadmium sulfide (CdS) semiconducting nanoparticles, which absorb light and use the energy to carry out photosynthesis. By feeding reducing equivalents directly into the WJP, solar energy can be converted into acetyl-CoA with 90% selectivity to acetate and 10% selectivity to biomass. This could be an efficient process to convert solar energy into liquid energy (i.e., fuels); however, it is in very early-stage development and requires biological optimization.

The advantages discussed above can be applied generally to autotrophic microorganisms. Acetogens are particularly attractive for commercialized gas fermentation due to their native ability to synthesize useful products such as ethanol, butanol, and 2,3-butanediol. Since they are anaerobes and does not require sugar and oxygen, they avoid flammability and biological contamination issues with combustion and dirty gases. Acetogens are found in over 20 different genera and over 100 different species [219]. Table 4.4 provides an overview of the most noteworthy acetogenic species.

What makes the biology of acetogens particularly effective is the WLP for CO_2 fixation. The WLP, also known as the reductive acetyl-CoA pathway, is the only linear CO_2 fixation pathway to acetyl-CoA [220] and considered as the most efficient nonphotosynthetic carbon fixation mechanism [221]. There are several excellent reviews on the detailed mechanism and enzymes of the WLP [210,222]. Briefly, the WLP consists of two branches, a methyl (Eastern) and a carbonyl (Western) branch. In the methyl branch, CO_2 is reduced to formate which undergoes numerous further reactions to form acetyl-CoA. When grown autotrophically on CO, the CO_2 required for the methyl branch is generated by the CODH-catalyzed water–gas shift reaction. Likewise, during autotrophic growth on CO_2, the CO is formed from CO_2 by CODH in the carbonyl branch. The WLP is also active during heterotrophic growth, where released CO_2 can be re-assimilated [222]. This ability is also exploited in a concept called acetogenic or anaerobic, nonphotosynthetic mixotrophy (ANP) to maximize yield in fermentation of sugar with additional hydrogen [223]. WLP path also requires conversion of acetyl-CoA to acetate.

A number of studies have examined two-stage processes to generate different chemicals from CO_2 via acetate route. Liew et al. [210] noted that gas fermentation process using steel mill off-gas as a feedstock resulted in acetate. In the initial stage of fermentation, CO was transformed to CO_2 and H_2 by *Thermococcus onnurineus*. Later, the mixtures of CO, CO_2, and H_2 were converted to acetate by *Thermoanaerobacter kivui*. When the initial step was absent, *T. kivui* was unable to generate acetate due to CO inhibition and lack of CO_2 as a substrate. In a multi-step process, acetate can be used to generate ethanol and other value-added products. Richter et al. [224,225] operated a two-stage system for generation of ethanol with CO_2 as a feedstock by using *C. ljungdahlii*. The first stage of the reactor was employed for acetate; then, the broth containing acetate was transferred to a second fermenter by lowering the pH for efficient conversion of acetate to ethanol. Martin et al. (2016) [226] showed that among different types of clostridium strains, *C. ljungdahlii* PETC

TABLE 4.4
Overview of Typical Acetogens [210]

Organism	Substrates	Products	Opt. Temp. (°C)	Opt. pH
	Mesophilic Microorganisms			
Acetobacterium woodii	H_2/CO_2, CO	Acetate	30	6.8
Acetonema longum	H_2/CO_2	Acetate, butyrate	30–33	7.8
Alkalibaculum bacchi	H_2/CO_2, CO	Acetate, ethanol	37	8–8.5
Butyribacterium methylotrophicum	H_2/CO_2, CO	Acetate, ethanol butyrate, butanol	37	6
Clostridium aceticum	H_2/CO_2, CO	Acetate	30	8.3
Clostridium autcethanogenum	H_2/CO_2, CO	Acetate, ethanol 2,3-butanediol,lactate	37	5.8–6.0
Clostridium carboxidivorans or "P7"	H_2/CO_2, CO	Acetate, ethanol, butyrate, butanol, lactate	38	6.2
Clostridium coskatii	H_2/CO_2, CO	Acetate, ethanol	37	5.8–6.5
Clostridium difficile	H_2/CO_2, CO	Acetate, ethanol, butyrate	35–40	6.5–7.0
Clostridium drakei	H_2/CO_2, CO	Acetate, ethanol, butyrate	25–30	3.6–6.8
Clostridium formicoaceticum	CO	Acetate, formate	37	NR
Clostridium glycolicum	H_2/CO_2	Acetate	37–40	7.0–7.5
Clostridium ljungdahlii	H_2/CO_2, CO	Acetate, ethanol, 2,3-butanediol, lactate	37	6.0
Clostridium magnum	H_2/CO_2	Acetate	30–32	7.0
Clostridium mayombei	H_2/CO_2	Acetate	33	7.3
Clostridium methoxybenzovorans	H_2/CO_2	Acetate, formate	37	7.4
Clostridium ragsdalei or "P11"	H_2/CO_2, CO	Acetate, ethanol, 2,3-butanediol, lactate	37	6.3
Eubacterium limosum	H_2/CO_2, CO	Acetate, butyrate	38–39	7.0–7.2
Oxobacter pfennigii	H_2/CO_2, CO	Acetate,butyrate	36–38	7.3
Blautia productus	H_2/CO_2, CO	Acetate	37	7
	Thermophilic Microorganisms			
Moorella thermoacetica	H_2/CO_2, CO	Acetate	55	6.5–6,8
Moorella thermoautotrophic	H_2/CO_2, CO	Acetate	58	6.1

NR, Not reported; GEM, genome-scale network reconstruction.

was the most optimal producer for ethanol. Hu et al. [227] developed a two-stage system for generation of long-chain fatty acids (C_{16}–C_{18}). This was carried out by *M. thermoacetica* in the first stage for acetate formation with syngas as a feedstock. The effluent broth was subjected to an aerobic reactor containing genetically modified *Yarrowia lipolytica*. The system could generate up to 18 g/L of C_{16}–C_{18} lipids at a rate of 0.19 g/L/h. The system once again proves the conversion of carbon in syngas to complex substrates with a multistage system.

4.7.2.1 Commercialization and Lifecycle Analysis of Gas Fermentation

Gas fermentation makes use of microorganisms known as acetogens. These convert CO_2 and other C1 molecules such as carbon monoxide into fuel, chemical, and nutrient products. Commercial gas fermentation processes are now being pioneered using industrially robust microbial strains and specialized bioreactor designs for the continuous production of low-carbon-transport fuel from industrial waste gas streams. Gas-fermenting microbes use the highly efficient one-carbon Wood-Ljungdahl pathway to harness the carbon and energy in mixtures of CO, or CO_2/H_2, or CO, CO_2, and H_2 gases for both microbial growth and the synthesis of fuels (such as ethanol) and chemicals, such as 2,3-butanediol. Converting gases to fuels has been traditionally achieved via thermo-catalytic processes at high temperature (150°C–350°C), and pressure (>30 atm) reactions [209]. Gas fermentation offers numerous advantages over these chemical processes as it takes place at ambient temperatures and low pressures, which allows significant energy and cost savings while achieving greater selectivity to the target product.

The efficiency, diversity of substrates, and product selectivity advantages of gas fermentation have led to scaling up the fermentation process for commercial-scale production of low-carbon fuels using acetogens. Three companies, Coskata, INEOS Bio, and LanzaTech, have operated pilot and demonstration plants for extended periods of time. Coskata's technology formed the basis of a new company Synata Bio [228] but has not been scaled up further. INEOS Bio and LanzaTech, on the other hand, are currently scaling up their processes to commercial scale.

INEOS Bio built an eight million gallon per year (Mgy) semi-commercial facility in Vero Beach, FL as a joint venture with New Planet Energy Holdings, LLC. Commissioned in 2012, the facility uses lignocellulosic biomass and MSW for generating gas fermentation substrates and generates 6 mW of electrical power. In July 2013, the company announced successful production of ethanol in its facility [229]. In September 2014, operational changes were imposed to optimize the technology and de-bottleneck the plant to achieve full production capacity [230].

LanzaTech has successfully operated two 0.1 Mgy pre-commercial plants in different locations in China with two steel companies, BaoSteel and Shougang Steel (see Figure 4.12). Both plants, the first at one of BaoSteel's mills in Shanghai in 2012 and the second at a Shougang steel mill near Beijing in 2013, used steel mill off-gases as substrates for gas fermentation. In 2015, both China Steel Corporation of Taiwan and ArcelorMittal of Luxembourg approved commercial projects with LanzaTech. The former was a 17 Mgy facility with the intention to scale up to 34 Mgy [231]. The latter 9.8 Mgy facility was built at ArcelorMittal's flagship steel plant in Ghent, Belgium with the intention to construct further plants across ArcelorMittal's operations [232]. In its full scale-up operation, the technology could enable the production of around 104 Mgy, which would displace 1.6 million barrels of fossil fuel-derived gasoline on a BTU basis. In addition to these two projects, Aemetis, Inc. acquired a license from LanzaTech for the conversion of agricultural waste, forest waste, dairy waste, and construction and demolition waste (CDW) to ethanol in California. Aemetis added an 8 Mgy gas fermentation unit to its existing 60 Mgy first-generation biofuel facility in Keyes. This technology enables Aemetis to produce advanced ethanol that is valued up to approximately $3 per gallon more than traditional ethanol [233].

FIGURE 4.12 A gas fermentation facility at the Shougang Steel mill in Caofeidian, China. Commercial-scale biological gas fermentation processes are currently being deployed to convert carbon-rich gas waste streams produced through steel manufacturing into low-carbon fuels. Pioneer plants are able to produce over 45,000 tons of fuel per annum. (Image courtesy of LanzaTech Inc [3].)

Lanzatech, in association with invista and SK innovation also proposed production of 1,3-butadiene in a two-step process, which is combination of biological and chemical steps. The recent reviews published by Liew et al. [210,234] give excellent overviews on the advances of synthetic biology on syngas-fermenting bacteria. These technologies are limited by low concentration of substrate uptake, productivity, and selectivity in terms of product formation.

In 2015, both China Steel Corporation of Taiwan and ArcelorMittal of Luxembourg approved commercial projects with LanzaTech. The former will be a 17 Mgy facility with the intention to scale up to 34 Mgy [231]. The latter 9.8 Mgy facility will be built at ArcelorMittal's flagship steel plant in Ghent, Belgium with the intention to construct further plants across ArcelorMittal's operations [232]. If scaled up to its full potential in Europe, the technology could enable the production of around 104 Mgy, which would displace 1.6 million barrels of fossil-fuel-derived gasoline on a BTU basis. In addition to these two projects, Aemetis, Inc. acquired a license from LanzaTech for the conversion of agricultural waste, forest waste, dairy waste, and construction and demolition waste (CDW) to ethanol in California. In a first phase, Aemetis plans to adopt the process by adding an 8 Mgy gas fermentation unit to its existing 60 Mgy first-generation biofuel facility in Keyes. This technology enables Aemetis to produce advanced ethanol that is valued up to approximately $3 per gallon more than traditional ethanol [233].

Several studies have carried out lifecycle analysis of gas fermentation. In a recently published cradle-to-grave LCA by Handler et al. [235], the production of ethanol from gas fermentation in the USA is estimated to result in 67% greenhouse gas (GHG) reduction (using blast oxygen furnace off-gas from steel manufacturing) and 88%–98% GHG reduction (utilizing gasified biomass), when compared to conventional fossil gasoline. In both feedstock scenarios, 20%–40% of the carbon in feed-gas is converted into ethanol. Conclusions drawn from LCA are often highly geographically dependent. A separate and older study showed that approximately 50% GHG savings can be attained from a microbial gas-to-ethanol platform based in China, relative to fossil gasoline [236].

Value-added products can also be generated by integration of biological and chemical processes. Lanzatech, in association with invista and SK innovation proposed production of 1,3-butadiene in a two-step process, which is combination of biological and chemical steps. The recent review published by Liew et al. [210,234] gives an excellent overview on the advances of synthetic biology on syngas-fermenting bacteria. These technologies are limited by low concentration of substrate uptake, productivity, and selectivity in terms of product formation.

4.8 NATURAL BACTERIA AND MICROBES FOR CO_2 CONVERSIONS

Several microorganisms and enzymes are very effective in conversion of CO_2. Some of these are described by Kondaveeti et al. [1]. Proteobacteria are Gram-negative bacteria, mainly composed of lipopolysaccharides in their outer membrane. These are found to be a geologically, environmentally, and evolutionarily important group of microorganisms. The members of this bacterial phylum exhibit an enormous metabolic diversity. Most of these bacteria have an industrial, medical, and agricultural significance [237]. The photosynthetic proteobacteria are known as purple bacteria, concerning their reddish pigments. One of the most common and well-known photosynthetic β-proteobacteria is *Ralstonia eutropha*. This bacterium can grow on variable carbon environments, such as heterotrophic, autrotrophic, and mixotrophic. In the absence of a heterotrophic carbon environment as an electron source, this bacterium has proven to utilize hydrogen as an energy source for fixing CO_2 via the Calvin–Bassham cycle (CBB) [238].

R. eutrohpa is a convenient photosynthetic bacterium to handle because of its aerobic H_2-oxidizing capabilities over obligate anaerobes to generate chemicals and other value-added products. The interest in *R. eutrohpa* has led to generation of bioplastics (PHAs/poly hydroxyalkanoates) by storing carbon in cytoplasm. PHAs are generally composed of PHB (poly 3-hydroxybutyrate) and PHBV (poly 3-hydroxybutyrate-co-3-hydroxyvalerate) [238,239]. These bioplastics are gaining high importance due to credentials such as biodegradability over conventional plastics and their green synthesis with CO_2 sink. These polymers with desired length can be pursued with genetic modifications [240]. Few studies are pursued in genetic modification of *R. eutropha* for the generation of products with variable properties (mechanical strength) and production yield. Voss et al. [241] modified *R. eutropha* and generated a cyanophycin, a protein-like polymer, with an increased production quantity. A few other studies have synthesized several useful compounds, such as ferulic acid (precursor to vanillin biotransformation) and 2 methyl citric acid, by genetic modification [1,242]. The limitation in these studies is use of organic carbon as an energy source rather than inorganic CO_2.

Muller et al. [1,243] engineered *R. eutropha* and generated methyl ketones of diesel range with productivity in a range of 50–180 mg/L by using CO_2 and H_2 as energy and electron sources, respectively. They also described an integrated approach for production of isobutanol and 3-methyl-1-butanol by using a microbial electrochemical process. In this system, initially, CO_2 was converted to formate and tailed by isobutanol and 3-methyl-1-butanol generation. This system achieved a final concentration of 850 and 570 mg/L of isobutanol and 3-methyl-1-butanol, respectively.

Idenolla species is another type of photosynthetic bacteria similar to *R. eutropha* in generation of polymers by using CO_2. An additional superiority of *Idenolla* over *R. eutropha* is the generation of products using carbon monoxide (CO) as an energy source. These bacterial species have an additional advantage by synthesizing products from exhaust gas [244].

Most clostridia are anaerobic and Gram-positive bacteria. They are found to have great importance in (a) human and animal health and physiology, (b) anaerobic mortification of simple and complex carbohydrates, (c) the carbon cycle, and (d) bioremediation/degradation of complex organic chemicals [245]. Many strains of clostridia are capable of autotrophic fixation of CO_2 or CO by using H_2 as an electron donor [172]. Clostridia also can use simple chain organic carbon molecules, such as formate or methanol, as an energy source. They can achieve this by the WLP pathway in which two molecules of CO_2 are scaled down to generate one molecule of Acetyl-CoA with CO or H_2 being used as a reducing proportionate [246].

Additionally, *Clostridium* species provide many fascinating characteristics for biotechnological utilization: (a) ability to use simple as well as complex substrates such as H_2, CO_2, and CO; (b) equipped with diversified pathways for generation of value-added products; and (c) tolerant to toxic metabolites and substrates. However, as they are strictly anaerobic, the cultivation of *Clostridium* is very difficult [172]. Normal atmospheric and facultative conditions are lethal to *Clostridium* species. Early work with *Clostridium* was mostly carried out for the production of acetic acid and other related products. The synthetic biology/genetic modification of *Clostridium* are being pursued at a slow pace due to their difficulty in lack of genetic tools as found in *R. eutropha*. However, the repertoires of genetic tools are applicable for *Clostridium* with development in plasmid DNA and chromosomal manipulation technologies [246]. The uses of mobile group II introns are generally targeted for gene disruption and genetic engineering of *Clostridium*. Cooksley et al. [247] demonstrated a ClosTron application through genetic engineering in *Clostridium acetobutylicum* for acetone-butanol-ethanol (ABE) fermentation and for expression of cellulosome. Leang et al. [248] employed double crossover homologous recombination technology and an improved electroporation protocol to the process in a proof-of-concept gene deletion study on the species of *C. ljundahlii*. Likewise, recent studies by the D.R. Lovley group demonstrated introduction of foreign genes and deletion of certain genes in *C. ljungdahlii* to enhance the production of biocommodities. Here, deletion of the ethanol-production pathway led to an increase in acetate production. The homolog recombination method described in their study can be applicable for the introduction of desired metabolic genes or reporters into the chromosomes of *C. ljungdahlii* [1,248].

Archaea are generally found in abnormal ecological niches, such as high and low temperatures, acidic and high-saline environments, and anaerobic atmospheres. The hydrogenotrophic methanogens of Archaea use H_2 as an energy carrier and CO_2 as a carbon source for CH_4 generation. Furthermore, several strains of Archaea belonging to the haloarchael group are equipped with a metabolic pathway for accumulation of PHA from CO_2. As most of the Archaea are thermophilic in nature, this becomes an excellent source of the carbonic anhydrases (CA) enzyme, which is thermostable and convenient for industrial CO_2 capture and a platform for gas fermentation [172]. The recognition of genetic tools for Archaea might enhance the product yield in the gas fermentation process. Recently, genetic reengineering of Archaea for valuable product formation has been rising. Keller et al. [249] reported a heterologous expression of five genes from a carbon fixation pathway in *M. sedula* into hyper theromophilic *Pyrococcus furious*, which grows on organic carbon at 100°C. Engineered *P. furious* is capable of incorporating CO_2 for generation of 3-hydroxypropionte-4-hydroxybut yrate (3BHP) at an optimum condition of *M. sedula*; i.e., at 70°C. This temperature-dependent strategy for generation of valuable product formations is established in the work of Basen et al. [250].

Whereas cyanobacteria derive their energy and carbon from the reduction of CO_2, chemolithotrophs derive their energy from the oxidation of reduced inorganic compounds and their carbon from CO_2 [251]. This allows chemolithotrophs to perform light-independent CO_2 fixation, eliminating photosynthetic production issues like cell shading. However, cultivating chemolithotrophs is more complex as two inputs are required instead of one. Previous studies have characterized and tested the CO_2 fixation capabilities and commercial applications of selected chemolithotrophs [252].

Acetogens are a well-studied subset of mixotrophic chemolithotrophs that operate strictly under anaerobic conditions [253]. One unique characteristic of these bacteria is the native Wood–Ljungdahl pathway (WLP), which fixes two CO_2 molecules into one acetyl-CoA with less energy than other carbon-fixation pathways [254]. This gives acetogens an advantage as microbial production platforms because they can bypass carbon loss under certain fermentation conditions.

During traditional heterotrophic fermentation, the conversion of sugar into acetyl-CoA results in one-third of all carbon lost to CO_2 production, limiting the maximum theoretical yield of products to 67% or less. During nonphotosynthetic mixotrophic fermentation, the WLP reassimilates the CO_2 generated from sugar conversion, resulting in three acetyl-CoA and one adenosine triphosphate (ATP) for every molecule of glucose or hexose sugar. The amount of CO_2 reassimilated is dependent on the availability of NAD(P)H and, therefore, is inversely dependent on the target product's degree of reduction. If the product is highly reduced, less NAD(P)H is available for CO_2 fixation and vice versa.

Mixotrophic production has many advantages relevant to industrial viability. Using syngas in addition to sugar fermentation provides added reducing power for CO_2 reassimilation. Sugar catabolism, specifically glycolysis, provides ample ATP, which would otherwise be severely limited in WLP gas-only production. Mixotrophic production also potentially enables larger product yields by dividing carbon utilization between biomass and product formation.

As pointed out by national academy report [2], carbon catabolite repression (CCR), in which one feedstock is preferred over the other [255], could present challenges for mixotrophic production. The concern with acetogens is that the sugar feedstock is preferred, thus reducing or eliminating the utilization of syngas for energy production or carbon fixation. However, nonphotosynthetic mixotrophic fermentation with syngas has been demonstrated in a variety of acetogenic microbes without CCR [256]. In that study, carbon labeling fermentation resulted in *Clostridium ljungdahlii* with 73%–80% and *C. autoethanogenum* with 51%–58% of its acetate derived from syngas, demonstrating concurrent utilization of both feedstocks with little CO_2 lost from glycolysis. The study was also able to engineer *Clostridium ljungdahlii* to produce acetone, a commodity with a market of $8 billion. Acetone anabolism does not require additional NAD(P)H downstream of acetyl-CoA, allowing for greater CO_2 fixation. Mixotrophic production from a high-density continuous fermenter resulted in a titer of 10 g/L and a productivity of 2 g/L/h, 92% of the theoretical mixotrophic maximum compared to 138% of the theoretical heterotrophic maximum. However, metabolic engineering was not applied to the host strain, and reactor conditions were not extensively optimized, especially for scale-up. More research is necessary to overcome the low degree of reduction required for the target products. The capability to engineer acetogens is currently limited by knowledge gaps and insufficient engineering tools [255].

4.8.1 Perspectives on Microbial Synthesis of CO_2 Conversion

Commercializing microbial production has always been an economic challenge due to the high cost of carbon feedstocks and low product yields. While significant work has been done to increase yields in recent years, the need to feed cultures with high-cost sugars is still an issue. Photosynthetic organisms like algae and cyanobacteria mitigate this problem by utilizing CO_2 as their feedstocks, but they are slow growing and it is difficult to achieve industrially relevant productivity and scale up the production systems. Nonphotosynthetic organisms that can convert gases like CO_2 or methane have become prime targets for microbial production due to their wide diversity of pathways and growth rates.

Nonphotosynthetic biological systems possess a number of potential advantages over photosynthetic systems. These include a wide variety of organisms, a larger range of potential target chemicals, and the ability to avoid the inefficiency of photosynthesis. Aerobic systems also have the advantage of high productivity, capacity for continuous cultivation, and compatibility with artificial photosynthesis. Some nonphotosynthetic organisms can take advantage of low-cost, low-emissions electrons from renewable energy sources, and many can be cultivated with cheap and ubiquitous gaseous carbon feedstocks comprised of hydrogen, CO, and CO_2, such as industrial waste gas, biogas, or syngas. Some of the products generated from nonphotosynthetic processes include acetogens, acetate, succinate, alcohols, and pyruvate. These products are formed by different approaches. For example, acetate and succinate are formed by direct electron transfer to microorganisms while alcohols and pyruvate are formed by indirect electron transfer via electrochemically synthesized electron donors. Acetogens are produced by either CO_2 or CO fixations or by two-stage integrated process. Various pathways for CO_2 fixation and metabolic conversion are described in the next section.

The two-stage integrated process was created by Hu et al. [227] in which the first-stage CO_2 and CO or H_2 are converted anaerobically to acetic acid by the acetogen *Moorella thermoacetica*. Acetic acid is then fed into the second stage where it is converted aerobically into lipids by *Yarrowia lipolytica*. Accounting for the fact that certain products are not readily formed under anaerobic conditions, this two-stage bioreactor process allows for the nonphotosynthetic fixation of CO_2 followed by acetate-dependent production of aerobic products. The fermentation conditions of *M. thermoacetica* and *Y. lipolytica* were optimized for acetate and lipid production, respectively. *M. thermoacetica* growth and production were divided into two phases. A CO-dependent growth phase was established under CO_2/CO conditions; then, the reactor was switched to H_2/CO_2 to increase production of acetate. In this process, the CO_2-fixation rate exceeded the CO_2-generation rate, indicating the feasibility of a two-stage bioprocess to convert syngas into useful commodities.

There are also significant challenges that need to be addressed in order for technologies based on nonphotosynthetic organisms to reach maturity. These include many of the same challenges associated with photosynthetic systems, such as poor solvation of CO_2 and H_2, a lack of genetic tools and metabolic understanding, a limited number of strains that have been explored and limited techniques for DSP of products. Nonphotosynthetic applications are at an earlier maturity level than photosynthetic ones, and support is warranted on these applications to advance understanding and the maturity of these systems.

4.9 SYNTHETIC BIOLOGY AND GENETIC ENGINEERING FOR THE CONVERSION OF CO_2

Direct use of microbes in generation of value-added products is typically hindered by several characteristics/properties, such as low product yield, expensive cultivation condition, and inadequate growth rate. In this aspect, synthetic biology has gained importance due to its interest in product formation by altering biochemical pathways or by introduction of heterologous pathways into microbes for favorable features [257]. With consideration of recent advances in metabolic and protein engineering, the synthetic biology proves to be an excellent tool for boosting up biological process and to improve economic viability.

4.9.1 Synthetic Biology for Single-Carbon Compounds

Synthetic biology enables improvements over natural systems by combining components that do not co-exist in nature (e.g., linking two pathways, localizing a pathway to an organelle; see Figure 4.13a). The primary molecular intermediates along the path between CO_2 and biomass are the one-carbon compounds depicted in Figure 4.13b, which serve as stepping stones and electron donors in the reduction and oxidation steps. Various naturally autotrophic (use CO_2 as sole carbon source while using light or chemicals for reducing power) and methylotrophic (use reduced one-carbon compounds such as formate or methanol both as carbon and reducing power sources) microbes have been considered and used in biotechnological industrial settings. Some of them can even be genetically engineered to a certain extent to suit bioproduction processes. For example, companies such as LanzaTech use anaerobic autotrophic bacteria called acetogens to produce simple compounds such as acetone, ethanol, and lactate from waste-derived syngas and flue gas. The input stream contains CO_2, CO, and H_2 as carbon and electron sources. The productivity of acetogenic CO_2 fixation can also be increased using elaborate cofeeding strategies. Acetogenic bacteria have also been mentioned as a possible platform for food production from electricity and CO_2 [258]. Recent work has shown that an engineered consortium of bacteria can collectively produce alcohol biofuels from electrochemically produced H_2 and CO with high energetic efficiency. Another naturally autotrophic model organism that can be cultivated aerobically with CO_2 as the carbon source and H_2 and/or formate as

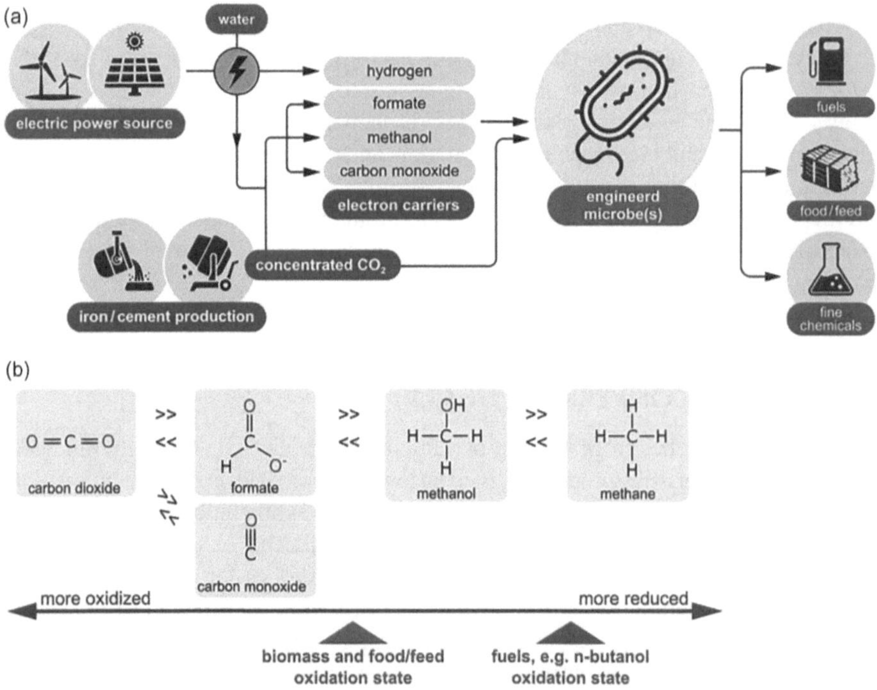

FIGURE 4.13 Framework for microbial production of value-added goods from renewable energy and CO_2. (a) Autotrophic organisms transfer electrons to (reduce) CO_2 to produce biomass. To do so, they require electron carriers. These electron carriers could be produced and recycled using electron input from an energy source (e.g., photovoltaic cells) and an electron source (water). Together with CO_2 concentrated from different industries (e.g., metallurgy), it is possible to provide these feedstocks to engineered microbes to produce various desired end products (e.g., fuel). (b) Oxidation states of one-carbon molecules. To assimilate CO_2, organisms must reduce it to a similar oxidation state as cellular biomass. This is done by transferring electrons (e) from water via electron carriers to CO_2. Reduced one-carbon molecules can be oxidized to produce reducing power, some of which can be invested to produce energy in the form of ATP [259].

the reducing power source is Cupriavidus necator. Two studies showed the efficient conversion of CO_2 to biofuels using this microbe combined with an electrical system that generated the reducing power source feedstock [259,260]. However, common model heterotrophs like *Escherichia coli* and *Saccharomyces cerevisiae* are much simpler to grow and modify than any naturally autotrophic organism known, even model systems like *C. necator.*

Recent studies have demonstrated dramatic advances in the ability to use CO_2 and other one-carbon compounds as carbon sources for two tractable and biotechnologically relevant microorganisms: the bacteria *E. coli* and the yeast *Pichia pastoris* (Figure 4.14). The energy sources used in these studies are one-carbon compounds that can be produced by the electrochemical reduction of CO_2 [259,261]. If a renewable source such as solar or wind power is used to drive this electrochemical reduction, then the overall bioproduction balance could be carbon negative.

Gleizer et al. [259] used a combination of rational design and adaptive evolution to enable *E. coli* to use CO_2 as its sole source of carbon (see Figure 4.14) and formate as

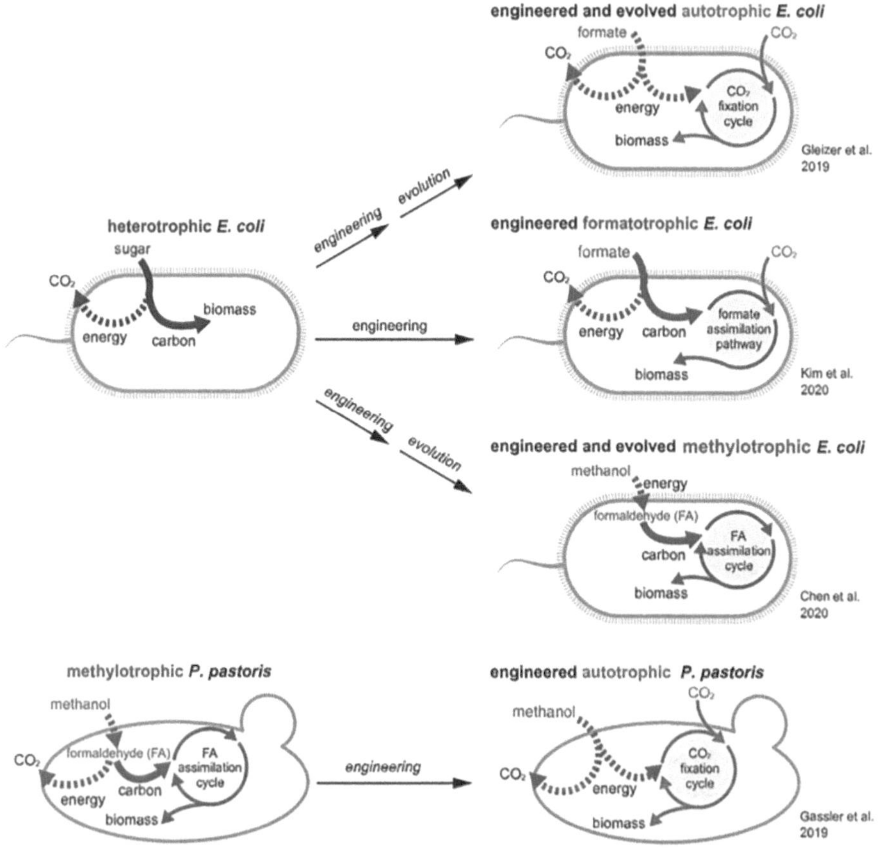

FIGURE 4.14 Recent developments in microbial engineering for one-carbon molecule conversion to biomass [259].

the sole source of energy. Kim et al. [263] engineered *E. coli* to use formate as both the energy and carbon source for growth (Figure 4.14). Expression of an additional gene encoding a methanol dehydrogenase enabled *E. coli* to use methanol as the carbon and energy source. Chen et al. [262] used rational design along with adaptive laboratory evolution to facilitate the growth of *E. coli* on methanol as both single energy and carbon sources (Figure 4.14) via the energetically efficient ribulose monophosphate pathway. Finally, in the study by Gassler et al. [264], the methylotrophic yeast *P. pastoris* was engineered to use its native carbon source, methanol, to provide only energy, while using CO_2 as the sole carbon source for biomass (Figure 4.14).

In the studies by Kim et al. [263] and Gassler et al. [264], the initial engineering was sufficient to enable slow growth, with subsequent adaptive evolution improving the growth rates by tenfold for *E. coli* and by twofold for *P. pastoris*. The resulting microorganisms in all but the study by Chen et al. [262] require elevated concentrations of CO_2 (>5% CO_2) to support growth, likely due to thermodynamic or kinetic constraints associated with most carbon fixation pathways. Despite recent advances, the reported strains are not yet suitable for industrial use, due to slow growth rates, incomplete characterization, and insufficient testing at industrial scale.

E. coli is given special attention as it is the most often used microbe in the field of synthetic biology [265]. Despite being heterotrophic in nature, the Gram-negative *E. coli* gained its place at the table of genetic engineering due to its simple cultivation, low cost, easy transformation, fermentation, and high production for new technologies, which include using CO_2 as a carbon source [172]. Heterologous expression of carbonic anhydrase (CA) in *E. coli* is extensively reported. CA expression of *Methanobacterium thermoautotrophicum* in periplasm of *E. coli* resulted in a whole-cell biocatalyst for CO_2 hydration. These systems successfully hydrated the CO_2 with a catalytic efficiency and productivity similar to free enzymes [266]. Zhuang et al. [267] examined heterologous expression of RuBisCo and phosphoribulokinase (PRK) in *E. coli* by utilizing different fermentation conditions. Jo et al. [268] observed additional benefits of *E. coli* like increase in stability and by passing the protein purification whole cells are isolated and recycled. Similarly, as *M. thermoautotrophicum*, the periplasmic expression of CA is noticed in *Neisseria gonorrhoeae*. These advances noted in *E. coli* are not confined in the expression of individual enzymes. Bonacci et al. [269] studied the expression of efficient carboxysomes from *Halothiobacillus neapolitanus* and encoded 10 genes in *E. coli* and noted that these bacteria are capable of carbon fixation.

Synthetic biology of *E. coli* has provided a possibility of encoding photobacterial genetics for the generation of value-added products. For instance, brown algae are well-known feedstock for the generation of carbon-neutral biofuel. However, from the point of economic industrial viability, these are limited by their inability to utilize alginate. Wargacki et al. [270] reported the DNA fragments in *Vibrio splendidus* that are responsible for secretion of enzymes that are capable of utilization of alginate. This enzyme expression in *E. coli* resulted in a bacterial platform that is capable of metabolizing alginate with a yield of 0.281 weight ethanol/dry weight of microalgae that is equivalent to 80% of the theoretical basis of total polysaccharide composition.

When coupled with the electrochemical reduction of CO_2 to one-carbon molecules, the ability of these engineered strains to harvest energy from one-carbon molecules raises the possibility of a circular one-carbon economy. Such coupled biotic-abiotic technologies can be called "hybrid systems" and these designs can ultimately surpass photosynthesis in productivity and energetic efficiency. For example, using photovoltaic cells to produce H_2 and CO as feedstocks for archaea was shown to have higher energy conversion efficiency than photosynthesis. Such designs are inherently modular, allowing the independent choice of biological host and energy source. The use of genetically tractable organisms is attractive as it simplifies the introduction of new pathways—for example, to use H_2 as an energy source in *E. coli*; H_2 can be produced more efficiently than formate and is compatible with microbial growth [271]. Genetically tractable hosts are also useful for introducing novel biosynthetic pathways (e.g., to produce a fuel such as n-butanol or more complex molecules such as dyes, drugs, and fragrances) [272]. Microbes feed for livestock or as food for human consumption (for the latter, social, and regulatory acceptance are preconditions). Since approximately half of the global agricultural calories are used as feed, the partial decoupling of livestock feed production from agriculture could greatly improve the sustainability of meat production, increase effective food yields, and free land for other uses. Finally, electrical current for the reduction of CO_2 can be provided by solar, wind, hydropower, or any other renewable source. However, to properly evaluate the industrial relevance of these approaches, techno-economic analyses are urgently required.

4.9.2 Electrofuel Host Development for Autotroph and Heterotroph

The development of electrofuel hosts requires the availability of genetic tools, as well as a sound knowledge of gene regulation and metabolism in the target host. While this effort is being developed for a number of genetic systems, current electrofuel projects are taking advantage of an established body of knowledge for genetic manipulation in well-studied organisms such as *R. eutropha* and *Clostridium* spp. There are two basic avenues for electrofuel host development: engineer a natural H_2-utilizing autotroph to produce the desired electrofuel, or engineer a suitable heterotroph to utilize CO_2/H_2 for electrofuel production. In the latter case, choice of a suitable host depends not only on the availability of a genetic system but also tools for systems-level analysis and the organism's suitability for bioprocess production. In the case of potentially suitable autotrophs, it may be that these tools are not available or are still in the early stages of development. Diverting carbon flow to an electrofuel rather than to native cellular metabolism may prove to be problematic [273–276].

The opposite is true for the heterotroph approach, where preventing CO_2-derived carbon from entering the host's native metabolism may be key to establishing an efficient process. It is not clear whether the 'autotrophic' or the 'heterotrophic' strategy is better suited for the generation of electrofuels. *R. eutropha* serves as an example of the 'autotrophic' approach. A variety of tools and techniques exist for genetic engineering of this H_2-utilizing autotrophic organism and there are both plasmid-based expression systems and chromosomal modifications via homologous recombination, as well as systems-level analysis tools including microarrays. While some species of *Clostridium* are also natural H_2-utilizing autotrophs, genetic tools for these organisms are still being developed. New efforts should enable clostridial hosts to be flexible electrofuel producing platform organisms in the near future [273–276].

The second avenue for electrofuel development using heterotrophs was recently demonstrated for the first time in *P. furiosus* [273–276]. Heterologous expression of the first five genes from the 3HP/4HB pathway allowed *P. furiosus* to utilize H_2 and incorporate CO_2 into 3HP, a crucial intermediate in the carbon fixation pathway and a valuable industrial chemical building block. This was accomplished by using a highly competent strain that enables chromosomal modification. The new genetic system has also been successfully utilized for overexpression of cytoplasmic hydrogenase I (SHI) thereby potentially allowing facile H_2 oxidation and NADPH production. Based on these promising preliminary results, efforts are underway for the construction of the electrofuels production host utilizing the complete *M. sedula* 3HP/4HB pathway for biological activation of carbon dioxide into a variety of chemical and fuel molecules [273–276].

4.9.3 Synthetic Biology Tools for Green Algae

The tools to facilitate genetic manipulation of green algae are far less advanced than tools for genetic manipulation of bacteria, yeast, and vascular plants. Key challenges include poor genome insertion, gene silencing, and unoptimized promoter systems. Nevertheless, attempts have been made for genetic modification of green algae for the purposes of improving photosynthetic efficiency, decreasing photodamage of the light harvesting complex, and optimizing the efficiency of carbon uptake and incorporation. Several products have also been sought from transgenic expression, including high-value therapeutic proteins, biohydrogen, lipids, and terpenoids [277].

In green algae, the majority of tools and techniques developed are for *Chlamydomonas reinhardtii* [278]. Green algae have three separate genomes: nuclear, chloroplast, and mitochondrial. In *C. reinhardtii*, each of these genomes has been sequenced, and tools and protocols for genetic manipulation have been developed. Commonly, nuclear and chloroplast genomes have been primarily targeted for transgene expression, and each location involves special tools, advantages, and challenges. Tools and methodologies for manipulating gene expression in the chloroplast have achieved significant advances in recent years. Just like nuclear expression, chloroplast expression at present, however, remains limited in terms of gene size and number. National Academy Report [2] points out that the developments of genetic insertion (homologous recombination) technologies, identification of robust promoters for gene expression, development of synthetic operons for multiple gene incorporation, tools for engineering large genes and pathways, and novel selection methods are needed to further advance CO_2 conversion by green algae [1–3].

4.9.4 Gene Expression and Its Constraints for Microalgae and Cyanobacteria Molecular Cell Physiology

The basic genetic construct used to express a specific gene must consider the proper recognition of the DNA structures related to transcription and translation by the host cell, that is, promoters, ribosome binding sites, the codon usage to translate the target gene, and transcription terminators. Therefore, there are many constraints that need to be considered during the planning of genetic modifications. These include transcriptional and transitional controls.

Promoters are important for transcriptional control. The selection of the right promoter will depend on diverse variables such as the organism to be modified (cyanobacteria, microalgae), the gene expression levels required (constitutive, induced expression, strong expression), or the target DNA that can belong to different cellular compartments in eukaryotes.

In eukaryotes, native promoters have been extensively described for nuclear and organelle transformation. Their main advantage is the correct recognition by the enzymatic transcriptional machinery of the microalgae. Heterologous promoters have also been used in different types of microalgae such as viral promoters 35S, CaMV35S, SV40, and CMV. One of the advantages of these promoters is that they can be recognized in some microalgal systems. For cyanobacteria, there is a set of constitutive and inducible native promoters used in biotechnological studies to express heterologous genes or increase the expression of native ones. However, efficient controllable promoters in cyanobacteria are scarce [117,279]. Additionally, the effectiveness of heterologous promoters in cyanobacteria is very low and successful examples using lac-regulated and *tetR*-regulated promoters need to include critical modifications to fit the sequences to the transcriptional machinery of cyanobacteria [280].

For translational control, codon usage is helpful. Different organisms usually bear particular codon usage patterns. Therefore, when a gene from one species is cloned and expressed in another, some codons might be rare in the new host, leading to poor translation efficiency or starvation for certain amino acids and consequently lose the cell's fitness [281]. Specific variations in codon usage are often cited as one of the major factors impacting protein expression levels. The presence of rare codons that correlate with low levels of their endogenous tRNA species in the host cell can decrease the translation rate of target mRNAs.

Genes in cyanobacteria show a bias in the use of synonymous codons [279,282]. The importance of codon optimization in algal genetic applications also been seen in microalgae such as *C. reinhardtii* [281], *Gonium pectorale* [283], and *Porphyra yezoensis* [284]. Sequencing studies and expression profiles of microalgae have clarified these issues, thereby allowing optimization of codon usage at the level of expression in the nucleus [283] and in the chloroplasts [279,285].

4.9.4.1 Barriers for Transfer Foreign DNA into Microalgae and Cyanobacteria

There are several obstacles to be faced when trying to insert exogenous DNA into eukaryotic and cyanobacterial systems. The first is physical barriers such as cell wall, plasma membrane, nuclear membrane, mitochondrial membrane, and chloroplast membrane. To overcome these "natural barriers," diverse experimental strategies have been designed, such as the generation of spheroplasts (cells without cell wall) for eukaryotic microalgae by mechanical disruption or enzymatic treatments [279,286]. Other strategies include the application of multi pulse or high electrical voltage for the disruption of the algae extremely rigid cell walls (e.g., *Nannochloropsis*sp.) and the entry of exogenous DNA [279,287]. Another obstacle is the transient insertion of the exogenous DNA and the eventual loss of the phenotype of interest over time.

Cyanobacteria have a special envelope formed by the outer membrane, a thick peptidoglycan layer, the s-layer, and frequently an exopolysaccharide EPS envelope [288]. Several protocols such as growth under constant agitation, insertion of low concentrations of saline sodium citrate, pumping the cells out of sheath using syringe [289] and treatment by agitation with concentrated NaCl and several washing steps [290] have been tried. In addition, cyanobacteria have mechanisms of defense against exogenous DNA that hinders the transformation process. The majority of filamentous cyanobacteria and some unicellular representatives contain restriction enzymes that can degrade the transferred DNA [291]. Strategies to improve the DNA transfer efficiency include the following: recipient strains with inactivated endonucleases [292]; deletion of the restriction sites from the foreign DNA [293]; or more often, the protection of the foreign DNA by methylation, using specific methylases encoded in helper plasmids, prior to transformation of the cyanobacteria [279,294].

4.9.4.2 Transformation of Eukaryotic Microalgae

Transformation in eukaryotic systems has several advantages when compared to cyanobacteria. The integration of genes for nuclear transformation of microalgae occurs at random locations, with homologous recombination occurring at very low frequency [295]. On the contrary, plastidial transformations are based on homologous recombination similar to what is described for cyanobacteria. Different strategies for inserting foreign DNA into microalgae have been described (biolistics, electroporation, natural transformation, random mutation) [279,296]. However, each microalgal species has its own morphological, structural, and physiological characteristics, making the development of standard protocols for all species difficult.

Biolistics is one of the most effective methods of transformation, which has been mainly used for the insertion of genes into plastids such as chloroplast. Electroporation has been used for many years for the transformation of numerous prokaryotic and eukaryotic cells. The great advantages of this technique are its versatility, simplicity, and high efficiency. Mutation followed by selection of favorable phenotypes has been used for crop plants, and some promising strategies are now beginning to emerge for microalgae. Random mutation strategies applied to microalgae include the use of mutagenic chemicals and radiation.

The scientific literature shows many examples of genetically modified cyanobacteria, especially unicellular cyanobacteria, mainly through natural transformation, conjugation, and electroporation. DNA bombardment (biolistics) is less frequent, but it has been used [279,290]. Some cyanobacteria are naturally competent, being able to be transformed by natural incorporation of foreign DNA from the environment. This ability is a simple and rapid way to introduce exogenous DNA into cyanobacteria and has been commonly used to transform cyanobacterial strains through gene replacement by double crossover homologous recombination. As natural transformation triggers DNA fragmentation [279,297], this procedure is not recommended for single recombination or intact replicating plasmids. For these, conjugation or electroporation is preferred. Electroporation is commonly used for transferring plasmids into diverse cyanobacterial cells. Electroporation promotes single rather than double recombination events. The most common technique for transforming cyanobacteria has been conjugation. Conjugation is the transfer of plasmid DNA from a donor (commonly *E. coli*) to a recipient cell through direct contact. Because filamentous cyanobacteria contain endogenous restriction endonucleases, the design of the plasmids must avoid predicted target sites of native restriction enzymes [293], or must be previously methylated to prevent the action of them [279,294].

Plasmids vectors are one of the most common tools for transferring foreign DNA containing the information for the genetic manipulation of cyanobacteria. They can be classified based on the type of transformation into (a) integrative vectors, which modify the genomic information of the target organism by recombination or transposition (*cis*-transgene expression); and (b) replicative or shuttle vectors, which are plasmids that can replicate themselves with the aim of expressing genes without any alteration to the host's genome (*trans*-transgene expression). Plasmids introduced by conjugation are circular and a single recombination allows integration of the selective marker. Gene expression can be enhanced through shuttle plasmids. The recognition of the replication origin by the host cell is essential for shuttle vectors, and in cyanobacteria, there are few studies using this type of plasmids, mainly based on the self-replicating base RSF1010 plasmid [279,298].

4.9.4.3 New Transformation Strategies for Microalgae and Cyanobacteria

While some target fuels and chemicals can be produced through exploitation of naturally occurring cyanobacteria traits, manipulation of organisms' genetic material can enable production of many other target molecules. An understanding of areas where desired genetic material may be inserted without interrupting essential functions (gene integration and plasmids), ways to control production of the desired chemical target (promoters and riboswitches), and ways to tell the cell where to start and stop generating the target molecule (ribosomal binding sites) are necessary in order to conduct genetic manipulations.

The insertion of exogenous genetic material using organisms such as *Agrobacterium tumefaciens* has also been reported in some microalgae. This strategy allows efficient insertion of genetic material into the nuclear DNA and expression of the respective reporter genes. The CRISPR/Cas system, a heritable adaptive immunity system [279,299], has been adapted for targeted gene editing in mammalian, plant, fungal, and bacterial hosts [300]. The use of the CRISPRi system for gene silencing in *Synechocystis* sp. PCC 6803 was described recently [301]. Additional studies are needed to evaluate the functionality of these promising systems as tools for genetic engineering in others cyanobacteria and eukaryotic microalgae.

In addition to an enhanced ability to introduce and control novel pathways in cyanobacteria, a thorough understanding of the flow of carbon through the cell's metabolism is key to creating an industrially relevant production system. Techniques for improving carbon flux have included the use of carbon sinks, disrupting side pathways, removing inhibitors, and protein fusions to improve rate. While these techniques have pushed cyanobacteria toward industrial relevance, there is still much that is unknown and needs further study.

One of the primary challenges of chemical production in cyanobacteria is the low titers that result from most pathways. While a great deal of effort has gone into engineering RuBisCO, its improvement and alterations still remain below the required threshold for wide-scale chemical production. Cyanobacteria can also be supplemented with alternate carbon sources including glucose, glycogen, acetate, and xylose. While these carbon sources effectively improve titers, this additional carbon is not always directed toward chemical production and is often lost as biomass. These alternate carbon sources can also increase the risk of contamination. As a slow-growing organism, cyanobacteria in culture can quickly become overcome by competing organisms that can utilize sources such as glucose. Additional strategies to prevent contamination and alternate carbon source can be expensive and take away the advantages of using cyanobacteria as a host strain.

4.9.4.4 Selection and Reporter Markers Genes

The use of selectable marker genes is normally required in all experiments that aim to generate stable transgenic algae because only a very low percentage of treated organisms are successfully transformed. Selectable markers are often antibiotic resistance genes, which are dominant markers as they confer a new trait to any transformed target strain of a certain species, independent of the respective genotype. Numerous genes that confer resistance to various antibiotics and herbicides have been reported in microalgae and cyanobacteria (antibiotics such as zeocin, hygromycin, bleomycin, chloramphenicol, kanamycin, phleomycin and the herbicides sulfounylurea, norflurazon) [302] allowing easy selection of transformed transgenic organisms (dominant selection markers). Among the known disadvantages of such markers is that their sensitivity is specific to each microalgae.

Another class of transformation markers are reporter genes that allow selecting transformant strains based on a particular phenotypic characteristic conferred by this gene. Among these are luminescent, fluorescent, and chromogenic proteins (such as sfCherry, mCherry, GFP, GUS, SHCP, and luciferase) are currently been used to evaluate gene expression levels in cyanobacteria and eukaryotic microalgae [279,303,304]. The use of reporter genes that provide detectability through optical methods has been recently implemented for the selection of successful transformants, especially in eukaryotic microalgae [279,305].

4.9.5 Genome Scale Models

Genome-scale models (GSMs) are important tools for assessing and engineering metabolic systems. The models may be used to describe an organism's entire metabolism utilizing genomic information [2,306]. GSM-directed engineering has been successfully used to improve a variety of production platforms in *E. coli*, including 1,4-butanediol [307], lycopene [308], lactic acid [309], and succinate [168]. While GSM are useful, the transfer of GSMs from heterotrophs to photoautotrophs is a difficult process. The recent development of two GSMs [2,310] for 7942 allows for greater predictive power when making modifications to metabolism. New insights have been gained using a GSM developed for 7002 [311], which could help engineer the strain [312].

Construction of a comprehensive GSM for 6803 has attempted to resolve many of the problems that are seen in 7942 and 7002 with GSM development in cyanobacteria

[2,313]. GSM analysis uses 3,167 genes to study gene function, carbon metabolism, photosynthesis, and chemical production [2,314]. While traditional GSMs are limited to the native genes found in 6803, recent iterations incorporate non-native metabolic reactions to construct hybrid phototrophic and heterotrophic cyanobacteria models [2,314]. These expanded models can predict new strategies for metabolic pathway construction and increase yields in targeted chemical production.

4.9.6 Limitations of Synthetic Biology

Synthetic biology tools are essential to utilize a strain as a production host. These tools can range from sites for the expression of nonnative genes to systems for controlling gene expression or tools for genetic manipulation [2,315]. While heterotrophic hosts such as yeast and *E. coli* have well-developed synthetic biology toolboxes, these tools are often incompatible with photoautotrophic hosts such as cyanobacteria. In fact, tools developed for cyanobacteria are often difficult to transfer between strains, necessitating the development of a unique set of tools for gene integration and controlling gene expression on a transcriptional and translational level for each strain. Many of these tools have been developed for cyanobacteria [2,117]. In addition to the difficulty of transferring tools between hosts, it can also be difficult to transfer pathways between host organisms. Gene expression, intermediates, and final products have varying toxicity across strains. Each strain can have a unique codon preference, altered enzyme activities, and constitutive promoters. One of the greatest challenges to metabolic engineering efforts for cyanobacteria is the careful tailoring of tools to each strain.

4.10 FUTURE PROSPECTS

Conversion of CO_2 to valuable products by bioprocesses is now well accepted by researchers and plant developers [316]. While there are still issues with bio-based technologies such as large land and water requirements, high energy needs, expensive catalysts, and significant amounts of chemical needs, well-optimized processes have significant commercial potential. Bioprocesses fit well in the concept of green technology. While not all bio-based processes are currently competitive on economical basis, chemicals and fuels produced from CO_2 conversion will have carbon neutral basis and high degree of sustainability. With further technical and productivity improvements, biological conversion of CO_2 has bright future. Bio-based processes can also be integrated to other chemical or waste remediation processes to serve multiple objectives [317]. The integration of a biological CO_2-mitigation process with conventional wastewater treatment technologies can minimize the economics related to waste remediation [318]. The conversion of CO_2 to high-value products through improvements of product selectivity, process optimization, genetic manipulations of microbes, and smart biological and chemical process integrations along with mathematical modeling can provide further momentum to this strategy [317,318].

According to Kondaveeti et al. [1], two technologies that have stood out in recent years are MES technology and gas fermentation. These technologies do not use biomass that can be a source for food. They both generate a wide variety of high-value chemicals and fuels. There are, however, few issues that need to be resolved. For MES, the low electron intake by bacteria at the cathode and low productivity, stability, selectivity, etc. needs to be addressed. The synergistic effects of the substrate, electrode materials, and bacterial cell should be studied and well-understood for scale-up. The conventional anaerobic fermentation system necessitates ~3 kg of glucose for production of 1 kg of butanol, which seems to be nonoptimal in spite of existing large-scale processes [319]. In this regard, alcohol generation in MES using organic-rich wastewater seems to be a promising choice [141]. More specifically, influence of current and electrical potential on the microbial metabolism and microbial strains related to electrofermentation or MES are required to consider for industrial application. Microbial carbon capture is likely to be a standalone process that could be competitive to conventional technologies.

Complex raw biomass sources, such as cellulosic, wood, and straw materials, cannot be easily degraded or fermented. Gas fermentation is very versatile and attractive alternative and its combination with chemical processes such as biomass gasification, biomass reforming, and partial oxidation of biomass along with water gas shift reaction can produce syngas, which is easy to process and ferment for a variety of chemicals and fuels. Gas fermentation also uses carbon content of the entire biomass (both cellulose and lignin), and it is not affected by the heterogeneity of the biomass. H_2 and CO_2 in syngas can also be utilized by microbes for product formation. The conversion of syngas to products with a higher productivity rate can be achieved with bioelectrochemical and photobiological processes [320]. Further research would be required in order to enhance biofuel generation with low energy utilization and a minimal amount of toxic by-product generation [1,141,317].

Biological conversion has a significant advantage over other technologies in that it does not require purified CO_2 streams; however, co-localization with CO_2 sources would be required due to barriers posed by transportation. Photosynthetic algal biomass production has some significant advantages compared with conventional crops in terms of its land footprint, water use footprint, and protein content. Photosynthesis is inherently inefficient though, requiring compensation with scale, making it land intensive. When developing any type of photosynthesis-based system, it is important to consider the impact of the cultivation method, restrictions in the use of genetically modified organisms, the degree of CO_2 solvation, nutrient requirements and downstream burdens, impacts on water and land use, and availability and suitability of CO_2 waste streams.

Biofuels have potential to advance the circular carbon economy and reduce reliance on and environmental impacts of fossil resource extraction. Renewable fuel production has been limited in its implementation due to its high cost compared with fossil sources and the capital infrastructure necessary to catalytically hydrogenate lipid extracts. In order to make biofuel economically more attractive, extractions and valorization of co-products are very important. Algal protein has the potential to supplement or replace conventional crops as a source of animal and/or human food, but this is not validated at large scale. Some carbon waste streams, such as those with heavy-metal contaminants, will be inappropriate for animal or human food applications. PUFAs are a promising and potentially lucrative product of algal biomass.

Pigments may be another valuable algae-based product. These applications may have the additional benefit of reducing pressure on conventional fish-based sources of these products.

In bioelectrochemical reaction, pathways using CO_2 as the feedstock are based on H_2 as a reaction companion, whereas algae can transform CO_2 into value-added products, such as omega fatty acids, proteins, and amino acids, and can simultaneously reduce pollutants. Likewise, other technologies, such as enzymes (e.g., formate dehydrogenase), can utilize/transform CO_2 by providing necessary energy. Figure 4.15 provides an overview of different CO_2-based product formation pathways and their current status of employment as outlined by Kondaveeti et al. [1]. Apart from urea, the leading CO_2-based product formations from other industrial processes involve the production of cyclic carbonates and salicylic acids, which are being produced at about 0.1 million tons/year [321]. Novomer in United States and Covestro (former Bayer Material Science) produce significant amounts of poly (propylene) carbonate and polycarbonate etherols polymrers using CO_2 as a building block. DNV (Det Norske Veritas) from Norway has constructed a pilot-scale operation plant for formic acid generation with a capability of 1 kg/day via electrochemical CO_2 reduction. Similarly, Mantra Energy Alternative Limited from Vancouver,

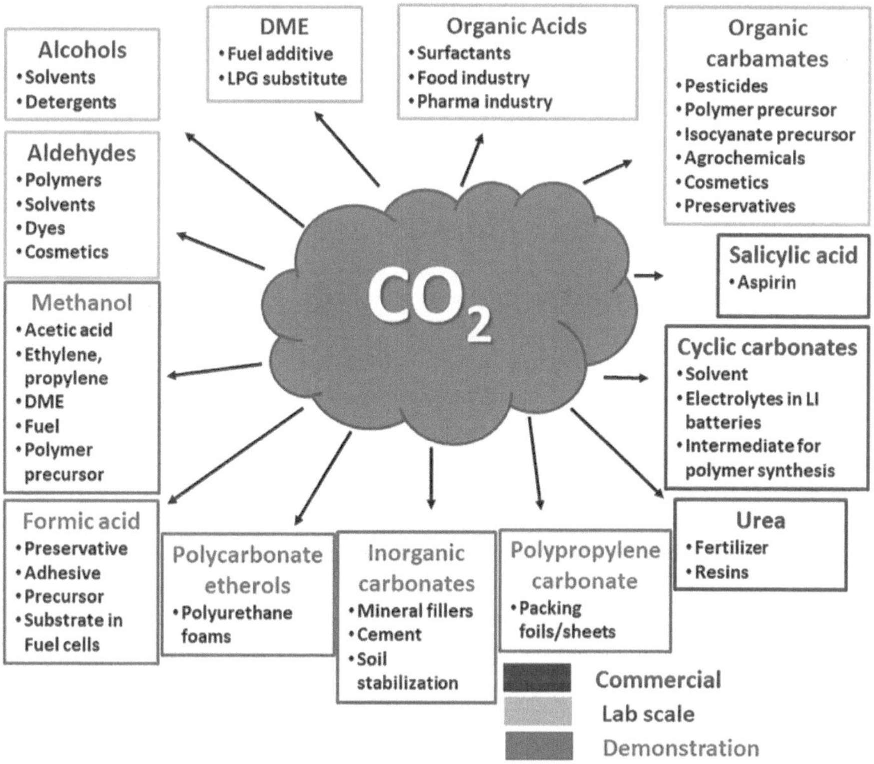

FIGURE 4.15 CO_2 conversion to target products and their position of implementation. [1].)

British Columbia, Canada, is constructing a plant for 100 kg/day formic acid generation by electro-reduction of CO_2 [321]. As shown earlier in Chapter 3, mineral carbonation is already being used in smaller and semi-commercial plants for various purposes, such as treatment of industrial waste, polluted soil, and generation of construction materials. Several applications are at laboratory or demonstration levels. One with high potential is the direct synthesis of dimethyl ether (DME) from CO_2. This method allows a CO_2 reduction of 0.125 t CO_2/t DME. Other ones are direct production of sodium acrylate from ethylene and CO_2 or electrocatalytic conversion of CO_2 to ethylene [322].

Stoichiometric conversion of CO_2 per ton of product is an important parameter in commercialization. Based on literature, acetic acid, ethanol, succinic acid, and caproic acid have stoichiometric production of 1.47, 1.91, 1.49, and 2.28 t CO_2/t, respectively. These chemicals have a projected market value of 16.4, 54.63, 0.237, and 0.27 billion dollars by 2026 [1]. Moreover, these products, such as acetate, can be used as building blocks for higher valued product generation, such as ethyl acetate, with a projected market value of 95 billion dollars by 2024 [1]. Also, the combination of ethanol and other compounds (e.g., volatile fatty acids) can be used for a carbon chain elongation reaction. On the basis of these projected market values, the mitigation of CO_2 to product formation can be beneficial. More lifecycle assessment (LCA) is, however, needed.

Algae and cyanobacteria have gained importance due to enhanced biomass generation for a feedstock and transesterification process. Cyanobacteria possess many advantages over other biological systems because they can easily be manipulated genetically to create different products, they are more efficient photosynthetically, and they can use CO_2 directly. Challenges with cyanobacteria include slow growth rate and limited availability of synthetic biology tools. Nonphotosynthetic microbial systems that can use CO_2 rather than sugars hold promise for utilization of gaseous carbon waste feedstocks. Of particular interest may be mixotrophic acetogens. However, β-proteobacteria, clostridia, archaea, and other bacterial species can overtake the photosynthetic microbes due to intense research and development [323]. Nonetheless, in utilization of various microbes for successful bioprocess development, one needs to pay attention to vital necessities, such as (a) suitable and appropriate strain selection for higher lipid generation (e.g., cell growth, autotrophic carbon fixation rate, product type and yield), (b) supplementation minerals/chemical in CO_2 fixation, (c) operation parameters (gas concentration, composition, etc.), (d) type of bioprocess operation (batch or continuous), and (e) reactor configuration (CSTR, fixed bed reactor, membrane reactor) [89].

A few decisive factors in biological processes and engineering of carbon capture and utilization include (a) safety (especially during the H_2 utilization as electron donor/source), (b) type of product selectivity and formation (concerning the purification and extraction economy), (c) optimized operational conditions, and (d) an increase in productivity by amplifying the cell density, etc.

In developing successful commercial operations, upstream and downstream processes are important. In fact, where ever possible, an integration of upstream, mainstream, and downstream processes can bring more savings and efficiency. As pointed out by Kondaveeti et al. [1], several issues to consider for successful commercialization are:

1. the technology should ideally cope with given impurities and/or CO_2 gas concentrations of their client's CO_2 waste streams at minimal pretreatment costs;
2. converting CO_2 in an efficient way (total CO_2 conversion and selectivity toward product of interest) at an appreciable productivity (space–time yield) toward product in order to minimize investment and operational costs and obtain a techno-economic feasible route;
3. obtain high product titers and high product selectivity in order to reduce further DSP steps or costs;
4. local value chains fit for given CO_2-to-product pathway via biological route;
5. legislation for the end product in case of high-volume application (ethanol, fuels) [324].
6. availability of renewable energy for hydrogen production and other electrical needs; for example, low potential photovoltaic energy prices in the Middle East, Southern Europe, and North Africa would enable the production of liquid CO_2 fuels that can be easily transported to Europe;
7. location of process and ease of transport of product to existing and potential new market;
8. proper scale and pricing for high-value products [1,325].

An integrated biorefinery approach can always help improving the economic viability of biological conversion processes.

This chapter has overviewed and discussed potential applications of biological processes for CO_2 fixation with generation of value-added products. While there is some progress in this strategy, there is still a need for finding suitable processes or microbial strains for CO_2 mitigation. For instance, the productivity of microbial CO_2 sequestration should be more than threefold higher than traditional CO_2 technologies, which seems to be a daunting goal with current methods. The combination of industrial processes with microbial CO_2 mitigation can be sustainable and economically viable for reduction of CO_2. The metabolism of CO_2 conversion by algae and photosynthetic processes can be equally viable as chemical and electrochemical technologies. Furthermore, market analysis and commercialization need to be further discussed by considering the real-field application. Bioelectrochemical CO_2 conversion and photo biological processes along with parts of gas fermentation for value-added product synthesis seem to be economical and sustainable approaches toward future technological development and commercialization. However, integration of suitable CO_2-converting processes with CO_2-generating processes helps for the development of a CO_2-based biorefinery that may lead to carbon-neutral industrialization. Finally, as pointed out by NAE report [2], future research should emphasize (a) bioreactor and cultivation optimization, (b) development of new analytical and monitoring tools, (c) bioprospecting, (d) valorizing of co-products, (e) development of genetic tools, and (f) pathways to new products.

REFERENCES

1. Kondaveeti, S., I. Abu-Reesh, G. Mohanakrishna, M. Bulut, and D. Pant 2020. Advanced routes of biological and bio-electrocatalytic carbon dioxide (CO_2) mitigation toward carbon neutrality. *Frontiers in Energy Research.* https://doi.org/10.3389/fenrg.2020.00094.
2. National Academies of Sciences, Engineering, and Medicine 2019. *Gaseous Carbon Waste Streams Utilization: Status and Research Needs.* Washington, DC: The National Academies Press. https://doi.org/10.17226/25232.
3. Buchanan, M. 2017. Accelerating breakthrough innovation in carbon capture, utilization and storage, a Report of the Carbon Capture, Utilization and Storage Experts' Workshop, September 26–28, Houston, Texas, Department of Energy, Washington, D.C.
4. Venkata Mohan, S., J. A. Modestra, K. Amulya, S. K. Butti, and Velvizhi, G. 2016. A circular bioeconomy with biobased products from CO_2 sequestration. *Trends in Biotechnology* 34:506–519. doi: 10.1016/j.tibtech.2016.02.012.
5. Hou, X., L. Huang, P. Zhou, F. Tian, Y. Tao, and G. Li Puma 2019. Electrosynthesis of acetate from inorganic carbon ($HCO3^-$) with simultaneous hydrogen production and Cd(II) removal in multifunctional microbial electrosynthesis systems (MES). *Journal of Hazardous Materials* 371:463–473. doi: 10.1016/j.jhazmat.2019.03.028.
6. Hunt, A. J., E. H. K. Sin, R. Marriott, and J. H. Clark 2010. Generation, capture, and utilization of industrial carbon dioxide. *ChemSusChem* 3:306–322. doi: 10.1002/cssc.200900169.
7. Singh, J., and S. Gu 2010. Commercialization potential of microalgae for biofuels production. *Renewable and Sustainable Energy Reviews* 14(9):2596–2610.
8. Vieira, E. D., G. Andrietta Mda, and S. R. Andrietta 2013. Yeast biomass production: A new approach in glucose-limited feeding strategy. *Brazilian Journal of Microbiology* 44(2):551–558.
9. Chisti, Y. 2007. Biodiesel from microalgae. *Biotechnology Advances* 25:294–306. doi: 10.1016/j.biotechadv.2007.02.001.
10. Demirbas, A. 2010. Use of algae as biofuel sources. *Energy Conversion and Management* 51:2738–2749. doi: 10.1016/j.enconman.2010.06.010.
11. Khan, M. I., J. H. Shin, and J. D. Kim 2018. The promising future of microalgae: Current status, challenges, and optimization of a sustainable and renewable industry for biofuels, feed, and other products. *Microbial Cell Factories* 17(1):36. https://doi.org/10.1186/s12934-018-0879-x
12. Clarens, A. F., E. P. Resurreccion, M. A. White, and L. M. Colosi 2010. Environmental life cycle comparison of algae to other bioenergy feedstocks. *Environmental Science & Technology* 44:1813–1819. doi: 10.1021/es902838n.
13. Shi, J., et al. 2015. Enzymatic conversion of carbon dioxide. *Chemical Society Reviews* 44:5981–6000. doi: 10.1039/C5CS00182J.
14. Asikainen, M., T. Munter, and J. Linnekoski 2015. Conversion of polar and non-polar algae oil lipids to fatty acid methyl esters with solid acid catalysts–A model compound study. *Bioresource Technology* 191:300–305.
15. Mubarak, M., A. Shaija, and T. V. Suchithra 2015. A review on the extraction of lipid from microalgae for biodiesel production. *Algal Research* 7:117–123.
16. Hazrat, M. A., M. G. Rasul, and M. M. Khan 2015. Lubricity improvement of the ultra-low sulfur diesel fuel with the biodiesel. *Energy Procedia* 75:111–117.
17. Al-Sabawi, M., and J. Chen 2012. Hydroprocessing of biomass-derived oils and their blends with petroleum feedstocks: A review. *Energy & Fuels* 26:5373–5399.
18. Davis, R., and M. Biddy 2013. Algal lipid extraction and upgrading to hydrocarbons technology pathway. National Renewable Energy Laboratory Technical Report NREL/TP-5100–58049, PNNL–22315.

19. Bhujade, R., M. Chidambaram, A. Kumar, and A. Sapre 2017. Algae to economically viable low-carbonfootprint oil. *Annual Review of Chemical and Biomolecular Engineering* 8:335–357.
20. Raab, A. M., G. Gebhardt, N. Bolotina, D. Weuster-Botz, and C. Lang 2010. Metabolic engineering of *Saccharomyces cerevisiae* for the biotechnological production of succinic acid. *Metabolic Engineering* 12:518–525.
21. Roy, J. J., L. Sun, and L. Ji 2014. Microalgal proteins: A new source of raw material for production of plywood adhesive. *Journal of Applied Phycology* 26:1415–1422.
22. Becker, E. W. 2007. Micro-algae as a source of protein. *Biotechnology Advances* 25:207–210.
23. Ahmed, F., Y. Li, and P. M. Schenk 2012. Algal biorefinery: Sustainable production of biofuels and aquaculture feed? In: Gordon, R. and Seckbach, J., eds., *The Science of Algal Fuels*. Netherlands: Springer, pp. 21–41.
24. Bimbo, A. P. 2007. Current and future sources of raw materials for the long-chain omega-3 fatty acid market. *Lipid Technology* 19:176–179.
25. Kaur, G., D. P. Singh, J. I. S. Khattar, and J. Nadda 2009. Microalgae: A source of natural colours. In: Khattar, J. I. S., Singh, D. P., and Kaur, G., eds., *Algal Biology and Biotechnology*. New Delhi: I. K. International Publishing House Pvt. Ltd., pp. 129–150.
26. Watanabe, Y., N. Ohmura, and H. Saiki 1992. Isolation and determination of cultural characteristics of microalgae which functions under CO_2 enriched atmosphere. *Energy Conversion and Management* 33:545–552. doi: 10.1016/0196-8904(92)90054-Z.
27. Kodama, M., H. Ikemoto, and S. Miyachi 1993. A new species of highly CO_2 tolerant fast growing marine microalga suitable for high-density culture. *Journal of Marine Biotechnology* 1:21–25.
28. Kratz, W. A., and J. Myers 1955. Nutrition and growth of several blue-green algae. *American Journal of Botany* 42:282–287. doi: 10.1002/j.1537-2197.1955.tb11120.x.
29. Allen, M. M., and R. Y. Stanier 1968. Selective isolation of blue-green algae from water and soil. *Microbiology* 51:203–209. doi: 10.1099/00221287-51-2-203.
30. Singh, S. P., and P. Singh 2014. Effect of CO_2 concentration on algal growth: A review. *Renewable and Sustainable Energy Reviews* 38:172–179.
31. Vocke, R. W., K. L. Sears, J. J. O'Toole, and R. B. Wildman 1980. Growth responses of selected freshwater algae to trace elements and scrubber ash slurry generated by coal-fired power plants. *Water Research* 14:141–150.
32. Kumar, K., C. N. Dasgupta, B. Nayak, P. Lindblad, and D. Das 2011. Development of suitable photobioreactors for CO_2 sequestration addressing global warming using green algae and cyanobacteria. *Bioresource Technology* 102:4945–4953. doi: 10.1016/j.biortech.2011.01.054.
33. Silva, C., et al. 2014. Commercial-scale biodiesel production from algae. *Industrial & Engineering Chemistry Research* 53:5311–5324. doi: 10.1021/ie403273b.
34. Csavina, J. L., B. J. Stuart, R. Guy Riefler, and M. L. Vis 2011. Growth optimization of algae for biodiesel production. *Journal of Applied Microbiology* 111:312–318. doi: 10.1111/j.1365-2672.2011.05064.x.
35. Georgianna, D. R., and S. P. Mayfield 2012. Exploiting diversity and synthetic biology for the production of algal biofuels. *Nature* 488:329–335. doi: 10.1038/nature11479.
36. Hanagata, N., T. Takeuchi, Y. Fukuju, D. J. Barnes, and I. Karube 1992. Tolerance of microalgae to high CO_2 and high temperature. *Phytochemistry* 31:3345–3348. doi: 10.1016/0031-9422(92)83682-O.
37. Maeda, K., M. Owada, N. Kimura, K. Omata, and I. Karube 1995. CO_2 fixation from the flue gas on coal-fired thermal power plant by microalgae. *Energy Conversion and Management* 36:717–720. doi: 10.1016/0196-8904(95)00105-M.
38. Tsuzuki, M., E. Ohnuma, N. Sato, T. Takaku, and A. Kawaguchi 1990. Effects of CO_2 concentration during growth on fatty acid composition in microalgae. *Plant Physiology* 93:851–856. doi: 10.1104/pp.93.3.851.

39. Radmann, E. M., F. V. Camerini, T. D. Santos, and J. A. Costa 2011. Isolation and application of SO_X and NO_X resistant microalgae in biofixation of CO_2 from thermoelectricity plants. *Energy Conversion and Management* 52:3132–3136.
40. Keffer, J. E., and G. T. Kleinheinz 2002. Use of *Chlorella vulgaris* for CO_2 mitigation in a photobioreactor. *Journal of Industrial Microbiology & Biotechnology* 29:275–280. doi: 10.1038/sj.jim.7000313.
41. Lee, J.-S., and J-P. Lee 2003. Review of advances in biological CO_2 mitigation technology. *Biotechnology and Bioprocess Engineering* 8:354. doi: 10.1007/BF02949279.
42. Sayre, R. 2010. Microalgae: The potential for carbon capture. *Bioscience* 60:722–727. doi: 10.1525/bio.2010.60.9.9.
43. Chen, H. W., T. S. Yang, M. J. Chen, Y. C. Chang, C. Y. Lin, I. Eugene, C. Wang, C. L. Ho, K. M. Huang, C. C. Yu, and F. L. Yang 2012. Application of power plant flue gas in a photobioreactor to grow Spirulina algae, and a bioactivity analysis of the algal water-soluble polysaccharides. *Bioresource Technology* 120:256–263.
44. de Morais, M. G., and J. A. Costa 2007. Isolation and selection of microalgae from coal fired thermoelectric power plant for biofixation of carbon dioxide. *Energy Conversion and Management* 48:2169–2173.
45. Wang, B., Y. Li, N. Wu, and C.Q. Lan 2008. CO_2 bio-mitigation using microalgae. *Applied Microbiology and Biotechnology* 79:707–718.
46. Ryan, C. 2009. *Cultivating Clean Energy. The Promise of Algae Biofuels*. Washington, DC: NRDC Publications.
47. Harun, R., M. Singh, G. M. Forde, and M. K. Danquah 2010. Bioprocess engineering of microalgaeto produce a variety of consumer products. *Renewable and Sustainable Energy Reviews* 14:1037–1047. doi: 10.1016/j.rser.2009.11.004.
48. Whitton, B. A. 2012. *Ecology of Cyanobacteria II: Their Diversity in Space and Time*. Dordrecht: Springer. doi: 10.1007/978-94-007-3855-3.
49. Searchinger, T., et al. 2008. Use of US croplands for biofuels increases greenhouse gases through emissions from land use change. *Science* 319:1238–1240. doi: 10.1126/science.1151861.
50. Hu, Q., et al. 2008. Microalgal triacylglycerols as feedstocks for biofuel production: Perspectives and advances. *Plant Journal* 54:621–639. doi: 10.1111/j.1365-313X.2008.03492.x.
51. Khan, S. A., Rashmi, M. Z. Hussain, S. Prasad, and U. C. Banerjee 2009. Prospects of biodiesel production from microalgae in India. *Renewable and Sustainable Energy Reviews* 13:2361–2372. doi: 10.1016/j.rser.2009.04.005.
52. Li, Y., M. Horsman, N. Wu, C. Q. Lan, and N. Dubois-Calero 2008. Biofuels from microalgae. *Biotechnology Progress* 24:815–820. doi: 10.1021/bp070371k.
53. Ho, S. H., W. M. Chen, and J. S. Chang 2010. Scenedesmus obliquus CNW-N as a potential candidate for CO_2 mitigation and biodiesel production. *Bioresource Technology* 101:8725–8730. doi: 10.1016/j.biortech.2010.06.112.
54. Sydney, E. B., et al. 2010. Potential carbon dioxide fixation by industrially important microalgae. *Bioresource Technology* 101:5892–5896. doi: 10.1016/j.biortech.2010.02.088.
55. Williams, P. J., and L. M. L. Laurens 2010. Microalgae as bio-diesel and biomass feedstocks: Review and analysis of the biochemistry, energetics and economics. *Energy & Environmental Science* 3:554–590. doi: 10.1039/b924978h.
56. Zhou, W., et al. 2017. Bio-mitigation of carbon dioxide using microalgal systems: Advances and perspectives. *Renewable and Sustainable Energy Reviews* 76:1163–1175. doi: 10.1016/j.rser.2017.03.065.
57. Pérez, A. T. E., M. Camargo, P. C. N. Rincón, and M. A. Marchant 2017. Key challenges and requirements for sustainable and industrialized biorefinery supply chain design and management: A bibliographic analysis. *Renewable and Sustainable Energy Reviews*. 69 350–359. doi: 10.1016/j.rser.2016.11.084.

58. Brown, M. 1991. The amino-acid and sugar composition of 16 species of microalgae used in mariculture. *Journal of Experimental Marine Biology and Ecology* 45:79–99. doi: 10.1016/0022-0981(91)90007-J.
59. Mata, T. M., A. A. Martins, and N. S. Caetano 2010. Microalgae for biodiesel production and other applications: A review. *Renewable and Sustainable Energy Reviews* 14:217–232. doi: 10.1016/j.rser.2009.07.020.
60. Singh, J., and D. W. Dhar 2019. Overview of carbon capture technology: Microalgal biorefinery concept and state-of-the-art. *Frontiers in Marine Science* 6:29. https://doi.org/10.3389/fmars.2019.00029URL=https://www.frontiersin.org/article/10.3389/fmars.2019.00029. doi: 10.3389/fmars.2019.00029, ISSN 2296-7745.
61. Choi, G.-G., B.-H. Kim, C.-Y. Ahn, and H.-M. Oh 2011. Effect of nitrogen limitation on oleic acid biosynthesis in *Botryococcus braunii. Journal of Applied Phycology* 23:1031–1037. doi: 10.1007/s10811-010-9636-1.
62. Gouveia, L., and A. C. Oliveira 2009. Microalgae as a raw material for biofuels production. *Journal of Industrial Microbiology & Biotechnology* 36:269–274. doi: 10.1007/s10295-008-0495-6.
63. Lam, M. K., K. T. Lee, and A. R. Mohamed 2012. Current status and challenges on microalgae-based carbon capture. *International Journal of Greenhouse Gas Control* 10:456–469. doi: 10.1016/j.ijggc.2012.07.010.
64. Gardner, R. D., et al. 2012. Use of sodium bicarbonate to stimulate triacylglycerol accumulation in the chlorophyte *Scenedesmus* sp. and the diatom *Phaeodactylum tricornutum. Journal of Applied Phycology* 24:1311–1320. doi: 10.1007/s10811-011-9782-0.
65. Gardner, R. D., J. Egan, E. J. Lohman, K. E. Cooksey, R. Robin Gerlach, and B. M. Peyton 2013. Cellular cycling, carbon utilization, and photosynthetic oxygen production during bicarbonate-induced triacylglycerol accumulation in a *Scenedesmus* sp. *Energies* 6:6060–6076. doi: 10.3390/en6116060.
66. Ghimire, A., et al. 2017. Bio-hythane production from microalgae biomass: Key challenges and potential opportunities for algal bio-refineries. *Bioresource Technology* 241:525–536. doi: 10.1016/j.biortech.2017.05.156.
67. Jankowska, E., S. Ashish, and O.-P. Piotr 2016. Biogas from microalgae: Review on microalgae's cultivation, harvesting and pretreatment for anaerobic digestion. *Renewable and Sustainable Energy Reviews* 75. doi: 10.1016/j.rser.2016.11.045.
68. Odjadjare, E. C., T. Mutanda, and A. O. Olaniran 2015. Potential biotechnological application of microalgae: A critical review. *Critical Reviews in Biotechnology* 37:37–52. doi: 10.3109/07388551.2015.1108956.
69. Jambo, S. A., et al. 2016. A review on third generation bioethanol feedstock. *Renewable and Sustainable Energy Reviews* 65:756–769. doi: 10.1016/j.rser.2016.07.064.
70. Markou, G., and E. Nerantzis 2013. Microalgae for high-value compounds and biofuels production: A review with focus on cultivation under stress conditions. *Biotechnology Advances* 31:1532–1542. doi: 10.1016/j.biotechadv.2013.07.011.
71. Yeong, T. K., K. Jiao, X. Zeng, L. Lin, S. Pan, and M. K. Danquah 2018. Microalgae for biobutanol production–Technology evaluation and value proposition. *Algal Research* 31 367–376. doi: 10.1016/j.algal.2018.02.029.
72. Cheng, H. H., et al. 2015. Biological butanol production from microalgae-based biodiesel residues by *Clostridium acetobutylicum. Bioresource Technology* 184:379–385. doi: 10.1016/j.biortech.2014.11.017.
73. Koller, M., A. Muhr, and G. Braunegg 2014. Microalgae as versatile cellular factories for valued products. *Algal Research* 6:52–63. doi: 10.1016/j.algal.2014.09.002.
74. 't Lam, G. P., M. H. Vermuë, M. H. M. Eppink, R. H. Wijffels, and C. van den Berg 2017. Multi-product microalgae biorefineries: From concept towards reality. *Trends in Biotechnology* 36:216–227. doi: 10.1016/j.tibtech.2017.10.011.

75. Vanthoor-Koopmans, M., R. H. Wijffels, M. J. Barbosa, and M. H. M. Eppink 2013. Biorefinery of microalgae for food and fuel. *Bioresource Technology* 135:142–149. doi: 10.1016/j.biortech.2012.10.135.
76. Han, W., C. Li, X. Miao, and G. Yu 2012. A novel miniature culture system to screen CO_2- sequestering microalgae. *Energies* 5:4372–4389. doi: 10.3390/en5114372.
77. Chen, B., C. Wan, M. A. Mehmood, J.-S. Chang, F. Bai, and X. Zhao 2017. Manipulating environmental stresses and stress tolerance of microalgae for enhanced production of lipids and value-added products – A review. *Bioresource Technology* 244(Pt 2):1198–1206. doi: 10.1016/j.biortech.2017.05.170.
78. Cheng, D., et al. 2017. Improving carbohydrate and starch accumulation in *Chlorella* sp. AE10 by a novel two-stage process with cell dilution. *Biotechnology Biofuels* 10:75. doi: 10.1186/s13068-017-0753-9.
79. Ng, I. S., S. I. Tan, P. H. Kao, Y. K. Chang, and J. S. Chang 2017. Recent developments on genetic engineering of microalgae for biofuels and bio-based chemicals. *Biotechnology Journal* 12:1600644. doi: 10.1002/biot.201600644.
80. Yang, B., et al. 2017. Genetic engineering of the Calvin cycle toward enhanced photosynthetic CO_2 fixation in microalgae. *Biotechnology for Biofuels* 10:229. doi: 10.1186/s13068-017-0916-8.
81. Kuo, C. M., et al. 2017. Ability of an alkali-tolerant mutant strain of the microalga *Chlorella* sp. AT1 to capture carbon dioxide for increasing carbon dioxide utilization efficiency. *Bioresource Technology* 244:243–251. doi: 10.1016/j.biortech.2017.07.096.
82. Kesaano, M., and R. Sims 2014. Algal biofilm based technology for wastewater treatment. *Algal Research* 5:231–240. doi: 10.1016/j.algal.2014.02.003.
83. Pires, J. C. M., M. C. M. Alvim-Ferraz, F. G. Martins, and M. Simoes 2012. Carbon dioxide capture from flue gases using microalgae: Engineering aspects and biorefinery concept. *Renewable and Sustainable Energy Reviews* 16:3043–3053. doi: 10.1016/j.rser.2012.02.055.
84. Koutra, E., C. N. Economou, P. Tsafrakidou, and M. Kornaros 2018. Bio-based products from microalgae cultivated in digestates. *Trends in Biotechnology* 36 819–833. doi: 10.1016/j.tibtech.2018.02.015.
85. Zhou, W., et al. 2012. Growing wastewater-born microalga *Auxenochlorella protothecoides* UMN280 on concentrated municipal wastewater for simultaneous nutrient removal and energy feedstock production. *Applied Energy* 98:433–440. doi: 10.1016/j.apenergy.2012.04.005.
86. Wang, J., W. Liu, and T. Liu 2017. Biofilm based attached cultivation technology for microalgal biorefineries—A review. *Bioresource Technology* 244:1245–1253. doi: 10.1016/j.biortech.2017.05.136.
87. Orr, V. C., and L. Rehmann 2016. Ionic liquids for the fractionation of microalgae biomass. *Current Opinion in Green and Sustainable Chemistry* 2:22–27. doi: 10.1016/j.cogsc.2016.09.006.
88. Wang, W.-N. 2014. Comparison of CO_2 photoreduction systems: A review. *Aerosol and Air Quality Research* 14:533–549. doi: 10.4209/aaqr.2013.09.0283.
89. Li, K., X. An, K. H. Park, M. Khraisheh, and J. Tang 2014. A critical review of CO_2 photoconversion: Catalysts and reactors. *Catalysis Today* 224:3–12. doi: 10.1016/j.cattod.2013.12.006.
90. Sorokin, C., and R. W. Krauss 1958. The effects of light intensity on the growth rates of green algae. *Plant Physiology* 33:109–113. doi: 10.1104/pp.33.2.109.
91. Xu, Y., I. M. Ibrahim, and P. J. Harvey 2016. The influence of photoperiod and light intensity on the growth and photosynthesis of *Dunaliella salina* (chlorophyta) CCAP 19/30. *Plant Physiology and Biochemistry* 106:305–315. doi: 10.1016/j.plaphy.2016.05.021.

92. Cloot, A. 1994. Effect of light intensity variations on the rate of photosynthesis of algae: A dynamical approach. *Mathematical and Computer Modelling* 19:23–33. doi: 10.1016/0895-7177(94)90038-8.
93. Anbalagan, A., et al. 2017. Continuous photosynthetic abatement of CO_2 and volatile organic compounds from exhaust gas coupled to wastewater treatment: Evaluation of tubular algal-bacterial photobioreactor. *Journal of CO_2 Utilization* 21:353–359. doi: 10.1016/j.jcou.2017.07.016.
94. Tang, D., W. Han, P. Li, X. Miao, and J. Zhong 2011. CO_2 biofixation and fatty acid composition of *Scenedesmus obliquus* and *Chlorella pyrenoidosa* in response to different CO_2 levels. *Bioresource Technology* 102:3071–3076. doi: 10.1016/j.biortech.2010.10.047.
95. Zelitch, I. 1975. Improving the efficiency of photosynthesis. *Science* 188:626–633.
96. Wilcox, J. 2012. *Carbon Capture.* New York, NY: Springer-Verlag.
97. Benedetti, M., V. Vecchi, S. Barera, and L. Dall'Osto 2018. Biomass from microalgae: The potential of domestication towards sustainable biofactories. *Microbial Cell Factories* 17:173. doi: 10.1186/s12934-018-1019-3.
98. Abishek, M. P., J. Patel, and A. P. Rajan 2014. Algae oil: A sustainable renewable fuel of future. *Biotechnology Research International* 2014:272814. doi: 10.1155/2014/272814.
99. Bleakley, S., and M. Hayes 2017. Algal proteins: Extraction, application, and challenges concerning production. *Foods* 6(5):33. doi: 10.3390/foods6050033.
100. Maurya, R., et al. 2016. Applications of de-oiled microalgal biomass towards development of sustainable biorefinery. *Bioresource Technology* 214:787–796. doi: 10.1016/j.biortech.2016.04.115.
101. Sarkar, O., M. Agarwal, A. Naresh Kumar, and S. Venkata Mohan 2015. Retrofitting hetrotrophically cultivated algae biomass as pyrolytic feedstock for biogas, bio-char and bio-oil production encompassing biorefinery. *Bioresource Technology* 178:132–138. doi: 10.1016/j.biortech.2014.09.070.
102. Zhang, Y., and A. Kendall 2019. Consequential analysis of algal biofuels: Benefits to ocean resources. *Journal of Cleaner Production* 231:35–42. doi: 10.1016/j.jclepro.2019.05.057.
103. Dong, T., E. P. Knoshaug, P. T. Pienkos, and L. M. L. Laurens 2016. Lipid recovery from wet oleaginous microbial biomass for biofuel production: A critical review. *Applied Energy* 177:879–895. doi: 10.1016/j.apenergy.2016.06.002.
104. Moreira, D., and Pires, J. C. M. 2016. Atmospheric CO_2 capture by algae: Negative carbon dioxide emission path. *Bioresource Technology*:215:371–379. doi: 10.1016/j.biortech.2016.03.060.
105. Schenk, P. M., Thomas-Hall, S. R., Stephens, E. et al. 2008. Second generation biofuels: high-efficiency microalgae for biodiesel production. *Bioenergy Research* 1:20–43. doi:10.1007/s12155-008-9008-8.
106. Nigam, P. S., and A. Singh 2011. Production of liquid biofuels from renewable resources. *Progress in Energy and Combustion Science* 37:52–68. doi: 10.1016/j.pecs.2010.01.003.
107. Carlsson, A. S., and D. J. Bowles 2007. *Micro- and Macro-algae: Utility for Industrial Applications : Outputs from the EPOBIO Project.* Speen; Newbury: CPL Press.
108. Kondaveeti, S., K. S. Choi, R. Kakarla, and B. Min 2014a. Microalgae *Scenedesmus obliquus* as renewable biomass feedstock for electricity generation in microbial fuel cells (MFCs). *Frontiers of Environmental Science & Engineering* 8:784–791. doi: 10.1007/s11783-013-0590-4.
109. Zhu, L. D., Z. H. Li, and E. Hiltunen 2016. Strategies for lipid production improvement in microalgae as a biodiesel feedstock. *BioMed Research International* 2016:8792548. doi: 10.1155/2016/8792548.
110. Xin, L., H. Hong-Ying, and Z. Yu-Ping 2011. Growth and lipid accumulation properties of a freshwater microalga *Scenedesmus* sp. under different cultivation temperature. *Bioresource Technology* 102:3098–3102. doi: 10.1016/j.biortech.2010.10.055.

111. Benemann, J. R., and W. J. Oswald 1996. Systems and economic analysis of microalgae ponds for conversion of CO_2 to biomass. Final report. Berkeley, CA: California Univ., Dept. of Civil Engineering. doi: 10.2172/493389.
112. Nagarajan, S., S. K. Chou, S. Cao, C. Wu, and Z. Zhou 2013. An updated comprehensive techno-economic analysis of algae biodiesel. *Bioresource Technology* 145:150–156. doi: 10.1016/j.biortech.2012.11.108.
113. Hatch, M., and C. Slack 1970. Photosynthetic CO_2-fixation pathways. *Annual Review of Plant Physiology* 21:141–162. doi: 10.1146/annurev.pp.21.060170.001041.
114. Pfennig, N. 1967. Photosynthetic bacteria. *Annual Review of Microbiology* 21:285–324. doi: 10.1146/annurev.mi.21.100167.001441.
115. Melis, A. 2009. Solar energy conversion efficiencies in photosynthesis: Minimizing the chlorophyll antennae to maximize efficiency. *Plant Science* 177:272–280.
116. Markley, A. L., M. B. Begemann, R. E. Clarke, G. C. Gordon, and B. F. Pfleger 2015. Synthetic biology toolbox for controlling gene expression in the cyanobacterium *Synechococcus* sp. strain PCC 7002. *ACS Synthetic Biology* 4:595–603.
117. Berla, B. M., R. Saha, C. M. Immethun, C. D. Maranas, T. S. Moon, and H. B. Pakrasi 2013. Synthetic biology of cyanobacteria: Unique challenges and opportunities. *Frontiers in Microbiology* 4:246.
118. Lau, N-S., M. Matsui, and A. A.-A. Abdullah 2015. Cyanobacteria: Photoautotrophic microbial factories for the sustainable synthesis of industrial products. *BioMed Research International* 2015:754934. doi: 10.1155/2015/754934.
119. Schirmer, A., M. A. Rude, X. Li, E. Popova, and S. B. Del Cardayre 2010. Microbial biosynthesis of alkanes. *Science* 329:559–562. doi: 10.1126/science.1187936.
120. van Haveren, J., E. L. Scott, and J. Sanders 2008. Bulk chemicals from biomass. *Biofuels, Bioproducts and Biorefining* 2:41–57.
121. Tran, A. V., and R. P. Chambers 1987. The dehydration of fermentative 2,3-butanediol into methyl ethyl ketone. *Biotechnology and Bioengineering* 29:343–351.
122. Dexter, J., and Fu, P. 2009. Metabolic engineering of cyanobacteria for ethanol production. *Energy & Environmental Science* 2:857–864. doi: 10.1039/b811937f.
123. Lambert, G. R., and Smith, G. D. 1977. Hydrogen formation by marine blue—green algae. *FEBS Letters* 83:159–162. doi: 10.1016/0014-5793(77)80664-9.
124. Tamagnini, P., R. Axelsson, P. Lindberg, F. Oxelfelt, R. Wünschiers, and P. Lindblad 2002. Hydrogenases and hydrogen metabolism of cyanobacteria. *Microbiology and Molecular Biology Reviews* 66:1–20. doi: 10.1128/MMBR.66.1.1-20.2002.
125. Benemann, J. R., P. Pursoff, and W. J. Oswald 1978. Engineering design and cost analysis of a large-scale microalgae biomass system, Final Report to the U.S. Energy Department, NTIS #H CP/T1605-01 UC-61, 91 p.
126. Martínez-Jerónimo, F., and F. Espinosa-Chávez 1994. A laboratory-scale system for mass culture of freshwater microalgae in polyethylene bags. *Journal of Applied Phycology* 6:423–425.
127. Richardson, J. W., M. D. Johnson, and J. L. Outlaw 2012. Economic comparison of open pond raceways to photo bio-reactors for profitable production of algae for transportation fuels in the Southwest. *Algal Research* 1:93–100.
128. Wagner, T., U. Ermler, and S. Shima 2016. The methanogenic CO_2 reducing-and-fixing enzyme is bifunctional and contains 46 [4Fe-4S] clusters. *Science* 354:114–117.
129. Ulmer, U., et al. 2019. Fundamentals and applications of photocatalytic CO_2 methanation. *Nature Communications* 10:3169. https://doi.org/10.1038/s41467-019-10996-2
130. Götz, M., et al. 2016. Renewable power-to-gas: A technological and economic review. *Renewable Energy* 85:1371–1390.
131. Zabranska, J., and D. Pokorna 2018. Bioconversion of carbon dioxide to methane using hydrogen and hydrogenotrophic methanogens. *Biotechnology Advances* 36:707–720.

132. Liao, J. C., L. Mi, S. Pontrelli, and S. Luo 2016. Fuelling the future: Microbial engineering for the production of sustainable biofuels. *Nature Reviews Microbiology* 14:288–304.
133. Shima, S., E. Warkentin, R. K. Thauer, and U. Ermler 2002. Structure and function of enzymes involved in the methanogenic pathway utilizing carbon dioxide and molecular hydrogen. *Journal of Bioscience and Bioengineering* 93:519–530.
134. Pang, H., T. Masuda, and J. Ye 2018. Semiconductor-based photoelectrochemical conversion of carbon dioxide: Stepping towards artificial photosynthesis. *Chemistry: An Asian Journal* 13:127–142.
135. Nichols, E. M., et al. 2015. Hybrid bioinorganic approach to solar-to-chemical conversion. *Proceedings of the National Academy of Sciences of the USA* 112:11461–11466.
136. Mersch, D., et al. 2015. Wiring of photosystem II to hydrogenase for photoelectrochemical water splitting. *Journal of the American Chemical Society* 137:8541–8549.
137. Bailera, M., P. Lisbona, L. M. Romeo, and S. Espatolero 2017. Power to gas projects review: Lab, pilot and demo plants for storing renewable energy and CO_2. *Renewable and Sustainable Energy Reviews* 69:292–312.
138. Sakimoto, K. K., Kornienko, N., and P. Yang 2017. Cyborgian material design for solar fuel production: The emerging photosynthetic biohybrid systems. *Accounts of Chemical Research* 50:476–481.
139. Haas, T., R. Krause, R. Weber, M. Demler, and G. Schmid 2018. Technical photosynthesis involving CO_2 electrolysis and fermentation. *Nature Catalysis* 1:32–39.
140. Logan, B. E. 2008. *Microbial Fuel Cells.* Hoboken, NJ: John Wiley and sons, Inc.
141. Rabaey, K., and R. A. Rozendal 2010. Microbial electrosynthesis — Revisiting the electrical route for microbial production. *Nature Reviews Microbiology* 8:706–716. doi: 10.1038/nrmicro2422.
142. Kadier, A., et al. 2020. Biorefinery perspectives of microbial electrolysis cells (MECs) for hydrogen and valuable chemicals production through wastewater treatment. *Biofuel Research Journal* 7:1128–1142. doi: 10.18331/BRJ2020.7.1.5.
143. Nevin, K. P., T. L. Woodard, A. E. Franks, Z. M. Summers, and D. R. Lovley 2010. Microbial electrosynthesis: Feeding microbes electricity to convert carbon dioxide and water to multicarbon extracellular organic compounds. *mBio* 1:e00103–10. doi: 10.1128/mBio.00103-10.
144. Nevin, K. P., S. A. Hensley, A. E. Franks, Z. M. Summers, J. Ou, T. L. Woodard, O. L. Snoeyenbos-West, and D. R. Lovley. 2011. Electrosynthesis of organic compounds from carbon dioxide is catalyzed by a diversity of acetogenic microorganisms. *Applied and Environmental Microbiology* 77(9):2882–2886. doi:10.1128/AEM.02642-10.
145. Zhang, T., et al. 2013. Improved cathode materials for microbial electrosynthesis. *Energy & Environmental Science* 6:217–224. doi: 10.1039/C2EE23350A.
146. Mohanakrishna, G., J. S. Seelam, K. Vanbroekhoven, and D. Pant 2015. An enriched electroactive homoacetogenic biocathode for the microbial electrosynthesis of acetate through carbon dioxide reduction. *Faraday Discuss* 183:445–462. doi: 10.1039/C5FD00041F.
147. Batlle-Vilanova, P., S. Puig, R. Gonzalez-Olmos, M. D. Balaguer, and J. Colprim 2016. Continuous acetate production through microbial electrosynthesis from CO_2 with microbial mixed culture. *Journal of Chemical Technology & Biotechnology* 91:921–927. doi: 10.1002/jctb.4657.
148. Marshall, C. W., E. V. LaBelle, and H. D. May 2013. Production of fuels and chemicals from waste by microbiomes. *Current Opinion in Biotechnology* 24:391–397. doi: 10.1016/j.copbio.2013.03.016.
149. Gildemyn, S., K. Verbeeck, R. Slabbinck, S. J. Andersen, A. Prévoteau, and K. Rabaey 2015. Integrated production, extraction, and concentration of acetic acid from CO_2 through microbial electrosynthesis. *Environmental Science & Technology Letters* 2:325–328. doi: 10.1021/acs.estlett.5b00212.

150. LaBelle, E. V., and H. D. May 2017. Energy efficiency and productivity enhancement of microbial electrosynthesis of acetate. *Frontiers in Microbiology* 8:756. doi: 10.3389/fmicb.2017.00756.
151. Bajracharya, S., et al. 2015. Carbon dioxide reduction by mixed and pure cultures in microbial electrosynthesis using an assembly of graphite felt and stainless steel as a cathode. *Bioresource Technology* 195:14–24. doi: 10.1016/j.biortech.2015.05.081.
152. Will, F. G. 1965. Hydrogen adsorption on platinum single crystal electrodes I. Isotherms and heats of adsorption. *Journal of the Electrochemical Society* 112:451–455. doi: 10.1149/1.2423567.
153. Mohanakrishna, G., K. Vanbroekhoven, and D. Pant 2016. Imperative role of applied potential and inorganic carbon source on acetate production through microbial electrosynthesis. *Journal of CO_2 Utilization* 15(Suppl. C):57–64. doi: 10.1016/j.jcou.2016.03.003.
154. Van Eerten-Jansen, M. C. A. A., et al. 2013. Bioelectrochemical production of caproate and caprylate from acetate by mixed cultures. *ACS Sustainable Chemistry & Engineering* 1:513–518. doi: 10.1021/sc300168z.
155. Vassilev, I., et al. 2019. Microbial electrosynthesis system with dual biocathode arrangement for simultaneous acetogenesis, solventogenesis and carbon chain elongation. *Chemical Communications* 55:4351–4354. doi: 10.1039/C9CC00208A.
156. Vassilev, I., et al. 2018. Microbial electrosynthesis of isobutyric, butyric, caproic acids, and corresponding alcohols from carbon dioxide. *ACS Sustainable Chemistry & Engineering* 6:8485–8493. doi: 10.1021/acssuschemeng.8b00739.
157. del Pilar Anzola Rojas, M., M. Zaiat, E. R. Gonzalez, H. De Wever, and D. Pant 2018. Effect of the electric supply interruption on a microbial electrosynthesis system converting inorganic carbon into acetate. *Bioresource Technology* 266:203–210. doi: 10.1016/j.biortech.2018.06.074.
158. Villano, M., F. Aulenta, C. Ciucci, T. Ferri, A. Giuliano, and M. Majone 2010. Bioelectrochemical reduction of CO_2 to CH_4 via direct and indirect extracellular electron transfer by a hydrogenophilic methanogenic culture. *Bioresource Technology* 101:3085–3090. doi: 10.1016/j.biortech.2009.12.077.
159. Villano, M., S. Scardala, F. Aulenta, and M. Majone 2013. Carbon and nitrogen removal and enhanced methane production in a microbial electrolysis cell. *Bioresource Technology* 130:366–371. doi: 10.1016/j.biortech.2012.11.080.
160. Yasin, M., Ketema, T., and Bacha, K. 2015. Physico-chemical and bacteriological quality of drinking water of different sources, Jimma zone, Southwest Ethiopia. *BMC Research Notes* 8:541. doi:10.1186/s13104-015-1376-5.
161. Christodoulou, X., T. Okoroafor, S. Parry, and S. B. Velasquez-Orta 2017. The use of carbon dioxide in microbial electrosynthesis: Advancements, sustainability and economic feasibility. *Journal of CO_2 Utilization.* 18(Suppl. C):390–399. doi: 10.1016/j.jcou.2017.01.027.
162. Blankenship, R. E., D. M. Tiede, J. Barber, G. W. Brudvig, G. Fleming, M. Ghirardi, M. R. Gunner, W. Junge, D. M. Kramer, A. Melis, T. A. Moore, C. C. Moser, D. G. Nocera, A. J. Nozik, D. R. Ort, W. W. Parson, R. C. Prince, and R. T. Sayre 2011. Comparing photosynthetic and photovoltaic efficiencies and recognizing the potential for improvement. *Science* 332:805–809.
163. Lovley, D. R. 2011. Powering microbes with electricity: Direct electron transfer from electrodes to microbes. *Environmental Microbiology Reports* 3:27–35.
164. Ross, D. E., J. M. Flynn, D. B. Baron, J. A. Gralnick, and D. R. Bond 2011. Towards electrosynthesis in *Shewanella*: Energetics of reversing the Mtr pathway for reductive metabolism. *PLoS One* 6(2):e16649. https://doi.org/10.1371/journal.pone.0016649.

165. Lovley, D. R., and K. P. Nevin. 2013. Electrobiocommodities: powering microbial production of fuels and commodity chemicals from carbon dioxide with electricity. *Current Opinion in Biotechnology* 24(3):385–390. doi:10.1016/j.copbio.2013.02.012.
166. Lee, H. J., J. Choi, S. M. Lee, Y. Um, S. J. Sim, Y. Kim, and H. M. Woo 2017a. Photosynthetic CO_2 conversion to fatty acid ethyl esters (FAEEs) using engineered cyanobacteria. *Journal of Agricultural and Food Chemistry* 65:1087–1092.
167. Lee, H. J., J. Lee, S. M. Lee, Y. Um, Y. Kim, S. J. Sim, J. I. Choi, and H. M. Woo 2017b. Direct conversion of CO2 to alpha-farnesene using metabolically engineered *Synechococcus elongatus* PCC 7942. *Journal of Agricultural and Food Chemistry* 65:10424–10428.
168. Lee, S. J., D. Y. Lee, T. Y. Kim, B. H. Kim, J. Lee, and S. Y. Lee 2005. Metabolic engineering of *Escherichia coli* for enhanced production of succinic acid, based on genome comparison and in silico gene knockout simulation. *Applied and Environmental Microbiology* 71:7880–7887.
169. Tashiro, Y., S. Hirano, M. M. Matson, S. Atsumi, and A. Kondo 2018. Electrical-biologcal hybrid system for CO_2 reduction. *Metabolic Engineering* 47:211–218.
170. 2021. Calvin Cycle, Wikipedia, The free encyclopedia, last visited April, 8, 2021.
171. Li, Z., et al. 2020. Engineering the Calvin–Benson–Bassham cycle and hydrogen utilization pathway of Ralstonia eutropha for improved autotrophic growth and polyhydroxybutyrate production. *Microbial Cell Factories* 19:228. https://doi.org/10.1186/s12934-020-01494-y.
172. Willey, J. M., L. M. Sherwood, and C. J. Woolverton 2014. *Prescott's Microbiology*. New York, NY: McGraw-Hill Education.
173. Cai, Z., G. Liu, J. Zhang, and Y. Li 2014. Development of an activity-directed selection system enabled significant improvement of the carboxylation efficiency of Rubisco. *Protein Cell* 5:552–562.
174. Myat, T. L., A. Occhialini, P. JohnAndralojc, A. J. P. Martin, and M. R. Hanson 2014. A faster Rubisco with potential to increase photosynthesis in crops. *Nature* 513:547–550.
175. Lee, H.-M., B.-Y. Jeon, and M.-K. Oh 2016. Microbial production of ethanol from acetate by engineered Ralstonia eutropha. *Biotechnology and Bioprocess Engineering* 21:402–407.
176. Marc, J., E. Grousseau, E. Lombard, A. J. Sinskey, N. Gorret, and S. E. Guillouet 2017. Over expression of GroESL in Cupriavidus necator for heterotrophic and autotrophic isopropanol production. *Metabolic Engineering* 42:74–84.
177. Chen, J. S., J. C. Way, B. Dusel, and J. P. Torella 2015. Production of fatty acids in *Ralstonia eutropha* H16 by engineering beta-oxidation and carbon storage. *PeerJ* 3:e1468.
178. Bi, C., P. Su, J. Müller, Y. C. Yeh, S. R. Chhabra, H. R. Beller, S. W. Singer, and N. J. Hillson 2013. Development of a broad-host synthetic biology toolbox for *Ralstonia eutropha* and its application to engineering hydrocarbon biofuel production. *Microbial Cell Factories* 12:1–10.
179. Kitadai, N., M. Kameya, and K. Fujishima 2017. Origin of the reductive tricarboxylic anoricid (rTCA) cycle-type CO_2 fixation: A perspective. *Life (Basel)* 7(4):39. doi: 10.3390/life7040039.
180. 2021. Reverse Krebs Cycle, Wikipedia, The free encyclopedia, last visited 2, Feb., 2021.
181. Ragsdale, S.W. 2003. Pyruvate ferredoxin oxidoreductase and its radical intermediate. *Chemical Reviews* 103:2333–2346.
182. Aoshima, M., M. Ishii, and Y. A. Igarashi 2004. Novel biotin protein required for reductive carboxylation of 2-oxoglutarate by isocitrate dehydrogenase in *Hydrogenobacter thermophiles* TK-6. *Molecular Microbiology* 51:791–798.

183. Weiss, M.C., F.L. Sousa, N. Mrnjavac, S. Neukirchen, M. Roettger, S. Nelson-Sathi, and W. F. Martin 2016. The physiologi and habitat of the last universal common ancestor. *Nature Microbiology* 1:16116.
184. Duclos, J. M., and P. Haake 1974. Ring opening of thiamine analogs. The role of ring opening in physiological function. *Biochemistry* 13:5358–5362.
185. Orgel, L.E. 2004. Prebiotic chemistry and the origin of the RNA world. *Critical Reviews in Biochemistry and Molecular Biology* 39:99–123.
186. Borrel, G., P. S. Adam, and S. Gribaldo 2016. Methanogenesis and the Wood–Ljungdahl pathway: An ancient, versatile, and fragile association. *Genome Biology and Evolution* 8(6):1706–1711. doi: 10.1093/gbe/evw114.
187. 2021. Wood-Ljungdahl pathway, Wikipedia, The free encyclopedia, last visited 5 Dec., 2020.
188. Ljungdahl, L., and H. G. Wood 1965. Incorporation of C^{14} from carbon dioxide into sugar phosphates, carboxylic acids, and amino acids by clostridium thermoaceticum. *Journal of. Bacteriology* 89:1055–1064.
189. Berg, I. A. 2011. Ecological aspects of the distribution of different autotrophic CO_2 fixation pathways. *Applied and Environmental Microbiology* 77:1925–1936. doi: 10.1128/AEM.02473-10.
190. Fuchs, G. 2011. Alternative pathways of carbon dioxide fixation: Insights into the early evolution of life? *Annual Review of Microbiology* 65:631–658 doi: 10.1146/annurev-micro-090110-102801.
191. Hattori, S., A. S. Galushko, Y. Kamagata, and B. Schink 2005. Operation of the CO dehydrogenase/acetyl coenzyme A pathway in both acetate oxidation and acetate formation by the syntrophically acetate-oxidizing bacterium Thermacetogenium phaeum. *Journal of Bacteriology* 187:3471–3476. doi: 10.1128/JB.187.10.3471-3476.2005.
192. 2020. 3-Hydroxypropionate bicycle, Wikipedia, The free encyclopedia, last visited 22, June 2020.
193. Hawkins, A. B., M. W. W._Adams, and R. M._Kelly 2014. Conversion of 4-hydroxybutyrate to acetyl coenzyme A and its anapleurosis in the Metallosphaera sedula 3-hydroxyp ropionate/4-hydroxybutyrate carbon fixation pathway [H. Nojiri, ed.], *Applied and Environmental Microbiology* 80(8):2536–2545. doi: 10.1128/AEM.04146-13.
194. Berg, I. A., D. Kockelkorn, W. Buckel, and G. Fuchs 2007. A 3-hydroxypropionate/4-hydroxybutyrate autotrophic carbon dioxide assimilation pathway in Ar- chaea. *Science* 318:1782–1786. http://dx.doi.org/10.1126/science.1149976.
195. Huber, H., M. Gallenberger, U. Jahn, E. Eylert, I. A. Berg, D. Kockelkorn, W. Eisenreich, and G. Fuchs 2008. A dicarboxylate/4-hydroxybutyrate autotrophic carbon assimilation cycle in the hyperthermophilic archaeum Ignicoccus hospitalis. *Proceedings of the National Academy of Sciences of the USA* 105:7851–7856. http: //dx.doi.org/10.1073/pnas.0801043105.
196. Jahn, U., H. Huber, W. Eisenreich, M. Hugler, and G. Fuchs 2007. Insights into the autotrophic CO_2 fixation pathway of the archaeon *Ignicoccus hospitalis*: Comprehensive analysis of the central carbon metabolism. *Journal of Bacteriology* 189:4108–4119. http://dx.doi.org/10.1128/JB.00047-07.
197. Estelmann, S., M. Hugler, W. Eisenreich, K. Werner, I. A. Berg, W. H. Ramos-Vera, R. F. Say, D. Kockelkorn, N. Gad'on, and G. Fuchs 2011. Labeling and enzyme studies of the central carbon metabolism in *Metallosphaera sedula*. *Journal of Bacteriology* 193:1191–1200. http://dx.doi.org/10.1128/JB.01155-10.
198. Loder, A. J., Y. Han, A. B._Hawkins, H._Lian, G. L._Lipscomb, G. J._Schut, M. W._Keller, M. W.W._Adams, and R. M._Kelly 2016. Reaction kinetic analysis of the 3-hydroxypro pionate/4-hydroxybutyrate CO_2 fixation cycle in extremely thermoacidophilic archaea. *Metabolic Engineering* 38: 446–463. doi: 10.1016/j.ymben.2016.10.009.

199. Lawton, T. J., and Rosenzweig, A. C. 2016. Biocatalysts for methane conversion: Big progress on breaking a small substrate. *Current Opinion in Chemical Biology* 35:142–149. doi: 10.1016/j.cbpa.2016.10.001.
200. Wang, B., S. Albarracín-Suazo, Y. Pagán-Torres, and E. Nikolla 2017. Advances in methane conversion processes. *Catalysis Today* 285:147–158. doi: 10.1016/j.cattod.2017.01.023.
201. Strong, P. J., M. Kalyuzhnaya, J. Silverman, and W. P. Clarke 2016. A methanotroph-based biorefinery: Potential scenarios for generating multiple products from a single fermentation. *Bioresource Technology* 215:314–323. doi: 10.1016/j.biortech.2016.04.099.
202. Bian, B., et al. 2018. Porous nickel hollow fiber cathodes coated with CNTs for efficient microbial electrosynthesis of acetate from CO_2 using *Sporomusa ovata*. *Journal of Materials Chemistry A* 6:17201–17211. doi: 10.1039/C8TA05322G.
203. Lee, J. C., J. H. Kim, W. S. Chang, and D. Pak 2012. Biological conversion of CO_2 to CH_4 using hydrogenotrophic methanogen in a fixed bed reactor. *Journal of Chemical Technology & Biotechnology* 87:844–847. doi: 10.1002/jctb.3787.
204. Scherer, P. A., G.-R. Vollmer, T. Fakhouri, and S. Martensen 2000. Development of a methanogenic process to degrade exhaustively the organic fraction of municipal "grey waste" under thermophilic and hyperthermophilic conditions. *Water Science and Technology* 41:83–91. doi: 10.2166/wst.2000.0059.
205. Shah, Y.T. 2014. *Water for Energy and Fuel Production*. New York, NY: CRC Press, Taylor and Francis CO.
206. Ahring, B. K., A. A. Ibrahim, and Z. Mladenovska 2001. Effect of temperature increase from 55 to 65°C on performance and microbial population dynamics of an anaerobic reactor treating cattle manure. *Water Research* 35:2446–2452. doi: 10.1016/S0043-1354(00)00526-1.
207. Conrad, R., and M. Klose 1999. Anaerobic conversion of carbon dioxide to methane, acetate and propionate on washed rice roots. *FEMS Microbiology Ecology* 30(2):147–155. https://doi.org/10.1111/j.1574-6941.1999.tb00643.x
208. Thauer, R. K. 2007. Microbiology – A fifth pathway of carbon fixation. *Science* 318:1732–1733. doi: 10.1126/science.1152209.
209. De Klerk, A., Y. W. Li, and R. Zennaro 2013. *Greener Fischer-Tropsch Processes for Fuels and Feedstocks*. Weinheim: John Wiley & Sons, Inc.
210. Liew, F., M. E. Martin, R. C. Tappel, B. D. Heijstra, C. Mihalcea, and M. Köpke 2016. Gas fermentation—A flexible platform for commercial scale production of low-carbon-fuels and chemicals from waste and renewable feedstocks. *Frontiers in Microbiology* 7:694. doi: 10.3389/fmicb.2016.00694.
211. Zheng, J. L., M. Q. Zhu, J. L. Wen, and R. C. Sun 2016. Gasification of bio-oil: Effects of equivalence ratio and gasifying agents on product distribution and gasification efficiency. *Bioresource Technology* 211:164–172. doi: 10.1016/j.biortech.2016.03.088.
212. National Energy Technology Laboratory 2013. *Water Gas Shift & Hydrogen Production*. Available at: http://www.netl.doe.gov/research/coal/energy-systems/gasification/gasifipedia/water-gas-shift (accessed March 30, 2016).
213. Klasson, K. T., M. D. Ackerson, E. C. Clausen, and J. L. Gaddy 1991. Bioreactor design for synthesis gas fermentations. *Fuel* 70 605–614. doi: 10.1016/0016-2361(91)90174-9.
214. Klasson, K. T., M. D. Ackerson, E. C. Clausen, and J. L. Gaddy 1992. Bioconversion of synthesis gas into liquid or gasesous fuels. *Enzyme and Microbial Technology* 14:602–608. doi: 10.1016/0141-0229(92)90033-K.
215. Michael, K., N. Steffi, and D. Peter 2011. The past, present, and future of biofuels – Biobutanol as promising alternative, In: dos Santos Bernades, M. A., ed., *Biofuel Production-Recent Developments and Prospects*. Rijeka: InTech, pp. 451–486.

216. Gaddy, J. L., et al. 2007. *Methods for Increasing the Production of Ethanol from Microbial Fermentation*. US7285402. Washington, DC: U.S. Patent and Trademark Office.
217. Tremblay, P.-L., and T. Zhang 2015. Electrifying microbes for the production of chemicals. *Frontiers in Microbiology* 6:1–10. doi: 10.3389/fmicb.2015.00201.
218. Sakimoto, Kelsey K., Andrew Barnabas Wong, and Peidong Yang. 2016. Self-photosensitization of nonphotosynthetic bacteria for solar-to-chemical production. *Science* 351(6268):74–77. doi:10.1126/science.aad3317.
219. Imkamp, F., and V. Müller 2007. Acetogenic bacteria, In *Encyclopedia of Life Sciences*. Chichester, UK: John Wiley & Sons, Ltd. doi: 10.1002/9780470015902.a0020086.
220. Drake, H. L., A. S. Gössner, and S. L. Daniel 2008. Old acetogens, new light. *Annals of the New York Academy of Sciences* 1125:100–128. doi: 10.1196/annals.1419.016.
221. Fast, A. G., and E. T. Papoutsakis 2012. Stoichiometric and energetic analyses of non-photosynthetic CO_2-fixation pathways to support synthetic biology strategies for production of fuels and chemicals. *Current Opinion in Chemical Engineering* 1:380–395. doi: 10.1016/j.coche.2012.07.005.
222. Drake, H. L., K. Küsel, and C. Matthies 2006. Acetogenic prokaryotes, In *The Prokaryotes*, Dworkin, M., Falkow, S., Rosenberg, E., Schleifer, K.-H., and Stackebrandt, E., eds., New York, NY: Springer, pp. 354–420. doi: 10.1007/0-387-30742-7_13.
223. Fast, A. G., E. D. Schmidt, S. W. Jones, and B. P. Tracy 2015. Acetogenic mixotrophy: Novel options for yield improvement in biofuels and biochemicals production. *Current Opinion in Biotechnology* 33:60–72. doi: 10.1016/j.copbio.2014.11.014.
224. Richter, H., S. E. Loftus, and L. T. Angenent 2013a. Integrating syngas fermentation with the carboxylate platform and yeast fermentation to reduce medium cost and improve biofuel productivity. *Environmental Technology* 34:1983–1994. doi: 10.1080/09593330.2013.826255.
225. Richter, H., M. Martin, and L. Angenent 2013b. A two-stage continuous fermentation system for conversion of syngas into ethanol. *Energies* 6:3987–4000. doi: 10.3390/en6083987.
226. Martin, M. E., Richter, H., Saha, S., and Angenent, L. T. (2016). Traits of selected clostridium strains and for syngas fermentation to ethanol. *Biotechnology and Bioengineering* 113:531–539. doi: 10.1002/bit.25827.
227. Hu, P., et al. 2016. Integrated bioprocess for conversion of gaseous substrates to liquids. *Proceedings of the National Academy of Sciences of the U.S.A.* 113:3773–3778. doi: 10.1073/pnas.1516867113.
228. Lane, J. 2016. *Coskata's Technology Re-emerges as Synata Bio*. Available at: http://www.biofuelsdigest.com/bdigest/2016/01/24/coskatas-technology-re-emerges-as-synata-bio/ (accessed March 30, 2016).
229. Schill, S. R. 2013. Ineos declares commercial cellulosic ethanol online in Florida, In *Ethanol Prod. Mag.* Available at: http://www.ethanolproducer.com/articles/10096/ineos-declares-commercial-cellulosic-ethanol-online-in-florida (accessed March 30, 2016).
230. INEOS Bio 2014. *INEOS Bio Provides Operational Update*. Available at: http://www.ineos.com/businesses/ineos-bio/news/ineos-bio-provides-operational-update (accessed January 27, 2016).
231. Lane, J. 2015b. China steel green-lights commercial-scale LanzaTech advanced biofuels project, In *Biofuels Dig.* Available at: http://www.biofuelsdigest.com/bdigest/2015/04/22/china-steel-green-lights-46m-for-commercial-scale-lanzatech-advanced-biofuels-project/ (accessed January 27, 2016).
232. Lane, J. 2015c. Steel's Big Dog jumps into low carbon fuels: ArcelorMittal, LanzaTech, Primetals Technologies to construct $96M biofuel production facility, in *Biofuels Dig.* Available at: http://www.biofuelsdigest.com/bdigest/2015/07/13/steels-big-dog-jumps-into-low-carbon-fuels-arcelormittal-lanzatech-primetals-technologies-to-construct-96m-biofuel-production-facility/ (accessed January 27, 2016).

233. LanzaTech 2016. *Aemetis Acquires License from LanzaTech with California Exclusive Rights for Advanced Ethanol from Biomass.* Available at: http://www.lanzatech.com/aemetis-acquires-license-lanzatech-california-exclusive-rights-advanced-ethanol-biomass-including-forest-ag-wastes/ (accessed March 25, 2016).
234. Liew, F. M., M. Köpke, and S. D. Simpson 2013. Gas fermentation for commercial biofuels production, In: Fang, Z., ed., *Biofuel Production-Recent Developments and Prospects.* Rijeka, Croatia: InTech, pp. 125–174.
235. Handler, R., D. Shonnard, E. Griffing, A. Lai, and I. Palou-Rivera 2015. Life cycle assessments of LanzaTech ethanol production: Anticipated greenhouse gas emissions for cellulosic and waste gas feedstocks. *Industrial & Engineering Chemistry Research* 55. doi: 10.1021/acs.iecr.5b03215.
236. Ou, X., X. Zhang, Q. Zhang, and X. Zhang 2013. Life-cycle analysis of energy use and greenhouse gas emissions of gas-to-liquid fuel pathway from steel mill off-gas in China by the LanzaTech process. *Frontiers in Energy* 7:263–270. doi: 10.1007/s11708-013-0263-9.
237. Sezenna, M. L. 2011. *Proteobacteria: Phylogeny, Metabolic Diversity and Ecological Effects.* New York, NY: Nova Science.
238. Jajesniak, P., H. E. Ali, and T. S. Wong 2014. Carbon dioxide capture and utilization using biological systems: Opportunities and challenges. *Journal of Bioprocessing & Biotechniques* 4:1. doi: 10.4172/2155-9821.1000155.
239. Luengo, J. M., B. Garcia, A. Sandoval, G. Naharro, and E. A. R. Olivera 2003. Bioplastics from microorganisms. *Current Opinion in Microbiology* 6:251–260. doi: 10.1016/S1369-5274(03)00040-7.
240. Reinecke, F., and A. Steinbüchel 2009. *Ralstonia eutropha* strain H16 as model organism for PHA metabolism and for biotechnological production of technically interesting biopolymers. *Journal of Molecular Microbiology and Biotechnology* 16:91–108. doi: 10.1159/000142897.
241. Voss, I., and A. Steinbüchel. 2006. Application of a KDPG aldolase gene-dependent addiction system for enhanced production of cyanophycin in *Ralstonia eutropha. Metabolic Engineering* 8:66–78. doi:10.1016/j.ymben.2005.09.003.
242. Overhage, J., A. Steinbüchel, and H. Priefert 2002. Biotransformation of eugenol to ferulic acid by a recombinant strain of *Ralstonia eutropha* H16. *Applied and Environmental Microbiology* 68:4315–4321. doi: 10.1128/AEM.68.9.4315-4321.2002.
243. Müller, J., et al. 2013. Engineering of *Ralstonia eutropha* H16 for autotrophic and heterotrophic production of methyl ketones. *Applied and Environmental Microbiology* 79:4433–4439. doi: 10.1128/AEM.00973-13.
244. Tanaka, K., K. Miyawaki, A. Yamaguchi, K. Khosravi-Darani, and H. Matsusaki 2011. Cell growth and P(3HB) accumulation from CO_2 of a carbon monoxide-tolerant hydrogen-oxidizing bacterium, *Ideonella* sp. O-1. *Applied Microbiology and Biotechnology* 92:1161–1169. doi: 10.1007/s00253-011-3420-2.
245. Kondaveeti, S., S-H. Lee, H-D. Park, and B. Min 2014b. Bacterial communities in a bioelectrochemical denitrification system: The effects of supplemental electron acceptors. *Water Research* 51:25–36. doi: 10.1016/j.watres.2013.12.023.
246. Tracy, B. P., S. W. Jones, A. G. Fast, D. C. Indurthi, and E. T. Papoutsakis 2012. Clostridia: The importance of their exceptional substrate and metabolite diversity for biofuel and biorefinery applications. *Current Opinion in Biotechnology* 23:364–381. doi: 10.1016/j.copbio.2011.10.008.
247. Cooksley, C. M., Y. Zhang, H. Wang, S. Redl, K. Winzer, and N. P. Minton 2012. Targeted mutagenesis of the *Clostridium acetobutylicum* acetone–butanol–ethanol fermentation pathway. *Metabolic Engineering* 14:630–641. doi: 10.1016/j.ymben.2012.09.001.
248. Leang, C., T. Ueki, K. P. Nevin, and D. R. Lovley 2012. A genetic system for *Clostridium ljungdahlii*: A chassis for autotrophic production of biocommodities and a model homoacetogen. *Applied and Environmental Microbiology* 79:1102–1109. doi: 10.1128/AEM.02891-12.

249. Keller, M. W., et al. 2013. Exploiting microbial hyperthermophilicity to produce an industrial chemical, using hydrogen and carbon dioxide. *Proceedings of the National Academy of Sciences of the U.S.A.* 110:5840–5845. doi: 10.1073/pnas.1222607110.
250. Basen, M., J. Sun, and M. W. W. Adams 2012. Engineering a hyperthermophilic archaeon for temperature-dependent product formation. *MBio* 3:e00053–e00012. doi: 10.1128/mBio.00053-12.
251. Kelly, D. P. 1981. Introduction to the chemolithotropihc bacteria, In: Starr, M. P., Stolp, H., Truper, H. G., Balows, A., and Schlegel, H. G., eds., *The Prokaryotes*. Berlin, Heidelberg: Springer.
252. Schiel-Bengelsdorf, B., and P. Durre 2012. Pathway engineering and synthetic biology using acetogens. *FEBS Letters* 586:2191–2198.
253. Ragsdale, S. W., and E. Pierce 2008. Acetogenesis and the Wood-Ljungdahl pathway of CO_2 fixation. *Biochimica et Biophysica Acta* 1784:1873–1898.
254. Ragsdale, S. W. 2008. Enzymology of the Wood-Ljungdahl pathway of acetogenesis. *Annals of the New York Academy of Sciences* 1125:129–136.
255. Bertsch, J., and V. Muller 2015. Bioenergetic constraints for conversion of syngas to biofuels in acetogenic bacteria. *Biotechnology for Biofuels* 8:210. doi: 10.1186/s13068-015-0393-x. eCollection 2015.
256. Jones, S. W., A. G. Fast, E. D. Carlson, C. A. Wiedel, J. Au, M. R. Antoniewicz, E. T. Papoutsaki, and B. P. Tracy 2016. CO_2 fixation by anaerobic non-photosynthetic mixotrophy for improved carbon conversion. *Nature Communications* 7:12800. doi: 10.1038/ncomms12800.
257. Ferry, M. S., J. Hasty, and N. A. Cookson 2012. Synthetic biology approaches to biofuel production. *Biofuels* 3:9–12. doi: 10.4155/bfs.11.151.
258. Mishra, A., J. N. Ntihuga, B. Molitor, and L. T. Angenent 2020. Power-to-protein: Carbon fixation with renewable electric power to feed the world. *Joule* 4:1142–1147.
259. Gleizer, S., Y. M. Bar-On, R. Ben-Nissan, and R. Milo 2020. Engineering microbes to produce fuel, commodities, and food from CO_2, Cell Press, open access report. *Cell Reports Physical Science* 1:100223. http://creativecommons.org/licenses/by/4.0/.
260. Li, H., P. H. Opgenorth, D. G. Wernick, S. Rogers, T.-Y. Wu, W. Higashide, P. Malati, Y.-X. Huo, K. M. Cho, and J. C. Liao 2012. Integrated electromicrobial conversion of CO_2 to higher alcohols. *Science* 335:1596.
261. Yishai, O., S. N. Lindner, J. Gonzalez de la Cruz, H. Tenenboim, and A. Bar-Even 2016. The formate bio-economy. *Current Opinion in Chemical Biology* 35:1–9.
262. Chen, F.Y.-H., H.-W. Jung, C.-Y. Tsuei, and J. C. Liao 2020. Converting *Escherichia coli* to a synthetic methylotroph growing solely on methanol. *Cell* 182:933–946.e14.
263. Kim, S., S.N. Lindner, S. Aslan, O. Yishai, S. Wenk, K. Schann, and A. Bar-Even 2020. Growth of *E. coli* on formate and methanol via the reductive glycine pathway. *Nature Chemical Biology* 16:538–545.
264. Gassler, T., M. Sauer, B. Gasser, M. Egermeier, C. Troyer, T. Causon, S. Hann, D. Mattanovich, and M. G. Steiger 2020. The industrial yeast Pichia pastoris is converted from a heterotroph into an autotroph capable of growth on CO_2. *Nature Biotechnology* 38:210–216.
265. Sawitzke, J. A., L. C. Thomason, N. Costantino, M. Bubunenko, S. Datta, and D. L. Court 2007. Recombineering: In vivo genetic engineering in *E. coli*, *S. enterica*, and beyond. *Methods Enzymology* 421:171–199. doi: 10.1016/S0076-6879(06)21015-2.
266. Patel, T. N., A-H. A. Park, and S. Banta 2013. Periplasmic expression of carbonic anhydrase in *Escherichia coli*: A new biocatalyst for CO_2 hydration. *Biotechnology and Bioengineering* 110:1865–1873. doi: 10.1002/bit.24863.
267. Zhuang, Z-Y., and S-Y. Li 2013. Rubisco-based engineered *Escherichia coli* for *in situ* carbon dioxide recycling. *Bioresource Technology* 150:79–88. doi: 10.1016/j.biortech.2013.09.116.

268. Jo, B. H., I. G. Kim, J. H. Seo, D. G. Kang, and H. J. Cha 2013. Engineered *Escherichia coli* with periplasmic carbonic anhydrase as a biocatalyst for CO_2 sequestration. *Applied and Environmental Microbiology* 79:6697–6705. doi: 10.1128/AEM.02400-13.
269. Bonacci, W., et al. 2012. Modularity of a carbon-fixing protein organelle. *Proceedings of the National Academy of Sciences of the U.S.A.* 109:478–483. doi: 10.1073/pnas.1108557109.
270. Wargacki, A. J., et al. 2012. An engineered microbial platform for direct biofuel production from brown macroalgae. *Science* 335:308–313. doi: 10.1126/science.1214547.
271. Claassens, N. J., I. Sá nchez-Andrea, D. Z. Sousa, and A. Bar-Even 2018. Towards sustainable feedstocks: A guide to electron donors for microbial carbon fixation. *Current Opinion in Biotechnology* 50:195–205.
272. Pontrelli, S., T.-Y. Chiu, E. I. Lan, F.Y.-H. Chen, P. Chang, and J. C. Liao 2018. *Escherichia coli* as a host for metabolic engineering. *Metabolic Engineering* 50:16–46.
273. Hawkins, A. S., Y. Han, H. Lian, A. J. Loder, A. L. Menon, I. J. Iwuchukwu, M. Keller, T. T. Leuko, M. W. W. Adams, and R. M. Kelly 2011. Extremely thermophilic routes to microbial electrofuels. *ACS Catalysts* 1:1043–1050. doi: 10.1021/cs2003017.
274. Hawkins, A., H. Lian, B. Zeldes, A. Loder, G. Lipscomb, G. Schut, M. Keller, M. Adams, and R. Kelly 2015. Bioprocessing analysis of Pyrococcus furiosus strains engineered for CO_2-based 3-hydroxypropionate production. *Biotechnology and Bioengineering* 112. doi: 10.1002/bit.25584.
275. Claassens, N., et al. 2016. Harnessing the power of microbial autotrophy. *Nature Reviews Microbiology* 14:692–706. https://doi.org/10.1038/nrmicro.2016.130
276. Hunt, K. A., R. M. Jennings, W. P. Inskeep, and R. P. Carlson 2018. Multiscale analysis of autotroph-heterotroph interactions in a high-temperature microbial community. *PLOS Computational Biology*, an open access paperBottom of FormPublished: September 27, 2018. https://doi.org/10.1371/journal.pcbi.1006431
277. Gimpel, J. A., E. A. Specht, D. R. Georgianna, and S. P. Mayfield 2013. Advances in microalgae engineering and synthetic biology applications for biofuel production. *Current Opinion in Chemical Biology* 17:489–495.
278. Rasala, B. A., P. A. Lee, Z. Shen, S. P. Briggs, and S. P. Mayfield 2012. Robust expression and secretion of Xylanasel in *Chlamydomonas reinhardtii* by fusion to a selection gene and processing with the FMDV 2A peptide. *PLoS One* 7(8):e43349. doi: 10.137/journal.pone.0043349.
279. Urtubia, H. O., L. B. Betanzo, and M. Vásquez 2016. Microalgae and cyanobacteria as green molecular factories: Tools and perspectives, an open access intech paper, http://dx.doi.org/10.5772/100261, Chapter June 2016 doi: 10.5772/63006.
280. Huang, H.-H., and P. Lindblad 2013. Wide-dynamic-range promoters engineered for cyanobacteria. *Journal of Biological Engineering* 7:10. doi: 10.1186/1754-1611-7-10.
281. Heitzer, M., H. Markus, E. Almut, F. Markus, and G. Christoph 2007. Influence of codon bias on the expression of foreign genes in microalgae, In: Aurora, L. R. G., ed., *Transgenic Microalgae as Green Cell Factories*, vol. 616. New York, NY: Springer, pp. 46–53. doi: 10.1007/978-0-387-75532-8_5.
282. Campbell, W. H., and G. Gowri 1990. Codon usage in higher plants, green algae, and cyanobacteria. *Plant Physiology* 92:1–11. doi: 10.1104/pp.92.1.1.
283. Lerche, K., and A. Hallmann 2009. Stable nuclear transformation of Gonium pectorale. *BMC Biotechnology* 9:64. doi: 10.1186/1472-6750-9-64.
284. Takahashi, M., T. Megumu, U. Toshiki, S. Naotsune, and M. Koji 2010. Isolation and regeneration of transiently transformed protoplasts from gametophytic blades of the marine red alga Porphyra yezoensis. *Electronic Journal of Biotechnology* 13. doi: 10.2225/vol13-issue2-fulltext-7.

285. Fuhrmann, M., A. Hausherr, L. Ferbitz, T. Schödl, M. Heitzer, and P. Hegemann 2004. Monitoring dynamic expression of nuclear genes in *Chlamydomonas reinhardtii* by using a synthetic luciferase reporter gene. *Plant Molecular Biology* 55:869–881. doi: 10.1007/s11103-004-2150-6.
286. Kwok, A. C. M., C. C. M. Mak, F. T. W. Wong, and J. T. Y. Wong 2007. Novel method for preparing spheroplasts from cells with an internal cellulosic cell wall. *Eukaryot Cell* 6:563–567. doi: 10.1128/ec.00301-06.
287. Vieler, A., et al. 2012. Genome, functionalgene annotation, and nuclear transformation of the heterokont oleaginous alga Nannochloropsis oceanica CCMP1779. *PLoS Genetics* 8:e1003064. doi: 10.1371/journal. pgen.1003064.
288. Hoiczyk, E., and A. Hansel 2000. Cyanobacterial cell walls: News from an unusual prokaryotic envelope. *Journal of Bacteriology* 182:1191–1199.
289. Guo, H., and X. Xu 2004. Broad host range plasmid-based gene transfer system in the cyanobacterium Gloeobacter violaceus which lacks thylakoids. *Progress in Natural Science* 14:31–35. doi: 10.1080/10020070412331343101.
290. Stucken, K., J. Ilhan, M. Roettger, T. Dagan, W. F. Martin 2012. Transformation and conjugal transfer of foreign genes into the filamentous multicellular cyanobacteria (subsection V) Fischerella and Chlorogloeopsis. *Current Microbiology* 65:552–560. doi: 10.1007/ s00284-012-0193-5.
291. Stucken, K., R. Koch, and T. Dagan 2013. Cyanobacterial defense mechanisms against foreign DNA transfer and their impact on genetic engineering. *Biological Research* 46:373–382. doi: 10.4067/ S0716-97602013000400009.
292. Onai, K., M. Morishita, T. Kaneko, S. Tabata, and M. Ishiura 2004. Natural transformation of the thermophilic cyanobacterium Thermosynechococcus elongatus BP-1: A simple and efficient method for gene transfer. *Molecular Genetics & Genomics* 271:50–59. doi: 10.1007/ s00438-003-0953-9.
293. Wolk, C. P., A. Vonshak, P. Kehoe, and J. Elhai 1984. Construction of shuttle vectors capable of conjugative transfer from *Escherichia coli* to nitrogen-fixing filamentous cyanobacteria. *Proceedings of the National Academy of Sciences of the USA* 81:1561–1565.
294. Thiel, T., and H. Poo 1989. Transformation of a filamentous cyanobacterium by electroporation. *Journal of Bacteriology* 171:5743–5746.
295. Parker, M. S., T. Mock, E. V. Armbrust 2008. Genomic insights into marine microalgae. *Annual Review of Genetics* 42:619–645. doi: 10.1146/annurev.genet.42.110807.091417.
296. Rasala, B. A., and S. P. Mayfield 2015. Photosynthetic biomanufacturing in green algae; production of recombinant proteins for industrial, nutritional, and medical uses. *Photosynthesis Research* 123:227–239. doi: 10.1007/s11120-014-9994-7.
297. Iwai, M., H. Katoh, M. Katayama, and M. Ikeuchi 2004. Improved genetic transformation of the thermophilic cyanobacterium, Thermosynechococcus elongatus BP-1. *Plant Cell Physiology* 45:171–175. doi: 10.1093/pcp/pch015.
298. Kreps, S., et al. 1990. Conjugative transfer and autonomous replication of a promiscuous IncQ plasmid in the cyanobacterium Synechocystis PCC 6803. *Molecular Genetics and Genomics* 221:129–133. doi: 10.1007/bf00280378.
299. Sorek, R., V. Kunin, and P. Hugenholtz 2008. CRISPR – A widespread system that provides acquired resistance against phages in bacteria and archaea. *Nature Reviews Microbiology* 6:181–186. doi: 10.1038/nrmicro1793.
300. Cong, L., et al. 2013. Multiplex genome engineering using CRISPR/Cas systems. *Science* 339:819–823. doi: 10.1126/science.1231143.
301. Yao, L., I. Cengic, J. Anfelt, and E. P. Hudson 2015. Multiple gene repression in cyanobacteria using CRISPRi. *ACS Synthetic Biology*. doi: 10.1021/acssynbio.5b00264.

302. Richmond, A., and Q. Hu 2013. *Handbook of Microalgal Culture: Applied Phycology and Biotechnology*. John Wiley & Sons. doi: 10.1002/9780470995280.
303. Peca, L., P. B. Kós, Z. Máté, A. Farsang, and I. Vass 2008. Construction of bioluminescent cyanobacterial reporter strains for detection of nickel, cobalt and zinc. *FEMS Microbiology Letters* 289:258–264. doi: 10.1111/j.1574-6968.2008.01393.x.
304. Kim, S-K, ed. 2015. *Handbook of Marine Microalgae: Biotechnology Advances*. London, UK: Academic Press.
305. Rasala, B. A., S.-S. Chao, M. Pier, D. J. Barrera, and S. P. Mayfield 2014. Enhanced genetic tools for engineering multigene traits into green algae. *PLos One* 9:e94028. doi: 10.1371/ journal.pone.0094028.
306. Kim, W. J., H. U. Kim, and S. Y. Lee 2017. Current state and applications of microbial genome-scale metabolic models. *Current Opinion in Systems Biology* 2:10–18.
307. Yim, H., R. Haselbeck, W. Niu, C. Pujol-Baxley, A. Burgard, J. Boldt, J. Khandurina, J. D. Trawick, R. E. Osterhout, R. Stephen, J. Estadilla, S. Teisan, H. B. Schreyer, S. Andrae, T. H. Yang, S. Y. Lee, M. J. Burk, and S. Van Dien 2011. Metabolic engineering of *Escherichia coli* for direct production of 1,4-butanediol. *Nature Chemical Biology* 7:445–452.
308 Alper, H., Y.-S. Jin, J. F. Moxley, and G. Stephanopoulos 2005. Identifying gene targets for the metabolic engineering of lycopene biosynthesis in *Escherichia coli*. *Metabolic Engineering* 7:155–164.
309. Fong, S. S., A. P. Burgard, C. D. Herring, E. M. Knight, F. R. Blattner, C. D. Maranas, and B. O. Palsson 2005. In silico design and adaptive evolution of *Escherichia coli* for production of lactic acid. *Biotechnology for Bioengineering* 91:643–648.
310. Triana, J., A. Montagud, M. Siurana, D. Fuente, A. Urchueguia, D. Gamermann, J. Torres, J. Tena, P. F. de Cordoba, and J. F. Urchueguia 2014. Generation and evaluation of a genome-scale metabolic network model of *Synechococcus elongatus* PCC7942. *Metabolites* 4:680–698.
311. Vu, T. T., E. A. Hill, L. A. Kucek, A. E. Konopka, A. S. Beliaev, and J. L. Reed 2013. Computational evaluation of *Synechococcus* sp. PCC 7002 metabolism for chemical production. *Biotechnology Journal* 8:619–630.
312. Hendry, J. I., C. B. Prasannan, A. Joshi, S. Dasgupta, and P. P. Wangikar 2016. Metabolic model of *Synechococcus* sp. PCC 7002: Prediction of flux distribution and network modification for enhanced biofuel production. *Bioresource Technology* 213:190–197.
313. Hucka, M., et al. 2003. The systems biology markup language (SBML): A medium for representation and exchange of biochemical network models. *Bioinformatics* 19:524–531.
314. Shirai, T., T. Osanai, and A. Kondo 2016. Designing intracellular metabolism for production of target compounds by introducing a heterologous metabolic reaction based on a *Synechosystis* sp. 6803 genome-scale model. *Microbial Cell Factories* 15:13.
315. Albers, S. C., V. A. Gallegos, and C. A. Peebles 2015. Engineering of genetic control tools in *Synechocystis* sp. PCC 6803 using rational design techniques. *Journal of Biotechnology* 216:36–46.
316. Kuk, S. K., et al. 2019. CO_2 reduction: Continuous 3D titanium nitride nanoshell structure for solar-driven unbiased biocatalytic CO_2 reduction. *Advanced Energy Materials* 9:1970097. doi: 10.1002/aenm.201970097.
317. ElMekawy, A., H. M. Hegab, G. Mohanakrishna, A. F. Elbaz, M. Bulut, and D. Pant 2016. Technological advances in CO_2 conversion electro-biorefinery: A step toward commercialization. *Bioresource Technology* 215:357–370. doi: 10.1016/j.biortech.2016.03.023.
318. Sadhukhan, J., et al. 2016. A critical review of integration analysis of microbial electrosynthesis (MES) systems with waste biorefineries for the production of biofuel and chemical from reuse of CO_2. *Renewable and Sustainable Energy Reviews* 56:116–132. doi: 10.1016/j.rser.2015.11.015.

319. He, A. Y., et al. 2016. Enhanced butanol production in a microbial electrolysis cell by *Clostridium beijerinckii* IB4. *Bioprocess and Biosystems Engineering* 39:245–254. doi: 10.1007/s00449-015-1508-2.
320. Jiang, Y., M. Su, Y. Zhang, G. Zhan, Y. Tao, and D. Li 2013. Bioelectrochemical systems for simultaneously production of methane and acetate from carbon dioxide at relatively high rate. *International Journal of Hydrogen Energy* 38:3497–3502. doi: 10.1016/j.ijhydene.2012.12.107.
321. Bazzanella, A., and F. Ausfelder 2017. *Low Carbon Energy and Feedstock for the European Chemical Industry.* Frankfurt am Main: DECHEMA; Gesellschaft für Chemische Technik und Biotechnologie eV.
322. Alexis Bazzanella, D. K. 2017. *Technologies for Sustainabilityand Climate Protection –Chemical Processes and Use of CO_2.* Available at: https://dechema.de/en/energyandclimate_CO2_Buch_engl.pdf (accessed February 15, 2020).
323. Wang, B., J. Wang, W. Zhang, and D. R. Meldrum 2012. Application of synthetic biology in cyanobacteria and algae. *Frontiers in Microbiology* 3:344. doi: 10.3389/fmicb.2012.00344.
324. Carr, M. 2005. Energy bill boosts industrial biotechnology; sweeping new legislation looks to bioproducts and biofuels to cut peak oil-dependency. *Industrial Biotechnology* 1:142–143. doi: 10.1089/ind.2005.1.142.
325. Illing, L., R. Natelson, M. Resch, I. Rowe, and D. Babson 2018. *Rewiring the Carbon Economy: Engineered Carbon Reduction Listening Day Summary Report.* Washington, DC: USDOE Office of Energy Efficiency and Renewable Energy (EERE). doi: 10.2172/1419624.

5 CO_2 Conversion to Fuels and Chemicals by Thermal and Electro-Catalysis

5.1 INTRODUCTION

Scientists have been aware of the potential economic and environmental benefits of using CO_2 as a feedstock for the synthesis of commodity chemicals and fuels for decades. It is generally an inexpensive waste product that contributes significantly to global warming. Nevertheless, despite the large amount of fundamental research that has been performed regarding the conversion of CO_2 into more valuable products, there are relatively few examples of industrially viable processes. The challenges associated with the conversion of CO_2 are primarily related to both its kinetic and thermodynamic stability. CO_2 cannot be converted into commodity chemicals or fuels without significant inputs of energy and contains strong bonds that are not particularly reactive. As a consequence, many of the available transformations of CO_2 require stoichiometric amounts of energy-intensive reagents. This can often generate significant amounts of waste and can result in large greenhouse gas footprints. The grand challenge for converting CO_2 waste streams into useful products is to develop processes that require minimal amounts of nonrenewable energy, are economically competitive, and provide substantial reductions in greenhouse gas emissions compared to existing technology.

In recent years, capturing carbon dioxide (CO_2), utilizing it, and storing it (CCUS) has emerged as an important method of mitigating its impact on environment. As shown in Chapter 2, in the past, major efforts were made to capture CO_2 streams, concentrate them, and store them underground or under the sea. This strategy was largely driven by fossil-fuel-based power industry and some energy-intensive industries like cement and steel manufacturing. There are, however, many challenges in the implementation of this strategy, which ultimately made them to be an expensive strategy. Some reasons include: the nature of CO_2 streams (volumetric flow, CO_2 concentrations, and other impurities) varies significantly; many waste streams are far away from their eventual storage place, thus requiring expensive transport of CO_2; and significant expenses are required to concentrate large-scale dilute CO_2 streams.

In recent years, more efforts are focused on the utilization strategy. As pointed out in Chapter 1, CO_2 can be utilized directly or indirectly by converting it into useful chemicals and fuels. While many of the direct utilization methods such as EOR,

DOI: 10.1201/9781003229575-5

EGR, and ECBM and others have been successful, they do not provide large markets needed for the vast amount of CO_2 emission produced by the use of fossil energy.

In Chapters 3, we examined conversion of CO_2 to useful construction materials by mineral carbonation process. Construction industry and the productions of aggregates, cement, and concrete are the second largest (behind water) industry in the world. Use of CO_2 to make synthetic concrete is a very valuable strategy. Significant efforts are being made to convert different types of feedstock to concrete by mineral carbonation process at large scale. In Chapter 4, we examined the strategy of biological methods for conversion of CO_2 to useful products. Biological photosynthetic method is the nature's way of converting CO_2 and water into oxygen. The chapter examines both photosynthetic and non-photosynthetic approaches for conversion of CO_2 to useful products. Once again, parts of this strategy are also successful and being commercialized. While these two methods have some very positive features for CO_2 utilization, other efforts are needed to further convert CO_2 to useful chemical and fuels.

The indirect utilization of CO_2 requires its chemical conversion. As shown in Figure 5.1, CO_2 is a very stable molecule and requires a large amount of energy to overcome activation barrier to convert it into useful chemicals and fuels. This can be achieved by the use of heterogeneous or electrocatalysts. As shown in this figure, about 85% of current chemicals production involves catalysis. The present chapter examines the effectiveness of CO_2 conversion to chemicals and fuels by homogeneous, heterogeneous, and electrocatalysis. Since historically fossil resources were produced via natural carbon-hydrogenation during photosynthesis, synthetic CO_2 hydrogenation is likely the best way to regenerate combusted hydrocarbons.

Heterogeneous catalysis for CO_2 conversion to produce C_1, C_{2+} (including higher alcohols, higher acids, and fuel productions) and high-molecular-weight polymer products can be carried out in a number of different ways. The first set of steps involves basic C_1 chemistry with CO_2 conversion that consists of CO_2 dissociation to produce CO ($CO_2 \rightarrow CO + 1/2\ O_2$), CO_2 hydrogenation to produce CO ($CO_2 + H_2 \rightarrow CO + H_2O$, which is also called the reverse water-gas shift [RWGS] reaction), methane

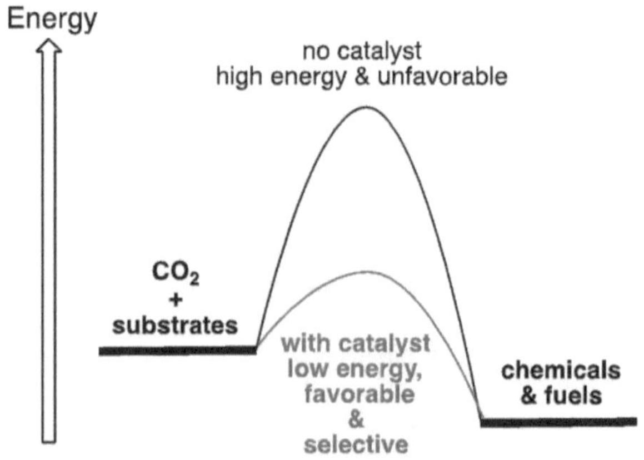

FIGURE 5.1 Role of catalysis in CO_2 conversion [1].

($CO_2 + 4H_2 \rightarrow CH_4 + 2H_2O$) or methanol ($CO_2 + 3H_2 \rightarrow CH_3OH + H_2O$), artificial photosynthesis of CO_2 (CO_2 + water) ($CO_2 + H_2O \rightarrow CO + H_2 + O_2$), and dry reforming of methane with CO_2 to produce syngas (H_2 + CO) ($CH_4 + CO_2 \rightarrow 2CO + 2H_2$). All of these reactions can be carried out in the gas phase, and they are facilitated by heterogeneous catalysis. The liquid-phase reaction includes the production of formic acid using CO_2 dissolved in aqueous phase (CO_2 (aq) + H_2 (aq) $\rightarrow$ COOH). Because the solubility of CO_2 is quite low in aqueous solution, the CO_2 conversion in liquid phase typically suffers from low productivity. Furthermore, CO_2 is thermodynamically the most stable carbon species, and its conversion into value-added products usually requires the expenditure of considerable energy. Several review papers about CO_2 hydrogenation can be found in the literature [2–7]. However, CO_2 hydrogenation or formic acid formation requires H_2, which is mainly produced from methane by steam reforming, which also produces a large amount of CO_2. In order to make CO_2 hydrogenation a green technology, hydrogen must be produced by electrolysis of water using solar or wind energy to provide required electricity. CO_2 dissociation to CO has been examined extensively particularly with the use of solar energy and ceria catalyst.

Efficient catalysts can minimize the energy needed for reactions by reducing the activation energy. Various catalysts are being actively investigated to enhance CO_2 conversion and to control selectivity toward specific target products. Various metals, metal oxides, metal carbides, and doped carbon materials have been used as catalysts for CO_2 conversion. Metal-based catalysts including precious metal or supported Ni catalysts have been used for the production of CO or CH_4. Not many good heterogeneous catalysts have been found for the production of formic acid; instead, homogeneous catalysts have been typically used. Robust heterogeneous catalyst producing formic acid might have high potential. As shown in subsequent chapters, catalysts that use both heat and light as energy sources have been developed, to minimize total energy use. Synthetic strategies have been developed to endure high reaction temperatures and minimize coke formation for dry methane reforming.

Considerable progress has been made toward converting CO_2 to single carbon (C_1) products (e.g., formic acid, carbon monoxide (CO), methane, methanol and syngas) [8,9]. Thermocatalytic hydrogenation of CO_2 to methane can be achieved easily at atmospheric pressure and high gas hourly space velocity (GHSV), and has been shown to achieve CO_2 conversion and CH_4 selectivity close to theoretical equilibrium values [10]. CO_2-to-methanol (CTM) has already been industrialized in Reykjavik, Iceland using heterogeneous catalysis and geothermal energy. Syngas produced from artificial photosynthesis or dry reforming of methane can be used to produce additional products by well-known FT synthesis.

Recently, significant progress has also been achieved in heterogeneous catalytic hydrogenation of CO_2 to various high-value and easily marketable fuels and chemicals containing two or more carbons (C_{2+} species), including dimethyl ether (DME) [11], olefins [12], liquid fuels [13], and higher alcohols and acids [14]. Compared to C_1 products, C_{2+} product synthesis is more challenging due to the extreme inertness of CO_2 and the high C–C coupling barrier, as well as many competing reactions leading to the formation of C_1 products. However, C_{2+} hydrocarbons and oxygenates possess higher economic values and energy densities than C_1 compounds and has been given significant attention.

The hydrogenation of CO_2 to C_{2+} products mainly occur via a methanol-mediated route or a CO_2 modified Fischer–Tropsch mechanism [13]. CO_2 can also be converted to polymers and other higher hydrocarbons by suitable catalysts. There are a variety of questions that need to be answered for each of these reaction mechanisms: what kinds of catalysts are beneficial for each route? How do the catalysts regulate product selectivity? What can be done to further enhance catalytic performance? What is the central challenge with catalysts for CO_2 hydrogenation to C_{2+} products? To answer these questions, the catalysts for CO_2 hydrogenation to C_{2+} species will be discussed based on the methanol-mediated route and the CO_2-modified Fischer-Tropsch route. Some experimental guidelines are provided to improve CO_2 conversion and to reduce C_1 byproducts. In addition, challenges and an outlook of transformational technologies for developing new catalysts are also briefly discussed in this chapter. Artificial intelligence (AI) is expected to guide the design and discovery of catalysis [15], while 3D-printing technologies are anticipated to be used to manufacture them on a large scale [16]. Finally, CO_2 can also be converted to high-molecular-weight polymer products by suitable catalysts. Several of the heterogeneous catalysis processes are commercialized.

The chapter is divided into two parts. In Sections 5.2–5.5, CO_2 conversion by thermal homogeneous or heterogeneous catalysis is examined. In Sections 5.6–5.10, the role electrocatalysis on CO_2 conversion is examined.

5.2 C_1 CHEMISTRY FOR CO_2 CONVERSION BY THERMAL CATALYSIS

While as shown in Chapter 6, CO_2 can be dissociated to CO using ceria catalyst with required high energy provided by solar heating, more favorable route to produce CO is the CO_2 hydrogenation. The hydrogenation of CO_2 can produce, CO, methane, or methanol depending on the reaction stoichiometry, operating conditions, and the nature of the catalyst. The reduction of carbon dioxide into carbon monoxide by RWGS is one of the more popular routes for CO_2 conversion studies mainly due to its low hydrogen consumption (as compared to methanation) and high equilibrium conversion rates [17,18]. Furthermore, the formation of CO forms an important first step for the subsequent synthesis of carbon-based chemicals and/or fuels [17,18].

The mechanisms suggested for RWGS reaction fall into two main categories: (a) the redox pathway [19] and (b) the formate decomposition mechanism [20]. The redox mechanism is most often described over a Cu-based catalyst, the metallic Cu^0 atoms are partially oxidized to form Cu_2O and carbon monoxide, and the oxidized copper is later reduced by the hydrogen present in the system, reforming it to its metallic state accompanied by the formation of a water molecule.

In contrast to the redox mechanism, the formate decomposition pathway suggests that the hydrogen molecule first hydrogenates the carbon dioxide into a formate intermediate. It is then from this intermediate that the cleaving of the carbon-oxygen double bond occurs to release a CO molecule [21]. The formate-mediated route is also reported to occur for the methanation pathway, but has often been debated regarding the extent of its contribution toward product formation. Using spectro-kinetic analysis of a Pt/CeO_2 catalyst for CO_2 hydrogenation, Goguet et al. [22] proposed that the Pt metal maintains the active (i.e. reduced) ceria surface via H_2-spillover. The resulting vacancies on the ceria surface then bond with CO_2 molecules to form surface carbonates, which are the main intermediates that facilitate the final formation of CO species.

One of the most apparent advantages methanol has over products such as CO and CH_4 is its liquid state at normal atmospheric temperature and pressure. This is desirable as it provides a practical alternative for energy storage and transport. Furthermore, much like CO, it can also serve as an important starting constituent for olefins and aromatics [17,23]. One of the major challenges for the methanol synthesis process is its less selectivity compared to methane and CO under low-pressure conditions. High pressures (50–100 bars) are often required to suppress CO formation via the RWGS pathway [17].

Similar to the Sabatier case, arguments on the CO_2 to methanol reaction pathway are dominated by the formate route and the CO route factions. The pathway suggested for the latter case proposes that CO_2 first undergoes RWGS-like hydrogenation before the CO formed is sequentially hydrogenated into formyl (HCO), formaldehyde (H_2CO), methoxy (H_3CO), and finally methanol [24]. The study by Grabow and Mavrikakis [25] suggested the formate pathway being more likely, whereby the hydrogenation of CO_2 into methanol occurs via the following sequence: $CO_2* \rightarrow HCOO* \rightarrow HCOOH* \rightarrow CH_3O_2* \rightarrow CH_2O* \rightarrow CH_3O* \rightarrow CH_3OH*$ where the symbol * indicates an adsorbed species.

A mechanistic study by Studt et al. [26], which compared CO_2 to methanol pathways on different Cu-based catalysts, shed further light on the debate regarding the importance of differences in catalyst composition impacting on the resulting mechanism experienced by the CO_2 molecules. In addition to the known effects of variation in reaction conditions (e.g., feed gas composition, temperature, and pressure), their calculations demonstrated that the promoting effect of Zn in a Cu–ZnO containing catalyst was not only kinetically based, but also stemmed from a deviation in the reaction mechanism due to the different sites at which the key reactions occur.

CO_2 hydrogenation requires hydrogen. There are two major ways to produce hydrogen. One method is to dissociate water by electrolysis. The energy required for this process can be obtained from excess solar, wind, or nuclear power. This method has been successfully used in a number of operations. The second method is to reform methane. Nowadays, more than 95% of the hydrogen for refinery use is produced via hydrocarbon steam reforming [27,28]. Industrial hydrogen production through methane steam reforming exceeds 50 million tons annually and accounts for 2%–5% of global energy consumption [27,28]. Both electrolysis and numerous methods for steam reforming, including with the use of solar energy, are extensively described in the literature and in my previous books [27,28].

5.2.1 CO_2 Conversion to Carbon Monoxide—Reverse Water-Gas Shift Reaction

As mentioned earlier, one method to convert CO_2 to CO is the RWGS reaction in which CO_2 and H_2 are converted into CO and H_2O [29]. The RWGS reaction is an endothermic reaction, and consequently, high temperatures (~500°C) are typically utilized to favor the formation of CO. Even then a large excess of either CO_2 or H_2 or the constant removal of products are often used as strategies to increase conversion. RWGS can also produce syngas, which is useful for synthesizing valuable chemicals such as ammonia, methanol, and fuels, from CO_2 [30]. However, the RWGS reaction is endothermic, requiring high temperature, and CO is the dominant product above 600°C as can be seen in Figure 5.2 [31]. Developing catalysts with high activity and durability are essential to obtain a maximum yield.

A range of heterogeneous catalysts have been used for the RWGS reaction, including copper-, iron-, or ceria-based systems, but in general they have poor thermal stability, and methane is commonly formed as an undesired side product [29]. Fluidized bed reactors give greater conversion to CO than fixed bed reactors, and this is one approach to improve conversion. Catalysts for RWGS should contain active sites to dissociate hydrogen and adsorb CO_2. Precious metal-based catalysts have been widely used for RWGS because of their superior activity for hydrogen dissociation. The effect of Pt particle size on the selectivity of CO and CH_4 was identified using rutile TiO_2 as a support. Small Pt particles preferred CO formation. Pt-CO species were the key intermediate deciding CO selectivity in the RWGS reaction [32].

The Au@UIO-67 catalyst has also been synthesized for the RWGS reaction. UIO-67, which is porous metal-organic framework, was used to disperse Au nanoparticles, which enhanced metal and support interaction. Au@UIO-67 showed high activity and CO selectivity for RWGS [33]. A Pd-In/SiO_2 showed 100% CO selectivity without

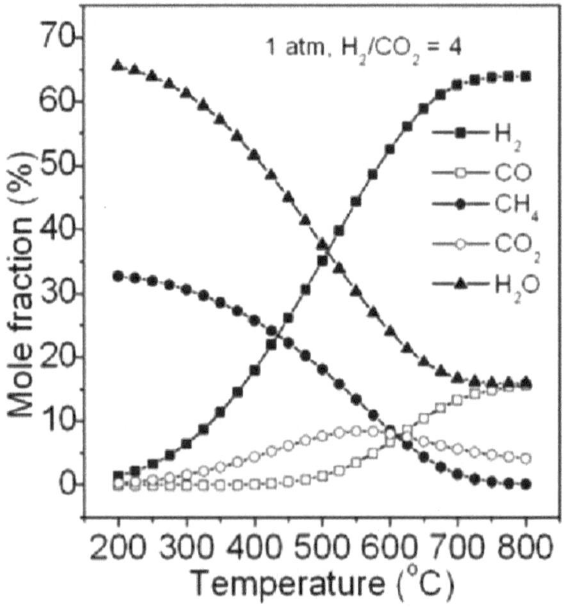

FIGURE 5.2 Thermodynamic equilibrium of CO_2 hydrogenation at 1 bar and a H_2/CO_2 molar ratio of 4. [3].

CH_4 formation. For the large-scale conversion of CO_2, non-precious metal catalysts are preferred due to cost and scarcity of precious metals. Ni- and Cu-based catalysts are promising because they also show high activity and selectivity for RWGS. But they usually suffer from sintering at high reaction temperatures.

Various ways of improving thermal stability have been investigated. Rossi et al. prepared highly dispersed Ni nanoparticles supported on SiO_2 using magnetron sputtering deposition, which showed better activity at T > 600°C and stability than a conventional catalyst [34]. Ni/Ce-Zr-O catalysts also showed high activity, stability, and selectivity for the conversion of CO_2 to CO [35]. Cu/ZnO catalysts with various Cu/Zn ratios were also used [30]. Zheng et al. prepared CeCu composite catalysts with different Ce/Cu mole ratios, and $Ce_{1.1}Cu_1$ catalyst showed the highest catalytic performance for the RWGS reaction [31]. Transition-metal carbides (TMCs) are also promising. Mo_2C showed higher CO_2 conversion and CO selectivity in the RWGS than other TMCs such as TiC, TaC, ZrC, WC, and NbC [36]. CO_2 adsorption was enhanced, and CO_2 dissociation barriers were reduced on the K-promoted Mo_2C catalyst [37]. Polycrystalline α-Mo_2C catalyst also showed better CO_2 conversion to CO [38]. Ma et al. [39] showed that Cu/β-Mo_2C showed high activity and stability compared with Pt- and Cu-based catalysts. Recently, Ajayan et al. reported a metal-free carbon-based catalyst for CO_2 hydrogenation [40]. Pyridinic N was doped at the edge sites of graphene quantum dots. With this catalyst, CO was produced dominantly at lower temperatures and CO selectivity increased to 85% at 300°C.

Carbon monoxide (CO) is an important feedstock in the synthesis of many chemicals and fuels [41]. For example, hydrogenation of CO can generate methanol, or through the Fischer–Tropsch process higher order hydrocarbons can be produced (Eq. 5.1).

$$(2n+1)H_2 + nCO \rightarrow C_nH_{2n+2} + nH_2O \tag{5.1}$$

5.2.2 CO_2 Methanation

CO_2 hydrogenation can also produce methane. The CO_2 methanation has recently received much attention as a way to store intermittent electricity, which is produced from solar cell or wind power [42]. The surplus electricity can electrolyze water, producing H_2. Then, CO_2 and H_2 can have reaction to produce methane, which can be used as a fuel. Various precious metals, such as Ru, Rh, and Pd, have been used as catalysts. Ni deposited on various supports has also been used for the CO_2 methanation [4]. Amal and Dai et al. produced methane from CO_2 and H_2 using porous perovskite materials [43,44]. The perovskite materials with ABO_x crystalline structure have been particularly interesting materials in solar cells or catalytic applications [45]. Ni-Rh nanoalloy nanoparticles on mesoporous $LaAlO_3$ perovskite showed good performance for CO_2 methanation [44]. Similarly, mesoporous Ni/Co_3O_4 also showed high activity for CO_2 methanation at low temperature [46].

Methane is widely used as a fuel and to make syngas. The Sabatier reaction, which hydrogenates CO_2 to methane using a nickel catalyst, has been known for more than a hundred years (Eq. 5.2) [47].

$$CO_2 + 4H_2 \rightarrow CH_4 + 2H_2O \tag{5.2}$$

Small-scale pilot plants in Norway generate CH_4 using CO_2 from flue gas and low-carbon H_2, and in Japan and the United States using H_2 from water electrolysis [48]. Currently, Germany has several small-scale plants, including a pilot plant run by Audi [47]. Specific challenges that need to be addressed for methanation include the development of catalysts that operate at lower temperatures (<200°C) where the reaction is more favorable and the sintering and oxidation-induced deactivation in nickel-based catalysts can be prevented. Although ruthenium catalysts have shown advantages compared to nickel systems, they are expensive. At this stage, the hydrogenation of CO_2 to methane is not practical on a large scale and is unlikely to be so in the near future given the low price and abundant availability of methane from natural gas. Additionally, there will be a significantly greater economic value in converting CO_2 to many other chemicals compared with methane.

The most widely used catalysts for the CO_2 methanation are noble metals, such as Rh, Ru, and Pd, and Ni-based catalysts. The noble metals are highly active toward CO_2 methanation at lower temperature and more resistant to the carbon formation than other transition metals; however, they are expensive. In particular, the noble metals are also used to promote the Ni catalysts to enhance their catalytic activities. The noble metals are generally supported by Al_2O_3, TiO_2, and CeO_2. Support plays an important role on activity and selectivity to methane [6]. The structure of support is also important. Ru dispersion is significantly influenced by the crystal phase structure of the TiO_2 supports [6]. The reaction rate of the Ru/r-TiO_2 was found to be 2.4 times higher than that of the Ru/a-TiO_2, which mainly originated from the different particle sizes of ruthenium [6]. A sequence of the surface-based activities (TOF): Ru/Al_2O_3 > Ru/$MgAl_2O_4$ > Ru/MgO > Ru/C reported in the literature is almost identical to that of electron-deficiencies of the metal, determined by the Lewis acidities of the supports [49]. An addition of second metal can substantially change the catalyst behavior. The literature indicates that when the alkaline salts were added to Ru/Al_2O_3 catalysts, a synergetic effect can be detected. The amount of metal loading can also affect metal particle size and turnover frequency, but as shown for Rh/y-Al_2O_3, the effect depends on temperature [6]. Furthermore, the metal loading can also change the product selectivity of CO_2 hydrogenation, as shown in Figure 5.3 for Rh/SiO_2 [50].

In recent years, significant efforts have been made to improve performance of Ni catalysts because of its lower cost compared to other noble metal catalysts. The literature results indicate significant dependence of CO_2 conversion on support and catalyst preparation method for Ni catalyst. Support is also important because a highly active catalyst (with high surface area) for CO_2 methanation requires a highly uniform dispersed active species over the support. In general, the support usually plays a very important role in the interaction between the Ni and the support. The nickel compounds on different support surfaces result in different "metal-support effects" [51], which implies that catalysts would exhibit different performance toward activity and selectivity for a given process. For example, a highly active catalyst Ni/MOF-5 showed unexpected activity at low temperature for CO_2 methanation [6]. The embedding of Ni nanoparticles onto the Al_2O_3 matrix enhances the metal-support interaction and prevents the sintering and/or the aggregation of the active

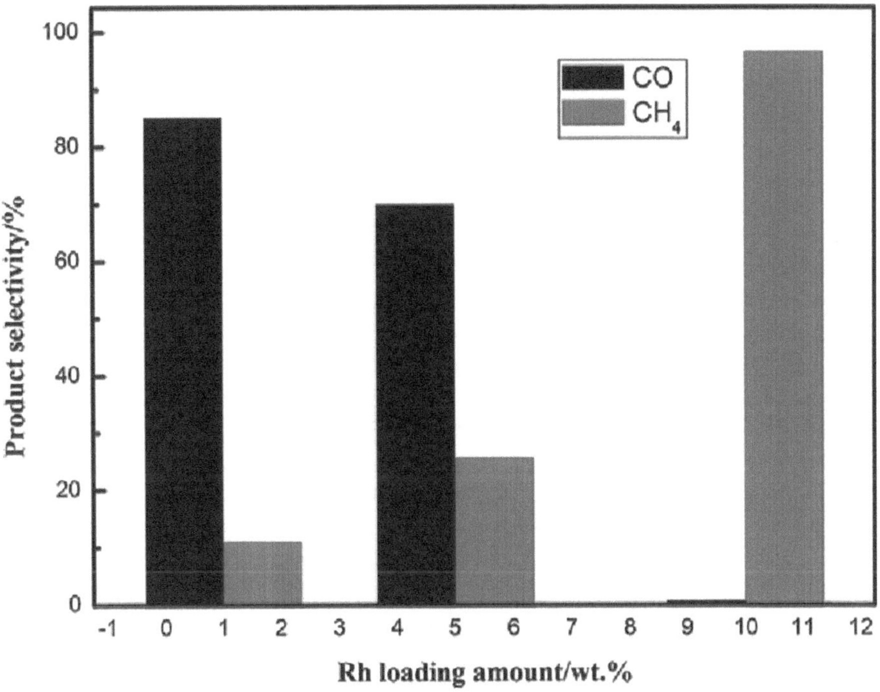

FIGURE 5.3 Effect of Rh loading on the distribution of CH_4 and CO [50]. Reaction conditions: temperature = 473 K, pressure = 5 MPa, H_2/CO_2 ratio = 3, flow rate = 100 cm^3/min [6].

nickel species, resulting in high dispersion degree, high stability, and high activity (less coking) during the long-term use. The performance of Ni catalyst can also be improved by adding promoters. Many efforts have been made to enhance the catalytic activity, including selection of appropriate supports and addition of catalytic promoters such as Ce, Zr, La, Mg, V, and Co [6,42]. Besides Ni, the Co-based Fischer–Tropsch catalysts exhibit a superior catalytic performance with respect to low-temperature CO_2 methanation [6,45]. A higher CH_4 selectivity was observed in the Fischer–Tropsch synthesis when the Co catalysts were not completely reduced or when the catalysts contain smaller Co_3O_4 particles [45].

5.2.3 CO_2 Hydrogenation to Methanol and Formic Acid

Methanol (CH_3OH), produced globally on a scale of approximately 70 million tons in 2015, typically is synthesized from syngas (H_2 + CO) obtained directly from fossil fuels [52,53]. A small amount of CO_2 (up to 30%) is generally added to the feed to improve performance [54]. This is successful in part because the mechanism of methanol production involves the initial conversion of CO and H_2O to CO_2 and H_2 via the water gas-shift reaction (Eq. 5.3). In fact, the development of methods to increase the amount of CO_2 in the syngas feed without causing a large decrease in methanol yield represents an opportunity to utilize waste CO_2 that is produced during syngas production. Although this strategy is only viable if excess H_2 is available, it could improve current technology and increase plant efficiency.

$$CO + H_2O \rightarrow H_2 + CO_2 \quad (5.3)$$

The *direct hydrogenation* of CO_2 to methanol could provide a more sustainable synthetic route if coupled with low-carbon methods for the production of H_2 [52,53]. Furthermore, the development of a practical method for the synthesis of methanol from CO_2 could also facilitate a transition toward a methanol economy, in which methanol is used either directly as a fuel or as a source of H_2 [55].

Researchers have developed several catalysts and reactors for direct hydrogenation of CO_2 to methanol, but high rates and high methanol selectivity have only been possible using high pressures (>300 bar) [52,53,56]. The cost of this technology presently is not competitive with the cost of methanol synthesis from syngas. Nevertheless, due to special circumstances related to location, presently two large pilot plants for direct methanol production from CO_2 are in operation. The Mitsui Chemical Company in Japan produces around 100 tons of methanol per year from CO_2 [48] located adjacent to petrochemical industry. The close links to the petrochemical industry make the process economically viable. Second, Carbon Recycling International, located in Iceland, produces approximately 4,000 tons of methanol from CO_2 each year [56]. The plant uses hydro- and geothermal energy to produce H_2 and uses CO_2 captured from the flue gas of a geothermal power plant, which is located next to the CO_2-to-methanol facility. The process is economically feasible because of the availability of low-cost electricity, required to generate the H_2, in Iceland, and because the composition of the flue gas is 85%–90% CO_2. This substantially lowers the cost of CO_2 compared with more traditional flue gas streams which contain lower amounts of CO_2.

Ye et al. [2] have elegantly described significant progress made to convert CO_2 to methanol by copper (Cu)- and indium (In)-based catalysts. They indicated that recent publications have dealt with Cu–ZnO composites, whose CH_3OH selectivity ranged from 30% to 70% at CO_2 conversions less than 30% under typical reaction conditions (temperature: 220°C–300°C, pressure <5 MPa, $H_2/CO_2 = 3$). Also by increasing the pressure to 36 MPa and H_2/CO_2 molar ratio to 10, the single-pass CO_2 conversion can be increased to 95.3% with 98.2% methanol selectivity over a Cu–ZnO–Al_2O_3 catalyst. When Cu nanoparticles (NPs) are encapsulated in metal organic frameworks (MOFs) and strongly interact with their secondary structural units, the resultant metal NPs@MOFs (e.g., Cu ⊂ UiO-66 and CuZn@UiO-bpy) showed enhanced activity and 100% CH_3OH selectivity, while preventing the agglomeration of the Cu NPs. In addition, ZrO_2 is a noted promoter or support in Cu-based catalysts for CTM hydrogenation [2,57].

Indium-based materials have shown promise as alternatives for CO_2 conversion to methanol. Pure In_2O_3 can convert 7.1% of CO_2 with 39.7% selectivity to CH_3OH at 330°C and 5 MPa [2]. A Pd/In_2O_3 catalyst with many interfacial sites and oxygen vacancies enhances CO_2 adsorption, achieving CO_2 conversions of above 20% [2]. In_2O_3/ZrO_2 catalysts significantly boost the methanol selectivity to 99.8%, with a

CO_2 conversion of 5.2% and long-term stability 40. Besides the Cu- and In-based catalysts, great progress has also been achieved with ZnO–ZrO_2 solid solution catalysts, as well as Pd/Pt-based catalysts [2,58,59].

The thermal hydrogenation of CO_2 to formic acid is thermodynamically unfavorable when starting from gas-phase reactants but becomes slightly exergonic when performed under aqueous conditions [52,53]. On a laboratory scale, a vast number of homogeneous and heterogeneous catalysts for CO_2 hydrogenation to formic acid have been developed [52,53]. In some cases, these catalysts give high turnover numbers and frequencies but typically only in the presence of a base, which reduces atom economy and increases cost. Challenges for future research into the development of systems for the hydrogenation of CO_2 to formic acid include (a) the discovery of catalyst systems which give high turnover numbers in the absence of base or with recycling of the base, (b) the need for cheaper ligands to stabilize homogeneous catalysts, and (c) the refinement of strategies to separate formic acid from the reaction media, which will be a crucial part of successful catalyst commercialization. Improved catalysts are critically needed if the direct hydrogenation of CO_2 to methanol is to replace methanol production from syngas. Methanol can be formed from CO_2 hydrogenation by both homogeneous and heterogeneous catalyses.

5.2.3.1 Homogenous Catalytic Conversion

Tominaga et al. [60] reported an example of direct CO_2 conversion to methanol using homogeneous catalysts. They showed that the performance of $Ru_3(CO)_{12}$–KI for CO_2 conversion was much better than the other transition metal carbonyl catalysts. Recently, Huff and Sanford [61] used three-step cascade process to reduce CO_2 to methanol instead of six electrons process. Three steps involve hydrogenation of CO_2 to formic acid; then, the esterification of formic acid to formate esters; and finally, hydrogenation of the formate ester to produce methanol. Different catalysts are used in each step of this approach under specific reaction conditions which are high temperature (135°C) and pressure (40 bars). Wesselbaum et al. [62] reported that the [(triphos)Ru-(TMM)] (TMM = trimethylenemethane, Triphos = 1,1,1-tris(diphenylp hosphinomethyl) ethane) can be used in the hydrogenation process to covert formate esters to methanol. Also, silanes and hydrides are the main reducing agents used in the homogeneous chemical reduction of CO_2 to methanol in the presence of organo catalysts such as N-heterocyclic carbenes (NHC). Although the cost of the silanes is high, it was proved that the NHC catalyst has the ability to reduce CO_2 to methoxides under ambient conditions [63]. The application of frustrated Lewis pairs to reduce CO_2 to methanol is considered to be another example of the metal-free catalysis [63].

5.2.3.2 Heterogeneous Catalytic Conversion

At this stage, significant amounts of research into the direct hydrogenation of CO_2 to methanol have focused on using heterogeneous copper-based catalysts that are closely related to those used for CO conversion to methanol [64]. In recent years, there have also been a number of reports of catalysts for CO_2 hydrogenation to methanol, which use metals other than copper and show promising activity [65,66]. Two general challenges for catalyst development are product inhibition by water (the byproduct of CO_2 hydrogenation) and poor selectivity because of the competing RWGS reaction between CO_2 and H_2 to generate CO and H_2O. Once more efficient catalysts are developed, further attention can be given to factors such as stability, cost, sustainability, and scale-up potential. It is generally accepted that a large-scale catalyst for direct methanol hydrogenation will almost certainly be heterogeneous.

The catalytic hydrogenation of CO_2 with H_2 is considered to be the most straightforward way for methanol from CO_2, as shown in Eq. (5.4).

$$CO_2 + 3H_2 \leftrightarrow CH_3OH + H_2O \qquad \Delta H298\,K = -11.9\,kcal\,mol \tag{5.4}$$

Heterogeneous catalysts have many advantages which includes easy separation of fluid from solid catalyst, convenient handling in different types of reactors (i.e., fixed-, fluidized-, or moving-bed), and the used catalyst can be regenerated. Many studies have proven that the Cu-based catalysts with different additives such as ZrO_2 and ZrO play an important role to improve the stability and activity of the heterogeneous catalyst. Cu and Zn and their oxides have been developed to be used as an efficient heterogeneous catalyst for the conversion of CO_2 to methanol [63,67]. This type of catalyst is similar to $Cu/ZnO/Al_2O_3$-based catalysts that are used to produce methanol in the industry. Various reviews have discussed the different factors that may affect the methanol production from syngas such as catalyst preparation, catalyst design, reaction kinetics, reactor design, and catalyst deactivation [63,68]. The future research works should be focused on the methanol production from CO_2 and H_2 in which the amount of produced methanol is high (greater than at least 10%). Furthermore, in order to sustain high plant output, the catalyst should remain active for several years. Improving the activity and stability of catalyst over time is very important in the economics of any methanol plant [63]. Recently, Lurgi has been collaborated with Süd-Chemie to use a high-activity catalyst (C79-05-GL, based on Cu/ZnO) to convert CO_2 and H_2 into methanol [63,67]. The Lurgi methanol reactor is a tube-based converter that contains the catalysts in fixed tubes and uses a steam pressure control to achieve the controlled temperature reaction. This type of reactor is able to achieve low recycle ratios and high yield. Lurgi has also developed a two-stage converter system, which uses two combined Lurgi reactors for high methanol capacities at lower energy costs [63]. Even though the operating temperature of the Lurgi system is around 260°C which is higher than that used for conventional catalysts to produce methanol, the methanol selectivity of this system is excellent. The activity decline of this catalyst is the same as current commercially used catalyst.

Mitsubishi Gas Chemical, Sinetix, and Haldor Topsøe are commercializing high stable catalysts for methanol production. Arena et al. [69] studied the solid-state interactions, functionality, and adsorption sites of $Cu–ZnO/ZrO_2$ catalysts and its ability for the conversion of CO_2 to methanol. Characterization data indicated that the strong Cu–ZnO interaction effectively promotes the dispersion and reactivity of metal copper to oxygen. The metal/oxide interface in $Cu–ZnO/ZrO_2$ catalysts plays an important role in hydrogenation of CO_2 to methanol.

During CO_2 hydrogenation, the formation of the undesired CO through RWGS is a competitive reaction to methanol synthesis. While decreasing the reaction temperature favors methanol synthesis reaction, and raising the space velocity decreases the CO selectivity, low temperature and/or high space velocity usually result in low single-pass CO_2 conversion. Thus, it remains a great challenge to simultaneously obtain high CO_2 conversion and high methanol selectivity. Very recently, significant progress has been made in developing more efficient catalysts for CO_2 hydrogenation to CH_3OH, including metal-supported catalysts, bimetallic systems, and reducible metal oxides.

Industrial methanol production from CO_2-containing syngas uses the well-known Cu-ZnO-Al_2O_3 catalysts. Currently, CH_3OH synthesis from catalytic CO_2 hydrogenation has been implemented at the pilot-plant level by Lurgi, Mitsui, and CRI, among others [63]. These processes mainly used modified Cu-ZnO-Al_2O_3 catalysts and were carried out under conditions similar to syngas-based methanol synthesis. Supported copper materials have attracted much attention and have been extensively investigated for CO_2 hydrogenation to CH_3OH. The methanol synthesis reaction is known to show strong support effects. Using a suitable support material can enhance CH_3OH selectivity, attributed to structural, electronic, and chemical promotional effects. For example, CH_3OH selectivity is usually high (>70%) at 200°C–260°C when ZrO_2 is used as the support. To simultaneously improve the intrinsic activity and catalyst stability, researchers directly used Cu–Zn-based LDH as the precursor to synthesize an efficient methanol synthesis catalyst with a confined structure, in which the active metallic Cu phase is highly dispersed and partially embedded in the remaining oxide matrix [2,3,7,63,70–74].

Co catalysts are widely used and extensively studied for FTS and can easily catalyze the CO_2 methanation reaction. Novel Co-based catalysts are also examined for methanol synthesis. The high performance for CH_3OH synthesis over Co-based catalysts was attributed to the formation of a new active phase, rather than the conventional metallic Co phase. Various bimetallic materials including Pd–Cu, Pd–Ga, Pd–In, Pd–Zn, Ni-Ga, In–Rh, In–Cu, In–Co, In–Ni, and Ni–Cu have also been examined for methanol production from CO_2 hydrogenation. Among these catalysts, some non-noble metal-based bimetallics were designed for efficient CO_2 hydrogenation to CH_3OH at low pressure. For the $GaPd_2/SiO_2$ system, a much higher intrinsic activity and CH_3OH selectivity than Cu-ZnO-Al_2O_3 have been observed at above 200°C and atmospheric pressure. However, under high reaction pressure (above 3.0 MPa), the performance advantage of the bimetallic catalyst is not obvious, even lower than conventional Cu-based catalysts [2,3,7,63,70–74].

In recent years, reducible oxides have received considerable attention due to their excellent performance with high CH_3OH selectivity in a wide range of temperatures (200°C–320°C). Nearly 100% selectivity can be attained over cubic In_2O_3 nanomaterial and In_2O_3 supported on monoclinic ZrO_2. Because of the lower H_2 splitting ability of In_2O_3 compared with metal catalyst, palladium (Pd) was introduced to enhance H_2 activation and facilitate oxygen vacancy formation and thereby substantially promote the activity of In_2O_3 [63]. The ZnO-ZrO_2 solid solution is another reducible oxide system efficient for methanol synthesis from CO_2 hydrogenation. Compared with Cu-based catalysts, the biggest advantage of the reducible metal oxides (In_2O_3-based oxides or ZnO-ZrO_2 solid solution catalysts) is that it can effectively inhibit the undesired RWGS reaction even if the reaction temperature is as high as 320°C [2,3,7,63,70–74].

Rungtaweevoranit and coauthors [74] reported a catalyst where Cu nanocrystal (Cu NC) was encapsulated in a metal–organic framework UiO-66 [$Zr_6O_4(OH)_4(BDC)_6$, BDC = 1,4-benzenedicarboxylate] to form catalyst Cu→UiO-66, which could catalyze the generation of methanol via the hydrogenation of CO_2. It was found that the performance of the catalyst exceeded that of the Cu/ZnO/Al_2O_3 catalyst at relative lower temperatures (<250°C), which can steadily increase the conversion rate by eight times and with 100% methanol selectivity.

Various carbon materials including carbon nanofibers, carbon nanotubes, biochar, and carbon felt have also been employed as carriers for CO_2 hydrogenation catalysts, taking advantage of their high hydrogen storage capacity, high thermal conductivity, and high specific surface area of carbon carriers. Nanosized materials are used to define nanoscale catalyst structures, in which the composition of catalysts and their surface structures can be adjusted and may bring to dare widespread applications [2,3,7,63,70–74].

The direct hydrogenation of CO_2 to methanol is as yet commercially not viable. There are still great challenges in developing catalysts with high catalytic performance and long-term stability, reducing the size of thermal catalytic reactors and decreasing the production costs. In addition, more effective and economical methods to produce H_2 are urgently needed. The primary challenge in converting CO_2 to value-added products is low selectivity due to excessive formation of CO and the large amount of water (H_2O) generated. CO and CH_4 are both formed due to partial reduction and cracking, respectively, during dehydration at the elevated temperatures (≥400°C). To date, it has limited conversion (<30%) toward desired products. In addition, H_2O produced from CH_3OH synthesis and dehydrative coupling further complicates separations and, in some cases, deactivates the catalyst. However, altering the zeolite structure has been shown to alter product distributions and increase selectivity to desired products, which was realized by increasing the space velocity during catalytic CO_2 hydrogenation over In_2O_3 combined with zeolites [2,3,7,63]. Through the introduction of CO conversion promoters, such as In–FeK, In–CoNa, or CuZnFe/zeolites, bifunctional catalysts can be upgraded to reduce CO formation. More work is, however, needed on the required zeolite properties and the structure of the bimetallic catalysts. An analysis of the integration between methanol synthesis catalysts and zeolites remains a challenge [2,3,7,63,70–74].

While due to intense research, various highly efficient and novel catalysts (such as Cu, Co, bimetallic systems, and reducible metal oxides) for CO_2 hydrogenation to methanol have been developed, considering the catalytic performance, catalyst cost, scale-up preparation feasibility, and other factors, Cu-based catalysts still hold the greatest prospect for large-scale industrial applications of methanol synthesis from pure CO_2 [63].

5.3 METHODS TO PRODUCE C_{2+} HYDROCARBONS, ACIDS, ALCOHOLS, OLEFINS, AROMATICS, AND FUELS FROM CO_2

There are four fundamental methods to produce C_{2+} hydrocarbons from CO_2.

1. Direct hydrogenation of CO_2 to C_{2+} products by FTS-based catalysis. This can produce C_2–C_5 olefins and other hydrocarbons, C_{5+} hydrocarbons, higher alcohols, and liquid fuels like gasoline, diesel, and jet fuel.

2. Productions of syngas by artificial photosynthesis or dry reforming of methane (which can be accompanied by steam reforming of methane and water gas shift reaction and partial oxidation of CH_4 to make tri-reforming of gas). Syngas of different H_2/CO compositions or CO can be transformed to C_{2+} hydrocarbons and fuels by FT, oxo, iso, and other types of syntheses.
3. Productions of higher alcohols, acids, and hydrocarbons from methanol what is generally known as methanol-based economy.
4. Insertions of CO_2 with other chemicals.

Here, we briefly examine each of these methods.

5.3.1 CO_2 Conversion to C_{2+} Products by FTS-Based Catalysis

Ethylene is a high-volume commodity chemical that is used to produce a large number of other chemicals and is an important monomer in a range of different polymers. In 2016, over 150 million tons of ethylene were produced, almost exclusively from fossil fuel–derived precursors. Although the conversion of CO_2 to ethylene requires a large energy input, the prospect of using CO_2 as a carbon source for a large fraction of commodity chemical production has motivated many research efforts. The production of light olefins by CO_2 hydrogenation by FTS-based catalysis can be challenging. For example, the selectivity of C_2 to C_4 can be above 80% for catalysts with methanol as the reactive intermediate, while only about 50% selectivity to C_2 to C_4 is achieved on the catalysts for CO_2 conversion via FTS.

The major obstacles of FTS-based CO_2 hydrogenation lie in the thermal stability of CO_2 and the complicated reaction mechanism with a wide product distribution. With the RWGS process being endothermic and FTS being exothermic, both need to be efficiently catalyzed under the same conditions, which set a very strict requirement for the catalytic system. The designed catalysts should be effective for the RWGS reaction and also active enough for the subsequent FTS reaction. Different active sites must be precisely tuned and carefully dispersed on the support materials. The key to improve the C_{2+} selectivity is to develop catalysts that have compatible bifunctional active sites for the two reaction steps, namely, *CO generation and subsequent *CO hydrogenation. The appropriately engineered reaction window is extremely important. The consequence of the incompatible activity of these two different types of active sites could be high CH_4 selectivity, as is suggested by the activation energy of RWGS being higher than that of CH_4 formation (81.0 and 59.3 kJ/mol, respectively) for Fe-based catalytic CO_2 hydrogenation [75].

The reaction temperature (~320°C) for the CO_2-FTS process is lower than that for the methanol-mediated route (~380°C). Thus, the CO selectivity (~30%) is dramatically reduced due to relatively low temperatures for the RWGS reaction, leading to improved hydrocarbon selectivity. Many researchers have tried to provide possible solutions for improving C_{2+} selectivity by adjusting the structures and compositions of the catalysts. Specifically, development of enhanced promoters, supports, and experimental conditions has been attempted. For example, bulk Fe is favorable for methane formation; however, olefin, and long-chain hydrocarbon production can be enhanced with the addition of promoters (i.e., alkaline promoters). Potassium, as an electronic promoter, can regulate the phase proportion of $Fe^0/FeO_x/FeC_x$ to maintain an optimum balance. The dissociative adsorption of *CO can be improved, while surface *H is decreased by donated electrons to the vacant d-orbital of Fe. The resulting higher C/H ratio favors chain growth and chain termination by forming unsaturated hydrocarbons [2,28]. Thus, a very important challenge is to increase the surface C/H ratio. Competitive adsorption between H_2 and CO occurs on the catalyst surface. The partial pressure of CO is always limited due the difficulty of converting CO_2 via the RWGS reaction. Since a high C/H ratio is desired to form unsaturated hydrocarbons, a possible strategy would be to enhance CO_2 activation and CO adsorption, while weakening H_2 adsorption [2,28]. In addition, the H_2 ratio in the feed gas should also be carefully chosen. To achieve good results from FTS-based CO_2 hydrogenation, more consideration is needed based on the characteristics of those reactions instead of imitation of the CO hydrogenation process.

Fe- and Co-based catalysts with alkali metal promoters are reported to be highly active for FTS to produce C_{2+} products [2] with as high as 57% selectivity at 40%–60% CO_2 conversion. The most effective promoters are alkali metals, especially Na and K, as they limit the formation of methane while improving the selectivity to C_{2+} products [2]. Meiri et al. [76] pointed out that the introduction of potassium can stabilize the texture of the Fe–Al–O spinel, increase the surface content of Fe_5C_2, and strengthen CO_2 adsorption. Martinelli et al. [77] concluded that the K-loading did not influence the CO_2 conversion, but increased the olefin/paraffin ratio and the average molecular weight of the products. Besides the alkali metals, other metals such as Cu, Zn, Ni, Zr, Mn, and Pt can also be used to modify Fe-based catalysts [2]. For example, bimetallic Fe–Cu/Al_2O_3 catalysts can suppress CH_4 formation and thus exhibit higher amounts of C_2–C_7 production compared with pure Fe/Al_2O_3 catalysts [78].

Support material is another important factor that can influence catalyst activity and product selectivity by affecting the dispersion of active metals and the interactions between reactive intermediates and the support material. The support materials for Fe-based FT catalysts can be divided into metal oxides (e.g., ZrO_2, CeO_2, and Al_2O_3) and carbonaceous materials (e.g., MOFs, mesoporous carbon, carbon nanotubes (CNTs), graphene, and organic precursors to mesoporous carbon). Carbon materials are naturally considered as good support materials for Fe catalysts. Carbon support materials can improve the dispersion of the active metals and lead to higher selectivity to olefins [79]. CNTs and graphene are also good carbon supports with superior thermal and chemical stability [2,79].

To increase selectivity to light olefins, further hydrogenation must be suppressed. Thus, synergy between the promoter and the support should be used. For example, an Al_2O_3 support can interact with a K promoter to form a $KAlO_2$ phase under calcination temperatures above 500°C, which has been shown to suppress further hydrogenation of olefins [80]. To increase the synergistic effect, the active metal can be mixed with supports. A promising support for Fe-based catalysts is MOF-derived carbon materials. Some researchers have recently developed methods of using MOFs to fabricate carbon-supported Fe-based catalysts [2,81]. Two Fe-MOF-derived catalysts have been reported for CO_2 hydrogenation, with high stability and selectivity to light olefins and liquid fuels [2,81]. However, the challenge of separating light olefins from unreacted CO_2 and H_2 remains. Increasingly advanced techniques and materials are required for their separation.

Heterogeneous FTS-based CO_2 hydrogenation can be realized in either one or two reactors, although the single-reactor process dominates. The direct one-reactor conversion system has drawn much attention due to its ease of operation and thus lower CO_2 conversion cost. This process integrates the reduction of CO_2 to CO via the RWGS reaction and hydrogenation of CO to hydrocarbons via FTS. An efficient catalyst for generating C_{2+} products, which generally refer to light olefins, liquid fuels, and higher alcohols, should be active for both RWGS and FTS under the same conditions. The product distribution can be wide, depending on the structure and composition of the catalysts. Iron (Fe)-, cobalt (Co)-, and ruthenium (Ru)-based supported catalysts, with appropriate promoters, are predominantly used in this area.

Prior to FTS, the RWGS reaction converts CO_2 to CO to form dissociatively adsorbed *CO precursor molecules on the catalyst surface. Therefore, an understanding of the reaction conditions and catalytic performance for the RWGS reaction is necessary. Since RWGS is endothermic, it requires high temperatures for reasonable conversions. CO selectivity up to 100% can be achieved at 200°C–600°C using the primary RWGS catalysts, such as Cu- and noble metal (Pt, Pd, Rh)-based catalysts, with CO_2 conversions up to 50% [82]. Recently, significant progress has been achieved in heterogeneous catalytic hydrogenation of CO_2 to various high-value and easily marketable fuels and chemicals containing two or more carbons (C_{2+} species), including dimethyl ether (DME), olefins, liquid fuels, and higher alcohols [2]. Compared to C_1 products, C_{2+} product synthesis is more challenging due to the extreme inertness of CO_2 and the high C=C coupling barrier, as well as many competing reactions leading to the formation of C_1 products.

5.3.2 C_{5+} Products by Direct Hydrogenation of CO_2 with FTS-Based Catalysis

The second set of challenges are in the production of C_{5+} products. These challenges are similar to those for the synthesis of light olefins, due to the shared fundamental reaction mechanism. The selectivity of C_{5+} products is also comparable to that of light olefins synthesis. CO_2 hydrogenation to C_{5+} products consist of multiple steps, including the RWGS reaction, C–C coupling, and acid-catalyzed reactions (oligomerization, isomerization, or aromatization). Therefore, cooperation between steps is required to develop an efficient and multifunctional catalyst. As zeolites are often used to incorporate the active sites, they generally undergo deactivation as a result of coke deposition. Thus, a big challenge for the synthesis of C_{5+} products is to reduce coke deposition in the zeolites and to enhance the catalyst stability. As heavier hydrocarbons contain many components, it is difficult to precisely control a specific type of product. Zeolites with different framework topologies are suggested to regulate product distribution. In addition, reformation of CO_2 from CO by the WGS reaction needs to be avoided. One method to decrease the reformation of CO_2 from CO is to develop better multifunctional catalysts by employing appropriate promoters to facilitate the formation of iron carbide, which is known as the active catalyst for heavy hydrocarbon formation in FTS96, and to enhance the chemisorption and dissociation of CO_2 [83]. Also, by cycling reactants, CO_2 conversion is increased. Dual-promoter systems are usually designed to improve the RWGS reaction activity, while increasing FTS activity and C_{5+} selectivity [2].

5.3.3 CO_2 Hydrogenation to Produce Higher Alcohols Based on FTS Catalysis

For higher alcohols synthesis, the major obstacles are the formation and insertion of the hydroxyl group during the C–C coupling process in the presence of parallel and consecutive reactions and usually higher energy barriers for CO insertion compared with CH_x hydrogenation [7]. This is made even more challenging since the formation pathways are different over Fe- and Co-based catalysts and the mechanisms are controversial. The synthesis of higher alcohols requires precise coordination between C–C coupling and OH formation; otherwise, more methanol or long-chain hydrocarbons would be produced. Also, the synergy of high metal dispersion and a high density of hydroxyl groups on the supports can promote the selectivity to ethanol because the hydroxyls are able to stabilize formate species and protonate methanol [2,7]. Noble metal (Au, Pd)-based catalysts were developed for direct synthesis of ethanol from CO_2 hydrogenation with high selectivity in a batch reactor [7,84]. Han's group reported water-promoted CO_2 hydrogenation to C_{2-4} alcohols, and the selectivity over Pt/Co_3O_4 is up to 88.1% at 220°C and 8.0 MPa in the water/1,3-dimethyl-2-imidazolidinone mixed solvent [85]. Recently, non-noble metal-based catalysts were found to also enable highly efficient liquid-phase ethanol synthesis from direct CO_2 hydrogenation [7].

Very recently, Ding's group reported the Cu@Na-Beta catalyst with Cu nanoparticles enclosed in the Na-Beta zeolite crystal particles that enable highly selective conversion of CO_2 to ethanol in a fixed-bed reactor due to the synergistic effects among irregular Cu nanoparticles and surrounding of zeolitic frameworks [86]. New routes using multifunctional catalysts have also been developed for higher alcohol synthesis (HAS) from syngas with high selectivity. Sun's group combined Co–Mn oxides (FTS catalyst) with Cu–Zn–Al–Zr oxides (methanol synthesis catalyst) and markedly increased the oxygenates selectivity to 58.1 wt% with the $C_{2+}OH$ fraction of 92.0 wt% [87].

5.3.4 CO_2 Hydrogenation Followed by FT Synthesis to Produce Hydrocarbon Fuels

Liquid fuels (e.g., gasoline and diesel) and other value-added chemicals (e.g., aromatics and isoparaffins) are also desired products from CO_2 hydrogenation via FTS. Liquid hydrocarbon (C_{5+}) fuels including gasoline (C_{5-11}), jet fuel (C_{8-16}), and diesel (C_{10-20}) play an instrumental role in the global energy supply chain and are wildly used as transportation fuels around the world. Jiang et al. [88] prepared Fe- and FeCo-based catalysts and achieved, when promoted with an appropriate amount of K, ≥54.6% CO_2 conversions and ≥47.0% C_{5+} selectivity. A $Na–Fe_3O_4$/HZSM-5 catalyst with three types of active sites was developed by Wei et al., demonstrating that 22% of CO_2 can be directly converted to gasoline with a selectivity of 78% [89].

When HZSM-5 was changed to HMCM-22, aromatization was suppressed, while isomerization was promoted due to the appropriate Brønsted acid properties of HMCM-22 and the special lamellar structure consisting of two independent pore systems. The octane number would be enhanced by developing catalysts that increase the fraction of isoparaffins in the gasoline. Similarly, aromatics are important feedstocks with applications in the synthesis of various polymers, petrochemicals, and medicines. $ZnFeO_x$–nNa/HZSM-5 catalysts can achieve 75.6% selectivity to total aromatics at a CO_2 conversion of 41.2% [2].

Owing to superior ability for chain growth, stability, and lower activity for the WGS reaction, Co-based catalysts have also been applied to produce long-chain C_{5+} hydrocarbons. Recently, a pure Co-based catalyst without promoters was reported to display high performance for both CO-FTS and CO_2-FTS [2,90]. The addition of promoters also helps. For example, with the introduction of Na as a promoter to a CoCu/TiO_2 catalyst, the CH_4 selectivity significantly decreased from 89.5% to 26.1%, while the C_{5+} selectivity increased from 4.9% to 42.1%, with an excellent stability of more than 200 hours [2]. He et al. [73] reported the bimetallic Co_6/MnO_x nanocatalyst with synergism between Co and Mn for 15.3% CO_2 conversion and 53.2% C_{5+} selectivity.

Syngas (CO/H_2) and CH_3OH are the most important C_1 platform molecules, and their conversions to value-added products via the Fischer–Tropsch synthesis (FTS) and methanol to hydrocarbons (MTH) processes, respectively, have been extensively applied in industry [2,7]. Therefore, combining the RWGS with FTS and combining high-temperature methanol synthesis with MTH over bifunctional/multifunctional catalysts are two efficient strategies for direct CO_2 hydrogenation to C_{2+} hydrocarbons including liquid hydrocarbons.

As CO_2 can be easily converted to CO via RWGS, modified CO-FTS catalysts were widely used for direct CO_2 hydrogenation to long-chain (C_{5+}) hydrocarbons. Methane and gaseous hydrocarbons (C_{2-4}) are also formed through this classical CO-FTS reaction. Compared with CO-FTS, more H_2 is usually needed in CO_2-based FTS, and the concentration of the CO intermediates is also lower during CO_2 hydrogenation, which results in a higher H/C ratio on the catalyst surface. As CO_2 can be directly hydrogenated to CH_4, the high H/C ratio favors methane formation and leads to a decrease in the chain growth probability. Therefore, CO_2–FTS catalysts to be developed should have high activity for both RWGS and FTS reactions but inhibit the CO_2 methanation reaction. Iron (Fe)- and Co-based catalysts are industrially adopted CO-FTS catalysts, though Co has little activity for RWGS, and thus a second component is needed for CO_2 conversion to CO. The combined selectivity of CH_4 and light hydrocarbons are usually very high (>55%), but the introduction of promoters such as Cu, Mo, and alkali metals to the Co catalyst enhances the selective formation of C_{5+} hydrocarbons from CO_2 hydrogenation [7,91]. For example, the optimized Na–Co–Mo/SiO_2–TiO_2 catalyst shows the C_{5+} selectivity of 27.3% with CH_4 selectivity of 40% at CO_2 conversion of 26.9% under the very mild reaction condition of 200°C and 0.1 MPa [91]. Nevertheless, Fe catalysts seem to be more suitable for direct CO_2 conversion because it has high RWGS reactivity but relatively lower CO_2 methanation activity.

To decrease CH_4 selectivity and enhance chain growth ability of Fe catalysts, promoters such as K, Na, Cu, Zn, Mn, and/or Ce are introduced. Simultaneously, some uniquely structured Fe catalysts were fabricated with improved activity and stability [7,92]. Zn serves as the structural promoter to improve the dispersion of Fe species [7,93]. Gasoline and diesel range hydrocarbons can also be obtained over the reduced $ZnFe_2O_4$ [93]. Under the same reaction condition, C_{5+} selectivity over the Fe–Zn catalyst is lower than the Fe–Cu catalyst (58.5% vs 66.3%) with much higher CO_2 conversion (34.0% vs 17.3%) and lower CO selectivity (11.7% vs 31.7%) [93,94]. K-promoted Fe–Co/Al_2O_3 with Co/(Co+Fe) = 0.17 also enables CO_2 hydrogenation to liquid hydrocarbons with jet fuel-range α-olefins as the main products [95].

The excellent performance for the synthesis of heavier hydrocarbons using these Co or Fe catalysts is usually related to the presence of alkaline metals (K or Na) because these promoters can enhance the RWGS reaction while suppressing the methanation reaction [7,93]. Additionally, the high temperature applied for traditional Fe-based FTS favors the endothermic RWGS reaction, and the traditional Co-based catalyst usually operates at a lower temperature of 180°C–240°C. Therefore, compared to catalysts with metallic Co as the active site, it is easier to obtain heavier hydrocarbons from CO_2, and the product distribution is closer to that derived from CO-FTS over the Fe catalyst with iron carbides as active sites at a higher reaction temperature (300°C).

However, the CO_2 hydrogenation product distribution over the modified Fe-based FTS catalysts still follows the well-known Anderson–Schulz–Flory (ASF) distribution because CO_2- and CO-FTS have similar reaction mechanisms. In addition, the hydrocarbons are mainly olefins and normal-paraffins (*n*-paraffins) for Fe-based FTS [7,89]. Recently, Wei et al. combine Fe catalysts with HZSM-5 zeolites to significantly increase the fraction of high-octane gasoline-range isoparaffins and aromatics in the CO_2 hydrogenation product [89].

Very recently, more researchers are considering the direct synthesis of value-added C_{2+} liquid products from CO_2 hydrogenation. Direct CO_2 hydrogenation to oxygenate possesses a better atom economy and also a higher efficiency in hydrogen utilization than hydrocarbon production [2]. On the other hand, worldwide consumption of liquid hydrocarbon fuels is greater, due to their higher energy content. Currently, high selectivity of gasoline fuels (above 75%) can be achieved over Na–Fe/HZSM-5 and In_2O_3/HZSM-5 bifunctional catalyst systems albeit via very different reaction mechanisms. The modified FTS catalysts are expected to be more suitable for the synthesis of heavier hydrocarbons (such as jet fuels and diesel) and $C_{2+}OH$ alcohols from CO_2 due to the great C–C coupling ability. For CO_2 hydrogenation to liquid hydrocarbons over Fe-based catalysts, better understanding of the iron carbide active sites is required for further improving the stability and decreasing the selectivity of light hydrocarbons as well as the industrial implementation of this process. To further enhance the performance of the Co-based CO_2–FTS catalyst, it is necessary to introduce a highly active low-temperature RWGS site. For the production of higher alcohols, catalyst development is still at a very early stage, where the problem of low alcohol yield remains to be solved. In addition to learning from the new strategy using multifunctional catalysts in HAS from syngas, using the confinement or modulation effects of zeolites on the reactive centers is an effective strategy to increase the selectivity of higher alcohols in CO_2 hydrogenation.

If the CO_2 source is coal- or biomass-combustion flue gas, the CO_2 conversion process becomes more complicated due to the copresence of CO, O_2, SO_x, and/or NO_x [2,7]. Cost associated with cleaning the flue gas especially via deep desulfurization will decrease the economic viability of CO_2 conversion [56]. In this regard, the ZnO-ZrO_2 catalyst mentioned above is more suitable due to its excellent sulfur tolerance. Although little attention has been paid to the effect of sulfur-containing molecules on the CO_2 hydrogenation reaction, it can be speculated that other oxide catalyst systems with the oxygen vacancy as the active site also have good sulfur resistance owing to their similar reaction mechanism. Therefore, for CO_2 sources from flue gas, partially reduced oxides and oxides/zeolites bifunctional catalysts are industrially more relevant for methanol production and liquid hydrocarbons synthesis via the methanol intermediate, respectively. The In_2O_3-based catalysts have received especially great attention due to their high performance and relatively simple active site structures. In addition, these catalyst systems usually show excellent selectivity and high stability, although their single-pass CO_2 conversions are usually much lower than those over supported metal catalysts (such as Cu- and Fe-based catalysts) and are generally below 20% even with noble metal modifiers and at relatively high reaction temperature (>280°C). A very high recycle ratio of the unconverted gas is needed in industrial applications of CO_2 hydrogenation, which may decrease its energy efficiency and economic value. Thus, it is necessary to further increase the amount of oxygen vacancies and enhance the H_2 splitting ability to improve the intrinsic activity of these oxide catalysts.

Recent years have seen emerging experimental and computational technologies for more efficient search and design of catalysts and other materials. Experimental technologies such as high-throughput catalyst synthesis and performance evaluation, 3D printing, and in situ characterizations and monitoring are increasingly being employed for the rapid discovery of novel catalysts and materials [7]. On the other hand, a similar array of computational technologies including high-throughput and automated computational simulations and reaction modeling coupled with machine-learning algorithms also start to enable the theoretical understanding and prediction of new catalysts [7]. The applicability of the above experimental and computational technologies in designing industrially relevant heterogeneous catalysts varies due to their greater complexity and stringent requirements. Nevertheless, these new technologies hold great potential in revolutionizing the way that industrial catalysts have been traditionally developed, and thus great progress can be expected in this research. Apart from further developing more efficient catalysts for milder reaction conditions, we should also pay sufficient attention to reactor design and optimization. For example, a membrane reactor can shift the equilibrium-limited CO_2 hydrogenation to liquid fuels by selective and continuous *in situ* removal of the byproduct water, leading to a substantial increase in CO_2 conversion as well as the yield of liquid fuels [7].

The use of bifunctional catalysis is another avenue to increase aromatics like benzene, toluene, xylene in liquid fuels. A combination of a metal/metal oxide and a zeolite for the heterogeneous gas-phase hydrogenation of CO_2 to value-added compounds (especially olefins and aromatics) is one of the most attractive technologies of the chemical industry. The greatest advantage of this bifunctional metal-zeolite configuration is that the product selectivity can be shifted to either light olefins (C_2–C_4) or aromatics by selecting the appropriate zeolite, overcoming many limitations of the standalone metal catalyst (i.e., thermodynamic limitations and/or product distribution) [2,3,7,63,71,72,96]. However, multiple reactions are involved (oligomerization, cracking, dehydrogenation, cyclization, alkylation, isomerization, etc.) and many intermediates can serve as reactants for other reaction pathways, limiting the total olefin or aromatics selectivity. In order to tackle this, more complete understanding of the global reaction mechanism is required. Additionally, despite some promising results, there are important issues to overcome, like the high CO selectivity (frequently higher than 50% [2,82]) and the zeolite selection and fine-tuning.

5.3.5 Syngas Formation by Dry Reforming of Methane

The dry reforming of methane (DRM) (Eq. 5.5) can consume two greenhouse gases (CO_2 and CH_4) simultaneously and produce synthesis gas (abbreviated syngas, a mixture of CO and H_2). The ratio of H_2 to CO products is 1, which is much lower than other reforming reactions, such as the steam reforming of methane (SRM) and partial oxidation of methane (POM) (Eqs. 5.6 and 5.7). The low ratio of H_2/CO is useful for synthesizing long-chain hydrocarbons via the Fischer-Tropsch reaction [3].

$$CH_4 + CO_2 \rightarrow 2CO + 2H_2, \qquad \Delta H^\circ = 247\,\mathrm{kJ/mol} \tag{5.5}$$

$$CH_4 + H_2O \rightarrow CO + 3H_2, \qquad \Delta H^\circ = 246\,\mathrm{kJ/mol} \tag{5.6}$$

$$CH_4 + 1/2\,O_2 \rightarrow CO + 2H_2, \qquad \Delta H^\circ = -36\,\mathrm{kJ/mol} \tag{5.7}$$

The DRM reaction is endothermic, requiring a high reaction temperature of >700°C, which results in high energy cost [3,97]. Figure 5.4 shows the DRM equilibrium plots at various temperatures and 1 atm [3]. The high temperature is required to attain high product yield. Side reactions such as the RWGS reaction (Eq. 5.8) or carbon gasification reaction (Eq. 5.9) can occur. The RWGS reaction affects the H_2/CO ratio; because the RWGS reaction happens more, the H_2/CO ratio decreases and CO_2 conversion increases. The steam produced from the RWGS reaction can react with carbon and produce syngas.

$$CO_2 + H_2 \rightarrow CO + H_2O, \qquad \Delta H^\circ = 41\,\mathrm{kJ/mol} \tag{5.8}$$

$$C + H_2O \rightarrow CO + H_2, \qquad \Delta H^\circ = 131\,\mathrm{kJ/mol} \tag{5.9}$$

Also, deactivation occurs easily due to sintering and coke deposition, which degrades long-term durability [3,98]. Carbon is produced due to the Boudouard reaction and methane cracking (Eqs. 5.10 and 5.11). The carbon is thermodynamically the main product at temperatures lower than 570°C [3,99]. DRM can proceed above 640°C with methane cracking. The Boudouard reaction cannot occur above 820°C. Carbon can be produced by the Boudouard reaction, and methane cracking occurs from 570°C to 700°C [100]. Coke deposition is thermodynamically favored at low temperature.

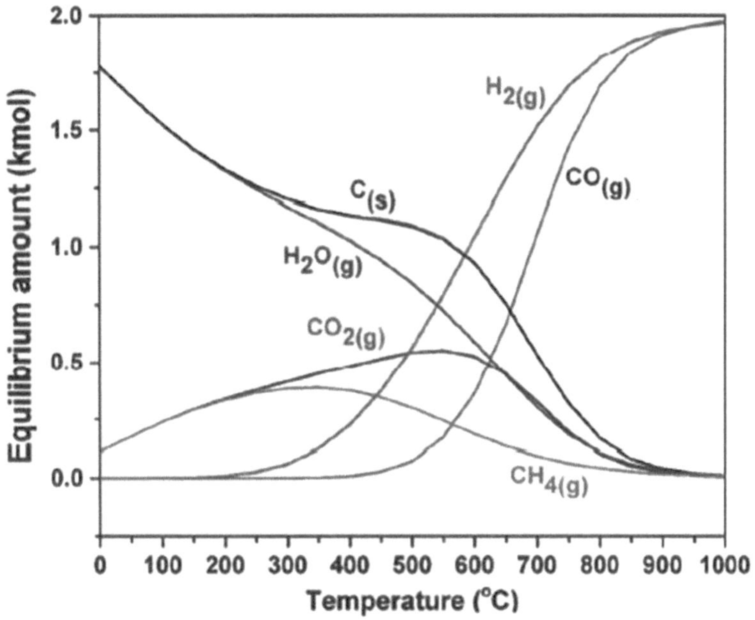

FIGURE 5.4 Thermodynamic equilibrium plots for DRM at 1 atm, 0°C–1,000°C and at a reactant feed ratio of $CO_2/CH_4 = 1$. These equilibrium calculations were conducted using the Gibbs free energy minimization algorithm on HSC Chemistry 7.1 software. [3].

The carbon produced by methane cracking is likely to react with steam or carbon dioxide at high temperature. Thus, the DRM reaction should be performed at high temperature.

$$2CO + C \rightarrow C + CO_2, \qquad \Delta H^\circ = -172\,\text{kJ/mol} \tag{5.10}$$

$$CH_4 \rightarrow C + 2H_2, \qquad \Delta H^\circ = 75\,\text{kJ/mol} \tag{5.11}$$

5.3.5.1 Role of Catalysts, Supports, and Promoters

Various heterogeneous catalysts have been developed, including metal supported catalysts, perovskites, and solid solution catalysts [3,101]. Precious metals (Pt, Rh or Ru) are known to have high activity and durability, although at high price. Non-precious metals (Ni or Co) have been widely investigated as well. Precious metals such as Pt, Rh and Ru are highly active for DRM and also resistant against coke deposition [3,98,102]. The precious metal-based catalysts show high activity in spite of the very small amount of metal catalyst used. Non-precious metals such as Ni or Co have been used more for DRM due to their cheap price and abundance [3,103]. Ni catalysts have shown a level of activity comparable to precious metals [102]. Alloyed metal catalysts are widely used for DRM because they have a different electronic structure than monometallic materials [104]. Monometallic Ni or Fe catalysts have shown poor durability, because the monometallic Ni catalyst is easily deactivated by coke deposition, and Fe is inactive for DRM.

Supports can affect catalytic activity and durability for DRM [3,98]. The interaction between a support and metal can change the dispersion of the metal, resistance against sintering, and the electronic structure of the metal [3,105]. The strong-metal support interaction (SMSI) causes a high dispersion of metals resulting in small-sized metal nanoparticles with enhanced DRM activity. In addition, it increases resistance against sintering with enhanced durability [9,10,106]. The acidity and basicity of the support also affects catalytic properties. CO_2 is activated by forming formate with surface hydroxyl on an acidic support. CO_2 is activated by forming oxy-carbonate on a basic support.

Promoters have been widely used for DRM. Alkali metal oxides such as K_2O or CaO have been used to increase the basicity of the catalyst enhancing CO_2 adsorption [3,107]. CeO_2 or ZrO_2 were used to increase oxygen mobility [3,108]. The transition between Ce^{3+} and Ce^{4+} can occur easily on CeO_2. The coke formed on Ni sites could be oxidized by using the redox cycle of CeO_2 with enhanced coke-resistivity [108]. The surface hydroxyl group can participate in the DRM surface reaction.

Various strategies to control the nanostructures of catalysts to prevent coke deposition have been reported. In order to prevent the sintering of nanoparticles, an inorganic oxide overlayer was formed on Ni nanoparticles [3,109]. The effects of Ni size and support have been evaluated independently. The Al_2O_3 and MgO overlayers showed enhanced CH_4 turnover frequency [109]. Similarly, core-shell structures have been investigated; an outer shell such as SiO_2 separates the catalytic core material such as Ni nanoparticles while preventing sintering [3,110]. The core-shell structures have been widely used in DRM [3,111]. Mesoporous materials have been widely used as supports because they provide a large surface area and facile mass transfer [3,112]. Mesoporous Al_2O_3 in particular has been widely used because it can have a strong interaction with a metal active phase, and it has high thermal stability [3,113]. An extensive review of dry reforming is given by Shah and Gardner [98]. Syngas can also be produced by artificial photosynthesis involving CO_2 and H_2O. This reaction is covered in a number of other chapters.

5.3.6 Fuel (Hydrocarbon) Production from Syngas by FT Synthesis

Syngas (CO/H_2) and CH_3OH are the most important C_1 platform molecules, and their conversions to value-added products via the Fischer–Tropsch synthesis (FTS) and methanol to hydrocarbons (MTH) processes, respectively, are extensively applied in industry [2,4,28,52,114]. The Fischer–Tropsch process is used to convert CO and H_2 into liquid fuels and has been commercialized on a large scale [115]. In addition to large-scale units for Fischer-Tropsch, smaller units, which may be better suited for use with waste carbon streams, are also being developed. Small-scale FT process can also be applied to CO generated from CO_2, electrochemically reacting with sustainably generated H_2.

Significant amount of research is currently being performed to develop systems that can perform Fischer-Tropsch chemistry starting from CO_2 in a single reactor using a single catalyst. Both Sunfire and INERATEC have developed small-scale demonstration plants for CO_2-based Fischer-Tropsch chemistry. However, these rely on water electrolysis to supply H_2, which is not economically competitive on a large scale. One of the challenges associated with the Fischer-Tropsch process using CO_2 is that there is only a small concentration of CO present during the reaction. This limits chain growth and consequently the product distribution is normally rich in light hydrocarbons, which are not suitable as liquid fuels. To date, iron-based catalysts, which are active for both the RWGS reaction and Fischer-Tropsch chemistry, have been the most extensively explored [89]. Additions of various transition metal–based promotors to iron-based catalysts improved product distributions, but mechanisms for these successes are not understood. Similarly the reasons for improved performance by support such as SiO_2 are also not understood. Further research to understand the mechanisms, development of new catalysts, and the optimal design of reactors is required.

5.3.7 Methanol-Based Economy

Methanol (CH_3OH), produced globally on a scale of approximately 70 million tons in 2015, typically is synthesized from syngas (H_2 + CO) obtained directly from fossil fuels [52,53]. As shown earlier, methanol can also be produced by CO_2 hydrogenation. While methanol can be used as transportation fuel, nowadays most of the production companies around the world use methanol as a raw material to produce different products. Methanol is used in producing solvents like the acetic acid, which represents 10% of the global demand [63]. Methanol can also be used in direct methanol fuel cells (DMFC), which is used for the conversion of chemical energy in methanol directly to electrical power under ambient conditions [63]. The term "Methanol Economy" includes an anthropogenic carbon cycle for methanol production as shown in Figure 5.5, which can be used as a renewable fuel or to produce nearly all products that are derived from fossil fuels [63,116]. Carbon Recycling

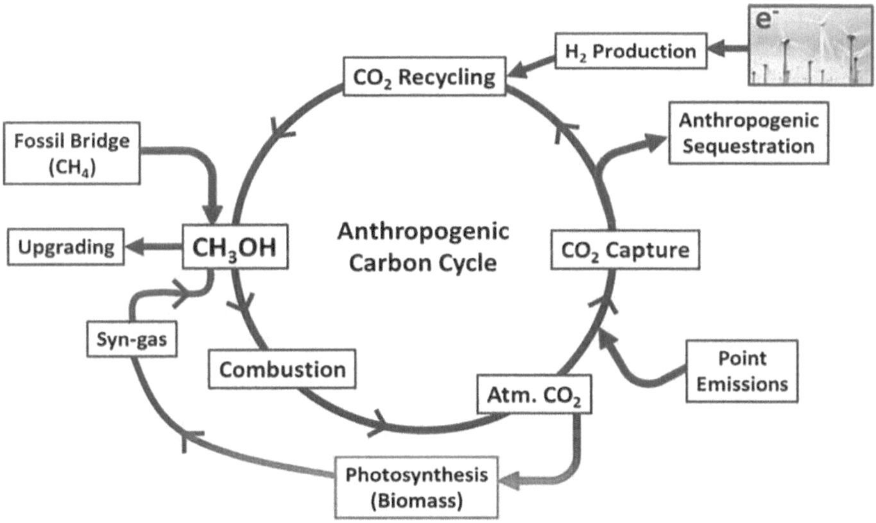

FIGURE 5.5 Anthropogenic carbon cycle for methanol production [63].

International's (CRI) George Olah plant is considered to be the world's largest CO_2 methanol plant. In 2015, Carbon Recycling International (CRI) scaled up the plant from a capacity of 1.3 million liters of methanol per year to more than five million liters a year. The plant now recycles 5.5 thousand tons of CO_2 a year. All energy used in the plant comes from the Icelandic grid that is generated from geothermal and hydro energy [117].

Methanol (CH_3OH) reaction-based CO_2 hydrogenation can be realized by coupling two sequential reactions over a bifunctional catalyst. First, CO_2 and H_2 are converted to CH_3OH over a partially reduced oxide surface (e.g., Cu, In, and Zn) or noble metals via a CO or formate pathway. Then, methanol is dehydrated or coupled over zeolites or alumina. Accordingly, bifunctional or hybrid catalysts are composed of a CH_3OH synthesis catalyst and a CH_3OH dehydration/coupling catalyst, which can convert CO_2 into high-value C_{2+} compounds, including DME, hydrocarbons like gasoline, and light olefins. An efficient catalyst for these high-value C_{2+} products should be active for both CH_3OH synthesis and dehydration/coupling under the same conditions (Eqs. 5.12–5.15).

$$CO_2 + 3H_2 \rightarrow CH_3OH + H_2O \tag{5.12}$$

$$2CH_3OH \rightarrow CH_3OCH_3 + H_2O \tag{5.13}$$

$$nCH_3OH + H_2 \rightarrow CH_3(CH_2)_n - 2CH_3 + nH_2O \tag{5.14}$$

$$nCH_3OH \rightarrow CH_2 = CH(CH_2)_n - 3CH_3 + nH_2O \tag{5.15}$$

In CO_2 hydrogenation to C_{2+} compounds, the reactions of CO_2 to CH_3OH and CH_3OH to C_{2+} compounds take place at 200°C–300°C and 400°C, respectively, over bifunctional catalysts. Therefore, investigation of the reaction conditions, catalyst properties, and catalytic performance for CO_2 to CH_3OH and CH_3OH to C_{2+} products is necessary.

5.3.8 Use of Methanol for Productions of Higher Hydrocarbons and Fuels

Methanol from direct hydrogenation of CO_2 or from syngas can be a raw material for a number of other chemicals and fuels. Dimethyl ether, which is closely related to methanol (dehydration of methanol gives dimethyl ether), is a platform chemical widely used as an alternative to liquefied petroleum gas as a clean fuel [48]. Currently, dimethyl ether is synthesized either via methanol dehydration or from syngas. The Korea Gas Corporation has developed an indirect strategy for dimethyl ether synthesis from CO_2 that is performed on a scale of 100 tons per day. It involves the tri-reforming of methane, CO_2, and H_2O into syngas. The syngas is then converted into dimethyl ether. There has been rapid development in DME synthesis from CO_2 hydrogenation using CH_3OH synthesis catalysts hybridized with CH_3OH coupling catalysts [2]. The effect of promoters, supports, and synthesis conditions has been explored [2]. For example, the acidic sites on γ-alumina surfaces and the $CuAl_2O_4$

spinel phase can be regulated by promoters like gallium or zinc oxides, resulting in higher stability for Cu NPs during CO_2-to-DME. To date, CO_2 conversion and DME selectivity mostly vary between 35% and 80% and 5%–50%, respectively, but CO_2 conversion can reach up to 97% at 280°C over a Cu–Zn–Al/HZSM-5 catalyst by drastically increasing the reaction pressure (to 36 MPa) [2].

Formic acid is used as a preservative, insecticide, or reducing agent primarily in the food, textile, and pharmaceutical industries [56]. In 2013, the global production of formic acid was 620 kilotons. The most common method for the synthesis of formic acid is a two-step process involving the initial reaction of CO with methanol to form methyl formate, followed by the conversion of methyl formate to formic acid either through reaction with ammonia to generate formamide and subsequent acidification with H_2SO_4, or through direct hydrolysis with water.

Dimethylcarbonate is used to synthesize polycarbonates, as a mild methylating agent in organic chemistry, and as a solvent [52,118]. It is currently produced on a scale of 90 kilotons per year [52]. Potential applications for dimethylcarbonate also exist in the fuel industry where it could be blended with gasoline as an oxygenate. Historically, dimethyl-carbonate was synthesized from phosgene and methanol, but now it is prepared either via transesterification of ethylene carbonate or propylene carbonate and methanol or using CO, methanol, and O_2 as feedstocks. While there has been considerable research into the development of catalysts for the conversion of CO_2 and two equivalents of methanol into dimethyl carbonate and water [119], results so far are not very successful. An alternative strategy to make dimethylcarbonate from CO_2 involves using urea as a CO_2 carrier [52,120]. In this approach, the reaction of urea and methanol generates dimethylcarbonate and ammonia.

The transformation of methanol to hydrocarbons (MTH reaction) is a domain of zeolite catalysis, usually HZSM-5 and SAPO-34. The mechanism of the MTH reaction has been the subject of vast amount of research. MTH reaction goes through two phases: (a) a short induction period followed by the (b) autocatalytic dual cycle mechanism. In the induction period, formation of the direct C-C bond takes place (i.e. the coupling of two methanol and/or dimethylether (DME). Out of many proposed, dual hydrocarbon-pool (HCP) mechanism is nowadays the most accepted one. This hydrocarbon pool, with an initially specified overall stoichiometry $(CH_2)_n$, represents an adsorbate (containing several polycondensed aromatic species), which has many characteristics in common with ordinary coke and contains less hydrogen than indicated. Product distribution depends on the zeolite topology, acidity and the operating conditions, causing either the aromatic or the olefinic cycle to propagate more than the other cycle. Nevertheless, the mechanism of the methanol to olefins (MTO) process is still a matter of investigation and discussion.

In the synthesis of carbamates, the aprotic organic bases can function as CO_2 absorbents and transcarboxylation agents. The initial attempt was made by Rossi group, in which CO_2 is trapped by a methanol solution of commercially available tetraethylammonium hydroxide. The resulting tetraethylammonium hydrogen carbonate can be used as a surrogate of CO_2 in the synthesis of carbamate. Meanwhile, the presence of tetraethylammonium ion as counterion increases the nucleophilicity of carbamate anion [121].

5.3.9 Role of Bifunctional Catalysts for CO_2 to Higher Hydrocarbons by RWGS or Methanol Route

Fujimoto and coworkers [122] were the first authors to combine a methanol synthesis catalyst with a zeolite in order to produce hydrocarbons from CO_2. Their focus was on the production of C_2–C_5 hydrocarbons by combining a Cu-Zn catalyst with a dealuminated Y zeolite. Analogously, Kuei and coworkers [123] were among the first authors to try to produce hydrocarbons from CO_2 via the RWGS route using multi-functional catalysis. They combined a fused Fe catalyst with an H-ZSM-5 zeolite and produced aromatics directly from CO_2. They were the first authors to report product yields of olefinic, paraffinic, and aromatic components from CO_2 hydrogenation using bifunctional catalysis. Fujiwara and coworkers [124] also reported the olefinic, paraffinic, and aromatic components of their products using the MeOH route for hydrocarbon production from CO_2. Their catalyst was a combination of the multi-metallic Cu-Zn-Cr catalyst and the proton form of zeolite Y. They reported high conversion of CO_2 (33.5%) but poor hydrocarbon yields due to high CO selectivities (>80%).

MeOH can not only be dehydrated to DME but also serve as an intermediate to synthesize hydrocarbon chains, $(CH_2)_n$, and final products, such as olefins, aromatics, and gasoline. In_2O_3 mixed with SAPO-34 is attractive for efficient CO_2 conversion to CH_3OH and subsequent selective C–C coupling of CH_3OH to form light olefins [2]. Similarly, a high yield of light olefins can also be achieved by composite catalysts, such as ZnZrO, ZnGaO, and CuZnZr mixed with SAPO-34 [2]. The selectivity of light olefins can be as high as 90%, while the CH_4 selectivity is less than 5% of the hydrocarbon products with 15%–30% CO_2 conversions over most bifunctional catalysts tested, which deviates from the Anderson–Schulz–Flory (ASF) distribution [2]. When the zeolite is changed from SAPO-34 to HZSM-5, more C_{5+} compounds are produced than light olefins. A 78.6% selectivity of gasoline-range hydrocarbons, with only 1% CH_4 selectivity, was obtained from the tandem In_2O_3/HZSM-5 catalyst26. When In_2O_3 is replaced by $ZnAlO_x$ or ZnZrO, CH_3OH is synthesized on the metal oxide surface and then converted to olefins and aromatics inside the HZSM-5 pores with an aromatic selectivity of 73%, which is attributed to a shielding of the external Brønsted acid sites of HZSM-5 by $ZnAlO_x$ [2]. Therefore, the type of product is affected by both the character of the metal oxides and the geometries of the zeolites that determine the confinement of the hydrocarbons [2].

Kim and coworkers [85] presented a very interesting configuration of a bifunctional catalyst. They impregnated alkali ion exchanged zeolite Y with potassium-doped Fe to hydrogenate CO_2 via the RWGS route. They reported quite low selectivity to CO (26.5%) and among the highest selectivity to light olefins (23.2%) available in open literature using the RWGS route. Again, due to milder temperatures (300°C), they did not achieve very high conversion of CO_2. Their follow-up works achieved similar results with an Fe-impregnated alkali ion exchanged zeolite Y (using Na instead of K in this case) and an Fe-Ce-impregnated alkali ion exchanged zeolite Y. Tan and coworkers [125] combined a methanol synthesis catalyst with a zeolite for hydrogenation of CO_2. Their focus, however, was on the production of isobutane from CO_2 and therefore they did not obtain high olefin or aromatics selectivity.

They studied various multi-metallic catalysts (combining Cr, Al, Ga, and Zr with Fe-Zn) and concluded that the Fe-Zn-Zr catalyst combined with HY zeolite gave the best yield for isobutane.

Li and coworkers [126] used a combination of Zn-modified ZrO_2 and Zn-modified HSAPO-34 in order to achieve selective conversion to lower olefins from CO_2 via the MeOH route. These authors were able to further enhance the selectivity of lower olefins from 80% to 93% in the hydrocarbon fraction by increasing the space velocity from 3,600 to 20,000 mL/gcat/h, but the conversion is drastically reduced. Liu and coworkers [127] obtained similar results by using a combination of $ZnGa_2O_4$ and SAPO-34. Gao and coworkers [128] also worked on production of hydrocarbons from CO_2 with a focus toward gasoline fraction selectivity. They, however, chose to take the MeOH route using a combination of In_2O_3 and HZSM-5. Although they achieved good selectivity toward the gasoline fraction in the hydrocarbon portion of products (~78%), owing to the high selectivity toward CO (~45%) their overall selectivity toward the gasoline fraction was limited (~43%). Additionally, CO_2 conversion was much lower when compared with the RWGS route.

Liu et al. [127] studied the MeOH route to selectively produce light olefins (C_2–C_4) from CO_2 using a combination of In_2O_3-ZrO_2 and HSAPO-34. Although they achieved a high proportion of light olefins in the hydrocarbon fraction (~90% and ~80%, respectively), the selectivity to CO was too high (>80%), limiting therefore the applicability of these catalysts. Dang et al. [129] improved the performance of its bifunctional catalyst by varying the In-Zr ratio. They reported their best yield of light olefins at an In:Zr ratio of 4:1. Moreover, they significantly improved their selectivity toward hydrocarbons by reducing CO selectivity from ~90% to ~64%. Finally, Ni and coworkers [130] reported a combination of ZnAlOx and HZSM-5, which yielded high aromatics selectivity (73.9%) in the hydrocarbon fraction by following the MeOH route. However, the CO_2 conversion was only 9.6% and the CO selectivity was more than half of the total products (57.4%).

The RWGS route has proven to be more productive with higher yields of olefins and/or aromatics, as high methanol selectivity from CO_2 can only be achieved at low reaction temperatures (usually lower than 300°C), which in turn limits the CO_2 conversion and consequently the yield of products *via* the MeOH route. Moreover, the RWGS route has led to significantly lower selectivities to CO as compared to the MeOH route due to CO undergoing FTS reaction. However, methane production is much more pronounced in the case of RWGS route owing to the mechanism of the FTS reaction, alongside the wider product distribution in comparison with the MeOH route. In addition, we have to take into consideration that aromatization reactions require higher temperatures (300°C–400°C olefins, 500°C–600°C paraffins) than Fisher-Tropsch (~300°C) limiting the aromatics selectivity in the RWGS route. Therefore, the development of catalytic materials that can work at higher temperatures seems to be mandatory to achieve commercial implementation for both routes (*i.e.*, to achieve higher conversion in the MeOH route and to couple aromatization with FTS in the RWGS route).

The selection of the most suitable zeolite is another key parameter. H-ZSM-5 has proven to be unbeatable in aromatic production, while SAPO-34 is the preferred choice for olefin production. However, a very limited number of different zeolites have been studied, pointing to the possibility of there being a hidden winner among the more than 200 types of discovered zeolites. Moreover, surface modification, acidity tuning, and incorporation of secondary metal are also features to take into consideration. Surface modification and acidity tuning can affect the product distribution, as it has been reported that BTX selectivity can be enhanced by suppressing isomerization, disproportionation, and dealkylation/alkylation reactions via surface/acidity modification.

Incorporation of metals into zeolites can also assist in aromatization, involving a bifunctional process where metal species provide dehydrogenation sites leading to olefins, which are further converted to aromatics. For example, over gallium-doped zeolites, gallium catalyzes the dehydrogenation of paraffins, while the protonic sites catalyze olefin oligomerization and diene cyclization. However, gallium species can also catalyze undesired reactions like cracking. Some authors have reported that the increasing the proximity of the two components shifts the equilibrium toward the desired products and hence increases the conversion. On the other hand, Wei and coworkers [13] reported that when their Na-Fe_3O_4 and HZSM-5 catalyst were integrated by powder mixing, the close proximity between iron-based sites and zeolite acid sites turned out to be detrimental due to zeolite acid sites poisoning the alkali sites on the iron catalyst. The spatial arrangements of the metal oxide/zeolite bifunctional catalyst investigated have ranged from mortar mixing (representing the closest proximity between metal catalyst and zeolite acidic sites) to dual bed configuration [13] (representing the furthest proximity) with powder mixing and hydrogenation of olefins [131].

The configuration of the bifunctional catalyst also plays a major role in the product selectivity. Granule stacking as the intermediate proximity options very well depend upon the hydrogenation route followed (RWGS or MeOH). Although not all the authors have explored the various spatial arrangements, very close proximity (i.e., mortar mixing) is usually detrimental to the catalytic performance regardless of the route. On the other hand, granule stacking and dual bed configurations seem to provide optimal site separation with the conversion and product distribution depending slightly on the configuration chosen. In summary, more comparative studies varying the spatial arrangement need to be carried out [131].

The literature also indicates that experimental and theoretical investigations involving model reactants in an atmosphere with CO_2, CO, H_2, and H_2O are needed, as these compounds can drastically change the classical MTH or aromatization networks. For example, CO not only functions as a reactant in the FTS pathway but also as a hydrogen acceptor to accelerate MTH and FTS. If this role develops further, strategic bifunctional catalyst development should also become more feasible. Bifunctional catalysis is a unique model; it requires at least double the effort for catalyst design, and there are obviously more parameters to consider compared to a single catalyst system. The literature also indicates that CO fraction should be included in the estimation of overall selectivities (selectivity should not be on CO free basis). If efficient routes foe green generation of H_2 are developed, CO_2 hydrogenation, particularly by methanol route can be a formidable alternative for higher hydrocarbon production [131].

5.3.10 CO_2 Insertion with Other Chemicals

A number of chemicals are produced by insertion of CO_2 molecules in other chemicals. Diphenylcarbonate is used as a feedstock for the synthesis of aromatic polycarbonates [132]. It is prepared either using toxic phosgene and phenol or via the transesterification of dimethylcarbonate. Recently, Asahi Kasei developed an indirect method to produce diphenylcarbonate from CO_2 and phenol. The direct synthesis of diphenylcarbonate from phenol and CO_2 is also possible but suffers from similar problems to the reaction of methanol with CO_2. In 2011, Shell opened a 500 ton per annum demonstration plant for the synthesis of diphenylcarbonate from phenol and CO_2 using propylene oxide as a water scavenger [52].

Carboxylic acids comprise a broad class of commodity chemicals that are used as solvents, reagents, and monomers for polymer production, among other applications. They are typically synthesized by aerobic oxidation of hydrocarbons which requires expensive reactors due to corrosive conditions [52]. Conceptually, inserting CO_2 into a C–H bond is an attractive alternative synthesis of carboxylic acids that would eliminate the need for a difficult oxidation step. However, CO_2 insertion is thermo-dynamically disfavored depending on the type of C–H bond. As a result, catalytic reactions either need to be performed in the presence of base (to drive the reaction forward by carboxylate formation) or need to be combined with a very effective separation process to remove the carboxylic acid in situ [52].

Acrylic acid and methacrylic acid are large commodity-scale chemicals used in the synthesis of polymers and water superabsorbers [133]. Acrylic acid can also be converted into acrylonitrile, which is a feedstock for carbon fiber synthesis. In principle, acrylic acid and methacrylic acid could be formed directly from CO_2 insertion into a C–H bond of ethylene or propylene, respectively [133]. As noted above, the thermodynamics of these processes are unfavorable, which has led to a stepwise approach in which the carboxylate is formed first and subsequently protonated in a second step. Furan-2,5-dicarboxylic acid (FDCA) is a monomer that has attracted strong commercial interest for polyester synthesis [134]. One potential method for FDCA synthesis involves edible fructose as a feedstock and requires difficult oxidation and purification steps. The carboxylation of furoic acid is potentially an advantageous route because furoic acid is produced from furfural, which is made industrially from inedible biomass [52].

Benzoic acid is a relatively small-scale commodity chemical used as an intermediate in phenol synthesis and in the production of preservatives (alkali benzoates), plasticizers (benzoate esters), and solvents. The preparation of benzoic acid by inserting CO_2 into a C–H bond of benzene has been investigated as an alternative strategy in a few early-stage research efforts [52]. These methods have relied on very energy-intensive stoichiometric reagents to activate benzene, such as $Al_2(CH_3)_3(OCH_2CH_3)_3$ or a combination of $AlCl_3$ and Al [52,135]. While these examples have provided insight into chemical strategies for activating aryl C–H bonds, the CO_2 footprint associated with the production of strong Lewis acids far outweighs the CO_2 consumed in the benzoic acid synthesis. New strategies to avoid stoichiometric reagents altogether are needed [52].

Franco group successfully identified the DBU-CO_2 complex via reacting CO_2 with DBU (1,8-diazabicyclo[5.4.0]undec-7-ene) in anhydrous acetonitrile, implying that DBU can be used as CO_2 trap reagent [136]. The resulting reactive DBU-CO_2 adduct can be utilized as transcarboxylating reagent for synthesis of *N*-alkyl carbamates. The combination of organic base and alcohol is an efficient CO_2 capture system and the absorbed CO_2 can be *in situ* transformed. Yang et al. [137] developed PEG/superbase system in which in the capture step, the superbase can be used as a proton acceptor and almost equimolar CO_2 per mole superbase can be absorbed. The resulting liquid amidinium carbonate can directly react with *n*-butylamine at 110°C to afford dibutyl urea in almost quantitative yield (96%) without any other additives. This protocol can also be used in the synthesis of other symmetrical urea derivatives.

Yoshida and coworkers use DBU to enrich and activate CO_2 in air and perform the first example of directly transforming atmospheric CO_2 into the substituted 5-vinylideneoxazolidin-2-ones using propargylic substrate 4-(benzylamino)-2-butynyl carbonates or benzoates as a substrate [138]. In subsequent work, they further improved the reaction efficiency by utilizing $AgNO_3$ as catalyst and propargylic amines as substrates, Yang et al. [137] designed a series of novel CO_2 capture and activation systems in which by employing ammonium iodide as catalyst, the cycloaddition reaction of various aziridines with the captured CO_2 by $NH_2PEG_{150}NH_2$ gave rise to oxazolidinones at 40°C in >94% yield and selectivity. *N*-isopropylglycinate is found to be the best absorbent for the rapid and reversible capture of almost equimolar CO_2. Crucially, the captured CO_2 can be activated simultaneously and the resulting carbamic acid can react with either aziridine or propargyl amine to afford oxazolidinones in the presence of NH_4I and AgOAc as a catalyst, respectively. Zhang et al. [139] further developed potassium phthalimide as absorbent to realize equimolar CO_2 capture in PEG_{150} and product is used as *in situ* transcarboxylating reagent to synthesize oxazolidinone derivatives.

Yu et al. [140] used carbamate salts generated from CO_2 and primary amines as substrates. The captured CO_2 not only acted as a reactant but also acted as a protecting reagent for the amine to avoid poisoning of the copper catalyst. By using 5 mol% of CuI as catalyst, carbamate salts can react with aromatic aldehydes and aromatic terminal alkynes, affording the important oxazolidin-2-ones. Song et al. [141] used ammonium carbamates as surrogates of carbon dioxide and secondary amines in the three-component synthesis of β-oxopropylcarbamates from propargylic alcohols, secondary amines, and CO_2. Catalyzed by silver (I) catalyst, ammonium carbamates can react with propargylic alcohols to generate β-oxopropylcarbamates under atmospheric pressure.

The transcarboxylation is an important transformation strategy for the CO_2 adducts to valuable chemicals. However, for the CO_2 adducts formed by base and CO_2, the transcarboxylation is still limited to the substrates including amines, propargylamines, and aziridines. Therefore, novel CO_2 absorbents and extended substrates are expected to facilitate the application of CO_2 adducts as transcarboxylation reagent in CCU strategy. Besides transcarboxylation, the integral transformation of CO_2 capture products is another attractive option in CCU strategy, wherein the ammonium carbamates derived from CO_2 and amines is a promising raw material [142].

5.3.11 Polymer Production

CO_2 can be used as either a direct or indirect feedstock for the synthesis of polymers. In the direct approach, CO_2 is used as a monomer unit, which is directly incorporated into the polymer. In the indirect approach, CO_2 is first converted into a different monomer, for example, methanol, organic carbonates, carbon monoxide, ethylene, dimethylcarbonate, or urea, which is then polymerized. Cheaper biodegradable polymers can be generated by the utilization of carbon oxides as a C_1 feedstock for the copolymerization with other monomers. The incorporation of carbon monoxide into organic molecules is a promising strategy toward the synthesis of new valuable materials. A large number of publications were focused on the copolymerization of reactive carbon monoxide with α-olefins (ethylene, propylene...) and vinyl arenes producing polyketone-based polymers. The copolymerization of CO with heterocycles such as epoxides and aziridines could be an interesting environmentally friendly alternative for the production of polyesters and polyamides [52].

The copolymerization of CO_2 with epoxides to produce polycarbonates is an alternative synthetic route. Numerous heterogeneous and homogeneous transition-metal catalysts have been developed which selectively form polycarbonates from a range of comonomers including ethylene oxide, propylene oxide, cyclohexene oxide, vinyl oxide, and styrene oxide, among others [52,143]. Typically homogeneous catalysts are preferred because they give higher selectivity. Additionally, by changing the nature of the catalyst, polymers can be produced which are either alternating and contain only carbonate groups (one molecule of CO_2 followed by one molecule of epoxide) or statistical (often referred to as polyether carbonates) and contain ether linkages that are formed when two ring-opened epoxides are adjacent to each other. Moreover, combining CO with imines can lead to polypeptides, which have broad applications in materials, catalysis and pharmaceuticals [52].

Despite the low reactivity of CO_2, it can be activated by suitable catalysis, and when copolymerized with epoxides, the overall reaction is thermodynamically favorable leading to polycarbonates. In the case of the use of basic aliphatic epoxides, polycarbonates susceptible to (bio)degradation can be obtained. The development of novel catalytic processes using commodity monomers could make the production of biodegradable (and also non-biodegradable) polymers comparably effective with current synthetic polymers. Using major petrol-based epoxides (ethylene oxide, propylene oxide) with CO_2 could lead to polymers, which are composed from 20% to 50% of renewable CO_2 building blocks (e.g., poly(propylene carbonate) (PPC)). One hundred percent bio-based polymers could be achieved when CO_2 would be effectively copolymerized with epoxides derived from, e.g., fatty acid or terpenes. Production of such materials could decrease the dependence of polymer industry on petrol [52,144,145].

A variety of catalysts have been reported to catalyze the copolymerization of epoxides with various substrates, such as anhydrides CO and CO_2. Among those, salen complexes of Al, Cr, and Co have been widely explored. The potential of salphen complexes in ring-opening copolymerization of epoxides with various substrates, which can afford biodegradable polycarbonates and polyesters is also high. Another possibility is copolymerization of epoxides with CO_2 with salphen chromium and cobalt complexes combined with organic base-cocatalyst leading to linear polycarbonates [52].

In general, alternating copolymers have low glass transition points, meaning that they will only be used in niche applications such as binders in ceramics and adhesives [146]. Nevertheless, Empower Materials currently sells poly(ethylenecarbonate) made from ethylene oxide and CO_2, and Econic manufactures polycarbonates that contain up to 50% CO_2 by weight [48,144,145]. A problem which companies such as Novomer are trying to address is the tendency of the polycarbonate to decompose to the cyclic carbonate, especially when electron-deficient epoxides are used. Both this problem and the tendency of polycarbonates with high CO_2 content to react with water can in principle be solved through the judicious use of additives but this still needs to be studied further [52]. At this stage, research is also still required to develop catalysts that are highly active with a wide range of epoxides, which may lead to alternating polymers with higher glass transition temperatures. Additionally, further development of strategies to incorporate another monomer like cyclohexene oxide into the copolymer to improve the properties of the polymer should be pursued. The reaction of CO_2 with epoxides is highly exothermic so finding catalysts that are thermally stable is also a key issue [144,145].

In contrast to the limited applications for alternating copolymers, polyether carbonates derived from epoxides and CO_2 are useful for a much wider range of applications [146]. In particular, they can be used as a component in polyurethanes. Novomer and Covestro are both selling polyether carbonates for use in polyurethane synthesis. Research has also begun on using epoxide starting materials which are derived from renewable feedstocks such as cyclohexadiene oxide, limonene oxide, and ⟨-pinene oxide, and catalysts have been developed for the copolymerization of CO_2 with these substrates [147]. Copolymerization of CO_2 with aziridines to produce polycarbamates and oxetanes may lead to polymers with different properties than those currently available [52].

5.4 MAJOR COMMERCIAL AND PILOT-SCALE CHEMICAL AND FUEL PRODUCTIONS BY HETEROGENEOUS CATALYSIS AND POSSIBLE BARRIERS

Several chemicals are produced commercially using heterogeneous catalysis. These include salicylic acid (30 kilotons per year), urea (112,000 kilotons per year), cyclic carbonates such as ethylene and propylene carbonate (40 kilotons per year), polycarbonate (600 kilotons per year), and propylene carbonate (10 kilotons per year). Here, two notable examples are the synthesis of salicylic acid and the synthesis of urea which was developed in 1922 [52,115]. CO_2 is combined with ammonia for urea synthesis. However, it should be noted that the CO_2 used in urea synthesis typically is produced from methane steam reforming, which also produces the H_2 required for ammonia synthesis. The use of CO_2 from a waste stream for sustainable urea synthesis would require using water electrolysis to make H_2 or an alternative ammonia synthesis.

Several fine and commodity chemicals are produced at the pilot scale. These include methanol by Carbon Recycling International (Iceland) at 4,000 tons per year and by Mitsui Chemical Company at 100 tons per year, methane by Audi (Germany) at 1,000 tons per year, carbon monoxide (via SOEC) by Haldor-Topsoe (Denmark)/Gas Innovations (USA) at 12 Nm^3/h, fuel via FT process by Sunfire (Germany) at 3 tons per year and INERATEC (Finland) at 200L per year, diphenyl carbonate by Shell (Singapore) at 500 tons per year and Asahi-Kasel at 1,000 tons per year, and oxalic acid by Liquid Light/Avantium (Netherland) at 2.4 tons per year.

While significant efforts are being made to commercialize valuable products from CO_2 by catalysis, there are barriers that need to be overcome for further progress. As shown earlier, methanol can be produced by direct hydrogenation of CO_2; however, the selectivity for methanol needs to be improved and catalyst inhibition by water needs to be removed. Similar challenges exist for the production of dimethyl ether and formic acid. For formic acid production by homogeneous hydrogenation of CO_2, stoichiometric addition of base is required for high turnover which makes process expensive. Separation of formic acid from reaction medium should also be carried out for base recycling. For dimethyl carbonate and diphenyl carbonate productions, alcohol/CO_2 condensation produces low pass per conversion and alcohol/urea condensation produces low selectivity along with low conversion per pass. There are also several barriers to various acid formations that need to be overcome. The turnover frequencies for acrylic and methacrylic acids are very and they require stoichiometric additives which make process expensive. For furan-2,5-dicarboxylic acid, the reaction rate is low and salt recycling process for carbonate generation is not proven on the large scale. For benzoic acid, the requirements for stoichiometric additives make the process expensive. Finally, oxalate and oxalic acid requires high potential and they have low selectivity. The production of hydrocarbon fuel is hindered by low selectivity and lack of understanding of carbon-carbon bond formation steps.

In the 1950s, methods for the commercial synthesis of cyclic carbonates from CO_2 were developed [148]. Specifically, the treatment of CO_2 with ethylene or propylene oxide in the presence of a basic catalyst generates ethylene carbonate or propylene carbonate, respectively. Similarly, styrene oxide, cyclohexene oxide, and 1,3-propylene oxide can also be used as substrates with CO_2 but the cyclic carbonates that are produced are made on a significantly smaller scale. Overall, approximately 80,000 tons of cyclic carbonates were produced worldwide from CO_2 in 2010 [149]. CO_2 has also been used as a feedstock for the synthesis of aromatic and aliphatic polycarbonates. The Asahi-Kasei process, which generates 600,000 tons of polycarbonate per year, uses CO_2, ethylene oxide, and bisphenol A as feedstocks [150]. Ethylene glycol, a commodity chemical, is produced as a stoichiometric by-product. In the production of polycarbonate polymers, products have the tendency to decompose into cyclic carbonate, and the process requires high purity of CO_2 which makes the process expensive. In the production of polyether carbonate, understanding of catalyst structure-polymer property relationships for tailored products is lacking. Finally, in the productions of carbon nanotubes, properties of currently produced carbon fibers are not suitable to act as replacements for carbon fibers [52].

Recently, Covestro developed a plant for the copolymerization of CO_2 with propylene oxide to generate polymeric polyols (polyether carbonates), branded as cardyon® [151]. These polyols are used to make polyurethanes, which are found in foam mattresses. The CO_2 is obtained from nearby ammonia production. Approximately

5,000 tons per year of polymeric polyols are produced in this facility. Novomer developed a related process, purchased by Saudi Aramco and branded Converge®, for the generation of polyols from CO_2 and propylene oxide, which is being performed on a scale similar to the Covestro system, using the facilities of Centauri Technologies in Texas. In the case of both the Converge® and cardyon® processes, the scale at which the CO_2-derived polyols are produced is approximately 60 times smaller than conventional polyol plants [52].

5.5 CHALLENGES AND INNOVATIONS IN CATALYST DEVELOPMENT

In general, catalysts for CO_2 conversion lack the required durability and stability. Even in cases where turnover frequencies are high, catalyst decomposition is a problem. The use of non-purified CO_2 streams would cause decomposition even more rapidly. Since purification can be expensive, it would be commercially more appealing if the CO_2 stream can be treated as is. Furthermore, the utilization of CO_2 from gaseous waste streams would be aided by improving the interface between CO_2 capture and conversion [52]. Depending on the application, this could include homogeneous, heterogeneous, and electrocatalysis systems. Where possible, an emphasis placed on developing catalysts that use sustainable raw materials would be beneficial. The addition of stoichiometric amounts of additives generate waste products, and increase the net carbon footprint of the process. Systems or reactor design that allows less than stoichiometric or no additives should be preferred. The identification of non-traditional targets and subsequent catalyst development could have transformative impacts. Improving the efficiency of CO_2 conversion chemistry would be aided by the integration of catalysts with the most efficient reactor technology. An efficient separation of product can make process more economical. Separation challenges can be mitigated by developing reactor designs that improve conversion per pass.

There are examples that the CO_2 from waste streams can be directly converted by designing catalysts that tolerate to the contaminants in the waste streams such as exogenous water, nitrogen, SO_2, amine, etc. For example, Williams et al. [152] reported the synthesis of polycyclohexylene and poly carbonate using stable homogeneous dinuclear Zn or Mg catalysts to treat CO_2 from contaminated power station gas steams. D'Elia and Basset group develops the combination of early transition metal halides (Y, Sc, Zr) and TBAB to quantitatively convert CO_2 from diluted streams and produce cyclic organic carbonates [94]. The features of metal-organic frameworks (MOFs) to selectively capture and catalyze CO_2 conversion make them a new type of platform for diluted CO_2 transformation. Recently, Hong group design and synthesize an acid-base resistant Cu(II)-MOF, which can convert CO_2 from simulated post-combustion flue gas into corresponding cyclic carbonates [153]. In direct conversion of diluted CO_2, the stability of the catalysts to the contaminants in the gas streams is crucial to the success of the process.

The keys for further progress are to discover alternative catalysts, modify current catalysts for CO_2 activation, develop methods to prepare specialized catalysts on a large-scale, and intelligently evaluate catalysts. More innovations in the design of the catalyst are needed. Innovations in catalyst preparation and modification, characterization, and AI-guided evaluation are needed (see Figure 5.6).

An innovative three-phase scheme for catalyst design was recently proposed by Ye et al. [2] for CO_2 hydrogenation. In this innovative scheme,

1. **Phase I is CO_2 hydrogenation catalyst preparation and modification.** Conventional labor-intensive lab-based CO_2 hydrogenation catalyst preparation may be replaced by low-cost 3D-printing approaches to include characteristics of high mechanical strength and surface-to-volume ratio, with precise control of porosity, size, and shape. As the functional components of active CO_2 hydrogenation catalysts are further studied (e.g., metals, promoters, and supports), more effort can be devoted to incorporating them into a fully integrated platform with diverse microstructures by 3D printing.
2. **Phase II is to characterize the CO_2 hydrogenation catalysts.** New material characterization technologies, including in situ scanning transmission electron microscopy and temporal analysis of product reactors, coupled with qualitative and/or quantitative species identification, such as gas chromatography–mass spectrometry, are suggested for characterization of the catalysts prepared with the state-of-the-art transformational technologies. AI-assisted CO_2 hydrogenation catalyst characterizations are needed. Machine-learning especially can bring advanced computational techniques to the forefront of characterization of heterogeneous catalysts.
3. **Phase III is to perform AI-guided evaluation of CO_2 hydrogenation catalysts.** The prospect of using AI for identification of reaction intermediates and pathways and establishing kinetic models would be promising. The AI-guided method is expected to predict catalytic performance and to discover promising catalyst candidates. AI-based catalyst evaluation might largely help predict and improve catalyst stability, which, however, will require more significant effort. This will include (a) in situ monitoring of the catalyst structure dynamics and transformation by AI, (b) more automatic, integrated, and flexible set-ups to evaluate catalysts, and (c) perform data analysis evaluation of kinetic models with the help of AI. Finally, predicting relationships among the characteristics of the catalysts, such as the Lewis acidity, CO_2 conversion efficiency, and product selectivity, with AI based on the available experimental data and DFT computation results will be helpful for discussing structure–function relationships and accelerating the discovery of catalytic mechanisms. This AI guided development of CO_2 hydrogenation catalyst and three phases of innovative scheme are graphically illustrated in Figure 5.6.

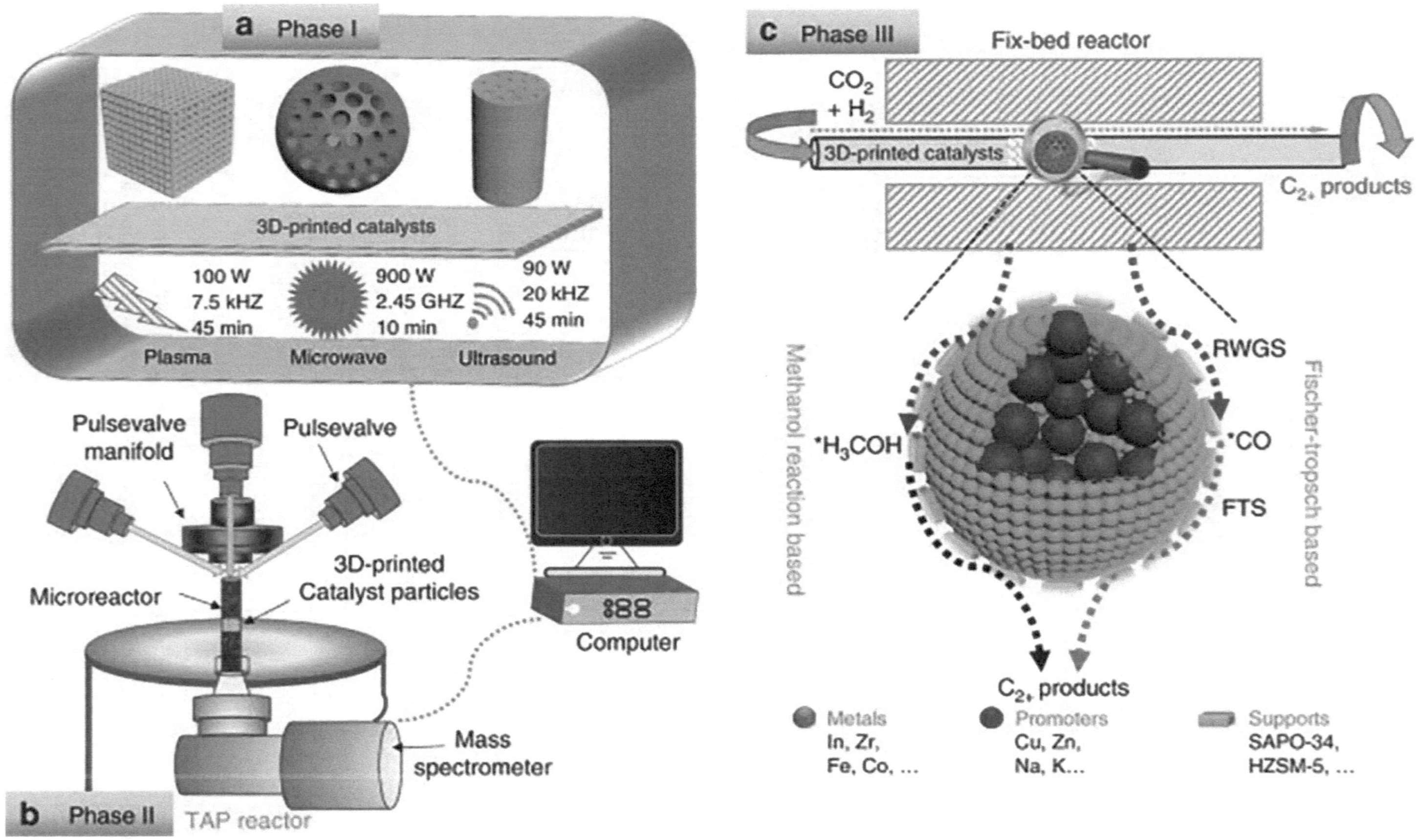

FIGURE 5.6 Scheme of AI-guided development of CO_2 hydrogenation catalysts [2]. Phase I is to prepare and modify catalysts using 3D printing technologies and new material modification technologies, such as plasma, microwave, and ultrasound modification. Phase II is to use advanced techniques to characterize the catalysts. Phase III is to perform AI-guided evaluation of the catalysts.

5.6 CARBON DIOXIDE CONVERSION BY ELECTROCHEMICAL CATALYSIS

As mentioned before, CO_2 is one of the most stable molecules in which carbon is in the highest valence state. It is difficult to have an electrophilic reaction because of its poor electron affinity. Hence, the conversion of CO_2 depends on nucleophilic attack of the carbon atom. As is known, the dissociation energy for breaking the C=O bond in CO_2 molecules is higher than 750 kJ mol^{-1} [154]. This is an uphill reaction from a thermodynamic point of view. To complete such a reaction, high temperature, high pressure environment, or highly efficient catalysts are typically required to provide the necessary energy. As is known, eight electrons are needed for each CO_2 molecule to complete the conversion to hydrocarbon compounds. This leads to various products during the reduction process, resulting in complicated purification procedures and poor yield of desired products.

Electrocatalysis [154] has been used to conduct the reduction of CO_2, in which electricity is used to supply essential energy for the reaction. In general, the electrocatalytic conversion of CO_2 to valuable chemicals is an attractive solution for reducing atmospheric CO_2 and storing energy. Using an external electric field as an energy source and water as the proton donor, various catalysts are applied to catalyze the reduction of CO_2. Compared with thermal catalysis, the electrocatalytic conversion is a more cost-effective method because water replacing H_2 is used as the proton donor. Electrocatalytic CO_2 reduction has attracted great attention due to its mild operating conditions (normal temperature and pressure), controllable reaction process conditions and reaction rate, recyclable catalyst and electrolyte, high energy utilization, simple equipment, and achievable conversion efficiency [155,156]. In the past few years, researchers have explored electrocatalytic reduction of CO_2 using different electrode materials, such as metals, transition metal oxides, transition metal chalcogenides, metal-free 2D materials, and metal–organic frameworks (MOFs), and various reduction products including CO, methane, formic acid, ethanol, and other compounds are obtained [154]. Homogeneous catalysis and immobilization of homogeneous catalysts are also considered.

The electrochemical CO_2 reduction reaction (CO_2RR) involves CO_2 conversion on an electrode powered by electrical energy. Because CO_2 is highly stable with a linear molecular shape having two C=O bonds, it requires a high overpotential to initiate the reaction. An overpotential is the difference between the actual applied potential and theoretical thermodynamic potential. Catalysts are used to reduce the overpotential. The performance and reaction pathway highly depend on the electrocatalyst. CO_2 is electrochemically reduced to chemicals or fuels at the cathode, and the water oxidation producing oxygen occurs at the anode which is typically known as oxygen evolution reaction (OER). A distinct feature of the electrochemical CO_2 reduction is its compatibility with renewable sources. Because CO_2 is thermodynamically stable, the required high energy can be obtained from renewable sources, such as solar or wind power, and the obtained electricity can drive the electrochemical CO_2RR. Additionally, electrochemical CO_2 conversion to fuels such as CH_4 can also be used for intermittent and undistributed renewable energy storage. Various chemicals, from several C_1 products to multicarbon products, can also be directly generated from electrochemical CO_2RR.

Many experiments with different conditions and electrocatalysts have been conducted for CO_2 reduction on metal electrodes [157]. The selection of catalyst and reaction conditions plays a significant role as compared to the potential in controlling between various reduced products. Table 5.1 shows several representative CO_2RR pathways with thermodynamic standard potentials versus a reversible hydrogen electrode (RHE) [157]. Neutral pH buffer electrolytes have been widely used for electrochemical CO_2RR studies. The CO_2RR typically competes with the hydrogen evolution reaction (HER). The hydrogen evolution reaction (HER) is very important during CO_2 electrocatalyst reduction in which H_2O is typically present as an electrolyte (and proton source). Suppressing HER is one of the most important issues in the development of electrocatalysts for CO_2RR. For this reason, the reported metals that can be used as an electrocatalyst for CO_2 reduction have relatively high HER overpotentials. A huge effort must be conducted in order to find the optimum electrode for CO_2 electrochemical reduction which will reduce the selectivity of CO_2 at low overpotentials and high rates without reducing water simultaneously [158].

As schematically illustrated in Figure 5.7, in this process, the electrochemical electrolyzer converts CO_2 and water into chemicals and fuels powered by renewable electricity. The resulting fuel is capable of long-term storage and can also be distributed or consumed, giving off CO_2 as the main waste, which will be captured and fed back to the reactor to close the loop. Moreover, the resulting small-molecule chemical feedstocks (e.g., carbon monoxide (CO) and formate) from ECR can be used as raw materials for more complicated chemical synthesis.

ECR in an aqueous electrolyte involves multi-electron/proton transfer processes together with a number of different possible reaction intermediates and products [157,159], making it highly complex. Table 5.1 summarizes the half electrochemical thermodynamic reactions of the main ECR products, including CO,

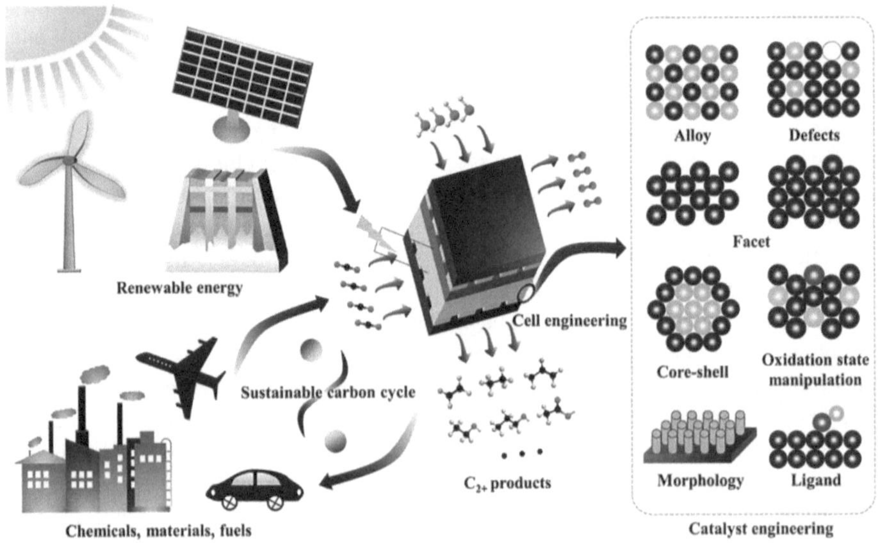

FIGURE 5.7 Schematic illustration of sustainable energy cycling based on ECR [157].

TABLE 5.1
Half Electrochemical Thermodynamic Reactions of the Main ECR Production, Together with Their Corresponding Standard Redox Potentials [V versus Reversible Hydrogen Electrode (RHE)] [157]

	Acid		Base	
Products	Equation	E (V)	Equation	E (V)
Hydrogen	$2H^+ + 2e^- \rightarrow H_2$	0.000	$2H_2O + 2e^- \rightarrow H_2 + 2OH^-$	−0.828
Carbon monoxide	$CO_2 + 2H^+ + 2e^- \rightarrow CO + H_2O$	−0.104	$CO_2 + H_2O + 2e^- \rightarrow CO + 2OH^-$	−0.932
Methane	$CO_2 + 8H^+ + 8e^- \rightarrow CH_4 + 2H_2O$	0.169	$CO_2 + 6H_2O + 8e^- \rightarrow CH_4 + 8OH^-$	−0.659
Methanol	$CO_2 + 6H^+ + 6e^- \rightarrow CH_3OH + H_2O$	0.016	$CO_2 + 5H_2O + 6e^- \rightarrow CH_3OH + 6OH^-$	−0.812
Formic acid/ formate	$CO_2 + 2H^+ + 2e^- \rightarrow HCOOH$	−0.171	$CO_2 + H_2O + 2e^- \rightarrow HCOO^- + OH^-$	−0.639
Ethylene	$2CO_2 + 12H^+ + 12e^- \rightarrow C_2H_4 + 4H_2O$	0.085	$2CO_2 + 8H_2O + 12e^- \rightarrow C_2H_4 + 12OH^-$	−0.743
Ethane	$2CO_2 + 14H^+ + 14e^- \rightarrow C_2H_6 + 4H_2O$	0.144	$2CO_2 + 10H_2O + 14e^- \rightarrow C_2H_6 + 14OH^-$	−0.685
Ethanol	$2CO_2 + 12H^+ + 12e^- \rightarrow CH_3CH_2OH + 3H_2O$	0.084	$2CO_2 + 9H_2O + 12e^- \rightarrow CH_3CH_2OH + 12OH^-$	−0.744
Acetic acid/acetate	$2CO_2 + 8H^+ + 8e^- \rightarrow CH_3COOH + 2H_2O$	0.098	$2CO_2 + 5H_2O + 8e^- \rightarrow CH_3COO^- + 7OH^-$	−0.653
n-Propanol	$3CO_2 + 18H^+ + 18e^- \rightarrow CH_3CH_2CH_2OH + 5H_2O$	0.095	$3CO_2 + 13H_2O + 18e^- \rightarrow CH_3CH_2CH_2OH + 18OH^-$	−0.733

methane (CH_4), methanol (CH_3OH), formic acid (HCOOH), ethylene (C_2H_4), ethanol (CH_3CH_2OH), and so on, together with their corresponding standard redox potentials [157]. In general, during an ECR process, CO_2 molecules first undergo adsorption and interaction with atoms on the catalyst surface to form $*CO_2^-$, followed by various stepwise transfer of protons and/or electrons toward different final products. For example, CH_4 is believed to form through the following pathways: $CO_2 \rightarrow *COOH \rightarrow *CO \rightarrow *CHO \rightarrow *CH_2O \rightarrow *CH_3O \rightarrow CH_4 + *O \rightarrow CH_4 + *OH \rightarrow CH_4 + H_2O$ [160].

Figure 5.8a summarizes the Faradaic efficiency (FE) under different production rates (current density) for the reported ECR electrocatalysts, which represents the product selectivity of the reaction [157,173]. Notably, while the state-of-the-art electrocatalysts can transform CO_2 into C_1 products (CO or formate) with over 95% FE under high production rate (>20 mA/cm² for H-type cell and >100 mA/cm² for flow cell) [157,173], the highly selective (>90%) and efficient production of more available multicarbon (C_{2+}) chemicals has not been realized so far. This is due to the fact that coupling to C_{2+} products requires arrival and adsorption of several CO_2

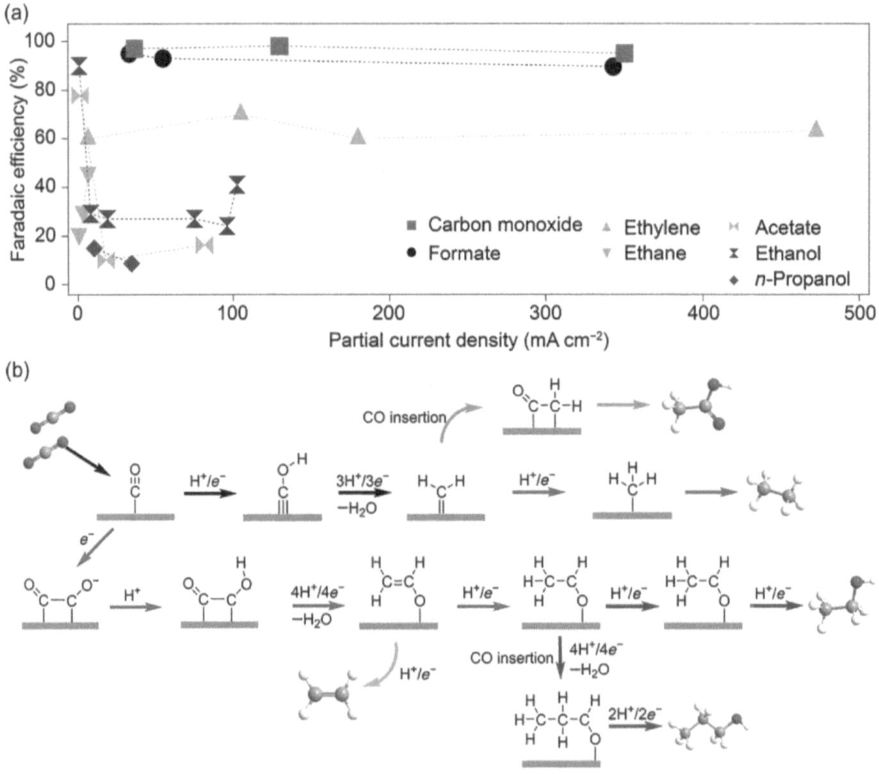

FIGURE 5.8 Summary of the state-of-the-art ECR performance and the C—C coupling mechanisms [157]. (a) The FE under different production rates (current density) for the reported ECR electrocatalysts [8–14,157,153,106,161–172]. (b) Most possible C_{2+} pathways during ECR.

molecules to the surface, stepwise transformation, and spatial positioning [174]. To be specific, as shown in Figure 5.8b, the subsequent reactions of *CO intermediates determine the final C_{2+} products of ECR. In general, C_2H_6 and CH_3COO^- share the same *CH_2 intermediate, which is generated from the proton-coupled electron transfer steps of *CO. Further protonation of *CH_2 gives *CH_3 intermediate, which leads to the formation of C_2H_6 via *CH_3 dimerization. Unlike C_2H_6 generation, CH_3COO^- is formed by CO insertion into *CH_2. The *CO dimerization is the rate-determining step for C_2H_4, CH_3CH_2OH, and *n*-propanol (*n*-C_3H_7OH) formation. After a series of electron transfer and protonation steps, the *CO—CO dimer forms the *CH_2CHO intermediate, which serves as the selectivity-determining step for C_2H_4 and C_2H_5OH. In addition, it was found that reducing *CH_2CHO to C_2H_4 has lower energy barrier than transforming *CH_3CHO to C_2H_5OH [175], which may explain the higher FE for C_2H_4 over C_2H_5OH on most copper catalysts. Furthermore, stabilized C_2 intermediates could transfer to *n*-C_3H_7OH via CO insertion. The complex and uncontrollable reaction pathways during C_{2+} chemical formation are mainly due to many more permutations to the protonation sites, along with the possible involvement of nonelectrochemical step [157,176]. As such, the design of highly selective electrocatalysts is a prerequisite for specific C_{2+} product formation at high yield.

Over the last four decades, there has been intense research to discover and tailor the selectivity and activity of catalysts for the electrochemical reduction of CO_2, and to unravel its underlying operating mechanisms. Significant work has also been carried out on the design of electrolytic cell, electrolytes and electrodes to optimize the electrocatalytic performance for CO_2 reduction. Works on scaled-up systems for electrochemical CO_2 reduction reaction have also been recently reported. Here we summarize recent advancements in the use of electrocatalysis for chemicals and fuel productions in three parts. First we examine details on chemicals that have been successfully produced by electrocatalytic process. Second, we examine the strategies to improve catalysts, electrolytic cell design/electrolyte and electrodes for better product selectivity. Finally, we examine barriers and opportunities for commercialization of CO_2RR for useful chemicals and fuels.

5.7 CHEMICALS FROM CO_2 BY ELECTROCATALYSIS

Although most commonly CO_2 is reduced to C_1 feedstocks such as CO, formic acid, methanol, and methane, there are a number of systems that can form products containing new carbon-carbon bonds. In particular, copper catalysts are effective in forming products containing C–C bonds such as ethylene, ethane, and higher-order hydrocarbons and alcohols [177]. Apart from selectively forming C_2 or higher products over C_1 products, major challenge is to inhibit the production of H_2 as a by-product. More research is needed to understand the elementary processes which facilitate C–C bond formation. As shown in Section 5.8, various strategies have been investigated to improve performances of catalyst, electrolytic cell/electrolytes, and electrodes to improve product selectivity. There are, however, significant barriers that need to be overcome to make this technology viable at commercial scale.

5.7.1 Carbon Monoxide

The direct electrochemical splitting of CO_2 into CO and O_2 can be performed either at high temperature using a solid oxide electrolysis cell (SOEC) or at low temperature using a solution-phase or gas diffusion electrolysis cell. SOECs for CO_2 splitting have recently been brought to market. SOECs are monolithic devices composed of two electrodes separated by a solid oxide- conducting (O^{2-}) electrolyte [52,178]. A commonly used combination consists of an yttria-stabilized zirconia (YSZ) solid oxide electrolyte, a composite Ni-YSZ cathode, and a perovskite-type anode such as strontium-doped lanthanum manganate. For CO_2 electrolysis, CO_2 is supplied to the cathode side and the anode side is swept with air or another gas. The cathode reduces CO_2 to CO and O, which migrates through the oxide-conducting electrolyte to the anode, where it is oxidized to O_2. High temperatures (700°C–900°C) are required to attain sufficient oxide conductivity in the electrolyte. SOECs operate at high current densities (0.2 up to 2 A/cm^2), achieve high energy efficiency (typically >95%), and can produce CO product streams of high purity (>99%). Researchers have demonstrated relatively stable performance for CO_2 electrolysis at current densities up to 0.5 A cm^2, with small cell voltage increases observed over several hundred hours of continuous operation [178,179]. Cell degradation is faster at higher current densities [52].

Haldor Topsoe currently sells a commercial SOEC system for onsite CO production that provides CO at 99.0%–99.999% purity and requires 6–8 kWh of power per normal cubic meter (Nm) CO produced [52]. The first commercial system with 12 Nm^3/h capacity began operating at Gas Innovations in Texas in 2016 [180]. There is room for substantial improvement in SOECs to increase deployment of this technology. Major challenges include developing new electrodes with enhanced stability at higher current densities, finding electrolytes that provide high oxide conductivity at lower temperatures, and increasing the tolerance to impurities [52].

In the past 20 years a large number of molecular and heterogeneous systems for low- temperature CO_2 electroreduction to CO have been developed [52,181]. Faradaic efficiencies in the range of 95%–98% can be achieved routinely in standard three-electrode or H-type cells using a variety of neutral or acidic electrolytes, typically using silver cathode catalysts [52,182]. Non–precious metal catalysts, such as N-doped carbon or carbon nanofibers, have achieved performance levels that are similar to those obtained with silver catalysts [182].

The electroreduction of CO_2 to CO is a two electrons transfer reaction. This makes the extent of CO_2 sequestration to CO per unit current higher than for products such as C_2H_4 which involves 12 electron transfers. Catalysts for highly selective CO synthesis with FEs close to 100% are already available. High selectivity is one of the highly desired performances because it helps to make the separation of products simpler [52]. Large scale CO generation has been attempted in PEM electrolyzers with Ag-based catalysts as well as Au, Cu, Pd, Pt and Co-phthalocyanine catalysts [163]. Masel et al. used imidazolium-functionalized Sustanion™ membranes with Ag-based catalysts and demonstrated continuous operation for 6 months with 98% FE CO at 50 mA cm^2 total current density [183].

Use of silver catalysts or supported gold catalysts in electrolyzers in which the electrodes are separated by a flowing liquid electrolyte, either alkaline or neutral, significantly increases the CO production rate to 150–450 mA/cm^2, while still maintaining Faradaic efficiencies of 60%–98% [184,185]. Although there is a trade-off between maximizing rate and energetic efficiency, the overpotentials for CO reduction using silver or gold catalysts are fairly low, which means that rates of 150 mA/cm^2 can be achieved at energetic efficiencies exceeding 50%. In all these cases, the oxygen evolution reaction takes place at the anode, typically using an IrO_2 catalyst. Based on these prior efforts that typically employ electrodes with a geometric area of 1–2 cm^2, Siemens performed experiments on a larger scale, using first 10 cm^2 and then 100 cm^2 gas diffusion electrodes in electrolyzer configurations with a flowing liquid electrolyte [52,168]. These cells were successfully operated for 200 hours using a neutral electrolyte at a rate of 150 mA/cm^2, with a Faradaic efficiency for CO formation of 60%. Subsequently, Siemens collaborated with Evonik to connect their 10 cm^2 CO_2-to-CO electrolyzer (running at 300 mA/cm^2 for over 1,200 hours) with a fermentation process in which the formed CO is combined with unreacted CO_2 to form butanol and hexanol, at close to 100% Faradaic efficiency [186]. Detailed economic analysis of this hybrid system highlights the promise for sustainable production of first CO and then other chemicals at scale using this approach. In addition to low-temperature and high-temperature electrochemical cells described here, intermediate-temperature carbon dioxide electrolysis, with cell temperatures ranging from 200°C to 500°C, is a possible new direction that could provide access to products that are not made efficiently at either temperature extreme currently [52].

Flow electrolyzers using Ag catalysts [187,188] have been studied intensively by several groups. Some groups have developed GDEs in-house, while others used commercial GDEs. Noteworthy effort has been made by Kenis et al. to develop carbon GDL (gas diffusion layer)-based GDEs with Ag catalysts. They investigated effects of composition of the micro porous layer and the substrate on performance [188]. They also studied the effect of electrolyte composition [184], and their best performing system showed CO partial current density as high as 440 mA cm^2 with FE above 90% [184]. Another outstanding work on GDE development was made by Sargent et al. [187]. They developed a new type of GDE by employing porous PTFE membrane as a hydrophobic GDL in place of traditional carbon GDL [187]. In such GDEs, catalysts were deposited on the PTFE membrane, and a porous carbon layer, for current collection, was deposited on top of catalyst layer. This new type of GDEs exhibited more than 100 hours of stable operation at 150 mA cm^2 total current density with >90% CO FE. It is believed that the hydrophobic PTFE membrane, which prevents electrolyte flooding, is responsible for long-term performance [52].

Commercially available Ag-based GDEs have been tested by at least two groups in flow electrolyzers. Haas et al. [186] obtained remarkable performance using Ag-based GDE from Covestro. They demonstrated 1,200 hours of stable operation at a total current density of 300 mA/cm^2, with 60% average FE for CO. Using a similar Ag-based GDE from Covestro, Fleischer et al. [168] also made significant progresses on the development of pilot-scale CO generation prototype system. They have shown over 600 hours of operation on a 100 cm^2 electrode with almost 60% FE for CO at 150 mA cm^2 total current density. Besides silver, Au, Co-phthalocyanine, and Ni-N-C-based catalysts have been investigated for CO generation using homemade GDEs in flow electrolyzers. Among these works, Strasser et al. reported an interesting result using Ni-N-C catalysts for CO generation. They achieved 20 hours operational duration and nearly 85% FE at a total current density of 200 mA/cm^2 [189].

While these electrolyzer systems exhibit encouraging performance levels for the reduction of CO_2 to CO, the use of gas diffusion electrodes, especially in combination with an alkaline electrolyte, can cause problems. Typically, CO_2 flow rates that greatly exceed the rate of CO_2 reduction are used to maximize the current density and Faradaic efficiency, which results in product streams that are diluted with a large excess of CO_2 (e.g., 3:1 CO_2:CO in the Siemens example above). In addition, some of the CO_2 is lost by diffusion through the gas diffusion electrodes into the electrolyte, where it can react with OH^- to form carbonates, which can precipitate on the electrode or migrate to the anode and release CO_2 into the O_2 stream. The rate of these undesired processes depends on the specific gas diffusion electrode used, the nature of the catalyst layer, as well as the flow rates and pressures of the CO_2 feed and the electrolyte. Achieving high CO_2 conversion with minimal loss of CO_2 into the electrolyte will be critical for advancing CO_2 electrolysis technology [52,163].

Dioxide Materials has reported a different electrolyzer configuration: an anion-conductive membrane-based electrolyzer with a silver cathode catalyst that sustains CO production rates in the range of 100–200 mA/cm^2 over more than 1,000 hours [183]. The use of a membrane as the electrolyte between the electrodes significantly reduces some of the afore- mentioned issues (CO_2 loss and electrolyte degradation) associated with the use of liquid electrolytes. In recent collaborative work Dioxide Materials and 3M explored the electrochemical generation of syngas from water and CO_2, either (a) by operating a flowing liquid-electrolyte CO_2 electrolyzer cell for CO production and a polymer electrolyte-based water electrolyzer for H_2 production in parallel, or (b) by performing co-electrolysis of CO_2 and H_2O in a single anion-exchange membrane-based electrolyzer [190]. Both systems produced CO and H_2 at industrially relevant rates (for example, CO_2 to CO at 100 mA/cm^2). **The various technoeconomic analyses reported to date suggest that the electroreduction of CO_2 to CO has promise to become economically feasible [52].** Remaining challenges for electrochemical conversion of CO_2 to CO include (a) achieving sufficient stability and durability for both catalysts and electrodes, (b) the need to develop strategies that minimize loss of CO_2 to the electrolyte, and (c) the need to further reduce the overall energy requirement under operating conditions at practical CO production rates [52].

5.7.2 Formic Acid

Electrochemical systems for the conversion of CO_2 and protons and electrons to formic acid have also been developed [52,181]. The electrocatalytic reduction of CO_2 to formate has been reported at Faradaic efficiencies (FEs) in the range of 80%–95% using tin-based catalysts [52,191] at moderate to high overpotential, both in standard three-electrode electrochemical cells and in liquid electrolyte-based electrolysis flow cells, the latter at a rate of 200 mA/cm^2 [192]. Palladium nanoparticle electrodes reduce CO_2 to formate with >90% FE at <200 mV overpotential, but the palladium requires periodic regeneration because of poisoning by trace CO, which is produced as a by-product [52,193]. Palladium has also been used recently as a cathode in a photoelectrochemical device for formate synthesis. The Joint Center for Artificial

Photosynthesis has demonstrated 10% efficient light-to-formate energy conversion using a Pd/C cathode wired to a tandem III-V GaAs/InGaP photoanode coated with a protective conductive TiO_2 layer [194].

The large-scale production of formic acid from CO_2 has also been studied intensively. Like CO, formic acid generation is a two-electron process, and catalysts with high selectivity are available. Formic acid generation in PEM electrolyzers has been studied using Sn catalysts on carbon paper-based GDE [11], and Sn-coated membrane electrode [12]. GDEs were developed in-house in all these publications. Among these, Masel et al. [11] developed a novel three-compartment PEM electrolyzer with Sustanion™ anion exchange membrane which is stable over 500 hours of operation. Their system shows up to 94% FE at a total current density of 140 mA/cm^2.

Flow electrolyzers for generating formic acid using homemade GDEs have been studied by a few groups. The majority of their works involved Sn-based catalysts on carbon GDL [52,195]. A significant improvement on GDE preparation method was reported by Kopljar et al. [195]. They developed a dry pressing method for GDE preparation which gives improved mechanical stability and high reproducibility. These improved GDEs gave close to 75% FE of formic acid at 400 mA/cm^2 total current density. Indium-, Pb-, and Bi-based catalysts have also been tested in flow electrolyzers [163]. Among these, Garcia et al. reported over 60 hours of operation and 90% FE at 200 mA/cm^2 total current density with Bi oxy-halide-derived catalysts [196]. Their results show that Bi oxy-halide-derived catalysts are superior to Bi catalysts. The preferential exposure of highly active Bi [134] facets is proposed to be responsible for the superior performance. Considering the above findings, it appears that both PEM electrolyzer and flow electrolyzer can achieve 90% formate FE at a total current density well above 100 mA/cm^2. In terms of stability, the three-compartment PEM electrolyzer based on Sustanion™ anion exchange membrane is superior [163].

More recently, dioxide materials were reported to reduce CO_2 to formic acid at a Faradaic efficiency of 94% at a rate of 200 mA/cm^2 in a three-compartment electrolyzer with a cation-anion exchange membrane that exhibited stable performance for 500 hours [197]. **Technoeconomic and life-cycle analyses suggest that electroreduction of CO_2 to formic acid has shown promise and this process is gradually moving towards scale-up and commercialization [198].** Remaining challenges for electrochemical conversion of CO_2 to formic acid include (a) achieving sufficient stability and durability for both catalysts and electrodes and (b) the refinement of energy-efficient strategies to separate formic acid from the product stream (typically a 5–20 wt% aqueous solution) [52,157,199].

5.7.3 Methane

Electrochemical reduction of CO_2 to methane is also a widely studied process with reported Faradaic efficiencies in the range of 80%–94% using N-doped carbon or copper-on-carbon catalysts in standard three-electrode or H cells [52,200]. Partial current densities for methane formation as high as 38 mA/cm^2 have been reported for a Cu catalyst electrodeposited on a carbon gas diffusion electrode [200]. **According to National Academy Report [52], despite continued progress in the development of more selective catalysts, electrocatalytic conversion of CO_2 to methane**

probably will not be pursued on a large scale given the global availability of low-cost methane derived from natural gas [52,157,199].

5.7.4 Ethylene

Large-scale synthesis of hydrocarbons by electrochemical reduction of CO_2 has been attempted by numerous groups. It is notable that all hydrocarbon synthesis reactions are multiple proton-electron transfer processes and involve a myriad of intermediates. Thus far, copper is an essential component of the catalysis. The selectivity (FE) for these products is typically far less than 90%. This makes the separation process non-trivial. The reaction requires a significantly larger overpotential than CO_2 reduction to CO using gold or silver catalysts. Extensive investigations of different copper nanostructures and operating conditions have led to systems that produce ethylene, but achieving selectivities exceeding 40% has been difficult. Strategies such as alloying silver with copper, in which the silver enhances formation of the needed CO intermediate [201], and precise engineering of the copper catalyst layer inside a sandwich-type gas diffusion electrode have increased Faradaic efficiencies for ethylene to 60%–70% at rates of 160–250 mA/cm^2 [202]. Hoang et al. [201] showed that copper-silver bimetallic wire catalysts could reduce CO_2 to ethylene with 60% FE and 180 mA/cm^2 partial current density.

Often a sizeable amount of ethanol (Faradaic efficiencies of 10%–30%) is co-produced with the ethylene. Co-production of ethylene and ethanol—a valuable commodity chemical itself (with an annual global production of approximately 80 million tons in 2016)—may bring economic feasibility for ethylene production closer as a result of the additional income generated from the produced ethanol. Additionally, given that ethylene is a gas and ethanol is a liquid, separation of the products should be relatively straightforward, although extracting ethanol from the electrolyte may be challenging. In related chemistry, a promising class of boron- and nitrogen-co-doped nanodiamond catalysts were reported, which can achieve Faradaic efficiencies for ethanol production as high as 93% [203]. It is unclear whether this catalyst can be integrated on a gas diffusion electrode in a liquid electrolyte or membrane-based electrolyzer in order to achieve practical production rates.

Despite the progress in the electroreduction of CO_2 to ethylene (and ethanol), a large number of challenges remain. The few studies that have explored the stability of the various copper catalysts suggest that catalyst degradation and remodeling is a serious issue. Indeed, sustained operation of an ethylene-producing electrolyzer for more than 2 hours has not been demonstrated. Furthermore, as described above, loss of CO_2 to the electrolyte has the potential to be a major drawback. **Despite these challenges, the prospect of converting CO_2 into a valuable and versatile chemical such as ethylene using low-carbon electricity is a powerful motivation to continue to develop better and more durable catalysts, electrodes, and electrochemical cell design [52,157,199].**

5.7.5 Oxalate and Oxalic Acid

Researchers continue to work on more efficient methods to synthesize oxalates and oxalic acid from CO_2. In early research (prior to 2000), Pb and other catalysts were used to generate oxalic acid in electrochemical cells with Faradaic efficiencies in the 70%–98% range, but this was only done at low rates in a standard electrochemical cell [52,204]. More recently, both mononuclear and binuclear copper complexes have shown promise for electrochemical reduction of CO_2 to oxalic acid [52,205]. **Nevertheless, further mechanistic understanding of this process is required to lower the overpotential and to improve selectivity.**

5.7.6 Methanol, Ethanol, and Propanol

Methanol is produced as the main source of energy using electricity [63]. There is a distinct advantage of directly converting the captured CO_2 into methanol of producing a useful product that can be used in many energy-consuming devices. This process allows for recycling captured CO_2 and produce methanol that could be used as a renewable energy instead of fossil fuel in energy-consuming devices. In other words, by electroreduction process, CO_2 could be reduced directly in the electrolysis cell back to methanol in one step. Different electrodes can be used to achieve methanol directly from CO_2 [63], as shown in Table 5.2. In 1983, Canfield and Frese [206] proved that some semiconductors such as *n*-GaAs, *p*-InP, and *p*-GaAs have the ability to produce methanol directly from CO_2, although at extremely low current densities and faradaic efficiencies (FEs). Many other researchers did some efforts to increase both the current density as well as faradaic efficiency of the process. Seshadri et al. [207] found that the pyridinium ion is a novel homogeneous electrocatalyst for CO_2 reduction to methanol at low overpotential. Recently, pyridine has been widely explored in which it is used to act as cocatalyst to form the active pyridinium species in situ [63,208]. Generally, the one-electron reduction products of CO_2 show lower current density than the two-electron reduction products such as CO. The direct electrochemical reduction of CO_2 to methanol is a promising process to reduce the amount of captured CO_2.

Popić et al. [209] showed that the Ru and Ru modified by Cd and Cu adatoms can be used as an electrode for CO_2 reduction at relatively small overpotentials.

TABLE 5.2
CO_2 Electrochemical Reduction to Methanol [63]

Electrode	Type of Electrode	E vs. NHE (V)	Current Density (mA/cm^2)	Faradaic Efficiency (%)	Electrolyte
p-InP	Semiconductor	−1.06	0.06	0.8	Sat. Na_2SO_4
n-GaAs			0.16	1.0	
p-GaAs			0.08	0.52	
CuO	Metal oxide	−1.3	6.9	28	0.5M $KHCO_3$
RuO_2/TiO_2 nanotubes		−0.6	1	60	0.5M $NaHCO_3$
Pt–Ru/C	Alloy	−0.06	0.4	7.5	Flow cell
n-GaP	Homogeneous catalyst	−0.06	0.27	90	10mM pyridine at pH = 5.2
Pd		−0.51	0.04	30	0.5M $NaClO_4$ with pyridine

The obtained results showed that on the surface of pure Ru, Ru modified by Cu and Cd adatoms, and RuOx + IrOx modified by Cu and Cd adatoms, the reduction of CO_2 was achieved to produce methanol during 8 hours of holding the potential at −0.8 V. Therefore, in case of CO_2 reduction on Ru modified by Cu and Cd adatoms, the production of methanol was depended on the presence of adatoms at the surface of ruthenium. RuO_2 is a promising material to be used as an electrode for CO_2 reduction to methanol due to its high electrochemical stability and electrical conductivity. For that reason, Qu et al. [210] prepared RuO_2/TiO_2 nanoparticles (NPs) and nanotubes (NTs) composite electrodes by loading of RuO_2 on TiO_2 nanoparticles and nanotubes, respectively. The obtained results showed that the current efficiency of producing methanol from CO_2 was up to 60.5% on the RuO_2/TiO_2 NTs modified Pt electrode. Therefore, RuO_2 and RuO_2/TiO_2 NPs composite electrodes showed lower electrocatalytic activity than RuO_2/TiO_2 NTs composite modified Pt electrode for the electrochemical reduction of CO_2 to methanol. In order to increase the selectivity and efficiency of CO_2 electrochemical reduction process, nanotubes structure is suggested to be used as an electrode as the studies proved.

Currently research is ongoing into the electrochemical reduction of CO_2 to methanol in which protons and electrons are used as the H_2 source. To date, however, most work reports the formation of methanol as a by-product, at selectivities less than 15% [181]. Several intermediates formed along the six-electron reduction pathway can release from the catalytic surface to form other products. Recently, it was reported that a molybdenum-bismuth bimetallic chalcogenide electrocatalyst could generate methanol with a Faradaic efficiency exceeding 70%, although this catalyst requires an acetonitrile/ionic liquid electrolyte solution [211]. Further exploratory and mechanistic research will be required to identify even more selective (and stable) catalysts that do not require organic electrolytes before electrocatalytic methanol production from CO_2 can be considered for large-scale application. Alternatively, methanol could be synthesized indirectly, via the initial electroreduction of CO_2 to CO followed by the conversion of CO to methanol [52,63,199].

Industrial synthesis of ethanol is an energy-intensive process, which also consumes a large amount of ethylene or agricultural feedstocks [163]. Thus, electrocatalytic production of ethanol or other C_{2+} oxygenates from CO_2 makes a lot of economic and environmental sense. Generation of ethanol and propanol in a flow electrolyzer has been demonstrated successfully by multiple groups. Publications are available using Cu nanoparticles, nanoporous copper film, nanoporous copper silver alloy, copper (I) chloride (CuCl) derived copper, metal organic framework (MOF)-based Cu catalysts, core-shell Cu nanoparticle, and other Cu-based catalysts [163]. Hoang et al. [201] reported close to 30% ethanol FE at 80 mA cm^2 partial current density on nanoporous Cu catalysts on carbon GDL. Zhuang et al. [212] reported over 30% FE for ethanol and propanol at 400 mA cm^2 total current density using copper-based core shell catalysts on carbon GDL. Using Cu-based PTFE membrane GDEs, their group reported nearly 10% ethanol FE at 750 mA cm^2 total current density [163,202].

Direct synthesis of alcohols from CO_2 electroreduction involves multiple electron transfer steps. As of now, the selectivity for alcohols generation is generally lower than that of ethylene. Methanol synthesis by electroreduction of CO_2 is generally difficult. Irabien et al. [213] demonstrated, using a flow electrolyzer, nearly 60% combined FE for methanol + ethanol + propanol at a current density of 10 mA/cm^2 with Cu_2O/

ZnO catalysts on carbon paper GDEs. In a very recent publication, 41% ethanol FE at 250 mA/cm^2 total current density using Cu/Ag catalysts was obtained [214]. The two-step process, CO_2 to CO and CO to alcohol synthesis, in a flow electrolyzer is quite promising [63,157,163,199]. Starting from a CO feed, Li et al. [215] demonstrated 23% FE (partial current density:11 mA cm^2) for propanol and 30% FE for ethanol with copper advanced dispersed particles catalysts on carbon paper-based GDEs. Jhonny et al. [216] demonstrated excellent performance using oxide-derived copper catalysts on carbon-based GDEs. They have shown ethanol and propanol can be produced from CO with FEs of 20.4% and 4.5%, respectively, at 1,020 mA/cm^2 total current density.

5.7.7 Carbon Nanotube Production

Recently, it has been demonstrated that carbon nanotubes can be produced electrochemically from CO_2 [52,217]. The process, which to date has only been performed on a laboratory scale, involves molten carbonate electrolysis in the presence of CO_2. Specifically, molten Li_2CO_3 can be reduced to generate carbon nanotubes and Li_2O. The Li_2O then reacts with CO_2 to regenerate Li_2CO_3, meaning that CO_2 is the ultimate carbon source for the nanotubes. If methods can be developed to produce more widely useful carbon nanotubes or carbon fibers from CO_2, this has the potential to be a disruptive technology.

5.7.8 Future Outlook

Various technoeconomic analyses of electrocatalysis of CO_2 have been presented in the literature [163,216]. While the FEs for CO and HCOOH formation from CO_2 reduction are approaching 100%, the optimized FE for ethylene, ethanol, and propanol are, respectively, around 70%, 40%, and 10%. An improvement in the FEs of these C_2–C_3 molecules is required to reduce the separation costs. Since it is primarily the catalysts that determine the FE of a product, future development should be aimed at improving their functionality, for example, through the mimicking of enzymatic catalytic steps for the selective synthesis of a product [163]. This could be accomplished via high-throughput catalyst screening as well as using operando techniques and high spatial resolution microscopy to identify the active sites on catalysts and their selectivity for CO_2 reduction.

Development of catalysts that requires low overpotentials for CO_2 reduction is also crucial for maximizing energy efficiency [163]. High current densities and longevity of a GDE are indispensable requirements for an industrial realization of electrochemical CO_2 reduction method. As of now, GDEs of 300–750 mA/cm^2 total current density have been commonly reported. Development of high-performance GDEs, which can operate well above 1,000 mA/cm^2 total current density with high FE toward a particular product, will drive electrochemical CO_2 reduction one step closer toward industrial deployment. Reported maximum GDE lifetimes, range between 150 hours (C_2H_4) and 4,200 hours (HCOOH), fall short of the desired lifetime of 30,000 hours used in technoeconomic analyses [163,218]. The deactivation of catalysts and flooding of electrodes are two major causes of GDE failure, and more work could be directed toward the understanding and prevention of these failure modes.

Recent studies indicate that carbon monoxide and formic acid electrosynthesis from CO_2 are most favorable. For these two products, above 90% selectivity (FE) at >100 mA/cm^2 total current density has been demonstrated by several groups. Furthermore, for both types of products, over 500 hours of operational stability has been achieved. Among hydrocarbons, ethylene generation appears to be promising, and total current density is already well above 100 mA/cm^2. However, the selectivity (FE = 70%) and electrode stability (150 hours) are still insufficient. Synthesis of alcohols is promising; however, their selectivity (FE = 30%–40%) remains far from practical realization. A significant improvement on selectivity and electrode stability is needed for alcohol synthesis [157,163,199].

5.8 STRATEGIES TO IMPROVE PRODUCT SELECTIVITY OF CATALYSTS, ELECTROLYTIC CELL/ELECTROLYTE, AND ELECTRODES

5.8.1 Strategies Used for Catalyst Improvement

Various strategies for new catalyst development are used to improve product selectivity of CO_2RR. While Sn, Bi, Hg, Cd, and Pb can produce formic acid (HCOOH) or formate ($HCOO^-$) in the CO_2RR [3,219], nanostructure of Sn or Bi can obtain formate with a better faraday efficiency of 60%–90% at an overpotential of 0.8–1.4 V [3,219]. On other metals, the selectivity of CO_2RR can be controlled by changing the reaction conditions to produce formic acid. Cu is the only element that can make a C–C bond directly from CO_2RR with meaningful selectivity [3,219]. It was reported that CO, formate, CH_4, C_2H_4, and H_2 could be obtained on a Cu electrode [120]. Furthermore, ethanol, *n*-propanol, allyl alcohol, a trace amount of methanol, glycolaldehyde, acetaldehyde, acetate, ethylene glycol, propionaldehyde, acetone, and hydroxyacetone were simultaneously observed in the large overpotential region, higher than 1 V [220]. Nanostructured Cu electrodes have also been studied [52,221], and the best CO_2RR performance on Cu electrode was 60%–70% faraday efficiency toward C_2H_4 production [52,221].

Oxide-derived metal electrodes improved catalytic performance compared to pristine metals [222]. Oxide-derived Cu produced more C_2 products of C_2H_4, C_2H_6, and ethanol than electro-polished Cu at the same overpotential [52,221]. Even C_3 and C_4 products like C_3H_7OH, C_3H_6, C_3H_8, and C_4H_{10} were observed from Cl ion-adsorbed oxide-derived Cu [223]. Oxide-derived Sn showed much higher current density and faraday efficiency than a pristine Sn electrode for formic acid production [3]. Oxide-derived Au [224] or Ag [225] showed superior faraday efficiency for CO production, with 90%–100% at only 0.3 V overpotential.

The intrinsic catalytic performance of metal can be modified, and high selectivity or activity can be obtained for CO_2RR by alloying with a secondary metal [3,219]. The presence of a secondary metal leads to modulation in the electronic structure of the active site. The binding energy between the active site and reaction intermediate could be tuned, changing the reaction pathway and the resulting selectivity [3,226]. The secondary metal can also change the surface geometric structure, with different atomic arrangements of reactant or intermediates. The electronic and geometric effect has been systematically studied for Au-Cu bimetallic nanoparticle catalysts with different compositions [3,226]. The alloy catalysts deviate from the general scaling relationship of binding strength to intermediates, and by controlling mixing patterns, they can exhibit unique selectivity. Well-designed bimetallic catalysts can improve the selectivity and activity for C_{2+} hydrocarbon production. A similar but not identical strategy has also been used to improve the electrocatalytic performance for C_{2+} oxygenates [3,227]. For instance, Ag-incorporated Cu-Cu_2O catalysts exhibited tunable ethanol selectivity, and the highest ethanol FE was 34.15% [3]. Besides ethanol, Cu-Ag bimetallic NPs have also been demonstrated to convert CO_2 to acetate with the addition of benzotriazole [227]. At −1.33 V versus RHE, the FE of acetate was 21.2%.

Carbon-based materials have been actively studied as promising electrocatalysts for the oxygen reduction reaction (ORR), OER, and HER [3,228]. The carbon materials possess high conductivity, high surface area, and good chemical and mechanical stability. Pure carbon materials are basically inert toward CO_2RR. However, if heteroatoms such as N are doped in the carbon matrix, electrocatalytic activity is greatly enhanced. Negatively charged N sites are considered active sites for CO_2RR [3,229]. N-doping introduces a Lewis base site to the catalyst, which is beneficial to stabilize CO_2 [3,229]. Elgrishi et al. [230] examined versatile functionalization of carbon electrode with a polypyridine ligand for CO_2 reduction. Single-atom catalysts (SAC) represent atomically dispersed metal catalysts on the surface of a support. They exhibit very distinct electronic structures and adsorption configurations of reactants and intermediates, with unique selectivity [231]. SACs have been used for electrochemical ORR [3,232], HER [233], formic acid oxidation reaction (FAOR) [234], and CO_2RR [235]. Ni single atoms on N-doped graphene can catalyze CO_2RR producing CO selectively. Other metal atoms of Fe, Co, Mn, and Cu with slightly different *d*-band structures have shown different selectivities. The single atomic Cu catalyzed CO_2RR, producing CH_4 with 58% faraday efficiency [236].

Morphology and/or structure regulation represents another alternative strategy to modulate catalytic selectivity and activity. Controlling the size, shape, and exposed facets of catalyst has been widely demonstrated for ECR performance improvement [157]. For example, the Cu(100) facet is intrinsically preferred for C_2H_4 generation, while the dominated product from the Cu(111) catalyst is methane (CH_4) [157]. In a study of Cu nanocrystals with various shapes and sizes, Loiudice et al. [237] revealed a nonmonotonic size dependence of the C_2H_4 selectivity in cube-shaped copper nanocrystals. Intrinsically, cubic Cu nanocrystals exhibited higher C_2H_4 activity and selectivity than spherical Cu nanocrystals owing to the predominance of the [69] facet. The smaller crystal size of cubic Cu could offer higher activity because of the increased concentration of low-coordinated surface sites, such as corners, steps, and kinks. However, the stronger chemisorption of low-coordinated sites was accompanied by higher H_2 and CO selectivity, resulting in lower overall hydrocarbon FE. On the other hand, the ratio of edge sites to plane sites decreased with the increase in particle sizes, which also affects the performance of C_2H_4 production.

Catalyst surface modification using small molecules is another well-known strategy to improve the electrochemical performance of ECR. This strategy can influence the microenvironment near the catalyst surface, which may stabilize the key intermediates due to the interaction between surface ligand and intermediate. Amine has been reported as a modifier to promote ECR [238]. Various amino acids, including glycine

(Gly), DL-alanine (Ala), DL-leucine (Leu), DL-tryptophan (Tyr), DL-arginine (Arg), and DL-tryptophan (Trp), have been investigated to study their effects on copper nanowires [238]. All amino acid–based ligands were capable of improving the selectivity of C_{2+} hydrocarbons. Selective exposure of crystal facets for electrocatalysts has been demonstrated as an effective and straightforward approach to achieving enhanced FE toward specific ECR products and an important way for fundamental understanding. A metal ion cycling method was developed to selectively expose the crystal facet of a Cu catalyst [239]. In situ morphology reconstruction has also been used to improve C_{2+} oxygenate FE. An active cube-like Cu catalyst was developed by Kim et al. [240], which showed improved C—C coupling performance.

Defects of electrocatalysts, such as atom vacancies and dopants, show the possibility of adsorbing unconventional ECR intermediates and, thus, selectively enhancing the corresponding pathway toward oxygenates [257,212]. Sargent and co-workers [212] studied the role of defects in a core-shell Cu electrocatalyst in detail and showed that the reaction energy barriers for ethylene and ethanol formation were similar in the early C—C coupling stage (0.5 V overpotential). Under such a condition, the introduction of copper vacancy would slightly increase the energy barrier for ethylene formation, yet it showed no influence on the ethanol generation. Furthermore, copper catalysts with vacancy and subsurface sulfur dopant would make ethylene route thermodynamically unfavorable with a negligible effect on the ethanol pathway.

Many effective and selective homogeneous metal complex catalysts have also been developed to promote CO_2 conversion. For example, a number of pincer complexes can reduce CO_2 into CO, CH_4, or other compounds. The ruthenium catalysts prepared by different methods are usually used in such systems. Hu et al. [241] investigated the electrocatalytic performance of cobalt meso-tetraphenylporphyrin (CoTPP) and its complex with carbon materials under both homogeneous and heterogeneous conditions. Their catalytic ability for CO_2 reduction was significantly increased by the strong interactions between CoTPP and carbon materials, when CoTPP was incorporated with carbon nanotubes (CNTs) or similar carbon materials. Wang and Weinstock [242] developed polyoxometalate-decorated nanoparticles The PMOFs, especially Co-PMOF, exhibited excellent electrocatalytic performance in CO_2 reduction. Davethu and Wiser [243] studied the electrochemical reduction of CO_2 to CO on an iron–porphyrin center using a computational modeling. The results showed that a ligand, rather than metal reduction, resulted in stable binding of CO_2 as an $[Fe^{III} (CO_2^{2}) (TPP)]^{2}$ complex during the reduction process.

Another strategy for catalyst improvement used is immobilization (or heterogenization) of molecular catalysts. There are numerous successful examples for CO_2 reduction by heterogenized molecular catalysts [244]. Meshitsuka et al. [245] successfully examined graphite electrodes that were immersed in suspensions of a range of metal phthalocyanines in benzene. Lieber and Lewis [246] immobilized Co phthalocyanine onto a carbon cloth from THF. The catalyst showed excellent selectivity for the production of CO at an overpotential of 300 mV in pH 5 citrate buffer. Enyo et al. [247] functionalized a glassy carbon electrode with pyridine. Electrocatalytic CO_2 reduction was observed in pH 7 phosphate buffer with an overpotential of 300 mV. The electrode displayed excellent selectivity and stability with FECO = 92% and TONCO = 107. Abruña and colleagues [248] reported the immobilization of a Co terpyridine (tpy) catalyst for CO_2 reduction to CO. A vinyl-functionalized tpy complex was electro polymerized on Pt electrodes and in dimethyl formamide (DMF) and it showed catalytic currents at 860 mV less negative potential than homogeneous $[Co(tpy)_2]^{2+}$. Kaneko and co-workers [249] followed this up and studied $[Co(tpy)_2]^{2+}$ immobilized in a Nafion membrane on carbon electrodes in aqueous electrolyte solution. At −0.85 V vs NHE in pH 7 phosphate solution, FEHCOOH = 51% and FEH_2 = 13% was observed with a low TONHCOOH = 11. Fontecave and colleagues [230] functionalized a carbon electrode with terpyridine through diazonium coupling. Electrolysis in DMF under a CO_2 atmosphere at a very negative potential (−1.73 V vs NHE) gave TONCO = 70.

Kaneko and co-workers [250] studied immobilized $[ReBr(bpy)(CO)_3]$ on graphite electrodes by casting a DMF/alcohol solution of the catalyst mixed with Nafion onto the surface of basal plane graphite. Electrolysis in pH 7 phosphate solution showed CO_2 reduction to a range of products, and the product distribution was dependent on the applied potential. The major product was H_2. Brunschwig, Gray and co-workers [251] functionalized a $[ReCl(bpy)(CO)_3]$ catalyst with pyrene groups for immobilization on graphite electrodes through π-π stacking. At applied potential of −1.67 V vs NHE in CH_3CN, FECO 70% and TONCO = 58 were obtained. A subsequent study [252] indicated that Mn analogues of the well-established $[ReX(bpy)(CO)_3]$ (L = Br^-, CH_3CN) catalysts are active for electrocatalytic CO_2 reduction. This is important from a scale-up perspective as Mn is 1.3 million times more abundant than Re in the Earth's crust [253].

The Mn catalyst operates at a lower overpotential relative to the Re analogue [254]. Cowan and colleagues [255] immobilized $[MnBr(bpy)(CO)_3]$ in a Nafion membrane onto a glassy carbon electrode and produced CO and H_2 in a ratio of 1:2 at −1.17 V vs NHE in pH 7 phosphate solution [255]. The Nafion immobilized system produced CO at 100 mV less negative potentials than for homogeneous $[MnBr(bpy)(CO)_3]$ (in CH_3CN/H_2O). This example highlights how immobilization can affect energy requirements and therefore the efficiency of the system. Other strategy to improve CO_2 reduction includes the use of enzymes. Protein film electrochemical studies of a molybdenum- and tungsten-containing formate dehydrogenase have shown that this class of enzyme can reversibly inter-convert CO_2 and formate at the thermodynamic potential [256]. Reversibility has also been demonstrated with a nickel-/iron-containing carbon monoxide dehydrogenase [257].

Molecular catalysts that mediate reactions with carbon dioxide often promote chemical reductions that form either carbon monoxide or formic acid (HCO_2H), but lack the activity and selectivity to reduce these compounds further to make other useful products, such as methanol, ethanol or methane. Wu et al. [258] indicated that when a complex called cobalt phthalocyanine is physically adsorbed to the surface of the carbon nanotubes as individual molecules, it has appreciable catalytic activity and selectivity for the electrochemical reduction of CO_2 to methanol. The key finding is that this mixed catalyst system not only activates CO_2 to produce carbon monoxide, but also, surprisingly, promotes further reduction to methanol when high voltages are applied in the electrochemical cell. The activity of the molecular catalysts and product selectivity in this situation was affected by the method used to immobilize the catalyst on the support; the specific carbon support chosen; the ratio of the concentration of the catalyst to that of the support; and the voltage used for the electrochemical reduction. The optimized system had significantly improved activity and selectivity compared with conventional molecular-catalyst systems. Long term stability is a major issue for molecular catalysts. Wu et al. [258] found that when they modified the ligand by appending amino (NH_2) substituents to it, their system's stability was enhanced. The study indicated that activity, selectivity and stability of the molecular catalyst–carbon nanotube hybrid system can be improved through judicious chemical manipulations of the catalyst and the support, and of the interactions between them.

With all these efforts, the FE of C_{2+} products is still far from practical application, where state-of-the-art catalysts allow production of C_2 products with around 60% FE [157], while the C_3 production is limited to less than 10% FE [157,259]. Reductive coupling of CO_2 to C_{2+} products requires heterogeneous catalysts with highly coordinated morphological and electronic properties [157,260,261]. The catalytic surface needs to break the scaling relations between the intermediates [220,262]. Moreover, to achieve C—C bond formation, the absorbed reaction intermediates at the catalyst surface must be in close proximity to one another. Furthermore, the pathway from the initially adsorbed intermediate toward a specific C_{2+} product needs to be well controlled because of the multiple proton-assisted electron transfer steps. Considering the high complexity of CO_2 reduction toward C_{2+} products, electrocatalysts should be carefully tailored to increase the selectivity. According to the intermediate species and chemical compositions, one can categorize C_{2+} products into multicarbon hydrocarbons and oxygenates [157,263]. To approach highly efficient electrocatalysts for specific C_{2+} molecule production, as shown above several catalyst design strategies including heteroatom doping, crystal facet regulation, alloy/dealloying, oxidation state tuning, and surface ligand control have been demonstrated [157,264]. Optimal design should rationally consider the aforementioned effects and maximize the benefits. While stability, resistivity to catalyst poisons, and activity on diluted CO_2 feed are also important for practical applications, they have not yet attracted much interest at this stage.

Improvements in catalyst alone has not allowed performance of ECR for CO_2 reduction to reach the performance obtained from gas-phase CO_2 reduction. To enhance electrochemical CO_2RR further, various other efforts have been tried. These include strategies to improve performances of electrolyzers/electrolytes and electrodes. To develop ECR technology to a commercially feasible level, these factors should be optimized.

5.8.2 Strategies to Improve Electrolyzer/Electrolyte Performance

While H-type cells are extensively used in lab-scale tests, they have low CO_2 solubility in aqueous electrolyte, limited current density, limited electrode surface area, and a large interelectrode distance to be very effective on large scale. For C_{2+} product generation, H-type cells usually show low selectivity under high overpotentials, e.g., 32% for ethylene at −0.98 V versus RHE, 13.1% for *n*-propanol at −0.9 V versus RHE, and 20.4% for ethanol at −0.46 V versus RHE, due to the seriously competitive hydrogen evolution [157].

In order to improve cell performance, the flow reactor was proposed [157]. In flow cells, gaseous CO_2 stream can be directly used as feedstock at cathode, thus leading to significantly improved mass diffusion and production rate [157,265]. Figure 5.9 shows the typical architecture of a flow cell, where a polymer electrolyte membrane (PEM) served as the electrode separator that is sandwiched between two flow channels. The catalyst is immobilized onto a gas diffusion electrode (GDE) to serve as the cathode electrode, in which gaseous CO_2 is directly fed. The catholyte, such as 0.5 M $KHCO_3$, is continuously flowed within the thin layer between the catalyst electrode and PEM. In addition, the anode side is typically circulated with an aqueous electrolyte for oxygen evolution reaction [157,265]. Compared with H-type cells, these membrane-based flow cells show much superior ECR performance. For example, Sargent and co-workers [212] evaluated the ECR performance of the Cu_2S-Cu-V catalyst in both H-type cell and flow cell. Using H-type cells, the maximum FE for C_{2+} products was 41% with a total current density of ~30 mA/cm^2 under −0.95 V versus RHE. However, the FE for C_{2+} products increased to 53% with a total current density

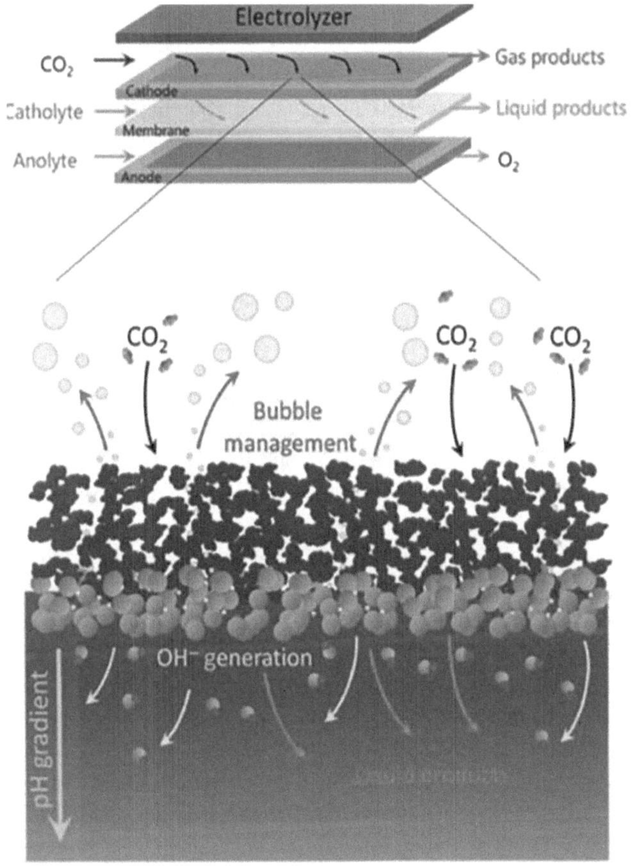

FIGURE 5.9 Diagram of the flow electrolyzer with a zoomed-in schematic of the electrode-electrolyte interface [157].

easily exceeding 400 mA/cm^2 under −0.92 V versus RHE in a flow system. Such a significant performance improvement using the flow reactor can be ascribed to the enhanced CO_2 diffusion and suppressed side reactions, mainly originating from the local gas-electrolyte-catalyst triple-interface architecture.

The zero gap cell is another emerging class of electrolyzers, which further removes the flow channels in flow cells and presses two electrodes together with an ion-exchange membrane in between. This configuration could significantly decrease mass transfer and electron transfer resistance and thus improve energy efficiency, making it more feasible in practical applications [265]. The reactants fed to the cathode can be either CO_2-saturated catholyte or humidified CO_2 stream. Water vapor or aqueous electrolyte is mandatorily fed to the anode for proton release to compensate the charge for the CO_2 reduction species [157]. Gutiérrez-Guerra et al. [266] evaluated the performance of the Cu-AC hybrid catalyst in the zero gap cell and reported that acetaldehyde is the main product with a high selectivity of 60%. As another advantage of this device, it is very easy to pressurize the reactant flow and significantly enhance the local CO_2 concentration, thus resulting in large current densities and high reaction rates [265]. However, the accelerated ion exchange rate in zero gap cells tends to acidify the catholyte, shifting the reaction toward H_2 evolution instead of CO_2 reduction [267]. To tackle this problem, Zhou and co-workers [157,267] inserted a buffer layer with a circulating aqueous electrolyte between the cathode and membrane to maintain the proper pH near the cathode for CO_2 reduction reaction. Although various C_{2+} products were detected on the basis of the zero gap cells, including acetone, ethanol, and *n*-propanol, the FEs are still relatively low. Most reported studies always focus on C_1 products that involve fewer numbers of proton and electron transfers during the reduction reaction. Therefore, the feasibility of the zero gap cell for C_{2+} products is still under debate [265].

Microfluidic electrolytic cells (MECs) are a kind of highly attractive electrolyzer configuration developed by Whipple et al. [192]. In this device, the membrane is replaced by a thin space (<1 mm in thickness) filled with flowing electrolyte stream to separate the anode and cathode. The CO_2 molecules could quickly diffuse into the electrode-electrolyte interface near cathode, and the two fixed GDEs are flushed by flowing electrolyte. Compared to membrane-based flow cells, MECs not only avoid the high membrane cost but also mitigate water management, which particularly refers to the anode dry-out and cathode flooding when operated at high current densities owing to the osmotic drag of water molecules along with proton transport from anode to cathode across the membrane [157]. Unfortunately, a minimal number of studies have achieved C_{2+} products in the original MECs. This is probably caused by the "floating" effect that protons formed in the anode are easily drained from the cathode vicinity or washed away by the flowing electrolyte, rather than participating in the multiple proton required C_{2+} formation reaction.

Lv et al. [268] synthesized a nanoporous Cu catalyst and then tested its ECR performance using different electrolytes ($KHCO_3$, KOH, K_2SO_4, and KCl) in a membrane-based MEC. They revealed that the CO_2 reduction in alkaline electrolyte (KOH) exhibits the highest C_{2+} selectivity and current density. At −0.67 V versus RHE in 1 M KOH electrolyte, the obtained FE for C_{2+} reaches up to 62% with a partial current density of 653 mA/cm^2, which is among the highest current densities that have ever been reported in electrochemical CO_2 reductions toward C_{2+} products. Ethylene (38.6%), ethanol (16.6%), and *n*-propanol (4.5%) are the main C_{2+} products with a small amount of acetate. They also pointed out that there is a strong correlation between the calculated surface pH and FE for C_{2+} products: The higher the surface pH, the higher current densities and C_{2+} products yield. The theoretical calculation proposed that the near-surface OH^- ions could strongly facilitate C—C coupling [157].

In addition to the electrolyzer configuration, the electrolyte applied in different electrolyzers could also substantially alter the final ECR products. Electrolyte compositions, pH, and concentrations affect activity and selectivity [3,269]. Ionic liquids help the activation of CO_2 [3,270]. As indicated above, highly alkaline KOH solutions are always used in flow cells with excellent performance. It is ascribed to the fact that KOH electrolyte could provide higher electrolyte conductivity, decrease ohmic resistance between the thin electrolyte coating on catalyst and bulk electrolyte, and further decrease the required overpotentials for C_{2+} formation [3]. An electric field near the electrode has shown to stabilize charged intermediates better [269]. The additives could control selectivity via electron shuttling or the formation of surface films [3,271]. The DFT results further confirm that the presence of OH^- ions could lower the energy barrier for CO dimerization, thus boosting the C_{2+} formation and suppressing the competition from C_1 and H_2 formation [157,268]. However, alkaline KOH could not be used as electrolyte in H-type cells. This is because CO_2 streams will rapidly react with KOH solutions and lastly create a bicarbonate solution with neutral pH in H-type cells [268]. In flow cells, however, once the CO_2 diffuses through the GDE, the CO_2 molecules will be consumed at the triple boundary phase (CO_2-catalyst-electrolyte) to form reduced products immediately. Besides, the poor buffering capacity of the electrolyte is able to rapidly increase the pH around the electrode in stationary electrolyzer configurations, whereas the flowing electrolyte will refresh the surface and minimize the pH fluctuation in the electrolyte [157,202]. A gas diffusion electrode (GDE) cell also allows unprecedentedly high current density [3,188].

Since ECR is a diffusion-controlled reaction, high reaction pressure could also significantly enhance the bulk and interface CO_2 concentration. The common high-pressure reactors are similar to the stainless steel autoclave, in which high-pressure CO_2 (up to 60 atm) could be introduced into the cell, leading to a remarkable increase in both the FE and the current density of C_{2+} [157,272]. Sakata and co-workers [273] showed that the current density could be improved to 163 mA/cm^2 under 30 atm on a Cu electrode with ethylene as the major product. Many metal catalysts (e.g., Fe, Co, and Ni), with no activity for C_{2+} production at ambient pressure, could reduce CO_2 to ethylene, ethane, propane, and other high-order C_{2+} products at elevated pressures. It has been demonstrated that the selectivity of the products markedly depends on the CO_2 pressure in the manner of altering CO_2 availability at the electrode surface [157,272]. The main reduced products are altered from H_2 to hydrocarbons (C_{2+} included) and lastly to CO/HCOOH with increased CO_2 pressure. Notably, the CO_2 pressure should be carefully monitored because excessive high or low CO_2 pressures would induce superfluous or limited CO_2 diffusion rate, which tends to favor the production of CO/HCOOH or H_2. Only a compatible amount of intermediate CO and current density that generated on the electrode surface could facilitate the C—C coupling reaction and enhance C_{2+} product selectivity [273].

5.8.3 Strategies for Improved Electrode Design

Designing a novel electrode with advanced structures is another important direction to enhance the selective C_{2+} production. In early stage, the working electrodes are nonporous metal foils and suffer from sluggish mass transfer [157,274]. As a result, GDE was proposed to alleviate the poor cell performance by providing hydrophobic channels that facilitate CO_2 diffusion to catalyst particles [275]. The conventional GDE usually comprises a catalyst layer (CL) and a gas diffusion layer (GDL) [202,268]. The gas-liquid-catalyst interface formed in GDE is crucial to improve the cell performance. The GDL assembled with porous materials (typically carbon paper) could provide abundant CO_2 pathways and ensure rapid electrolyte diffusion rate. It also acts as a low-resistance transportation medium for protons, electrons, and reduction products from the CL into the electrolyte [275]. Drop casting, airbrushing, and electrodeposition are the common technologies for preparation of GDEs [276]. Catalysts assembled with GDEs have been intensively investigated in CO_2 electroreduction to C_{2+} products [277]. Notably, the aforementioned flow cells with favorable performance are all coupled with GDEs. As early as 1990, Sammells and co-workers [278] reported that Cu-coated GDEs achieved high FE of 53% for ethylene with a high density of 667 mA/cm^2. Enhancing the selectivity of ethylene and ethanol is a major challenge that is always coproduced on Cu-based catalysts because of their very similar mechanistic reaction pathways. Moreover, it is important to point out that the elevated productivity and selectivity of ethylene compared to ethanol have been observed on Cu-based GDE [157,201]. Hoang et al. [201] showed an excellent FE of 60% for ethylene and a suppressed FE for ethanol of 25% on electrodeposited Cu-Ag GDE, when the total current density reached ~300 mA/cm^2 at −0.7 V versus RHE. It is a rare work that achieved such a high selectivity at a large current density. This finding suggests that a GDE-incorporated electrode provides a promising avenue for tuning the reaction pathways, in which the selectivity of reduced products can be obtained at high current densities. Reports are available on GDEs developed in-house using carbon GDL and Cu nanoparticles catalysts [279], nanoporous copper film [280], nanoporous copper silver alloy [281], copper (I) chloride derived copper [282], metal organic framework (MOF)-based Cu catalysts [283] and other Cu-based catalysts [284,285].

Efforts have been made to generate ethylene using CO feedstock [285]. Han et al. [286] demonstrated 50.8 mA/cm^2 partial current density for C_2H_4 from CO using a Cu-based GDE. Ripatti et al. [275] demonstrated an electrolyzer using a Cu-based GDE, which generated the acetate salt in addition to ethylene, and other carbonaceous products. In their system, the H_2O used for the reduction came from the anodic compartment (dragged through the membrane). This helped to increase the concentration of the acetate salt. They achieved almost 40% C_2H_4 FE over 24 hours at a total current density of 144 mA cm^2. The use of CO as feedstock has an inherent advantage, in that it does not react with alkaline electrolytes to give carbonates.

The stability of GDEs is also a significant issue that should be addressed because stable long-term operation is essential to realize practical application for flow cells. There are multiple factors that can decrease the lifetimes of GDEs. This includes catalyst deactivation as a result of poisoning from contaminants. Despite the outstanding

CO_2-to-C_{2+} performance achieved with GDEs, the stability is still poor due to the weak mechanical adhesion of the catalyst, GDL, and binder layers [287,288]. The carbon surface of GDL might change from hydrophobic to hydrophilic during the electrochemical reaction due to the oxidation reaction that occur at elevated overpotentials, which leads to the flooding in GDL and obstruct CO_2 diffusion pathways [202]. Rapid release of product gases can damage catalysts layer.

To solve this problem, researchers integrated hydrophobic scaffold of polytetrafluoroethylene (PTFE) into GDEs. Compared to hydrophilic Nafion, a hydrophobic PTFE layer renders a superior long-term stability [202]. Sargent and co-workers [202] assembled a Cu catalyst between the separated PTFE and carbon NPs, in which the hydrophobic PTFE layer could immobilize the NPs and graphite layers, thus constructing a stable electrode interface. As a result, the FE for ethylene production was increased to 70% in 7 M KOH solution at current densities of 75–100 mA/cm^2. The life span of this flow reactor was extended to more than 150 hours with negligible loss in ethylene selectivity, which is 300-fold longer than traditional GDEs. A possible reason behind the relatively long lifetimes of PTFE membrane type GDEs over carbon-based GDEs is the suppression of electrolyte flooding inside the GDE. Such a sandwich structure has been demonstrated to be an excellent GDE design. For example, Li et al. [288] designed a trilayer structure with an active electrode layer clipped by two hydrophobic nanoporous polyethylene films. The outer hydrophobic layers could slow down the electrolyte flux from the bulk solution, leading to stable, high local pH around the working electrode. Optimization of the interlayer space, which can improve CO_2 transport and adsorption, is also important in such a design [288]. Recently, carbon nanotubes have also been integrated into the GDEs because of their high porosity, good conductivity, and hydrophobicity, which could facilitate electron and mass transportation [287].

Modelling and simulation of GDEs and electrolyzers play an important role for the design and optimization of a system [278]. However, such modeling and simulation works are comparatively rare. A steady-state isothermal model, for CO generation, studying the effects of applied cell potential, feed in concentration, feed flow rates, channel length, and porosity of GDE was developed by Wu et al. [289]. A modeling work on GDEs to study the effects of catalysts layer, hydrophobicity, loading, porosity, and electrolyte flow rate was also presented by Weng et al. [290]. Technoeconomic analysis is also important to assess the economic feasibility and scalability of CO_2 reduction [167,291,292]. **One recent analysis [292] indicates that CO_2 reduction can become competitive against fossil fuel feedstock, when electrical to chemical energy conversion factor is above 60% for carbon monoxide, ethanol and ethylene, assuming electricity cost is less than 4 US cents per kWh.**

5.9 BARRIERS AND POSSIBILITIES FOR COMMERCIALIZATION OF ELECTROCATALYTIC REDUCTION OF CO_2

While electrocatalysis of CO_2 reduction is being aggressively pursued, there are some barriers that need to be overcome. For methane, methanol, oxalate, oxalic acid, and DME productions, high overpotentials and low selectivity need to be overcome. For formic acid, poor catalysts stability and separation of formic acid from reaction medium need to be facilitated. For CO production, high flux of CO_2 to cathode is required and conversion per pass needs to be improved. For ethylene and ethanol, low selectivity and poor catalyst stability are barriers. Low selectivity and lack of understanding of carbon-carbon bond formation steps are major issues with production of hydrocarbon fuel. Finally, for carbon nanotubes, properties of currently produced carbon fibers are not suitable to act as replacements for carbon fibers.

While the FEs for CO and HCOOH formation from CO_2 reduction are approaching 100%, the optimized FE for ethylene, ethanol, and propanol are, respectively, around 70%, 40%, and 10%. An improvement in the FEs of these C_2–C_3 molecules is required to reduce the separation costs. Since it is primarily the catalysts that determine the FE of a product, future development should be aimed at improving their functionality, for example, through the mimicking of enzymatic catalytic steps, for the selective synthesis of a product. This could be accomplished via high-throughput catalyst screening as well as using operando techniques and high spatial resolution microscopy to identify the active sites on catalysts and their selectivity for CO_2 reduction. Development of catalysts that requires low overpotentials for CO_2 reduction is also crucial for maximizing energy efficiency.

Malkhandi and Yeo [163] summarize the important works that produced more than 100 mA cm^2 current density. **These works indicate that carbon monoxide and formic acid electrosynthesis from CO_2 are most favorable. For these two products, above 90% selectivity (FE) at >100 mA cm^2 total current density has been demonstrated by several groups.** Furthermore, for both types of products, over 500 hours of operational stability has been achieved. Among hydrocarbons, ethylene generation appears to be promising and total current density is already well above 100 mA cm (FE = 70%), and electrode stability (150 hours) are still insufficient. Synthesis of alcohols is promising; however, their selectivity (FE = 30%–40%) remains far from practical application. **A significant improvement on selectivity and electrode stability is needed for alcohol synthesis.**

An effective cathode for CO_2 reduction is a major obstacle that has captured significant attention. An ideal cathode will reduce CO_2 to a single desired product at a high rate with minimal overpotential and maintain its activity for long periods of continuous operation. High synthesis rates require effective mass transport of CO_2 to the catalyst material while maintaining high ionic conductivity. These requirements have been met by using gas diffusion electrodes (GDE), which are used in commercial technologies such as fuel cells and electrolyzers. Electrolysis cells employing gas diffusion electrodes have been used successfully for electroreduction of CO_2 to products such as formic acid, CO, and ethylene at potentially commercially relevant rates exceeding 100 mA/cm^2. Most of these flow electrolysis systems, however, are operated such that (a) only a modest percentage of CO_2 (1%–20%) is converted to the desired product, that is, a low single-pass conversion, and (b) some of the CO_2 is captured by OH^- in the electrolyte to form carbonate. A low single-pass conversion implies that more energy is needed to separate gaseous products from the product stream, and to recycle the unreacted CO_2 back into the feed. Second, formation of carbonates represents a net loss of CO_2 and causes electrolyte degradation and electrode performance degradation (carbonate precipitation). Given the energy required to obtain CO_2 of sufficient purity to begin with, low CO_2 utilization per pass and loss of CO_2 in the electrolyte imposes additional energy penalties that are typically not accounted for in system analyses. In fact most experimental work reported does not explicitly acknowledge, let alone quantify, these parasitic losses. Achieving practical electrochemical systems for conversion of CO_2 to valuable intermediates such as CO and ethylene requires the development of systems that achieve high single-pass conversion of CO_2 to the desired product with minimal loss of CO_2 into the electrolyte.

High-performance electrodes are necessary to generate products with yields required for industrial scales. Electrodes such as single crystal metal surfaces, studied in the lab using an H-cell, are incapable of this due to mass transport limitation of CO_2. This arises because of the poor solubility of the reactant CO_2 in aqueous electrolyte (33 mM at 298 K and 1 atm). The mass transport constraint can be managed by either filling the H-cell with CO_2 at higher pressures [169] or by using gas diffusion electrodes (GDE). The latter types of electrodes are widely used in fuel cells and other electrochemical processes. GDEs can circumvent the mass transport bottleneck and improve the yield by two orders of magnitude or even more [161,162].

In the simplest terms, the structure of a typical GDE consists of an electrically conducting porous support layer (I), and a micro-porous gas diffusion layer (II) (GDL) [161,162]. The catalysts are coated on the interior wall of channels in the GDL. In an electrolyzer, the GDE stays between the gas and electrolyte compartments [106]; the electrolyte partially penetrates those catalysts-coated channels and forms a three-phase boundary. During a reaction, gaseous CO_2 directly approaches the catalytic centers at those three phase boundaries through tiny channels from the gas side of the electrode. Note that the bulk electrolyte and CO_2 will not be considerably mixed to give carbonates. Thus, alkaline electrolytes, which have been experimentally shown to improve C_{2+} product, can be used [164,165]. In contrast, in an electrochemical system housed in a H-cell, the gaseous CO_2 usually has to dissolve in the bulk electrolyte first before arriving at the catalytic centers. This arrangement gives rise to the CO_2 mass transport bottleneck, due to poor solubility of the CO_2. It also rules out the use of alkaline electrolytes owing to its rapid reaction with CO_2.

Here are some research directions that could be taken to accelerate practical deployment of these systems in an industrial setting, as discussed in various technoeconomic analyses [166,167]. High current densities and longevity of a GDE are indispensable requirements for an industrial realization of electrochemical CO_2 reduction method. As of now, GDEs of 300–750 mA cm^2 total current density have been commonly reported. Development of high-performance GDEs, which can operate well above 1,000 mA cm^2 total current density with high FE toward a particular product, will drive electrochemical CO_2 reduction one step closer toward industrial deployment. Reported maximum GDE lifetimes, range between 150 hours (C_2H_4) and 4,200 hours (HCOOH), fall short of the desired lifetime of 30,000 hours, used in technoeconomic analyses [168]. The deactivation of catalysts and flooding of electrodes are two major causes of GDE failure, and more work could be directed toward the understanding and prevention of these failures.

Another key factor to make the overall electrolysis process (more) economically feasible is to enhance energy efficiency and minimize overall energy requirements. Some gains can still be made by reducing the overpotential of the

various cathode catalysts. However, those potential gains are small compared to the energy requirements of an overall electrolysis process that typically couples the CO_2 electroreduction at the cathode to the highly energy intense oxygen evolution reaction (OER) at the anode. An analysis of the energy requirements based on only Gibbs free energies shows that about 90% of the energy needed to drive the CO_2 electrolysis process is required for OER at the anode, to form O_2, for which there is very limited commercial value at scale. Indeed efforts have started to identify alternative anode reactions that involve the oxidation of readily available chemicals. For example, glycerol (a by-product of biodiesel production), glucose, and even methane, for which the thermodynamic potentials are 0.8–1.1 V lower than the minimum potential needed for the OER, have potential in this regard. In fact, it has been demonstrated that glycerol oxidation can lead to a 37%–50% reduction in the energy requirement for the overall CO_2 electrolysis process, and with the cost of electricity often the prime cost-determining factor [167], this significantly changes the likelihood of achieving economic viability. Of course, no chemical is as abundantly available as water for the OER; however, vast amounts of (waste) chemicals like glycerol are available in various places around the world, for example, in Brazil's biofuel production. In those locations, CO_2 electrolysis to CO, ethylene, and/or ethanol has the potential to become a viable technology when coupled with glycerol oxidation.

Despite the exciting progresses on ECR, strategies for low-cost, large-scale C_{2+} product generation are rarely present [170]. At this stage, the challenges and opportunities are concurrent to understand the reaction mechanisms of ECR and commercialize this promising technology. Despite all the effort put into ECR, there are still many problems with the current catalysts and ECR system that must be addressed before commercializing ECR. First, as the dominating catalyst to realize efficient C—C coupling, Cu suffers from serious stability issues, especially in aqueous electrolyte, and can rarely survive for 100 hours due to their high atom mobility, particle aggregation, and structure deterioration under ECR conditions. Thus, how to achieve long-period stability using a Cu-based catalyst is still an open challenge. Anchoring the Cu-based catalyst on specific support with strong interaction might be a reliable strategy to preserve the catalyst structure/morphology and thus provides enhanced life span. Furthermore, using a polymer membrane electrolyte to replace the aqueous solution during ECR can probably further improve the stability of the Cu-based catalyst. In addition, from the perspective of catalysts, in situ/in operando characterization techniques and theoretical modeling should also be used to monitor and understand the catalyst performance decay, thus, in turn, suppressing the degradation and poisoning of catalyst to the lowest levels. Another important issue of ECR catalysts that should be addressed is to make the synthesis protocol viable for mass production. To this end, streamlining the synthetic procedures using widely available feedstocks is preferred.

Second, the generated C_{2+} oxygenated from ECR are usually mixed with solutes (e.g., $KHCO_3$ and KOH) in the electrolyte for traditional H- or flow-cell reactors, which, however, requires extra separation and concentration processes to recover pure liquid fuel solutions in practical applications. At the same time, the evolved C_{2+} hydrocarbons are also mixed with H_2 and residual CO_2. Thus, a costly separation process is indispensable for current ECR technology, which further hinders ECR from practical application. Therefore, how to directly and continuously produce pure liquid fuel solutions and pure gas hydrocarbons, particularly with high product concentrations, is highly desirable for the practical deployment of ECR. We thus predict the rising importance of direct generation of pure products via ECR in the near future, which may take the ECR technology much closer to market [171].

Third, while the formation of C—O and C—H bonds, such as ethanol, acetic acid, and ethylene, in ECR technology has been heavily studied, exploration of other types of products is also important for ECR technology and shows economical interest. For example, recently, Han and co-workers [172] reported the production of 2-bromoethnol by ECR. The in situ formation of C—Br bond transforms the product from ethanol to 2-bromoethnol, which is an important building block in chemical and pharmaceutical synthesis and shows higher added value. Thus, beyond the current well-studied C_{2+} products, targeting of other rarely explored products such as oxalic acid [104] and synthesis of more complex C_{2+} molecules such as cyclic compounds is another promising route for future ECR research.

Last but not least, novel electrode and reactor designs such as waterproof GDE, liquid-flow cells, and PEM cell should be widely adopted to boost the ECR production rate to commercial level (>200 mA/cm^2). However, the large discrepancy in electrocatalytic activity is always observed when electrocatalysts are applied to the full cell test. Therefore, more systematic studies should be performed to minimize the gap between half-cell studies and full-cell device application to bring the ECR from lab-scale test to practical use.

5.10 CCUS STRATEGY USING FUEL-CELL TECHNOLOGY

Fuel cell is an electrolytic cell where chemical energy is transformed into electrical energy as long as fuel and oxygen are supplied. As shown in my previous book [293], high-temperature MCFC fuel cells can take dilute CO_2 stream and convert it into power and heat and produce concentrated stream of Carbon dioxide which can be either sequestrated or converted to valuable chemicals by downstream electrocatalytic process. The book describes three processes that can be parts of CCUS strategy; (a) Exxon/Mobil-Fuel Cell technology approach to use MCFC fuel cell to convert CO_2 to power, (b) use MCFC fuel cell as trigeneration plant to produce power heat and hydrogen, and (c) use Audi-Sunfire-Climeworks approach to convert CO_2 to diesel fuel (blue-crude). Here we examine two other innovative and integrated approaches to convert CO_2 to power and/or valuable chemicals [252,294]

Rossi et al. [294] built a cylindrical MCFC with innovative stack design. The main technology benefits of this design were high electrical efficiency (up to 40%), thermal self-sustain conditions kept down to the kW-size stack, because of minimum heat losses due to the cylindrical geometry and gas recirculation, non-pressurized devices, long life (the proposed MCFC worked for 4,500 working hours), compact design and modularity. The proposed CO_2 capture chemical process by molten carbonate fuel cells is shown in Figure 5.10. The CO_2 capture chemical process was analyzed when integrated with a gas turbine plant. The turbine plant was based on an open Brayton cycle. The exhaust gas of the gas turbine plant was sent to the cathode of a molten carbonate fuel cell. CO_2 in exhaust gases was carried to the operative temperature of the MCFC by exchanging heat in a reformer and then it

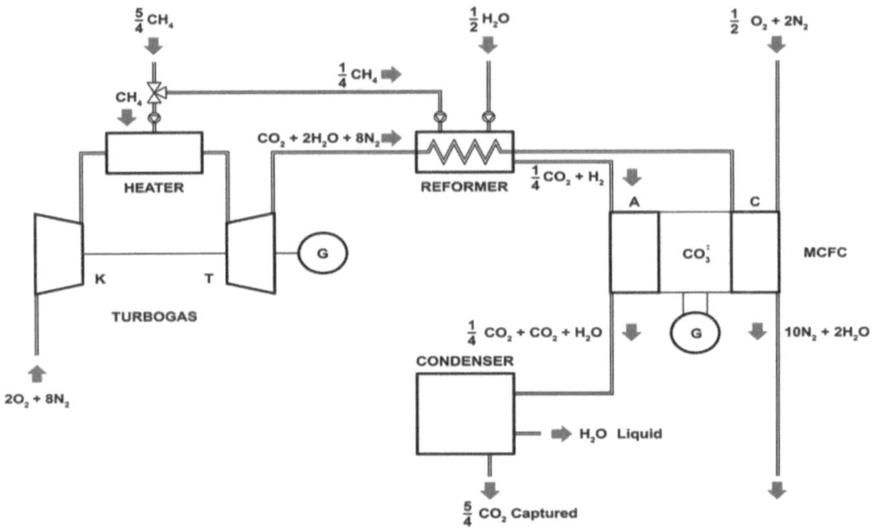

FIGURE 5.10 Novel combustion-reformer-MCFC integrated study by Rossi et al. [294].

was sent to the cathode. The reformer was used to generate the hydrogen which was sent at the anode of the fuel cell. Oxygen reacted with carbon dioxide at the cathode producing carbonate ion (CO_3^{2-}) and electrons ($2e^-$). At the anode, hydrogen was combined with the carbonate ion to generate water, carbon dioxide and electrons ($H_2O + CO_2 + 2e^-$). The exhausts of the fuel cell reactions were water and carbon dioxide ($1/4CO_2 + CO_2 + H_2O$). Exhausts were sent to a condenser, where CO_2 was separated from H_2O and finally captured. The energy consumption per mass unit of captured CO_2 (kJ/kgCO_2) was calculated as the difference between two contributions (see Figure 5.10): energy, in terms of methane energy content, to produce the hydrogen required to supply the MCFC and to capture carbon dioxide ($5/4CO_2$), and the electric energy produced by the MCFC fuel cell. The details of these calculations are given by Rossi et al. [294].

A comparison in terms of energy consumption between the proposed study and other CCS technologies is shown in Table 5.3.

TABLE 5.3
Comparison of Energy Required by Rossi et al. [294] with Other CCS Technologies

Method	Energy required (kJ/kgCO_2)
Chemical solvent absorption (e.g., amine)	4000–6000
Physical absorption/adsorption	4000–6000
Membrane separation processes	500–6000
Phase separation (e.g., cryogenic-clathrate hydrates)	6000–10,000
Innovative-CO_2 capture MCFC	About 500

5.10.1 Integrated Carbon Capture and Utilization

As shown in Figure 5.11, Mohsin et al. [252] evaluated an integrated carbon capture and utilization scenario. In this approach, an electrochemical conversion unit is directly coupled with a compatible CO_2 capture process, leading to in situ production of high-value products.

High-temperature fuel-cell-based carbon capture technology offers tremendous opportunities, as it offers simultaneous production of electricity and concentrated CO_2 (>97%) for subsequent electrochemical conversion. This fuel cell can be combined with natural gas combined cycle (NGCC) power plants. This approach is similar to one adopted by ExxonMobil and FuelCell Energy consortium [293]. Mohsin et al. [252] conducted three case studies on a 500 MW NGCC power plant, as illustrated in Figure 5.12. The study considered electrochemical (i.e., MCFC-based) carbon capture and CO_2 conversion (ECC) technologies, and where CO_2 is converted to products. The study assumed that the electrical energy required for the ECC will be supplied by the MCFC unit. In case I, a 500 MW NGCC power plant located in North

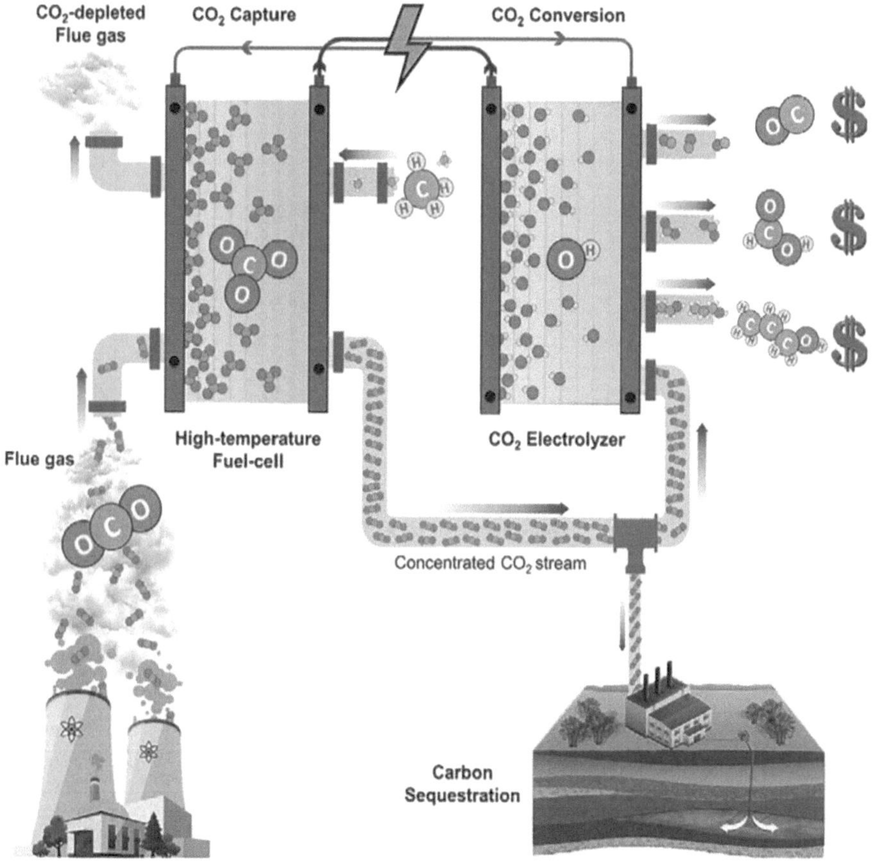

FIGURE 5.11 Integrated carbon capture and utilization [252].

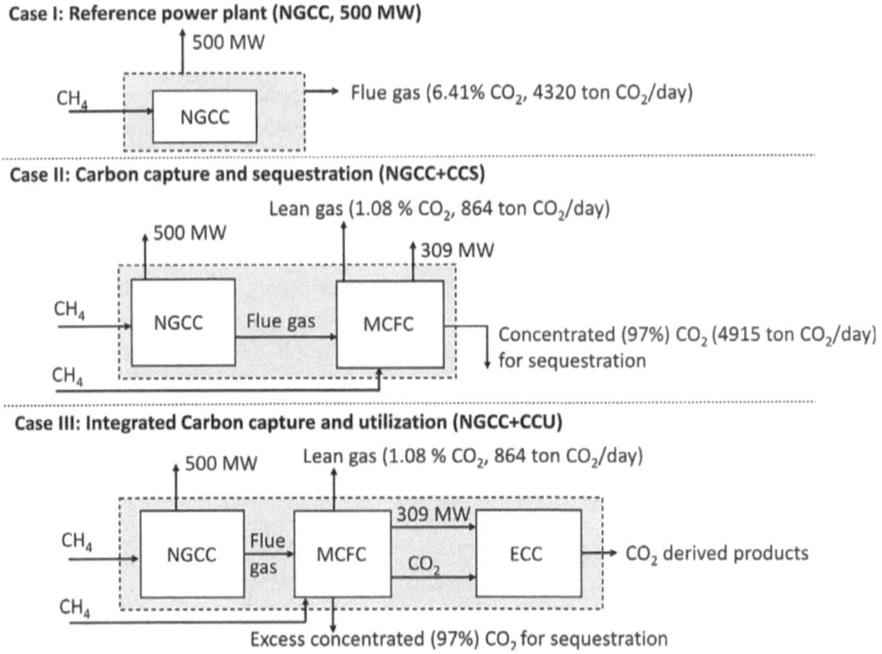

FIGURE 5.12 Three case studies assessed in work by Mohsin et al. [252].

America is used as a reference, based on an advanced (e.g., "F-class") gas turbine that produces flue gas containing 4 mol% CO_2. In case II (NGCC + CCS), MCFC is used to concentrate CO_2 (97 mol%) from flue gas and concomitantly produce electricity (309 MW) over and above the 500 MW produced by the NGCC plant. In this case, the study assumed that the incremental electricity generated (309 MW) would be sold to the grid and the CO_2 captured by the MCFC is transported 200 km by pipeline for sequestration in a geological formation. In case III (NGCC + CCU), the study assumed that the incremental electricity generated (309 MW) would be used in a co-located ECC unit to convert concentrated CO_2 into one of the six different products (see Table 5.4).

The study assumed the excess captured CO_2 (i.e., CO_2 captured minus CO_2 converted via ECC) would be transported 200 km by pipeline for sequestration in a geological formation. The study then applied technoeconomic analysis and cradle-to-gate life cycle assessment (LCA) under conservative, baseline, and optimistic scenarios to reveal the overall economic feasibility and environmental impact for each of these scenarios. This analysis revealed that under the baseline scenario, along the CCU route, only three carbon products would reap either economic (carbon monoxide [CO] and *n*-propanol 67% and 10% gross margin increase, respectively) or environmental benefits (formic acid net GHG reduction potential of 721 thousand ton CO_2e/year) relative to the CCS route. Under the optimistic scenario, while all of the CO_2-derived products are economically compelling over the CCS route, only formic acid would provide a net GHG reduction potential of 1,465 thousand ton CO_2e/year over CCS (575 thousand ton CO_2e/year).

5.10.1.1 Levelized Cost of Electricity

Figure 5.13 shows the gross and net levelized cost of electricity (LCOE) under three case scenarios. The cost incurred due to the addition of MCFC in case II is largely compensated for by additional power generation (i.e., 309 MW) relative to case I (i.e., an increase of 6% in case II). The net LCOE for case III showed that the choice of CO_2-derived product strongly influences the results, not only due to the different market prices but also because the production rates and underlying cost of production (e.g., product separation cost) vary between products. For example, the gross LCOE is the highest ($190–$225/MWh) when formic acid is considered to be the key CO_2-derived product, which is attributed to the expensive separation process required to produce formic acid of a salable purity [166]. However, formic acid has the highest production rate (along with CO), as it requires only a two electron transfer (Table 5.4), and, when combined with the market price, this results in formic acid having a net LCOE ($60/MWh) that is very similar to other products, but higher than CO ($20/MWh) and *n*-propanol (45$/MWh) under baseline scenario. There is uncertainty in the net LCOE for cases II and III that is associated with the market demand and price of the products of interest, whether they are chemical in nature (as in case III) or simply electricity (as in case II). Considering the uncertainty associated with the net LCOE in case III, the study concluded that under the baseline scenario, the only CO_2-derived products that would improve the overall economics of the plant (relative to case II) would be CO and *n*-propanol, with the highest gross margin of 67% for CO. In the optimistic scenario, all of the CO_2-derived products could provide improved system economics for the CCU route over CCS.

In case III, the CO_2-derived products were sold at market price (Table 5.4) to offset the cost of electricity production. Gross LCOE represents the cost without considering the revenue from CO_2 derived products. Net LCOE offsets cost with product revenue.

TABLE 5.4
Electrochemical CO_2 Reduction Reaction Products [252]

Products	No. Electrons per CO_2	Av. Market Price ($/kg)	Normalized Market Price ($/Electron) 3 10³	Annual Global Production (Million Tons)	Product Mol Wt/Electron
Carbon monoxide	2	0.6	8.0	*	14.0
Formic acid	2	0.74	16.1	0.6	23.01
Methanol	6	0.58	3.1	110	5.34
Ethylene	6	1.3	3.0	140	2.33
Ethanol	6	1.00	3.8	77	3.83
n-Propanol	6	1.44	4.8	0.2	3.33
Hydrogen (by product)	–	–	–	–	–

*CO is produced and used extensively in chemical processes; however, it is typically produced in captive facilities (e.g., SMR) and not marketed in bulk. Hence, we do not report estimated production volumes.

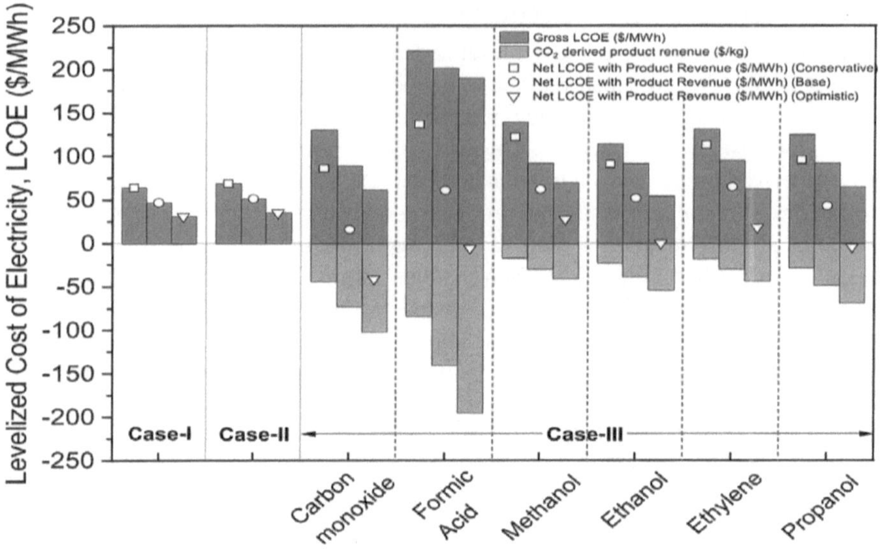

FIGURE 5.13 LCOE under three different cases [252].

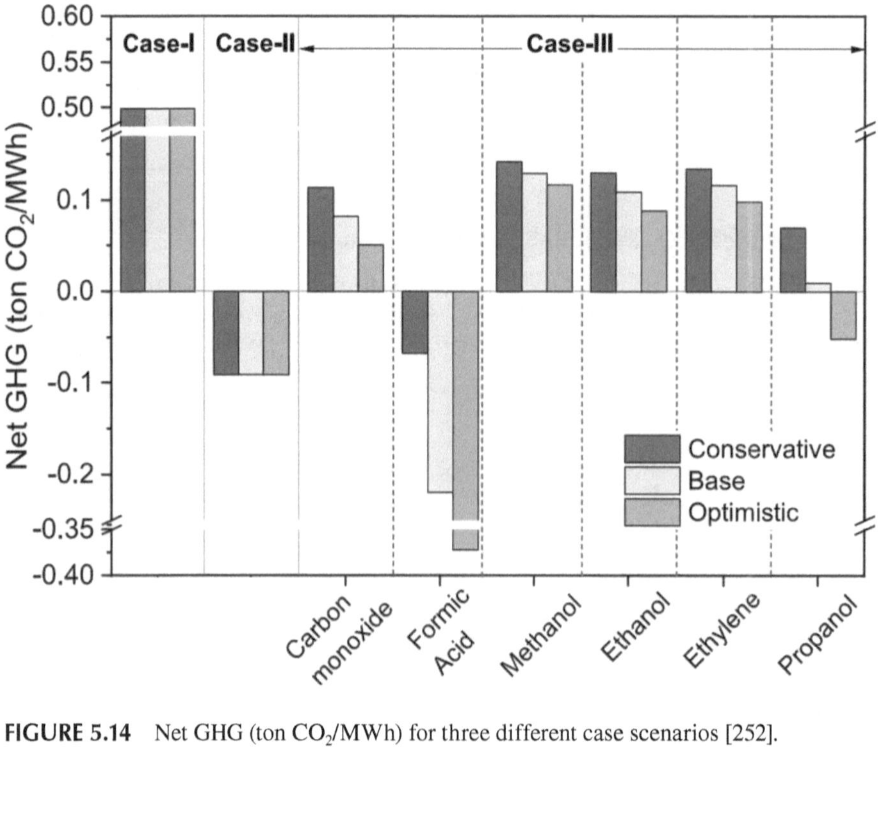

FIGURE 5.14 Net GHG (ton CO_2/MWh) for three different case scenarios [252].

5.10.1.2 Net GHG Emission

Results in Figure 5.14 show that the application of CO_2 capture via the MCFC and subsequent geological storage in case II results in a negative carbon footprint for electricity generated as a result of substitution. This negative value does not imply "negative emissions" (i.e., a physical reduction of CO_2 in the atmosphere), but simply that a reduction of emissions is occurring as a result of substitution. Electricity generated from all of the CO_2-derived product pathways (case III) has a carbon footprint that is lower than that of an unabated NGCC (case I), but higher than that from the CCS (case II). The production of formic acid results in the avoidance of emissions from fossil fuel-based formic acid production due to substitution, as does *n*-propanol when optimistic assumptions are applied. The negative carbon footprint of CO_2-derived products (formic acid and *n*-propanol) are attributed not only to their higher production rate but also to the high global warming potential (GWP) of these products via conventional routes. These results indicate compelling environmental incentives for the conventional power producer (i.e., NGCC) to adopt the CCU route (case III), with formic acid as the CO_2-derived product over the CCS route (case II), both under baseline and optimistic scenarios. More detailed analyses with other scenarios are described by Mohsin et al. [252].

REFERENCES

1. Buchanan, M. 2017. Accelerating breakthrough innovation in carbon capture, utilization and storage, a Report of the Carbon Capture, Utilization and Storage Experts' Workshop, September 26–28, Houston, TX, Department of Energy, Washington, D.C.
2. Ye, R. P., et al. 2019. CO_2 hydrogenation to high-value products via heterogeneous catalysis. *Nature Communications* 10:5698. https://doi.org/10.1038/s41467-019-13638-9.
3. Whang, H.S., et al. 2019. Heterogeneous catalysts for catalytic CO_2 conversion into value-added chemicals. *BMC Chemical Engineering* 1:9. https://doi.org/10.1186/s42480-019-0007-7.
4. Frontera, P., A. M. M. Ferraro, and P. Antonucci 2017. Supported catalysts for CO_2 methanation: A review. *Catalysts* 7:59. doi: 10.3390/catal7020059. www.mdpi.com/journal/catalysts.
5. Yao, Y., B. C. Sempuga, X. Liu, and D. Hildebrandt 2020. Production of fuels and chemicals from a CO_2/H_2 mixture. *Reactions* 1:130–146. doi: 10.3390/reactions1020011. www.mdpi.com/journal/reactions.
6. Qin, Z., Y. Zhou, Y. Jiang, Z. Liu, and H. Ji 2017. Recent advances of heterogeneous catalytic hydrogenation of CO_2 to methane. doi: 10.5772/65407. Downloaded from: http://www.intechopen.com/books/new-advances-inhydrogenation-processes-fundamentals-and-applications.
7. Gao, P., L. Zhang, S. Li, Z. Zhou, and Y. Sun 2020. Novel heterogeneous catalysts for CO_2 hydrogenation to liquid fuels. *ACS Central Science* 6:1657–1670.
8. Jin, F., et al. 2011. High-yield reduction of carbon dioxide into formic acid by zero-valent metal/metal oxideredox cycles. *Energy & Environmental Science* 4:881–884.
9. Sun, Y., et al. 2019. In situ hydrogenation of CO_2 by Al/Fe and Zn/Cu alloy catalysts under mild conditions. *Chemical Engineering & Technology* 42:1223–1231.
10. Yan, Y., Y. H. Dai, H. He, Y. B. Yu, Y. H. Yang 2016. A novel W-doped Ni-Mg mixed oxide catalyst for CO_2 methanation. *Applied Catalysis B: Environmental* 196:108–116.

11. Saravanan, K., H. Ham, N. Tsubaki, and J. W. Bae 2017. Recent progress for direct synthesis of dimethyl ether from syngas on the heterogeneous bifunctional hybrid catalysts. *Applied Catalysis B: Environmental* 217:494–522.
12. Li, Z., et al. 2017. Highly selective conversion of carbon dioxide to lower olefins. *ACS Catalysis* 7:8544–8548.
13. Wei, J., et al. 2017. Directly converting CO_2 into a gasoline fuel. *Nature Communications* 8:15174–15181.
14. Yang, C., et al. 2019. Hydroxyl-mediated ethanol selectivity of CO_2 hydrogenation. *Chemical Science* 10:3161–3167.
15. Tran, K., and Z. W. Ulissi 2018. Active learning across intermetallics to guide discovery of electrocatalysts for CO_2 reduction and H_2 evolution. *Nature Catalysis* 1:696–703.
16. Parra-Cabrera, C., C. Achille, S. Kuhn, and R. Ameloot 2018. 3D printing in chemical engineering and catalytic technology: Structured catalysts, mixers and reactors. *Chemical Society Reviews* 47:209–230.
17. Kho, E. T., T. H. Tan, E. Lovell, R. J. Wong, J. Scott, and R. Amal 2017. A review on photo-thermal catalytic conversion of carbon dioxide. *Green Energy and Environment* 2:204–217.
18. Kim, S. S., H. H. Lee, and S. C. Hong 2012. The effect of the morphological characteristics of TiO2 supports on the reverse water–gas shift reaction over Pt/TiO_2 catalysts. *Applied Catalysis B: Environmental* 119:100–108.
19. Fujita, S.-I., M. Usui, and N. Takezawa 1992. Mechanism of the reverse water gas shift reaction over Cu/ZnO catalyst. *Journal of Catalysis* 134:220–225.
20. Chen, C.-S., W.-H. Cheng, and S.-S. Lin 2000. Mechanism of CO formation in reverse water–gas shift reaction over Cu/A12O3 catalyst. *Catalysis Letters* 68:45–48.
21. Kung, H. H. 1989.*Transition Metal Oxides, vol. 45, 1st Edition: Surface Chemistry and Catalysuis*. Amsterdam, The Netherlands: Elsevier.
22. Goguet, A., F. C. Meunier, D. Tibiletti, J. P. Breen, and R. Burch 2004. Spectrokinetic investigation of reverse water-gas-shift reaction intermediates over a Pt/CeO_2 catalyst. *The Journal of Physical Chemistry B* 108:20240–20246.
23. Olah, G. A., G. S. Prakash, and A. Goeppert 2011. Anthropogenic chemical carbon cycle for a sustainable future. *Journal of the American Chemical Society* 133: 12881–12898.
24. Graciani, J., et al. 2014. Highly active copper-ceria and copper-ceria-titania catalysts for methanol synthesis from CO_2. *Science* 345:546–550.
25. Grabow, L., and M. Mavrikaki 2011. Mechanism of methanol synthesis on Cu through CO_2 and CO hydrogenation. *ACS Catalysis* 1:365–384.
26. Studt, F., et al. 2015. The mechanism of CO and CO_2 hydrogenation to methanol over Cu-based catalysts. *ChemCatChem* 7:1105–1111.
27. Shah, Y. T. 2014. *Water for Energy and Fuel Production*. New York, NY: CRC Press, Taylor and Francis.
28. Shah, Y. T. 2017. *Chemical Energy from Natural and Synthetic Gas*. New York, NY: CRC Press, Taylor and Francis.
29. Wang, W., S. Wang, X. Ma, and J. Gong 2011. Recent advances in catalytic hydrogenation of carbon dioxide. *Chemical Society Reviews* 40(7):3703–3727. doi: 10.1039/C1CS15008A.
30. Galvan, C. A., J. Schumann, M. Behrens, J. L. G. Fierro, R. Schlogl, and E. Frei 2016. Reverse water-gas shift reaction at the Cu/ZnO interface: Influence of the Cu/Zn ratio on structure-activity correlations. *Applied Catalysis B* 195:104–111.
31. Zhou, G. L., B. C. Dai, H. M. Xie, G. Z. Zhang, K. Xiong, and X. X. Zheng 2017. CeCu composite catalyst for CO synthesis by reverse water-gas shift reaction: Effect of Ce/Cu mole ratio. *Journal of CO_2 Utilization* 21:292–301.

32. Chen, X. D., X. Su, H. M. Duan, B. L. Liang, Y. Q. Huang, T. Zhang 2017. Catalytic performance of the Pt/TiO_2 catalysts in reverse water gas shift reaction: Controlled product selectivity and a mechanism study. *Catalysis Today* 281:312–318.
33. Xu, H. T., Y. S. Li, X. K. Luo, Z. L. Xu, and J. P. Ge 2017. Monodispersed gold nanoparticles supported on a zirconium-based porous metal-organic framework and their high catalytic ability for the reverse water-gas shift reaction. *Chemical Communications* 53(56):7953–7956.
34. Goncalves, R. V., L. L. R. Vono, R. Wojcieszak, C. S. B. Dias, H. Wender, E. Teixeira-Neto, L. M. Rossi 2017. Selective hydrogenation of CO_2 into CO on a highly dispersed nickel catalyst obtained by magnetron sputtering deposition: A step towards liquid fuels. *Applied Catalysis B* 209:240–246.
35. Sun, F. M., C. F. Yan, Z. D. Wang, C. Q. Guo, S. L. Huang 2015. Ni/Ce-Zr-O catalyst for high CO_2conversion during reverse water gas shift reaction (RWGS). *International Journal of Hydrogen Energy* 40(46):15985–15993.
36. Porosoff, M. D., S. Kattel, W. H. Li, P. Liu, and J. G. Chen 2015. Identifying trends and descriptors for selective CO_2 conversion to CO over transition metal carbides. *Chemical Communications* 51(32):6988–6991.
37. Porosoff, M. D., J. W. Baldwin, X. Peng, G. Mpourmpakis, and H. D. Willauer 2017. Potassium-promoted molybdenum carbide as a highly active and selective catalyst for CO_2 conversion to CO. *ChemSusChem* 10(11):2408–2415.
38. Liu, X. Y., C. Kunkel, P. R. de la Piscina, N. Homs, F. Vines, and F. Illas 2017. Effective and highly selective CO generation from CO_2 using a polycrystalline alpha-Mo_2C catalyst. *ACS Catalysis* 7(7):4323–4335.
39. Zhang, X., X. B. Zhu, L. L. Lin, S. Y. Yao, M. T. Zhang, X. Liu, X. P. Wang, Y. W. Li, C. Shi, and D. Ma 2017. Highly dispersed copper over beta-Mo_2C as an efficient and stable catalyst for the reverse water gas shift (RWGS) reaction. *ACS Catalysis* 7(1):912–918.
40. Wu, J. J., et al. 2017. Carbon dioxide hydrogenation over a metal-free carbon-based catalyst. *ACS Catalysis* 7(7):4497–4503.
41. Keim, W. 1989. Carbon monoxide: Feestock for chemicals, present and future. *Journal of Organometallic Chemistry* 372(1):15–23.
42. Hwang, S., U. G. Hong, J. Lee, J. H. Baik, D. J. Koh, H. Lim, and I. K. Song 2012. Methanation of carbon dioxide over mesoporous nickel-M-alumina (M = Fe, Zr, Ni, Y, and Mg) xerogel catalysts: Effect of second metal. *Catalysis Letters* 142:860–868. doi: 10.1007/s10562-012-0842-0.
43. Arandiyan, H., Y. Wang, H. Y. Sun, M. Rezaei, and H. X. Dai 2018. Ordered meso- and macroporous perovskite oxide catalysts for emerging applications. *Chemical Communications* 54(50):6484–6502.
44. Arandiyan, H., Y. Wang, J. Scott, S. Mesgari, H. X. Dai, and R. Amal 2018. In situ exsolution of bimetallic Rh-Ni nanoalloys: A highly efficient catalyst for CO_2 methanation. *ACS Applied Materials & Interfaces* 10(19):16352–16357.
45. Zhu, H., R. Razzaq, L. Jiang, and C. Li 2012. Low-temperature methanation of CO in coke oven gas using single nanosized Co_3O_4 catalysts. *Catalysis Communications* 23:43–47. doi: 10.1016/ j.catcom.2012.02.029.
46. Wang, Y., H. Arandiyan, J. Scott, H. X. Dai, and R. Amal 2018. Hierarchically porous network-like Ni/Co_3O_4: Noble metal-free catalysts for carbon dioxide methanation. *Advanced Sustainable Systems* 2(3):1700119.
47. Su, X., J. Xu, B. Liang, H. Duan, B. Hou, and Y. Huang 2016. Catalytic carbon dioxide hydrogenation to methane: A review of recent studies. *Journal of Energy Chemistry* 25(4):553–565. doi: 10.1016/j.jechem.2016.03.009.

48. Quadrelli, E. A., G. Centi, J.-L. Duplan, and S. Perathoner 2011. Carbon dioxide recycling: Emerging largescale technologies with industrial potential. *ChemSusChem* 4(9):1194–1215. doi: 10.1002/cssc.201100473.
49. Kowalczyk, Z., K. Stolecki, W. Rarog-Pilecka, E. Miskiewicz, E. Wilczkowska, and Z. Karpiniski 2008. Supported ruthenium catalysts for selective methanation of carbon oxides at very low COx/H_2 ratios. *Applied Catalysis A: General* 342: 35–39. doi: 10.1016/j.apcata. 2007.12.040.
50. Kusama, H., K. K. Bando, K. Okabe, and H. Arakawa 2000. Effect of metal loading on CO_2 hydrogenation reactivity over Rh/SiO_2 catalysts. *Applied Catalysis A: General* 197:255–268. doi: 10.1016/s0926-860x(99)00486-x.
51. Chang, F. W., M. S. Kuo, M. T. Tsay, and M. C. Hsieh 2003. Hydrogenation of CO_2 over nickel catalysts on rice husk ash-alumina prepared by incipient wetness impregnation. *Applied Catalysis A: General* 247:309–320. doi: 10.1016/s0926-860x(03)00181-9.
52. National Academies of Sciences, Engineering, and Medicine 2019. *Gaseous Carbon Waste Streams Utilization: Status and Research Needs*. Washington, DC: The National Academies Press. https://doi.org/10.17226/25232.
53. Álvarez, A., A. Bansode, A. Urakawa, A. V. Bavykina, T. A. Wezendonk, M. Makkee, J. Gascon, and F. Kapteijn 2017. Challenges in the greener production of formates/formic acid, methanol, and DME by heterogeneously catalyzed CO_2 hydrogenation processes. *Chemical Reviews* 117(14):9804–9838. doi: 10.1021/acs.chemrev.6b00816.
54. Jadhav, S. G., P. D. Vaidya, B. M. Bhanage, and J. B. Joshi 2014. Catalytic carbon dioxide hydrogenation to methanol: A review of recent studies. *Chemical Engineering Research and Design* 92(11):2557–2567. doi: 10.1016/j.cherd.2014.03.005.
55. Olah, G. A. 2005. Beyond oil and gas: The methanol economy. *Angewandte Chemie International Edition* 44(18):2636–2639. doi: 10.1002/anie.200462121.
56. Klankermayer, J., S. Wesselbaum, K. Beydoun, and W. Leitner 2016. Selective catalytic synthesis using the combination of carbon dioxide and hydrogen: Catalytic chess at the interface of energy and chemistry. *Angewandte Chemie International Edition* 55(26):7296–7343. doi: 10.1002/anie.201507458.
57. Larmier, K., et al. 2017. CO_2-to-methanol hydrogenation on Zirconia-supported copper nanoparticles: Reaction intermediates and the role of the metal-support interface. *Angewandte Chemie International Edition* 56:2318–2323.
58. Nie, X., et al. 2018. Mechanistic understanding of alloy effect and water promotion for Pd-Cu bimetallic catalysts in CO_2 hydrogenation to methanol. *ACS Catalysis* 8:4873–4892.
59. Yin, Y., et al. 2018. Pd@zeolitic imidazolate framework-8 derived PdZn alloy catalysts for efficient hydrogenation of CO_2 to methanol. *Applied Catalysis B: Environmental* 234:143–152.
60. Tominaga, K.-i., Y. Sasaki, M. Kawai, T. Watanabe, and M. Saito 1993. Ruthenium complex catalysed hydrogenation of carbon dioxide to carbon monoxide, methanol and methane. *Journal of the Chemical Society, Chemical Communications* (7):629–631.
61. Huff, C. A., and M. S. Sanford 2011. Cascade catalysis for the homogeneous hydrogenation of CO_2 to methanol. *Journal of the American Chemical Society* 133(45):18122–18125.
62. Wesselbaum, S., T. vom Stein, J. Klankermayer, and W. Leitner 2012. Hydrogenation of carbon dioxide to methanol by using a homogeneous ruthenium–phosphine catalyst. *Angewandte Chemie* 124(30):7617–7620.
63. Al-Saydeh, S. A., and S. J. Zaidi 2018. *Carbon Dioxide Conversion to Methanol: Opportunities and Fundamental Challenges*, Chapter 3, an Intech open access paper. doi: 10.5772/intechopen.74779.

64. Ganesh, I. 2016. Electrochemical conversion of carbon dioxide into renewable fuel chemicals—The role of nanomaterials and the commercialization. *Renewable and Sustainable Energy Reviews* 59(Supplement C):1269–1297. doi: 10.1016/j.rser.2016.01.026.
65. Martin, O., A. J. Martín, C. Mondelli, S. Mitchell, T. F. Segawa, R. Hauert, C. Drouilly, D. Curulla-Ferré, and J. Pérez-Ramírez 2016. Indium oxide as a superior catalyst for methanol synthesis by CO_2 hydrogenation. *Angewandte Chemie International Edition* 55(21):6261–6265. doi: 10.1002/anie.201600943.
66. Studt, F., I. Sharafutdinov, F. Abild-Pedersen, C. F. Elkjær, J. S. Hummelshøj, S. Dahl, I. Chorkendorff, and J. K. Nørskov 2014. Discovery of a Ni-Ga catalyst for carbon dioxide reduction to methanol. *Nature Chemistry* 6:320. doi: 10.1038/nchem.1873. Supplemental information available at https://www.nature.com/articles/nchem.1873#supplementary-information.
67. Goehna, H., and P. Koenig 1994. Producing methanol from CO, CHEMTECH, 24-6, Journal ID: ISSN 0009-2703, OSTI identifier 7157792
68. Cheng, W.-H. 1994. *Methanol Production and Use*. CRC Press, New York, New York
69. Arena, F., et al. 2008. Solid-state interactions, adsorption sites and functionality of Cu-ZnO/ZrO_2 catalysts in the CO_2 hydrogenation to CH_3OH. *Applied Catalysis A: General* 350(1):16–23.
70. Ronda-Lloret, M., Y. Wang, P. Oulego, G. Rothenberg, X. Tu, and N. Raveendran Shiju 2020. CO_2 hydrogenation at atmospheric pressure and low temperature using plasma-enhanced catalysis over supported cobalt oxide catalysts. *ACS Sustainable Chemistry & Engineering* 2020 8(47):17397–17407.
71. Zhang, X., G. Zhang, C. Song, and X. Guo 2021. Catalytic conversion of carbon dioxide to methanol: Current status and future perspective. *Frontiers in Energy Research.* https://doi.org/10.3389/fenrg.2020.621119.
72. Sharma, P., J. Sebastian, S. Ghosh, D. Creaser, and L. Olsson 2021. Recent advances in hydrogenation of CO_2 into hydrocarbons via methanol intermediate over heterogeneous catalysts. *Catalysis Science & Technology* 11:1665.
73. He, Z., M. Cui, Q. Qian, J. Zhang, H. Liu, and B. Han 2019. Synthesis of liquid fuel via direct hydrogenation of CO_2. *PNAS* 116(26):12654–12659. www.pnas.org/cgi/doi/10.1073/pnas.1821231116.
74. Rungtaweevoranit, B., J. Baek, J. R. Araujo, B. S. Archanjo, K. M. Choi, O. M. Yaghi1, and G. A. Somorjai 2016. Copper nanocrystals encapsulated in Zr-based metal-organic frameworks for highly selective CO_2 hydrogenation to methanol. *Nano Letters* 16(12):7645–7649. doi: 10.1021/acs.nanolett.6b03637.
75. Owen, R. E., D. Mattia, P. Plucinski, and M. D. Jones 2017. Kinetics of CO_2 hydrogenation to hydrocarbons over iron-silica catalysts. *ChemPhysChem* 18:3211–3218.
76. Meiri, N., et al. 2015. Novel process and catalytic materials for converting CO_2 and H_2 containing mixtures to liquid fuels and chemicals. *Faraday Discussions* 183:197–215.
77. Martinelli, M., et al. 2014. CO_2 reactivity on Fe-Zn-Cu-KFischerâ-Tropsch synthesis catalysts with different K-loadings. *Catalysis Today* 228:77–88.
78. Wang, W., X. Jiang, X. Wang, and C. Song 2018. Fe-Cu bimetallic catalysts for selective CO_2 hydrogenation to olefin-rich $C_2{}^+$ hydrocarbons. *Industrial & Engineering Chemistry Research* 57:4535–4542.
79. Mattia, D., et al. 2015. Towards carbon-neutral CO_2 conversion to hydrocarbons. *ChemSusChem* 8:4064–4072.
80. Numpilai, T., et al. 2017. Structure–activity relationships of Fe-Co/K-Al_2O_3 catalysts calcined at different temperatures for CO_2 hydrogenation to light olefins. *Applied Catalysis A: General* 547:219–229.

81. Liu, J., et al. 2018. Pyrolyzing ZIF-8 to N-doped porous carbon facilitated by iron and potassium for CO_2 hydrogenation to value-added hydrocarbons. *Journal of CO_2 Utilization* 25:120–127.
82. Kattel, S., P. Liu, and J. G. Chen 2017. Tuning selectivity of CO_2 hydrogenation reactions at the metal/oxide interface. *Journal of the American Chemical Society* 139:9739–9754.
83. Choi, Y. H., et al. 2017. Carbon dioxide Fischer-Tropsch synthesis: A new path to carbon-neutral fuels. *Applied Catalysis B: Environmental* 202:605–610.
84. He, Z., Q. Qian, J. Ma, Q. Meng, H. Zhou, J. Song, Z. Liu, and B. Han 2016. Water-enhanced synthesis of higher alcohols from CO_2 hydrogenation over a Pt/Co_3O_4 catalyst under milder conditions. *Angewandte Chemie International Edition* 55:737–741.
85. Kim, H., D.-H. Choi, S.-S. Nam, M.-J. Choi, and K.-W. Lee 1998. The selective synthesis of lower olefins (C_2–C_4) by the CO_2 hydrogenation over iron catalysts promoted with potassium and supported on ion exchanged(H, K) zeolite-Y. *Studies in Surface Science and Catalysis* 114:407–410 [Inui, T., Anpo, M., Izui, K., Yanagida, S., Yamaguchi, T., eds.]. Elsevier.
86. Ding, L., T. Shi, J. Gu, Y. Cui, Z. Zhang, C. Yang, T. Chen, M. Lin, P. Wang, N. Xue, L. Peng, X. Guo, Y. Zhu, Z. Chen, and W. Ding 2020. CO_2 hydrogenation to ethanol over Cu@Na-Beta. *Chem.* doi: 10.1016/j.chempr.2020.07.001.
87. Lin, T. J., X. Z. Qi, X. X. Wang, L. Xia, C. Q. Wang, F. Yu, H. Wang, S. G. Li, L. S. Zhong, and Y. H. Sun 2019. Direct production of higher oxygenates by syngas conversion over a multifunctional catalyst. *Angewandte Chemie International Edition* 58(14):4627–4631.
88. Jiang, F., B. Liu, S. Geng, Y. Xu, and X. Liu 2018. Hydrogenation of CO_2 into hydrocarbons: Enhanced catalytic activity over Fe-based Fischer–Tropsch catalysts. *Catalysis Science & Technology* 8:4097–4107.
89. Wei, J., Q. Ge, R. Yao, Z. Wen, C. Fang, L. Guo, H. Xu, and J. Sun 2017. Directly converting CO_2 into a gasoline fuel. *Nature Communications* 8:15174.
90. Tarasov, A. L., V. I. Isaeva, O. P. Tkachenko, V. V. Chernyshev, and L. M. Kustov 2018. Conversion of CO_2 into liquid hydrocarbons in the presence of a Co-containing catalyst based on the microporous metal-organic framework MIL-53(Al). *Fuel Processing Technology* 176:101–106.
91. Owen, R. E., P. Plucinski, D. Mattia, L. Torrente-Murciano, V. P. Ting, and M. D. Jones 2016. Effect of support of Co-Na-Mo catalysts on the direct conversion of CO_2 to hydrocarbons. *Journal of CO_2 Utilization* 16:97–103.
92. Hwang, S. M., C. D. Zhang, S. J. Han, H. G. Park, Y. T. Kim, S. Yang, K. W. Jun, and S. K. Kim 2020. Mesoporous carbon as an effective support for Fe catalyst for CO_2 hydrogenation to liquid hydrocarbons. *Journal of CO_2 Utilization* 37:65–73.
93. Choi, Y. H. E. C. Ra, E. H. Kim, K. Y. Kim, Y. J. Jang, K. N. Kang, S. H. Choi, J. H. Jang, and J. S. Lee 2017. Sodium-containing spinel zinc ferrite as a catalyst precursor for the selective synthesis of liquid hydrocarbon fuels. *ChemSusChem* 10(23):4764–4770.
94. Barthel, A., et al. 2016. Highly integrated CO_2 capture and conversion: Direct synthesis of cyclic carbonates from industrial flue gas. *Green Chemistry* 18:3116–3123. doi: 10.1039/c5gc03007b.
95. Satthawong, R., N. Koizumi, C. S. Song, and P. Prasassarakich 2013. Bimetallic Fe-Co catalysts for CO_2 hydrogenation to higher hydrocarbons. *Journal of CO_2 Utilization* 3–4:102–106.
96. Li, W., H. Wang, X. Jiang, J. Zhu, Z. Liu, X. Guo, and C. Song 2018. A short review of recent advances in CO_2 hydrogenation to hydrocarbons over heterogeneous catalysts. *RSC Advances* 8:7651.
97. Kim, H. Y., J. N. Park, G. Henkelman, and J. M. Kim 2012. Design of a highly nano dispersed $Pd\text{-}MgO/SiO_2$ composite catalyst with multifunctional activity for CH_4 reforming. *ChemSusChem* 5(8):1474–1481.

98. Shah, Y. T., and T. Gardner 2014. Dry reforming of hydrocarbon feedstocks. *Catalysis Reviews: Science and Engineering* 54:476–536. New York, NY: CRC Press, Taylor and Francis.
99. Pakhare, D., C. Shaw, D. Haynes, D. Shekhawat, and J. Spivey 2013. Effect of reaction temperature on activity of Pt- and Ru-substituted lanthanum zirconate pyrochlores ($La_2Zr_2O_7$) for dry (CO_2) reforming of methane (DRM). *Journal of CO2 Utilization* 1:37–42.
100. Wang, S. B., G. Q. M. Lu, and G. J. Millar 1996. Carbon dioxide reforming of methane to produce synthesis gas over metal-supported catalysts: State of the art. *Energy & Fuels* 10(4):896–904.
101. Kambolis, A., H. Matralis, A. Trovarelli, and C. Papadopoulou 2010. Ni/CeO_2-ZrO_2 catalysts for the dry reforming of methane. *Applied Catalysis A* 377(1–2):16–26.
102. Ferreira-Aparicio, P., A. Guerrero-Ruiz, and I. Rodriguez-Ramos 1998. Comparative study at low and medium reaction temperatures of syngas production by methane reforming with carbon dioxide over silica and alumina supported catalysts. *Applied Catalysis A* 170(1):177–187.
103. Park, J.-H., S. Yeo, I. HeO, and T.-S. Chang 2018. Promotional effect of Al addition on the Co/ZrO_2 catalyst for dry reforming of CH_4. *Applied Catalysis A* 562:120.
104. Paris, A., and A. B. Bocarsly 2019. High-efficiency conversion of CO_2 to oxalate in water is possible using a Cr-Ga oxide electrocatalyst. *ACS Catalysis* 9:2324–2333.
105. Yokota, S., K. Okumura, and M. Niwa 2002. Support effect of metal oxide on Rh catalysts in the CH_4-CO_2 reforming reaction. *Catalysis Letters* 84(1–2):131–134.
106. Vennekoetter, J.-B., R. Sengpiel, and M. Wessling 2019. Beyond the catalyst: How electrode and reactor design determine the product spectrum during electrochemical CO_2 reduction. *Chemical Engineering Journal* 364:89–101. ISSN 1385-8947. https://doi.org/10.1016/j.cej.2019.01.045.
107. Alipour, Z., M. Rezaei, and F. Meshkani 2014. Effect of alkaline earth promoters (MgO, CaO, and BaO) on the activity and coke formation of Ni catalysts supported on nanocrystalline Al_2O_3 in dry reforming of methane. *Industrial & Engineering Chemistry Research* 20(5):2858–2863.
108. Liang, T. Y., C. Y. Lin, F. C. Chou, M. Q. Wang, and D. H. Tsai 2018. Gas-phase synthesis of Ni-CeOx hybrid nanoparticles and their synergistic catalysis for simultaneous reforming of methane and carbon dioxide to syngas. *Journal of Physical Chemistry C* 122(22):11789–11798.
109. Han, J. W., J. S. Park, M. S. Choi, and H. Lee 2017. Uncoupling the size and support effects of Ni catalysts for dry reforming of methane. *Applied Catalysis B* 203:625–632.
110. Song, H. 2015. Metal hybrid nanoparticles for catalytic organic and photochemical transformations. *Accounts of Chemical Research* 48(3):491–499.
111. Cargnello, M., J. J. D. Jaen, J. C. H. Garrido, K. Bakhmutsky, T. Montini, J. J. C. Gamez, R. J. Gorte, and P. Fornasiero 2012. Exceptional activity for methane combustion over modular Pd@CeO_2 subunits on functionalized Al_2O_3. *Science* 337(6095):713–717.
112. Wang, K., X. J. Li, S. F. Ji, X. J. Shi, and J. J. Tang 2009. Effect of $CexZr_{1-x}O_2$ promoter on Ni-based SBA-15 catalyst for steam reforming of methane. *Energy Fuels* 23(1–2):25–31.
113. Wu, Z., Q. Li, D. Feng, P. A. Webley, and D. Zhao 2010. Ordered mesoporous crystalline γ-Al_2O_3 with variable architecture and porosity from a single hard template. *Journal of the American Chemical Society* 132(34):12042–12050.
114. Liu, M., J. Lan, L. Liang, J. Sun, and M. Arai 2017. Heterogeneous catalytic conversion of CO_2 and epoxides to cyclic carbonates over multifunctional tri-s-triazine terminal-linked ionic liquids. *Journal of catalysis* 347:138–147 https://doi.org/10.1016/j.jcat.2016.11.038 (March, 2017), http://hdl.handle.net/2115/72724.

115. Artz, J., T. E. Müller, K. Thenert, J. Kleinekorte, R. Meys, A. Sternberg, A. Bardow, and W. Leitner 2018. Sustainable conversion of carbon dioxide: An integrated review of catalysis and life cycle assessment. *Chemical Reviews* 118:434–504. doi: 10.1021/acs.chemrev.7b00435.
116. Olah, G. A., A. Goeppert, and G. K. S Prakash 2011. *Beyond Oil and Gas: The Methanol Economy.* John Wiley & Sons, Hoboken, New Jersey
117. Carbon Recycling International (CRI), World's Largest CO_2 Methanol Plant 2016. Available at: http://carbonrecycling.is/george-olah/2016/2/14/worlds-largest-co2-methanol-plant.
118. Ma, J., N. Sun, X. Zhang, N. Zhao, F. Xiao, W. Wei, and Y. Sun 2009. A short review of catalysis for CO_2 conversion. *Catalysis Today* 148(3):221–231. doi: 10.1016/j.cattod.2009.08.015.
119. Tamboli, A. H., A. A. Chaugule, and H. Kim 2017. Catalytic developments in the direct dimethyl carbonate synthesis from carbon dioxide and methanol. *Chemical Engineering Journal* 323:530–544. doi: 10.1016/j. cej.2017.04.112.
120. Shukla, K., and V. C. Srivastava 2017. Synthesis of organic carbonates from alcoholysis of urea: A review. *Catalysis Reviews* 59(1):1–43. doi: 10.1080/01614940.2016.1263088.
121. Inesi, A., V. Mucciante, and L. Rossi 1998. A convenient method for the synthesis of carbamate esters from amines and tetraethylammonium hydrogen carbonate. *Journal of Organic Chemistry* 63:1337–1338. doi: 10.1021/jo971695y.
122. Fujimoto, K., and T. Shikada 1987. Selective synthesis of C_2-C_5 hydrocarbons from carbon dioxide utilizing a hybrid catalyst composed of a methanol synthesis catalyst and zeolite. *Applied Catalysis* 31(1):13–23.
123. Kuei, C.-K., and M.-D. Lee 1991. Hydrogenation of carbon dioxide by hybrid catalysts, direct synthesis of aromatics from carbon dioxide and hydrogen. *Canadian Journal of Chemical Engineering* 69(1):347–354.
124. Fujiwara, M., H. Ando, M. Tanaka, and Y. Souma 1995. Hydrogenation of carbon dioxide over Cu-Zn- chromate/zeolite composite catalyst: The effects of reaction behavior of alkenes on hydrocarbon synthesis. *Applied Catalysis A* 130(1):105–116.
125. Tan, Y., M. Fujiwara, H. Ando, Q. Xu, and Y. Souma 1998. Selective formation of iso-butane from carbon dioxide and hydrogen over composite catalysts. *Studies in Surface Science and Catalysis* 114:435–438 [Inui, T., Anpo, M., Izui, K., Yanagida, S., Yamaguchi, T., eds.]. Elsevier.
126. Li, Z., J. Wang, Y. Qu, H. Liu, C. Tang, S. Miao, Z. Feng, H. An, and C. Li 2017. Highly selective conversion of carbon dioxide to lower olefins. *ACS Catalysis* 7(12):8544–8548.
127. Liu, X., M. Wang, C. Zhou, W. Zhou, K. Cheng, J. Kang, Q. Zhang, W. Deng, and Y. Wang 2018. Selective transformation of carbon dioxide into lower olefins with a bifunctional catalyst composed of $ZnGa_2O_4$ and SAPO-34. *Chemical Communications* 54(2):140–143.
128. Gao, P., et al. 2017. Direct conversion of CO_2 into liquid fuels with high selectivity over a bifunctional catalyst. *Nature Chemistry* 9:1019.
129. Dang, S., P. Gao, Z. Liu, X. Chen, C. Yang, H. Wang, L. Zhong, S. Li, and Y. Sun 2018. Role of zirconium in direct CO_2 hydrogenation to lower olefins on oxide/zeolite bifunctional catalysts. *Journal of Catalysis* 364:382–393.
130. Ni, Y., Z. Chen, Y. Fu, Y. Liu, W. Zhu, and Z. Liu 2018. Selective conversion of CO_2 and H_2 into aromatics. *Nature Communications* 9(1):3457.
131. Dokania, A., A. Ramirez, A. Bavykina, and J. Gascon 2018 Heterogeneous catalysis for the valorization of CO_2: Role of bifunctional processes in the production of chemicals. *ACS Energy Letters* 4: 167–176. http://dx.doi.org/10.1021/acsenergylett.8b01910.
132. Gong, J., X. Ma, and S. Wang 2007. Phosgene-free approaches to catalytic synthesis of diphenyl carbonate and its intermediates. *Applied Catalysis A: General* 316(1):1–21. doi: 10.1016/j.apcata.2006.09.006.

133. Limbach, M. 2015. Chapter four: Acrylates from alkenes and CO_2, the stuff that dreams are made of. In: P. J. Pérez, ed., *Advances in Organometallic Chemistry*. New York, NY: Academic Press, pp. 175–202.
134. Sousa, A. F., A. C. Fonseca, A. C. Serra, C. S. R. Freire, A. J. D. Silvestre, and J. F. J. Coelho 2016. New unsaturated copolyesters based on 2,5-furandicarboxylic acid and their crosslinked derivatives. *Polymer Chemistry* 7(5):1049–1058. doi: 10.1039/C5PY01702E.
135. Olah, G. A., B. Török, J. P. Joschek, I. Bucsi, P. M. Esteves, G. Rasul, and G. K. Surya Prakash 2002. Efficient chemoselective carboxylation of aromatics to arylcarboxylic acids with a superelectrophilically activated carbon dioxide–Al_2Cl_6/Al system. *Journal of the American Chemical Society* 124(38):11379–11391. doi: 10.1021/ja020787o.
136. Pérez, E. R., M. O. Silva, V. C. da Costa, U. P. Rodrigues-Filho, and D. W. Franco 2002. Efficient and clean synthesis of *N*-alkyl carbamates by transcarboxylation and O-alkylation coupled reactions using a DBU–CO_2 zwitterionic carbamic complex in aprotic polar media. *Tetrahedron Letters* 43:4091–4093. doi: 10.1016/S0040-4039(02)00697-4.
137. Yang, Z. Z., L. N. He, Y. N. Zhao, B. Li, and B. Yu 2011. CO_2 capture and activation by superbase/polyethylene glycol and its subsequent conversion. *Energy & Environmental Science* 4:3971–3975. doi: 10.1039/c1ee02156g.
138. Yoshida, M., Y. Komatsuzaki, and M. Ihara 2008. Synthesis of 5-vinylideneoxazolidin-2-ones by DBU-mediated CO_2-fixation reaction of 4-(benzylamino)-2-butynyl carbonates and benzoates. *Organic Letters* 10:2083–2086. doi: 10.1021/ol800663v.
139. Zhang, S., Li, Y. N., Zhang, Y. W., He, L. N., Yu, B., Song, Q. W., et al. (2014). Equimolar carbon absorption by potassium phthalimide and in situ catalytic conversion under mild conditions. *ChemSusChem* 7:1484–1489. doi: 10.1002/cssc.201400133.
140. Yu, B., et al. 2016. Atmospheric pressure of CO_2 as protecting reagent and reactant: Efficient synthesis of oxazolidin-2-ones with carbamate salts, aldehydes and alkynes. *Advanced Synthesis & Catalysis* 358:90–97. doi: 10.1002/adsc.201500921.
141. Song, Q. W., Z. H. Zhou, H. Yin, and L. N. He 2015. Silver(I)-catalyzed synthesis of β-oxopropylcarbamates from propargylic alcohols and CO_2 surrogate: A gas-free process. *ChemSusChem* 8:3967–3972. doi: 10.1002/cssc.201501176.
142. Fu, H.-C., F. You, H.-R. Li, and L.-N. He 2019. CO_2 capture and *in situ* catalytic transformation. *Frontiers in Chemistry* 7:525. doi: 10.3389/fchem.2019.00525.
143. Lu, X.-B., and D. J. Darensbourg 2012. Cobalt catalysts for the coupling of CO_2 and epoxides to provide polycarbonates and cyclic carbonates. *Chemical Society Reviews* 41(4):1462–1484. doi: 10.1039/C1CS15142H.
144. Calabrese, C., F. Giacalone, and C. Aprile 2019. Hybrid catalysts for CO_2 conversion into cyclic carbonates. *Catalysts* 9(4):325. https://doi.org/10.3390/catal9040325.
145. Alkordi, M. H., Ł. J. Weseliński, V. D'Elia, S. Barman, A. Cadiau, M. N. Hedhili, A. J. Cairns, R. G. AbdulHalim, J.-M. Basset, and M. Eddaoudi 2016. CO_2 conversion: The potential of porous-organic polymers (POPs) for catalytic CO2–epoxide insertion. *Journal of Materials Chemistry A* 4:7453–7460. doi: 10.1039/C5TA09321J.
146. Qin, Y., X. Sheng, S. Liu, G. Ren, X. Wang, and F. Wang 2015. Recent advances in carbon dioxide based copolymers. *Journal of CO_2 Utilization* 11(Supplement C):3–9. doi: 10.1016/j.jcou.2014.10.003.
147. Poland, S. J., and D. J. Darensbourg 2017. A quest for polycarbonates provided via sustainable epoxide/CO_2 copolymerization processes. *Green Chemistry* 19(21):4990–5011. doi: 10.1039/C7GC02560B.
148. Sakakura, T., J.-C. Choi, and H. Yasuda 2007. Transformation of carbon dioxide. *Chemical Reviews* 107(6):23652387. doi: 10.1021/cr068357u.
149. Alper, E., and O. Yuksel-Orhan 2017. CO_2 utilization: Developments in conversion processes. *Petroleum* 3(1):109–126. doi: 10.1016/j.petlm.2016.11.00351T.

150. Fukuoka, S., M. Tojo, H. Hachiya, M. Aminaka, and K. Hasegawa 2007. Green and sustainable chemistry in practice: Development and industrialization of a novel process for polycarbonate production from CO_2 without using phosgene. *Polymer Journal* 39:91. doi: 10.1295/polymj.PJ2006140.
151. Langanke, J., A. Wolf, J. Hofmann, K. Bohm, M. A. Subhani, T. E. Muller, W. Leitner, and C. Gurtler 2014. Carbon dioxide (CO_2) as sustainable feedstock for polyurethane production. *Green Chemistry* 16(4):18651870. doi: 10.1039/C3GC41788C.
152. Chapman, A. M., C. Keyworth, M. R. Kember, A. J. J. Lennox, and C. K. Williams 2015. Adding value to power station captured CO_2: Tolerant Zn and Mg homogeneous catalysts for polycarbonate polyol production. *ACS Catalysis* 5:1581–1588. doi: 10.1021/cs501798s.
153. Liang, L., et al. 2017. Carbon dioxide capture and conversion by an acid-base resistant metal-organic framework. *Nature Communications* 8:1233. doi: 10.1038/s41467-017-01166-3.
154. Xu, L., Y. Xiu, F. Liu, Y. Liang, and S. Wang 2020. Research progress in conversion of CO_2 to valuable fuels. *Molecules* 25:3653. doi: 10.3390/molecules25163653. www.mdpi.com/journal/molecules.
155. Lin, S., et al. 2015. Covalent organic frameworks comprising cobalt porphyrins for catalytic CO_2 reduction in water. *Science* 349:1208.
156. Duan, X., J. Xu, Z. Wei, J. Ma, S. Guo, S. Wang, H. Liu, and S. Dou 2017. Metal-free carbon materials for CO_2 electrochemical reduction. *Advanced Materials* 29:1701784.
157. Fan, L., C. Xia, F. Yang, J. Wang, H. Wang, and Y. Lu 2020. Strategies in catalyst and electrolyzer design for electrochemical CO_2 reduction towards C_{2+} products. *Science Advances* 6(8):eaay3111. doi: 10.1126/sciadv.aay3111.
158. Jiang, K., S. Siahrostami, T. Zheng, Y. Hu, S. Hwang, E. Stavitski, Y. Peng, J. Dynes, M. Gangisetty, D. Su, K. Attenkofer, and H. Wang 2018. Isolated Ni single atoms in graphene nanosheets for high-performance CO_2 reduction. *Energy & Environmental Science* 11:893–903.
159. Verma, S., B. Kim, H.-R. M. Jhong, S. Ma, and P. J. A. Kenis 2016. A gross-margin model for defining technoeconomic benchmarks in the electroreduction of CO_2. *ChemSusChem* 9:1972–1979.
160. Peterson, A. A., F. Abild-Pedersen, F. Studt, J. Rossmeisl, and J. K. Nørskov 2010. How copper catalyzes the electroreduction of carbon dioxide into hydrocarbon fuels. *Energy & Environmental Science* 3:1311–1315.
161. Burdyny, T. 2019. Smith WA: CO_2 reduction on gas-diffusion electrodes and why catalytic performance must be assessedat commercially-relevant conditions. *Energy & Environmental Science* 12:1442–1453.
162. Higgins, D., C. Hahn, C. Xiang, T. F. Jaramillo, and A. Z. Weber 2019. Gas-diffusion electrodes for carbon dioxide reduction: A new paradigm. *ACS Energy Letters* 4:317–324.
163. Malkhandi, S., and B. Yeo 2019. Electrochemical conversion of carbon dioxide to high value chemicals using gas-diffusion electrodes. *Current Opinion in Chemical Engineering* 26:112–121. www.sciencedirect.com.
164. Xiao, H., T. Cheng, W. A. Goddard, and R. Sundararaman 2016. Mechanisticexplanation of the pH dependence and onset potentials forhydrocarbon products from electrochemical reduction of COon Cu (111). *Journal of the American Chemical Society* 138:483–486.
165. Liu, X., et al. 2019. pH effects on the electrochemical reduction of CO_2 towards C_2 products on stepped copper. *Nature Communications* 10:32.
166. Spurgeon, J. M., and B. Kumar 2018. A comparative techno economicanalysis of pathways for commercial electrochemical CO_2 reduction to liquid products. *Energy & Environmental Science* 11:1536–1551.

167. Jouny, M., W. Luc, and F. Jiao 2018. General techno-economic analysis of CO_2 electrolysis systems. *Industrial & Engineering Chemistry Research* 57:2165–2177.
168. Jeanty, P., C. Scherer, E. Magori, K. Wiesner-Fleischer, O. Hinrichsen, and M. Fleischer 2018. Upscaling and continuous operation of electrochemical CO_2 to CO conversion in aqueous solutions on silver gas diffusion electrodes. *Journal of CO_2 Utilization* 24:454–462.
169. Ramdin, M., A. R. T. Morrison, M. de Groen, R. van Haperen, R. de Kler, L. J. P. van den Broeke, J. P. M. Trusler, W. de Jong, and T. J. H. Vlugt 2019. High pressure electrochemical reduction of CO_2 to formic acid/formate: A comparison between bipolar membranes and cation exchange membranes. *Industrial & Engineering Chemistry Research* 58:1834–1847.
170. Sánchez, O. G., O. G. Sánchez, Y. Y. Birdja, M. Bulut, J. Vaes, T. Breugelmans, and D. Pant 2010. Recent advances in industrial CO_2 electroreduction. *Current Opinion in Green and Sustainable Chemistry* 16:47–56.
171. Xia, C., P. Zhu, Q. Jiang, Y. Pan, W. Liang, E. Stavitski, H. N. Alshareef, and H. Wang 2019. Continuous production of pure liquid fuel solutions via electrocatalytic CO_2 reduction using solid-electrolyte devices. *Nature Energy* 4:776–785.
172. Zhong, S., Z. Cao, X. Yang, S. M. Kozlov K.-W. Huang V. Tung L. Cavallo, L.-J. Li, and Y. Han 2019. Electrochemical conversion of CO_2 to 2-bromoethanol in a membraneless cell. *ACS Energy Letters* 4:600–605.
173. Zheng, X., Y. Ji, J. Tang, J. Wang, B. Liu, H.-G. Steinrück, K. Lim, Y. Li, M. F. Toney, K. Chan, and Y. Cui 2019. Theory-guided Sn/Cu alloying for efficient CO_2 electroreduction at low overpotentials. *Nature Catalysis* 2:55–61.
174. Bushuyev, O. S., P. De Luna, C. T. Dinh, L. Tao, G. Saur, J. de Lagemaat, S. O. Kelley, and E. H. Sargent 2018. What should we make with CO_2 and how can we make it? *Joule* 2:825–832.
175. Ledezma-Yanez, I., E. P. Gallent, M. T. M. Koper, and F. Calle-Vallejo 2016. Structure-sensitive electroreduction of acetaldehyde to ethanol on copper and its mechanistic implications for CO and CO_2 reduction. *Catalysis Today* 262:90–94.
176. Handoko, D., F. Wei, B. S. Yeo, and Z. W. She 2018. Understanding heterogeneous electrocatalytic carbon dioxide reduction through operando techniques. *Nature Catalysis* 1:922–934.
177. Ren, D., Y. Deng, A. D. Handoko, C. S. Chen, S. Malkhandi, and B. S. Yeo 2015a. Selective electrochemical reduction of carbon dioxide to ethylene and ethanol on copper(I) oxide catalysts. *ACS Catalysis* 5(5):2814–2821. doi: 10.1021/cs502128q.
178. Ebbesen, S. D., and M. Mogensen 2009. Electrolysis of carbon dioxide in solid oxide electrolysis cells. *Journal of Power Sources* 193(1):349–358. doi: 10.1016/j.jpowsour.2009.02.093.
179. Ebbesen, S. D., S. H. Jensen, A. Hauch, and M. B. Mogensen 2014. High temperature electrolysis in alkaline cells, solid proton conducting cells, and solid oxide cells. *Chemical Reviews* 114(21):10697–10734. doi: 10.1021/cr5000865.
180. Mittal, C., C. Madsbjerg, and P. Blennow 2017. Small-scale CO from CO_2 using electrolysis. *Chemical Engineering World* 52(3):44–46.
181. Costentin, C., M. Robert, and J.-M. Saveant 2013. Catalysis of the electrochemical reduction of carbon dioxide. *Chemical Society Reviews* 42(6):2423–2436. doi: 10.1039/C2CS35360A.
182. Mistry, H., Y.-W. Choi, A. Bagger, F. Scholten, C. S. Bonifacio, I. Sinev, N. J. Divins, I. Zegkinoglou, H. S. Jeon, K. Kisslinger, E. A. Stach, J. C. Yang, J. Rossmeisl, and B. Roldan Cuenya 2017. Enhanced carbon dioxide electroreduction to carbon monoxide over defect-rich plasma-activated silver catalysts. *Angewandte Chemie International Edition* 56(38):11394–11398. doi: 10.1002/anie.201704613.

183. Kutz, R. B., Q. Chen, H. Yang, S. D. Sajjad, Z. Liu, and I. R. Masel 2017. Sustain ionimidazolium-functionalized polymers for carbon dioxideelectrolysis. *Energy Technology* 5:929–936.
184. Verma, S., X. Lu, S. Ma, R. I. Masel, and P. J. A. Kenis 2016. The effect of electrolyte composition on the electroreduction of CO_2 to CO on Ag based gas diffusion electrodes. *Physical Chemistry Chemical Physics* 18(10):7075–7084. doi: 10.1039/C5CP05665A.
185. Verma, S., Y. Hamasaki, C. Kim, W. Huang, S. Lu, H.-R. M. Jhong, A. A. Gewirth, T. Fujigaya, N. Nakashima, and P. J. A. Kenis 2018. Insights into the low overpotential electroreduction of CO_2 to CO on a supported gold catalyst in an alkaline flow electrolyzer. *ACS Energy Letters* 3(1):193–198. doi: 10.1021/ acsenergylett.7b01096.
186. Haas, T., R. Krause, R. Weber, M. Demler, and G. Schmid 2018. Technical photosynthesis involving CO_2 electrolysis and fermentation. *Nature Catalysis* 1(1):32–39. doi: 10.1038/s41929-017-0005-1.
187. Dinh, C-T, F. P. Garcı́ade Arquer, D. Sinton, and E. H. Sargent 2018. High rate, selective, and stable electroreduction of CO_2 to CO in basic and neutral media. *ACS Energy Letters* 3:2835–2840.
188. Kim, B., F. Hillman, M. Ariyoshi, S. Fujikawa, and P. J. A. Kenis 2016. Effects of composition of the micro porous layer and the substrate on performance in the electrochemical reduction of CO_2 to CO. *Journal of Power Sources* 312:192–198.
189. Mö ller, T., W. Ju, A. Bagger, X. Wang, F. Luo, T. Ngo Thanh, A. S. Varela, J. Rossmeisl, and P. Strasser 2019. Efficient CO_2 to CO electrolysis on solid Ni–N–C catalysts at industrial current densities. *Energy & Environmental Science* 12:640–647.
190. Liu, Z., R. I. Masel, Q. Chen, R. Kutz, H. Yang, K. Lewinski, M. Kaplun, S. Luopa, and D. R. Lutz 2016. Electrochemical generation of syngas from water and carbon dioxide at industrially important rates. *Journal of CO_2 Utilization* 15:50–56. doi: 10.1016/j.jcou.2016.04.011.
191. Kumar, B., V. Atla, J. P. Brian, S. Kumari, T. Q. Nguyen, M. Sunkara, and J. M. Spurgeon. 2017. Reduced SnO_2 porous nanowires with a high density of grain boundaries as catalysts for efficient electrochemical CO_2 into HCOOH conversion. *Angewandte Chemie International Edition* 56(13):3645–3649. doi: 10.1002/ anie.201612194.
192. Whipple, D. T., E. C. Finke, and P. J. A. Kenis 2010. Microfluidic reactor for the electrochemical reduction of carbon dioxide: The effect of pH. *Electrochemical and Solid-State Letters* 13(9):B109–B111. doi: 10.1149/1.3456590.
193. Min, X., and M. W. Kanan 2015. Pd-catalyzed electrohydrogenation of carbon dioxide to formate: High mass activity at low overpotential and identification of the deactivation pathway. *Journal of the American Chemical Society* 137(14):4701–4708. doi: 10.1021/ja511890h.
194. Zhou, X., R. Liu, K. Sun, Y. Chen, E. Verlage, S. A. Francis, N. S. Lewis, and C. Xiang 2016. Solar-driven reduction of 1 atm of CO_2 to formate at 10% energy-conversion efficiency by use of a TiO_2-protected III–V tandem photoanode in conjunction with a bipolar membrane and a Pd/C cathode. *ACS Energy Letters* 1(4):764–770. doi: 10.1021/acsenergylett.6b00317.
195. Kopljar, D., N. Wagner, and E. Klemm 2016. Transferring electrochemical CO_2 reduction from semi-batch into continuous operation mode using gas diffusion electrodes. *Chemical Engineering & Technology* 39:2042–2050.
196. Garcı́ade Arquer, F. P., et al. 2018. 2D metal oxyhalide-derived catalysts for efficient CO_2 electroreduction. *Advanced Materials* 30:1802858.
197. Yang, H., J. J. Kaczur, S. D. Sajjad, and R. I. Masel 2017. CO_2 conversion to formic acid in a three compartment cell with Sustainion™ membranes. *ECS Transactions* 77(11):1425–1431. doi: 10.1149/07711.1425ecst.

198. Pérez-Fortes, M., J. C. Schöneberger, A. Boulamanti, G. Harrison, and E. Tzimas 2016a. Formic acid synthesis using CO_2 as raw material: Techno-economic and environmental evaluation and market potential. *International Journal of Hydrogen Energy* 41(37):16444–16462. doi: 10.1016/j.ijhydene.2016.05.199.
199. Zhang, X., S.-X. Guo, K. A. Gandionco, A. M. Bond, and J. Zhang 2020. Electrocatalytic carbon dioxide reduction: From fundamental principles to catalyst design. *Materials Today Advances* 7:100074.
200. Qiu, Y.-L., H.-X. Zhong, T.-T. Zhang, W.-B. Xu, X.-F. Li, and H.-M. Zhang 2017. Copper electrode fabricated via pulse electrodeposition: Toward high methane selectivity and activity for CO_2 electroreduction. *ACS Catalysis* 7(9):6302–6310. doi: 10.1021/acscatal.7b00571.
201. Hoang, T. T. H., S. Verma, S. Ma, T. T. Fister, J. Timoshenko, A. I. Frenkel, P. J. A. Kenis, and A. A. Gewirth 2018. Nanoporous copper–silver alloys by additive-controlled electrodeposition for the selective electroreduction of CO_2 to ethylene and ethanol. *Journal of the American Chemical Society* 140(17):5791–5797. doi: 10.1021/jacs.8b01868.
202. Dinh, C.-T., T. Burdyny, M. G. Kibria, A. Seifitokaldani, C. M. Gabardo, F. Pelayo García de Arquer, A. Kiani, J. P. Edwards, P. De Luna, O. S. Bushuyev, C. Zou, R. Quintero-Bermudez, Y. Pang, D. Sinton, and E. H. Sargent 2018. CO_2 electroreduction to ethylene via hydroxide-mediated copper catalysis at an abrupt interface. *Science* 360(6390):783–787. doi: 10.1126/science.aas9100.
203. Yanming, L., Z. Yujing, C. Kai, Q. Xie, F. Xinfei, S. Yan, C. Shuo, Z. Huimin, Z. Yaobin, Y. Hongtao, and M. R. Hoffmann 2017. Selective electrochemical reduction of carbon dioxide to ethanol on a boron- and nitrogen-co-doped nanodiamond. *Angewandte Chemie International Edition* 56(49):15607–15611. doi: 10.1002/anie.201706311.
204. Rudolph, M., S. Dautz, and E.-G. Jäger 2000. Macrocyclic [N4P for the formation of oxalate by electrochemical reduction of carbon dioxide. *Journal of the American Chemical Society* 122(44):10821–10830. doi: 10.1021/ja001254n.
205. Angamuthu, R., P. Byers, M. Lutz, A. L. Spek, and E. Bouwman 2010. Electrocatalytic CO_2 conversion to oxalate by a copper complex. *Science* 327(5963):313–315. doi: 10.1126/science.1177981.
206. Canfield, D., and K. W. Frese Jr. 1983. Reduction of carbon dioxide to methanol on n-and p-GaAs and p-InP. Effect of crystal face, electrolyte and current density. *Journal of the Electrochemical Society* 130(8):1772–1773.
207. Seshadri, G., C. Lin, and A. B. Bocarsly 1994. A new homogeneous electrocatalyst for the reduction of carbon dioxide to methanol at low overpotential. *Journal of Electroanalytical Chemistry* 372(1–2):145–150.
208. Morris, A. J., R. T. McGibbon, and A. B. Bocarsly 2011. Electrocatalytic carbon dioxide activation: The rate-determining step of Pyridinium-catalyzed CO_2 reduction. *ChemSusChem* 4(2):191–196.
209. Popić, J. P., M. L. Avramov-Ivić, and N. B. Vuković 1997. Reduction of carbon dioxide on ruthenium oxide and modified ruthenium oxide electrodes in 0.5 M $NaHCO_3$. *Journal of Electroanalytical Chemistry* 421(1):105–110.
210. Qu, J., X. Zhang, Y. Wang, and C. Xie 2005. Electrochemical reduction of CO_2 on RuO_2/TiO_2 nanotubes composite modified Pt electrode. *Electrochimica Acta* 50(16):3576–3580.
211. Sun, X., Q. Zhu, X. Kang, H. Liu, Q. Qian, Z. Zhang, and B. Han 2016b. Molybdenum-bismuth bimetallic chalcogenide nanosheets for highly efficient electrocatalytic reduction of carbon dioxide to methanol. *Angewandte Chemie International Edition* 55(23):6771–6775. doi: 10.1002/anie.201603034.
212. Zhuang, T. T., et al. 2018. Steering post-C–C coupling selectivity enables high efficiency electroreduction of carbon dioxide to multi-carbon alcohols. *Nature Catalysis* 1:421–428. https://doi.org/10.1038/s41929-018-0084-7.

213. Albo, J., and A. Irabien 2016. Cu_2O-loaded gas diffusion electrodes for the continuous electrochemical reduction of CO_2 to methanol. *Journal of Catalysis* 343:232–239.
214. Li, Y. C., et al. 2019. Binding site diversity promotes CO_2 electroreduction to ethanol. *Journal of the American Chemical Society* 141:8584–8591.
215. Li, J., et al. 2018. Copper adparticle enabled selective electrosynthesis of n-propanol. *Nature Communications* 9:4614.
216. Jouny, M., W. Luc, and F. Jiao 2018. High-rate electroreduction of carbonmonoxide to multi-carbon products. *Nature Catalysis* 1:748–755.
217. Licht, S., A. Douglas, J. Ren, R. Carter, M. Lefler, and C. L. Pint 2016. Carbon nanotubes produced from ambient carbon dioxide for environmentally sustainable lithium-ion and sodium-ion battery anodes. *ACS Central Science* 2(3):162–168. doi: 10.1021/acscentsci.5b00400.
218. Fu, Q., C. Mabilat, M. Zahid, A. Brisse, and L. Gautier 2010. Syngas production via high-temperature steam/CO_2 co-electrolysis: An economic assessment. *Energy & Environmental Science* 3:1382–1397.
219. Vasileff, A., C. C. Xu, Y. Jiao, Y. Zheng, and S. Z. Qiao 2018. Surface and interface engineering in copper-based bimetallic materials for selective CO_2 electroreduction. *Chem* 4(8):1809–1831.
220. Kuhl, K. P., E. R. Cave, D. N. Abram, and T. F. Jaramillo 2012. New insights into the electrochemical reduction of carbon dioxide on metallic copper surfaces. *Energy & Environmental Science* 5(5):7050–7059.
221. Lee, S. Y., H. Jung, N. K. Kim, H. S. Oh, B. K. Min, and Y. J. Hwang 2018. Mixed copper states in anodized Cu electrocatalyst for stable and selective ethylene production from CO_2 reduction. *Journal of the American Chemical Society* 140(28):8681–8689.
222. Pander, J. E., D. Ren, Y. Huang, N. W. X. Loo, S. H. L. Hong, and B. S. Yeo 2018. Understanding the heterogeneous electrocatalytic reduction of carbon dioxide on oxide-derived catalysts. *ChemElectroChem* 5(2):219–237.
223. Lee, S., D. Kim, and J. Lee 2015. Electrocatalytic production of C_3-C_4 compounds by conversion of CO_2 on a chloride-induced bi-phasic Cu_2O-Cu catalyst. *Angewandte Chemie International Edition* 54(49):14701–14705.
224. Chen, Y. H., C. W. Li, and M. W. Kanan 2012. Aqueous CO_2 reduction at very low overpotential on oxide-derived Au nanoparticles. *Journal of the American Chemical Society* 134(49):19969–19972.
225. Ma, M., B. J. Trzesniewski, J. Xie, and W. A. Smith 2016. Selective and efficient reduction of carbon dioxide to carbon monoxide on oxide-derived nanostructured silver electrocatalysts. *Angewandte Chemie International Edition* 55(33):9748–9752.
226. Kim, D., J. Resasco, Y. Yu, A. M. Asiri, and P. D. Yang 2014. Synergistic geometric and electronic effects for electrochemical reduction of carbon dioxide using gold-copper bimetallic nanoparticles. *Nature Communications* 5:4948.
227. Lindstrom, C., and X.-Y. Zhu 2006. Photoinduced electron transfer at molecule–metal interfaces. *Chemical Reviews* 106(10):4281–4300.
228. Liu, X., and L. M. Dai 2016. Carbon-based metal-free catalysts. *Nature Reviews Materials* 1(11):16064.
229. Guo, D. H., R. Shibuya, C. Akiba, S. Saji, T. Kondo, and J. Nakamura 2016. Active sites of nitrogen-doped carbon materials for oxygen reduction reaction clarified using model catalysts. *Science* 351(6271):361–365.
230. Elgrishi, N., S. Griveau, M. B. Chambers, F. Bedioui, and M. Fontecave 2015. Versatile functionalization of carbon electrodes with a polypyridine ligand: metallation and electrocatalytic H^+ and CO_2reduction. *Chemical Communications* 51:2995.
231. Kim, J., H. E. Kim, and H. Lee 2018. Single-atom catalysts of precious metals for electrochemical reactions. *ChemSusChem* 11(1):104–113.

232. Yang, S., J. Kim, Y. J. Tak, A. Soon, and H. Lee 2016. Single-atom catalyst of platinum supported on titanium nitride for selective electrochemical reactions. *Angewandte Chemie International Edition* 55(6):2058–2062.
233. Deng, J., H. B. Li, J. P. Xiao, Y. C. Tu, D. H. Deng, H. X. Yang, H. F. Tian, J. Q. Li, P. J. Ren, and X. H. Bao 2015. Triggering the electrocatalytic hydrogen evolution activity of the inert two-dimensional MoS2 surface via single-atom metal doping. *Energy & Environmental Science* 8(5):1594–1601.
234. Kim, J., C. W. Roh, S. K. Sahoo, S. Yang, J. Bae, J. W. Han, and H. Lee 2018. Highly durable platinum single-atom alloy catalyst for electrochemical reactions. *Advanced Engineering Materials* 8(1):1701476.
235. Yang, H. B., et al. 2018. Atomically dispersed Ni(i) as the active site for electrochemical CO_2 reduction. *Nature Energy* 3(2):140–147.
236. Wang, Y. F., Z. Chen, P. Han, Y. H. Du, Z. X. Gu, X. Xu, and G. F. Zheng 2018. Single-atomic Cu with multiple oxygen vacancies on ceria for electrocatalytic CO_2 reduction to CH_4. *ACS Catalysis* 8(8):7113–7119.
237. Loiudice, A., P. Lobaccaro, E. A. Kamali, T. Thao, B. H. Huang, J. W. Ager, and R. Buonsanti 2016. Tailoring copper nanocrystals towards C2 products in electrochemical CO_2 reduction. *Angewandte Chemie International Edition* 55:5789–5792.
238. Xie, M. S., B. Y. Xia, Y. Li, Y. Yan, Y. Yang, Q. Sun, S. H. Chan, A. Fisher, and X. Wang 2016. Amino acid modified copper electrodes for the enhanced selective electroreduction of carbon dioxide towards hydrocarbons. *Energy & Environmental Science* 9:1687–1695.
239. Jiang, K., R. B. Sandberg, A. J. Akey, X. Liu, D. C. Bell, J. K. Nørskov, K. Chan, and H. Wang 2018. Metal ion cycling of Cu foil for selective C–C coupling in electrochemical CO_2 reduction. *Nature Catalysis* 1:111–119.
240. Kim, D., C. S. Kley, Y. F. Li, and P. D. Yang 2017. Copper nanoparticle ensembles for selective electroreduction of CO_2 to C2–C3 products. *Proceedings of the National Academy of Sciences of the USA* 114:10560–10565.
241. Hu, X.-M., M. Rønne, S. Pedersen, T. Skrydstrup, and K. Daasbjerg 2017. Enhanced catalytic activity of cobalt porphyrin in CO_2 electroreduction upon immobilization on carbon materials. *Angewandte Chemie International Edition* 56. doi: 10.1002/anie.201701104.
242. Wang, Y., and I. A. Weinstock 2012. Polyoxometalate-decorated nanoparticles. *Chemical Society Reviews* 41:7479. doi: 10.1039/c2cs35126a.
243. Davethu, P., and S. Visser 2019. CO_2 reduction on an iron-porphyrin center: A computational study. *The Journal of Physical Chemistry A* 123. doi: 10.1021/acs.jpca.9b05102.
244. Windle, C. D., and E. Reisner 2015. Heterogenised molecular catalysts for the reduction of CO_2 to fuels. *Chimia* 69:435–441. doi: 10.2533/chimia.2015.435.
245. Meshitsuka, S., M. Ichikawa, and K. Tamaru 1947. Electrocatalysis by metal phthalocyanines in the reduction of carbon dioxide. *Journal of the Chemical Society, Chemical Communications* 158.
246. Lieber, C. M., and N. S. Lewis 1984. Catalytic reduction of CO_2 at carbon electrodes modified with cobalt phthalocyanine. *Journal of the American Chemical Society* 106:5033.
247. Atoguchi, T., A. Aramata, A. Kazusaka, and M. Enyo 1991. Electrocatalytic activity of Co^{II} TPP-pyridine complex modified carbon electrode for CO_2 reduction. *Journal of Electroanalytical Chemistry and Interfacial Electrochemistry* 318:309.
248. Guadalupe, A. R., D. A. Usifer, K. T. Potts, H. C. Hurrell, A.-E. Mogstad, and H. D. Abruña 1988. Novel chemical pathways and charge-*transport* dynamics of electrodes modified with electropolymerized layers of [Co(v-terpy)2]2+. *Journal of the American Chemical Society* 110:3462.

249. Yoshida, T., T. Iida, T. Shirasagi, R.-J. Lin, and M. Kaneko 1993. Electrocatalytic reduction of carbon *dioxide* in aqueous medium by bis(2,2′: 6′,2″-terpyridine)cobalt(II) complex incorporated into a coated polymer membrane. *Journal of Electroanalytical Chemistry* 344:355.
250. Yoshida, T., K. Tsutsumida, S. Teratani, K. Yasufuku, and M. Kaneko 1993. Electrocatalytic reduction of CO2 in water by [Re(bpy)(CO)3Br] and [Re(terpy)(CO)3Br] complexes incorporated into coated nafion membrane (bpy = 2,2′-bipyridine; terpy = 2,2′;6′,2″-terpyridine). *Journal of the Chemical Society*:631.
251. Blakemore, J. D., A. Gupta, J. J. Warren, B. S. Brunschwig, and H. B. Gray 2013. Noncovalent immobilization of electrocatalysts on carbon electrodes for fuel production. *Journal of the American Chemical Society* 135:18288.
252. Mohsin, I., T. Al-Attas, K. Sumon, J. Bergerson, S. McCoy, and Md. Kibria 2020. Economic and environmental assessment of integrated carbon capture and utilization. *Cell Reports Physical Science* 1:100104. doi: 10.1016/j.xcrp.2020.100104.
253. Smieja, J. M., M. D. Sampson, K. A. Grice, E. E. Benson, J. D. Froehlich, and C. P. Kubiak 2013. Manganese as a substitute for rhenium in CO2 reduction catalysts: the importance of acids. *Inorganic Chemistry* 52:2484.
254. Bourrez, M., F. Molton, S. Chardon-Noblat, and A. Deronzier 2011. [Mn(bipyridyl)(CO)3Br]: An Abundant metal carbonyl complex as efficient electrocatalyst for CO_2 reduction. *Angewandte Chemie International Edition* 50:9903.
255. Walsh, J. J., G. Neri, C. L. Smith, and A. J. Cowan 2014. Electrocatalytic CO2 reduction with a membrane supported manganese catalyst in aqueous solution. *Chemical Communications* 50:12698.
256. Reda, T., C. M. Plugge, N. J. Abram, and J. Hirst 2008. Reversible interconversion of carbon dioxide and formate by an electroactive enzyme. *Proceedings of the National Academy of Sciences of the USA* 105:10654.
257. Parkin, A., J. Seravalli, K. A. Vincent, S. W. Ragsdale, and F. A. Armstrong 2007. Rapid and Efficient Electrocatalytic CO2/CO Interconversions by Carboxydothermus hydrogenoformans CO Dehydrogenase I on an Electrode. *Journal of the American Chemical Society* 129:10328.
258. Wu, Y., Z. Jiang, X. Lu, Y. Liang, and H. Wang 2019. Domino electroreduction of CO 2 to methanol on a molecular catalyst. *Nature* 575:639–642.
259. Hori, Y., I. Takahashi, O. Koga, and N. Hoshi 2003. Electrochemical reduction of carbon dioxide at various series of copper single crystal electrodes. *Journal of Molecular Catalysis A: Chemical* 199:39–47.
260. Arán-Ais, R. M., D. Gao, and B. R. Cuenya 2018. Structure- and electrolyte-sensitivity in CO_2 electroreduction. *Accounts of Chemical Research* 51:2906–2917.
261. Gao, D., R. M. Arán-Ais, H. S. Jeon, and B. R. Cuenya 2019. Rational catalyst and electrolyte design for CO_2 electroreduction towards multicarbon products. *Nature Catalysis* 2:198–210.
262. Liu, X., J. Xiao, H. Peng, X. Hong, K. Chan, and J. K. Nørskov 2017. Understanding trends in electrochemical carbon dioxide reduction rates. *Nature Communications* 8:15438.
263. Larrazábal, G. O., A. J. Martín, and J. Pérez-Ramírez 2017. Building blocks for high performance in electrocatalytic CO_2 reduction: Materials, optimization strategies, and device engineering. *Journal of Physical Chemistry Letters* 8:3933–3944.
264. Luo, W., X. Nie, M. J. Janik, and A. Asthagiri 2015. Facet dependence of CO_2 reduction paths on Cu electrodes. *ACS Catalysis* 6:219–229.
265. Endrődi, B., G. Bencsik, F. Darvas, R. Jones, K. Rajeshwar, and C. Janáky 2017. Continuous-flow electroreduction of carbon dioxide. *Progress in Energy and Combustion Science* 62:133–154.
266. Gutiérrez-Guerra, N., L. Moreno-López, J. Serrano-Ruiz, J. Valverde, and A. de Lucas-Consuegra 2016. Gas phase electrocatalytic conversion of CO_2 to syn-fuels on Cu based catalysts-electrodes. *Applied Catalysis B: Environmental* 188:272–282.

267. Wu, J., F. G. Risalvato, P. P. Sharma, P. J. Pellechia, F.-S. Ke, and X.-D. Zhou 2013. Electrochemical reduction of carbon dioxide II. Design, assembly, and performance of low temperature full electrochemical cells. *Journal of the Electrochemical Society* 160:F953–F957.
268. Lv, J.-J., M. Jouny, W. Luc, W. Zhu, J.-J. Zhu, and F. Jiao 2018. A highly porous copper electrocatalyst for carbon dioxide reduction. *Advanced Materials* 30:1803111.
269. Resasco, J., L. D. Chen, E. Clark, C. Tsai, C. Hahn, T. F. Jaramillo, K. Chan, and A. T. Bell 2017. Promoter effects of alkali metal cations on the electrochemical reduction of carbon dioxide. *Journal of the American Chemical Society* 139(32):11277–11287.
270. Rosen, B. A., A. Salehi-Khojin, M. R. Thorson, W. Zhu, D. T. Whipple, P. J. A. Kenis, and R. I. Masel 2011. Ionic liquid-mediated selective conversion of CO_2 to CO at low overpotentials. *Science* 334(6056):643–644.
271. Cole, E. B., P. S. Lakkaraju, D. M. Rampulla, A. J. Morris, E. Abelev, and A. B. Bocarsly 2010. Using a one-electron shuttle for the multielectron reduction of CO_2 to methanol: Kinetic, mechanistic, and structural insights. *Journal of the American Chemical Society* 132(33):11539–11551.
272. Hara, K., A. Kudo, and T. Sakata 1995. Electrochemical reduction of carbon dioxide under high pressure on various electrodes in an aqueous electrolyte. *Journal of Electroanalytical Chemistry* 391:141–147.
273. Hara, K., A. Tsuneto, A. Kudo, and T. Sakata 1994. Electrochemical reduction of CO_2 on a Cu electrode under high pressure factors that determine the product selectivity. *Journal of Electroanalytical Chemistry* 141:2097–2103.
274. Ma, S., R. Luo, J. I. Gold, A. Z. Yu, B. Kim, and P. J. A. Kenis 2016. Carbon nanotube containing Ag catalyst layers for efficient and selective reduction of carbon dioxide. *Journal of Materials Chemistry A* 4:8573–8578.
275. Ripatti, D. S., T. R. Veltman, and M. W. Kanan 2019. Carbon monoxide gas diffusion electrolysis that produces concentrated C2 products with high single-pass conversion. *Joule* 3:240–256.
276. Cook, R. L., R. C. MacDuff, and A. F. Sammell 1990. High rate gas phase CO_2 reduction to ethylene and methane using gas diffusion electrodes. *Journal of the Electrochemical Society* 137:607–608.
277. Burdyny, T., and W. A. Smith 2019. CO_2 reduction on gas-diffusion electrodes and why catalytic performance must be assessed at commercially-relevant conditions. *Energy & Environmental Science* 12:1442–1453.
278. Weng, L.-C., A. T. Bell, and A. Z. Weber 2019. Towards membrane-electrode assembly systems for CO_2 reduction: A modeling study. *Energy & Environmental Science* 12:1950–1968.
279. Ma, S., M. Sadakiyo, R. Luo, M. Heima, M. Yamauchi, and P. J. A. Kenis 2016. One-step electrosynthesis of ethylene and ethanol from CO_2 in analkaline electrolyzer. *Journal of Power Sources* 301:219–228.
280. Hoang, T. T. H., S. Ma, J. I. Gold, P. J. A. Kenis, and A. A. Gewirth 2017. Nanoporouscopper films by additive-controlled electrodeposition: CO_2 reduction catalysis. *ACS Catalysis* 7:3313–3321.
281. Hoang, T. T. H., S. Verma, S. Ma, T. T. Fister, J. Timoshenko, A. I. Frenkel, P. J. A. Kenis, and A. A. Gewirth 2018. Nanoporous copper–silver alloys by additive-controlled electrodeposition for the selective electroreduction of CO_2 to ethylene and ethanol. *Journal of the American Chemical Society* 140:5791–5797.
282. Kibria, M. G., et al. 2018. A surface reconstruction route to high productivity and selectivity in CO_2 electroreduction toward C_{2+} hydrocarbons. *Advanced Materials* 30:1804867.
283. Nam, D.-H., et al. 2018. Metal–organic frameworks mediate Cu coordination for selective CO_2 electroreduction. *Journal of the American Chemical Society* 140:11378–11386.
284. De Luna, P., R. Quintero-Bermudez, C.-T. Dinh, M. B. Ross, O. S. Bushuyev, P. Todorovic, T. Regier, S. O. Kelley, P. Yang, and E. H. Sargent 2018. Catalyst electro-redeposition controls morphology and oxidation state for selective carbon dioxidereduction. *Nature Catalysis* 1:103–110.

285. Romero Cuellar, N. S., K. Wiesner-Fleischer, M. Fleischer, A. Rucki, and O. Hinrichsen 2019. Advantages of CO over CO_2 as reactant for electrochemical reduction to ethylene, ethanol and n-propanol on gas diffusion electrodes at high current densities. *Electrochim Acta* 307:164–175.
286. Han, L., W. Zhou, and C. Xiang 2018. High-rate electrochemical reduction ofcarbon monoxide to ethylene using Cu-nanoparticle-basedgas diffusion electrodes. *ACS Energy Letters* 3:855–860.
287. Chang, Z., S. Huo, W. Zhang, J. Fang, and H. Wang 2017. The tunable and highly selective reduction products on Ag@Cu bimetallic catalysts toward CO_2 electrochemical reduction reaction. *Journal of Physical Chemistry C* 121:11368–11379.
288. Li, J., G. Chen, Y. Zhu, Z. Liang, A. Pei, C.-L. Wu, H. Wang, H. R. Lee, K. Liu, S. Chu, and Y. Cui 2018. Efficient electrocatalytic CO_2 reduction on a three-phase interface. *Nature Catalysis* 1:592–600.
289. Wu, K., E. Birgersson, B. Kim, P. J. A. Kenis, and I. A. Karimi 2015. Modeling and experimental validation of electrochemical reduction of CO_2 to CO in a microfluidic cell. *Journal of the Electrochemical Society* 162:F23–F32.
290. Weng, L.-C., A. T. Bell, and A. Z. Weber 2018. Modeling gas-diffusion electrodes for CO_2 reduction. *Journal of the Chemical Society, Faraday Transactions* 20:16973–16984.
291. Spurgeon, J. M., and B. Kumar 2018. A comparative techno economic analysis of pathways for commercial electrochemical CO_2 reduction to liquid products. *Energy & Environmental Science* 11:1536–1551.24.
292. De Luna, P., C. Hahn, D. Higgins, S. A. Jaffer, T. F. Jaramillo, and E. H. Sargent 2019. What would it take for renewably powered electrosynthesis todisplace petrochemical processes? *Science*:364eaav3506.
293. Shah, Y. T. 2021. *Hybrid Power.* New York, NY: CRC Press, Taylor and Francis.
294. Rossi, Federico, Andrea Nicolini, Massimo Palombo, Beatrice Castellani, Elena Morini, and Mirko Filipponi 2014. An innovative configuration for CO_2 capture by high temperature fuel cells. *Sustainability* 6:6687–6695. www.mdpi.com/journal/sustainability.

6 Carbon Dioxide Conversion Using Solar Thermal and Photo Catalytic Processes

6.1 INTRODUCTION

Solar energy can be used to convert CO_2 to useful products in a number of different ways. In nature, solar energy is used to convert CO_2 to oxygen by the process of photosynthesis. This biological process was described in great detail in Chapter 4. The subjects of hybrid biological processes which include (a) bio-photosynthesis/microbial bio-methation in the presence of solar energy, (b) bio-electrophotocatalysis and (c) microbial electro-synthesis are also treated in details in Chapter 4 and will not be discussed here. Solar thermal, solar PV, or their combination can also be used to generate power. This power can be used to generate hydrogen by electrolysis of water. As shown in Chapter 5, the generated hydrogen can be used to carry out CO_2 hydrogenation to produce CO, CH_4, or methanol. Solar power can also be used to drive electro-catalytic reaction as shown in Chapter 5. If harness properly, besides these approaches discussed in Chapters 4 and 5, there are a number of different ways solar energy can be used to reduce CO_2.

Solar energy carries heat and light, and both are important to induce chemical reactions. In this chapter, we evaluate different methods to convert CO_2 using both heat and light components of solar energy. The chapter examines solar thermal and solar photochemical conversions of CO_2. Because of its unique advantages and wide availability, the use of solar energy to convert CO_2 to useful chemicals and fuels (commonly known as solar fuels) is a hot topic of the current investigation. Unfortunately, while solar energy is abundant clean source, its efficient use faces several limitations, such as the low energy density, the intermittent and unstable supply, uneven spatial distribution of solar radiation, and the difficulty of direct storage. The chapter discusses seven different ways of using solar thermal and photochemical energy to convert CO_2. These are [1–13]:

1. Thermochemical dissociation of CO_2 using solar energy. This approach involves the use of thermochemical cycles using metal oxides (normally cerium oxide) as a catalyst at high temperatures (around 1,400°C). The process involves concentrated solar power and special types of reactor. If the process involves H_2O instead of oxygen, hydrogen can be produced along with CO. Novel technologies involve the use of membrane reactor

DOI: 10.1201/9781003229575-6

to simultaneously separate products (CO and O_2) or a pipe reactor. This method is still at the development stage. The method can also carry out methane reforming reaction.

2. Photothermal conversion of CO_2. This method uses photothermal catalysts or plasmon-driven approach to convert CO_2 to useful products. The use of photon energy to convert CO_2 using semiconductor materials has gained some momentum. Unfortunately, the process suffers from low efficiency.
3. Photocatalytic conversion of CO_2. This method can use a number of different catalyst and can be carried out along with photo-synthesis. Photocatalysis can be homogeneous or heterogeneous.
4. Photocatalytic conversion with enzymes for conversion of CO_2. Enzymes further accelerate the reactions and improve product selectivity.
5. Photoelectrochemical catalysis to convert CO_2 to valuable products like CO, CH_4, or methanol. Catalysts at electrodes accelerate the conversion. Photoelectrolysis approach is most convenient, but it is also limited by the conversion rate, and researchers are seeking for the catalysts, which can produce better performance.
6. Photoelectrochemical catalysis aided by enzymes to convert CO_2 to useful products. Once again enzymes improve performance and product selectivity.
7. PV-supported electrochemical catalysis to convert CO_2 to useful products. This novel method has significant commercial possibility. The electrolysis using photovoltaic (PV) materials and electrolyzer is the most mature approach for producing solar fuel. However, the PV materials can only utilize the light with a certain range of wavelength (usually short-wavelength light), and the other part of sunlight absorbed is converted into thermal energy, which is wasted as residual heat, leading to a limited PV cell efficiency (the commercial PV cell efficiency is about 15%; the highest multiple-junction PV cell efficiency in lab is higher than 40% with high cost). The total energy efficiency from solar energy to chemical energy is the product of solar power efficiency (e.g., PV cell efficiency) and electrolysis efficiency, so the total efficiency has the potential to be further improved. Thermo photo-voltaic cell (TPV cell) is another approach used to improve the efficiency. In PV/EC concept, the performance of PV cell is integrated with electrocatalysis.

Some of the important aspects of the strategies covered in this chapter as well as in Chapters 4 and 5 are graphically illustrated in Figure 6.1 and their classifications and definitions are outlined in Table 6.1.

Solar/thermal approach involves the use of metal oxide thermochemical cycles to dissociate CO_2.

The conversion of CO_2 can lead to several different chemical/fuel products depending on the materials and/or methods employed, including carbon monoxide (CO), formic acid (HCOOH), methane (CH_4), methanol (CH_3OH), ethylene (C_2H_4), ethane (C_2H_6), propane (C_3H_8), ethanol (CH_3CH_2OH), acetic acid (CH_3COOH),

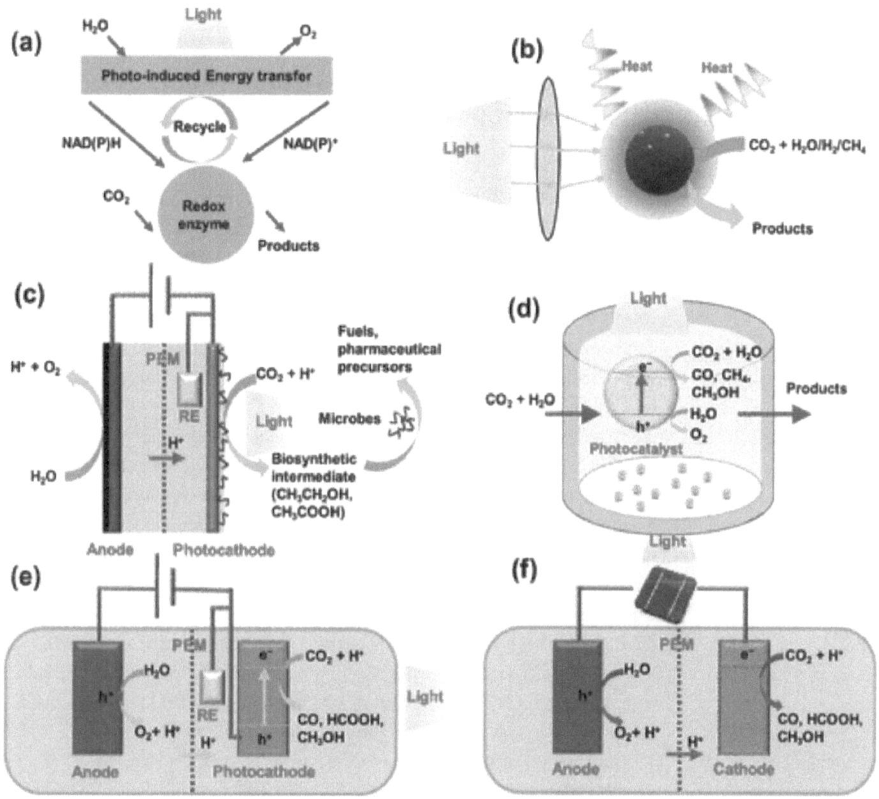

FIGURE 6.1 Schematic illustration of (a) biophotosynthetic, (b) photothermal, (c) microbial-photoelectrochemical, (d) photosynthetic and photocatalytic (PS/PC), (e) photoelectrochemical (PEC), and (f) photovoltaic plus electrochemical (PV+EC) approaches for CO_2 conversion [6].

acetone, *n*-propanol, acetaldehyde, allyl alcohol, dimethyl ether, glycolaldehyde, hydroxyacetone, ethylene glycol, propionaldehyde, and glycerol [14]. Although the conversion level and selectivity of these products and carbon content of these products, at the present time, accounts for only a fraction of the emitted CO_2 [15], the concept of solar-driven CO_2 can be further advanced to fuel production in the future (especially for aviation where high energy density is inevitable), which accounts for a much larger carbon footprint [16].

While CO_2 can be converted to numerous products as mentioned above, major reactions examined in the literature to demonstrate effectiveness of solar energy assisted CO_2 conversion are (a) CO_2 dissociation, (b) CO_2 hydrogenation which can be reverse water gas shift reaction to produce CO, hydrogenation to produce methane or methanol, (c) CO_2 reaction with water, and (d) dry reforming of CO_2 (CO_2 reaction with methane). In this chapter, we will use one or more of these reactions to illustrate effectiveness of various types of solar thermal, solar photothermal,

TABLE 6.1
Classifications and Definitions of Solar-Driven CO_2 Conversion: Solar–Thermal and Photo-Catalytic Approaches [6]

Category	Definition
Biophotosynthetic/ biophotocatalytic	An approach that mimics natural photosynthesis, which usually involves redox enzyme molecules as photocatalysts or artificial microbes for photosynthesis
Photothermal	An approach that uses high-temperature solar reactors, typically employing concentrated solar radiation, to split CO_2, potentially utilizing the entire solar spectrum and offering high product formation rate
Microbial photoelectrochemical	Combines the advantages of semiconductor photoelectrodes and the high-selectivity microbe-based biocatalysts, directly converting CO_2 into fuels or chemicals
Photosynthetic and photocatalytic (PS/PC)	Two sister approaches using particulate or molecular photocatalysts, either in solution or immobilized on a surface. This category includes both PC ($\Delta G < 0$) and PS ($\Delta G > 0$) processes, depending on the oxidation half-reaction. Because of many similarities, they are discussed together herein, but in the light-to-fuel efficiency comparison (Figure 6.9b), only PS processes were selected to ensure fair comparison.
Photoelectrochemical (PEC)	Either one or both electrodes of the electrochemical cell is/are semiconductor photoelectrode(s). Photogenerated charge carriers drive either one or both half-reactions. The studies using the "buried junction" concept are included here, where a solar cell is covered by one or more catalyst(s) (and possibly a protecting layer) and this whole assembly acts as a photoelectrode.
Photovoltaic plus electrochemical (PV +EC)	The combination of PV cells with CO_2 electrolysis in one device. This approach decouples the light-harvesting and the electrochemical conversion steps.

photo-catalytic, and photoelectrocatalytic (with or without enzymes) conversion strategies.

6.2 THERMOCHEMICAL CONVERSION OF CO_2 AND CH_4 USING SOLAR ENERGY

The overall objective of this process is to collect low density and distributed solar energy by solar collectors and use this concentrated solar energy to generate heat to drive chemical reaction. High-density solar fuel created by this reaction can be stored and transported where it is needed. Thus, the process transforms an intermittent and geographically distributed renewable source into more stable and reliable high-density chemical fuel for energy. The concentrated solar energy can be used for CH_4 and CO_2 conversions in two ways: (a) reforming of CH_4 by steam or CO_2 to produce syngas (H_2/CO mixture) that can be used for many downstream products through F-T, methanol, and other oxygenated syntheses, or (b) dissociation of CO_2 to CO or $CO_2 + H_2O$ to syngas by thermochemical cycles involving metal oxides such as CeO_2. Here, we briefly examine both approaches.

6.2.1 Solar Energy-Based Dry Reforming

The methane dry reforming is the reaction between methane and carbon dioxide for syngas generation, given as:

$$CH_4g + CO_2g \rightleftharpoons 2COg + 2H_2g, \qquad \Delta H°25°C = 247.0\,kJ/mol \qquad (6.1)$$

The reforming reaction, Eq. (6.1), is highly endothermic, and a large amount of heat is often provided by burning a supplemental amount of methane [17], which decreases the heat value of fuel gas generated by 22% for the same amount of methane consumed and releases large amounts of greenhouse gas CO_2 [18]. In recent years, these reactions are carried out using CSE [19] instead of fossil fuels. The process, thus, results in transforming solar energy in syngas. Syngas can be used to produce chemicals, fuels, and fuel additives using F-T, methanol, or other oxygenated syntheses. Numerous studies [20–22] have used novel reactor design, like fluidized bed reactors [20], solar tubular reactor [22], or specially designed reactor supported by heliostat to focus sunlight to improve the efficiency of reforming reactions.

6.2.2 Solar Energy Based CO_2 Dissociation to CO

CO_2 dissociation to CO using CSE can be carried out using two-step redox cycles process in which a concentrated and focused beam of sunlight heats ceria up to 1,400°C or more, driving its endothermic reduction and releasing oxygen as a result. The reduced ceria is then cooled (non-solar) to 1,000°C or below (known as a temperature-swing cycle), and re-oxidized under a flow of carbon dioxide, creating carbon monoxide. This process is graphically illustrated in Figure 6.2.

The efficiency of this process depends on the morphology of the ceria redox material and the solar thermochemical reactor design. High surface area material is important. The reactors can be operated as high flux solar simulators (HFSS) with furnaces heated by simulated solar light, or high flux solar furnaces (HSSF) utilizing actual (not simulated) solar energy, which can be used to study these processes. The largest solar furnaces, such as that at PROMES-CNRS in Odeillo, France, can offer a thermal power of 1 MW or more. The oxidation step is performed during cooling

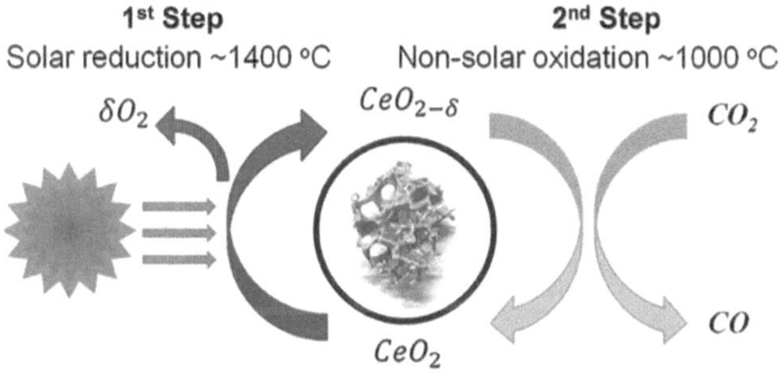

FIGURE 6.2 The two-step thermochemical redox process for the spitting of CO_2 using ceria [3].

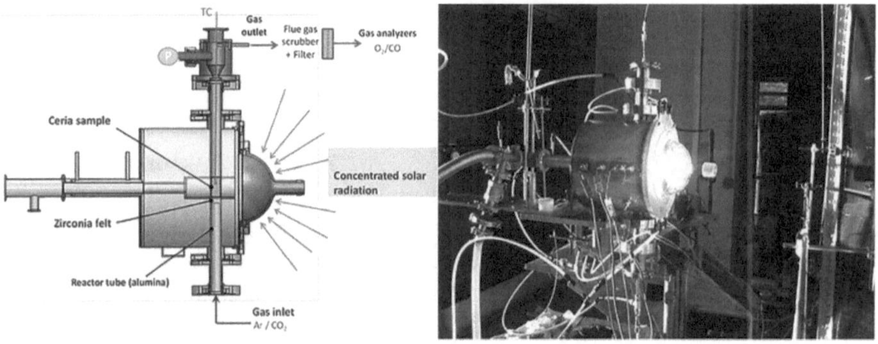

FIGURE 6.3 The actual concentrated solar light reactor used at PROMES [3].

without exposure to concentrated sunlight as an energy source. The actual concentrated solar light reactor used at PROMES is shown in Figure 6.3.

6.2.2.1 Ceria for Two-Step CO_2 Splitting Cycle

Cerium dioxide (ceria) is a versatile reducible oxide. It tolerates a considerable reduction without phase change, especially at elevated temperatures. CeO_2 has an unusually high entropy change associated with oxygen exchange compared to other non-stoichiometric redox materials [23–25], resulting in reduced temperature swings between the reduction and oxidation steps [26]. Ceria also has rapid reaction kinetics and oxygen diffusion rates [27], which is thermally stable and relatively resistant to sintering even at high temperatures due to its high melting point (~2,400°C). Ceria was first investigated as a material specifically for solar thermochemical CO_2 splitting in 2010 by Chueh et al. [28], Chueh and Haile [29], and Haussener and Steinfeld [30,31]. Since then, many studies have investigated ceria redox materials, as they showed higher oxygen ion mobility and rapid fuel production kinetics compared to ferrites and other non-volatile metal oxides [29,32,33]. The two-step splitting cycle of CeO_2 is based on:

1. The solar thermal reduction of CeO_2 (endothermic, high temperature) at low oxygen partial pressure in a neutral (typically Ar or N_2) atmosphere, to create oxygen-deficient non-stoichiometric ceria ($CeO_{2-\delta}$, where δ is the degree of oxygen deficiency) via the formation of oxygen vacancies and the subsequent release of O_2 gas; and
2. The non-solar oxidation of $CeO_{2-\delta}$ back to CeO_2 (exothermic, lower temperature), that will take oxygen from CO_2 as the temperature is decreased with CO_2 present, releasing CO gas and re-incorporating some oxygen into the ceria lattice.

This redox process is depicted in Figure 6.2. The fuel production yield is dependent on the degree of non-stoichiometry (δ), and is determined by temperature and oxygen partial pressure. Ceria can accommodate quite high amounts of oxygen non-stoichiometry, and can support high levels of oxygen storage/loss and mobility while maintaining the crystallographic fluorite structure [34]. As one molecule of O_2 from the reduction step should produce two molecules of CO in the oxidation step, the ideal ratio of O_2:CO production would equal 2.

The study by Chueh et al. [28] was carried out in the temperature range of 1,581°C–1,624°C and with porous monolithic ceria. After multiple cycles, production of CO remained constant but the rate declined due to considerable grain growth of ceria from 5 to 15 μm. This indicates that oxidation of ceria with CO_2 is a surface-limited process, as the oxidation rate is strongly linked to the SSA (specific surface area) [29]. An overall solar-to-fuel energy conversion of only 0.4% was achieved, but they predicted that values as high as 16%–19% should be attainable, even without sensible heat recovery [28]. While numerous experiments under different operating conditions were performed subsequently, the final conclusion was that optimum temperature during the oxidation step was about 1,000°C and CeO_2 reduction step was kinetic limited [35].

The literature research has shown that as well as grain size, stoichiometry and effects of dopants, variations in porosity and microstructure can also have a great effect on the ability of ceria to split CO_2. It is very difficult to directly compare results and attribute differences purely to microstructure, as there are many other variables between the CO_2 splitting experiments of different research groups, such as gas partial pressures, heating rates and dwell times, source of heating energy (solar, electrical, IR, TGA), quantity and form of sample, etc.

Different structures of ceria have been tested in the literature. These include porous monolithic ceria, compressed ceria fibers with and without Zr doping, 3-DOM ceria, reticulated porous ceria, mesoporous ceria nanoparticles, cork based and other bio-based templating materials and a biomimetic approach, use of waste materials like red mud, etc. The reported results have demonstrated that the material morphology clearly affects the oxidation kinetics, as well as the heat and mass transfer rates within the reactive porous structure. Identification of proper materials shaping strategies favoring kinetics, while offering suitable redox activity, high oxygen transport properties, and long-term cycling stability under concentrated solar irradiation, is thus required. Relevant morphologies such as 3D porous structures capable of absorbing concentrated solar radiation, while offering a large available geometrical surface area for the solid/gas reactions, are promising options. Using bio-based templating materials and a biomimetic approach to elaborate eco-ceramics has to be pursued to advance sustainable processes. More efforts are needed to improve conversion to make the process commercially viable [3].

6.2.2.2 Concept of Membrane Reactor

As shown above, solar thermochemistry for CO_2 splitting and methane reforming require 3,000°C and 700°C–1,000°C, respectively. These high temperatures result in high concentration ratio, large mirror area, and complex and expensive systems for CSE. The temperature can be reduced, and the reaction rate can be enhanced by the simultaneous separation of products from the reactant. Combination of membrane reactor and solar thermal collection offers unique advantages in many respects, such as the increment of conversion rate, decrease of reaction temperature, and emission reduction. Furthermore, the all-solid-state feature and isothermal operation enable compact design of solar fuel reactors with minimized thermal stress.

CO_2 dissociation using membrane reactor has been studied by a number of investigators [36–40]. Preferred materials for membrane are perovskites, ZrO_2, and CeO_2 (or doped ZrO_2 and CeO_2). The most notable studies were carried out by Steinfeld [38,39] who examined solar CO_2 splitting for CO generation by oxygen permeation membrane with 100% selectivity (e.g., $La_{0.6}Sr_{0.4}Co_{0.2}Fe_{0.8}O_{3-\delta}$ at 1,030°C [38], CeO_2 at 1,600°C [39]). The materials of the hydrogen permeation membrane are various, such as metal (e.g., palladium, nickel), perovskites, pyrochlores, fluorites, and polymers, which are usually used in the reaction of reforming, splitting, and partial oxidation of hydrocarbon. Methane reforming using membrane reactor was also carried out both experimentally [41,42] and theoretically [43]. Carbon dioxide permeation membrane includes mixed e^-/CO_3^{2-} conducting membrane, O^{2-}/CO_3^{2-} conducting membrane, OH^-/CO_3^{2-} conducting membrane (hydroxide/ceramic dual-phase membrane), etc. [44,45]. The combination of hydrogen permeation membrane and carbon dioxide permeation membrane has been proposed for methane steam reforming by way of an alternate H_2 and CO_2 separation driven by solar energy [46]. This results in high energy efficiency due to relatively high partial pressure and less required separation energy. While the solar membrane reactor has lots of advantages and immense potential for application, the efficient approach to lower the partial pressure of gas product (or avoid the relatively low pressure) is the main challenge to maintain a high energy conversion rate, and the improvement of stability and permeability of membrane material at corresponding reaction temperature is also significant.

An application of membrane reactor for CO_2 dissociation is graphically illustrated in Figure 6.4. While this method gives high conversion and efficiencies, the temperature/pressure swing commonly applied between reduction and oxidation steps incurs irreversible energy losses and severe material stresses. This single-step method under steady-state isothermal/isobaric conditions was, however, demonstrated

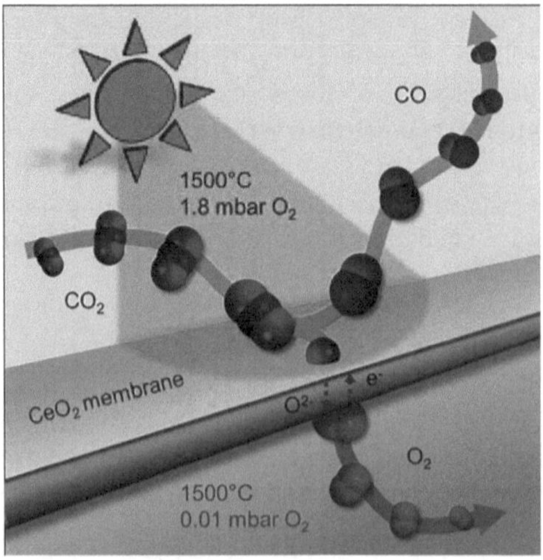

FIGURE 6.4 Illustration of a membrane reactor [39].

for ceria membrane by Tou et al. [39]. The study experimentally demonstrated for the first time the single-step continuous splitting of CO_2 into separate streams of CO and O_2 under steady-state isothermal/isobaric conditions. The system was operated with the solar reactor at 1,600°C, 3×10^6 bar p_{O2} and 3,500 suns radiation, yielding total selectivity of CO_2 to $CO + {}^1/_2O_2$ with a conversion rate of 0.024 mmolxs[1] per cm^2 membrane [39].

The novel solar reactor configuration consisted of a thermally insulated cavity-receiver with a small aperture for the access of concentrated solar radiation. The cavity geometry enabled efficient radiative capture by internal multiple reflections, approaching a blackbody absorber. The cavity contained a capped tubular membrane, made of ceria, enclosed by a coaxial alumina tube. CO_2 was supplied to the inner side (oxidation side) of the membrane, and a sweep inert gas Ar was supplied to the outer side (reduction side) to control the oxygen partial pressure $p_{O2.}$ Both sides were operated at ambient total pressure. The study also indicated that, the analogous net splitting of H_2O to produce H_2 and O_2 in separate streams can be achieved by supplying H_2O instead of CO_2 [39].

6.2.2.3 Syngas from CO_2-H_2O Reaction Using Solar Energy

Scientists with the SOLAR-JET Project demonstrated a novel process to make kerosene, the jet fuel used by commercial airlines. The technique uses a high-temperature thermal solar reactor (see Figure 6.5) to create syngas using two-step thermochemical cycle similar to the one shown in Figure 6.2 but for CO_2 and H_2O mixture using CeO_2 catalyst. Because of H_2O, the process created H_2 along with CO resulting in syngas production. Shell Global Solutions in Amsterdam refined the solar syngas into jet fuel using the new Fischer–Tropsch method. Researchers at the Department of Mechanical and Process Engineering at ETH Zurich, Switzerland, concentrated

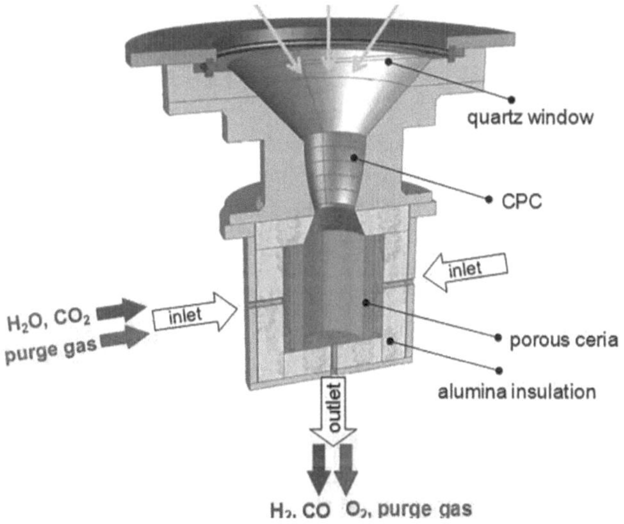

FIGURE 6.5 The scientists performed 295 consecutive cycles in a 4 kW solar reactor during the SOLAR-JET project, yielding 700 standard liters of syngas. (Credit: SolarPACES [2,5].)

3,000 "suns" of solar thermal energy into a solar reactor at 1,500°C for thermochemical splitting of H_2O and CO_2 into hydrogen and carbon monoxide (syngas), the precursor to kerosene and other liquid fuels. The paper, "Solar Thermochemical Splitting of CO_2 Into Separate Streams of CO and O_2 with High Selectivity, Stability, Conversion, and Efficiency," presented at the 2017 Solar PACES Conference in September details the successful solar-driven thermochemical splitting of CO_2 into separate streams of CO and O_2 with 100% selectivity, 83% molar conversion, and 5.25% solar-to-fuel energy efficiency [2,5].

The scientists performed 295 consecutive cycles in a 4kW solar reactor during the SOLAR-JET project, yielding 700 standard liters of syngas. Once the researchers thermochemically split the water and carbon dioxide into carbon monoxide and hydrogen (syngas), they sent the syngas to Shell Global Solutions in Amsterdam, where the Fischer–Tropsch process was applied to refine it into kerosene, the jet fuel used by airplanes. The solar thermochemical kerosene can be certified for commercial aviation by minor addendum to the existing D7566 specification for synthesized hydrocarbons to allow for kerosene produced by solar energy. The SOLAR-JET team was the first to have demonstrated the entire process using solar heat, to split the H_2O and CO_2 and to store, compress, and then process the resulting gas via the Fischer-Tropsch method into jet fuel. H_2O and CO_2 can supply the feedstock for all the liquid fuels currently used for transportation, which are just various molecular recombination of hydrogen and carbon [2].

A solar thermochemical reactor for splitting of H_2O and CO_2 requires temperatures up to 1,500°C and therefore much greater solar flux than today's CSP for electricity generation which operates at only up to 560°C. So solar reactors for thermochemistry use thousands rather than hundreds of suns, to get to these temperatures. The key component of the splitting process is a high-temperature solar reactor containing a reticulated porous ceramic (RPC) structure made of ceria (CeO_2) which facilitates molecule splitting.

The researchers have boosted the efficiency of the solar process from 2% to 5% by moving to vacuum extraction on the 4kW reactor. Their two-step solar thermochemical conversion with ceria shows a long-term efficiency potential of beyond 30%. Ultimately, industrial-scale solar fuel production systems would use megawatt-scale reactor systems on solar towers with heliostats (mirrors) concentrating solar energy on the receiver, running at much higher temperatures than current commercial solar power plants using solar tower. According to the SOLAR-JET Project Coordinator at Bauhaus Luftfahrt, Dr. Sizmann, a solar reactor with a 1 square kilometer heliostat field could generate 20,000 liters of kerosene a day. This output from one solar fuels refinery could fly a large 300-body commercial airliner for about 7hours. ETH spin-off Sun redox is to commercialize the technology [2].

6.2.3 Syngas Production by Thermochemical Conversion of CO_2 and H_2O Using a High-Temperature Heat Pipe-Based Reactor

Pearlman and Chen [1] used a simple multi-dimensional thermal resistance model to explain the low system efficiency of current solar thermochemical reactors, which are now at ~2% or less. They then designed a heat pipe-based reactor to convert CO_2

and H_2O into syngas using solar energy and thermochemical cycle similar to the one described above. The reactor incorporated the use of high-temperature heat pipe(s) that efficiently transfer the heat from a solar collector to a porous ceria metal oxide material. The focus was on increasing the heat transfer area and thereby reduce the thermal resistance to the working material. The thermal performance and O_2 release and CO production rates were demonstrated using CO_2 as the feedstock for select test conditions. Special attention was given to the thermal characteristics of the reactor, which are key factors affecting the overall system efficiency and amount of fuel produced.

The reactor used in the study was limited to operation at 1,100°C due to the creep stress of the nickel-based alloy material (Haynes 230). To further increase fuel productivity, the reactor temperature needed to be further increased (to 1,300°C–1,500°C) and additional work aimed at lowering the reduction temperature of metal oxides without adversely affecting their re-oxidation kinetics was also needed. For higher temperature operation, a refractory metal heat pipe reactor encapsulated in a quartz vacuum tube (to suppress oxidation) was developed in the study. In this new reactor design, heat was transferred solely by radiation from the heat pipe to the working material. The reactor design focused on reducing the thermal resistance and the timescale needed to uniformly heat the packed bed. To accomplish this, the new reactor had a large heat transfer surface area required for the incident solar energy to rapidly be transferred to the working material such that it can be uniformly reduced while minimizing re-radiation loss. As noted in the study, the heat pipes in these designs simply acted as a fin structure with nearly perfect fin efficiency to efficiently distribute the concentrated solar power to the metal oxide material in the reactor. The details of the heat pipe-based reactor are given by Pearlman and Chen [1].

6.3 PHOTOTHERMAL CATALYTIC CONVERSION OF CO_2 WITH HYDROGEN

6.3.1 Mechanisms for Photothermal Activations

Photocatalysis is initiated by the excitation of electrons by light absorption. There are three different ways; interband, intraband, and plasmonic excitation of electrons can occur depending on the energy of the incident photons and nature of catalyst [47–50]. Interband excitations occur between valence and conduction bands, intraband excitations occur to or from defect states within a band [47–49] and plasmonic excitations involve a collective excitation of conduction band electrons [50]. Besides these three, an increase in photon absorption across a plasmonic metal nanoparticle can also occur by so-called "Antenna effects", which further enhances light absorption [50]. As shown in Figure 6.6a, in photothermal catalysis, light absorption band is wider than the ones in photocatalysis using semiconductor materials. The energy captured by electronic excitation is facilitated by non-radiative electron relaxation in which energy is transferred from excited electrons to adjacent atomic nuclei and it gets converted to heat. This process is illustrated in Figure 6.6b. Non-radiative energy transfer is dominated by processes such as electron–phonon scattering [51], which is accomplished particularly well in plasmonic materials [52].

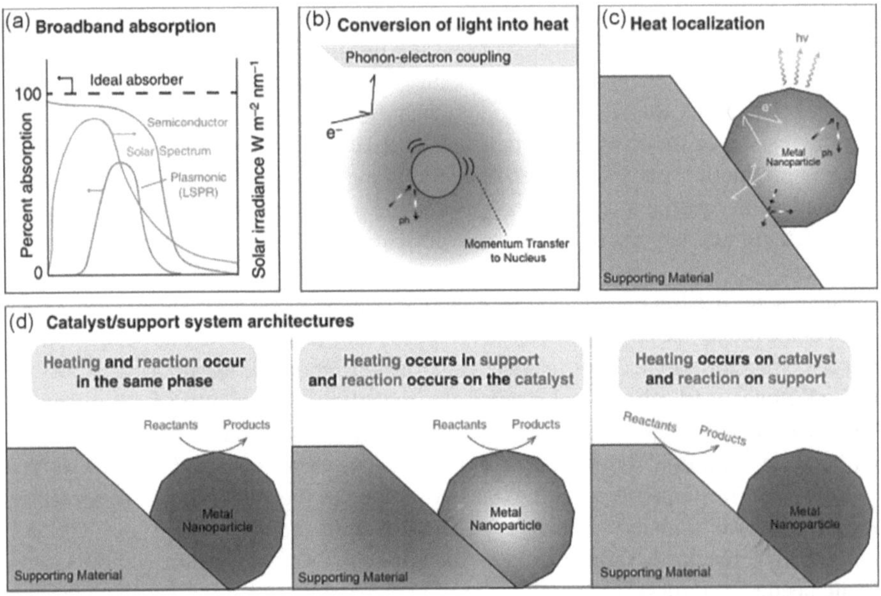

FIGURE 6.6 Key concepts and examples of photothermal methanation architectures. (a) Schematic representation of the light absorption spectra of semiconductors and plasmonic metals in comparison to the solar emission spectrum. (b) Light-to-heat conversion: photo-excited electrons (e^-) interact with atomic nuclei, possibly generating phonons (ph). (c) Mechanisms of heat transfer within a particulate catalyst. Heat can be localized in catalyst nanoparticles by inhibiting phonon and electron transfer from the nanoparticle to its support. Structural defects, as well as phase boundaries between the catalyst and support phases, inhibit phonon and electron transfer and are hence beneficial for heat localization. Properties such as thermal and electronic conductivity, size, and shape of the catalyst/support systems govern nanoscale heat transfer. (d) Selected potential photothermal catalyst architectures [7].

Nano-sized particles facilitate the concentration of generated heat by electron excitation. This concentration of heat makes photothermal catalyst very active. This heat is either transferred or retained depending on the catalyst structure [50], and it may require adequate heat transport strategies at the nanoscale. The support used to enhance catalyst activity and stability must also have electrical and thermal conductivities such that required heat transport is facilitated. Phase boundaries between catalyst and support can help in providing thermal energy carriers who scatter heat back into the active catalyst material, thus concentrating heat and optimizing photothermal effect as shown in Figure 6.6c and d. This concentrated photothermal heat along with a sustainable supply of H_2 drive the CO_2 methanation reaction in photothermal catalysis. Plasmonic nanometals can also initiate surface chemical reactions via "hot electron injection", wherein the energy of localized surface plasmons excite charge carriers (electrons and holes) on the metal surface. These charge carriers are then transferred to an adsorbed reactant or intermediate and create excited states that facilitate the chemical transformation of the adsorbed species [53].

Catalyst activity and light absorption depend significantly on both nanoscale active catalyst materials and support. Nanoscale metal catalysts such as Ni, Ru, Rh, Fe, Au, and Pd [54–64] are also often studied as photothermal catalysts, due to their strong broadband optical absorption. These catalysts may be supported on various materials including Al_2O_3, ZnO, TiO_2, Nb_2O_5, Si, and MOF [54–60] to enhance their dispersion and stability. Local electric fields generated by oscillating plasmonic electrons can be enhanced by metal nanostructure morphologies, especially edges and corners which can result in the amplification of the light absorption. Reaction pathways and rates can be modified [61–63] by injection of electrons into adsorbed reactants or intermediates which is facilitated by enhanced electric field in combination with the lower coordination of atoms found at edge and corner sites.

Both catalyst particle size and support are important for reaction rate [65–67]. Nanoscale catalyst particle dispersion on support provides high catalyst surface area. Along with providing mechanical and thermal stability to active catalyst particles, support can also play role in changing surface chemistry for the reaction. For CO_2 hydrogenation, CO_2 can also be adsorbed and get activated by the support. Such observations have been reported for Cu/ZnO catalyst for CO_2 hydrogenation in thermal catalysis. Similar investigations on surface defect-rich oxides such as TiO_2 CeO_2, Ga_2O_3, In_2O_3, and their combinations as well as for carbide materials support such as titanium carbide and molybdenum carbides have been carried out. With these supports, CO_2 conversions have been reported despite the absence of metals. Furthermore, when coupled with metals such as copper, nickel, or gold, a significant enhancement in activity (with varying results in product selectivity) has been observed [68]. Such synergism is often attributed to the interaction between the metal catalyst and the oxide or carbide surface via occurrences such as creation of active centers at the metal-support interface, which promotes CO_2 adsorption upon the oxide or carbide surface and charge redistribution at the metal-support interface.

Active catalyst materials can also play role in light absorption. For example, active materials like Cu, Ni, and Au for CO_2 reduction also exhibit plasmonic absorption that lies within the solar spectrum. This dual role played by the catalyst material is demonstrated in two reported studies [63,69]. The first study [69] used Cu/ZnO-impregnated aerogel containing microreactor for self-heating CO_2 reduction, while in the second study the effectiveness of laser-induced heating on Au/ZnO for a similar application was examined. Just as catalyst, support materials can also play a dual role of supporting catalyst activity and contribute to the light absorption. Transition metal oxides, such as those mentioned above, are semiconductors that are capable of light energy absorption due to their unique band structures. A combination of an active, plasmonic metal catalyst with a semiconducting material can result in enhanced charge carrier generation (due to thermo plasmonics or optical near-field enhancement) or hot electron injection into the semiconductor as illustrated by Figure 6.6d and as shown by Baffou and Quidant [70]. Many catalytic studies have observed the role of catalyst support on light absorption.

The dual role played by both catalyst and support like oxides and carbides can be used as selective absorber coatings for solar thermal collectors. For this purpose, aluminum oxide can be coupled with metals such as cobalt, silver, platinum, molybdenum, and most prevalently, nickel. The popularity of the nickel-alumina composite stems from its solar absorption properties – it possesses the ideal combination of high absorbance (for effective energy absorption) and low thermal emissivity (to minimize losses through thermal radiation) [71]. CAESAR (*CA*talytically *E*nhanced *S*olar *A*bsorption *R*eceiver) project [72] used a catalyst material in the form of a catalyst foam which acted as a receiver for concentrated solar radiation for CO_2 reforming of methane. Carbide materials such as hafnium, tantalum, and titanium carbides have also been used because of their high thermal resistance [73] which is very important for fabricating a catalyst/absorber foam with greater durability. Titanium or Mo carbides, are however, less active for CO_2 reduction. The excitation of plasmon resonance in Rh nanoparticles results in up to a seven-fold increase in selectivity for CH_4 over CO, relative to thermal reaction conditions [61]. Theoretical simulations suggest that, in the thermo catalytic reaction, phonons activate intermediates during both CH_4 and CO formation, resulting in comparable production rates of these two products. In the photocatalytic reaction, however, hot electrons selectively transfer to CH_4 intermediates, reducing the activation energy of CH_4 formation and increasing its production rate [61].

Novel catalyst/support systems are being developed to improve both catalyst activity and light absorption. Light-absorptive support materials, such as vertically aligned Si nanowires [57,74] and inverse opal Si photonic crystals [56,59] have been tested. The minimal reflection losses and strong broadband absorption made possible using these materials are instrumental in enhancing CO_2 methanation rates [56,57,74]. For example, photothermal methanation rates for RuO_2 dispersed on inverse opal Si photonic crystals (denoted as RuO_2/i–Si–o) were enhanced relative to RuO_2 deposited on a Si wafer. No photocatalytic effects were observed and the enhancement was attributed to increased temperatures resulting from improved light harvesting by the i–Si–o support. Thus, the choice of the supporting material can greatly influence light absorption by the catalyst/support system. Catalyst-support system should be optimized both with respect to CO_2 reduction rate as well as for their ability to absorb light so that they make the most effective photothermal catalyst/support system.

Since photothermal catalysts carry out both photochemical and thermochemical reactions, it is important to know the respective contributions by these two paths for reaction rates, selectivities, and turn-over numbers so that improved catalysts can be designed. Fortunately, these two paths can be differentiated based on their response to light. Plasmonically initiated reactions exhibit a super-linear ("power law") dependence on light intensity (i.e., rate$\propto$intensity) [75] and are characterized by a positive relationship between quantum efficiency and photon flux/temperature [67]. Thus, unlike traditional semiconductor-based photocatalysis, wherein quantum efficiency decreases with temperature, heat, and light work synergistically in plasmonic reactions: increased temperature yields increased efficiency [60,66,67].

There are several challenges that remain with the design and testing of photothermal methanation catalysts. These are

1. Find relationships between size, composition, and morphology of catalyst/support system with its light harvesting and catalyst properties,
2. Identifying the effects of light intensity and spectral distribution on light-harvesting properties, quantum efficiency, catalytic rate and selectivity, and

temperature evolution and distribution of these parameters within the catalyst bed,
3. Separating thermal and photocatalytic effects on the catalyst performance such as its activity and selectivity as a function of time.

6.3.2 Industrial Implications

There are a number of industrial applications of photothermal heat on CO_2 methanation.

1. Photothermal system can be used to heat the reactor instead of conventional heat jacket or other heating systems for methanation reaction. Since photothermal process can generate heat closed to the catalyst surface, it will be more efficient, reduce start-up times, and improve load flexibility.
2. Due to photocatalytic or plasmonic effects described earlier, photothermal heating can increase the reaction rate. This will allow either increase in reactants throughput and increase production rate or reduce the reactor size for the same reactants throughput.
3. Plasmonic photocatalysts can be used to change the product selectivity such as increase the production of higher-value chemicals like ethane instead of methane or other higher hydrocarbons. This will also change the caloric value of the products. These higher hydrocarbons can be targeted with the use of selective catalysts and/or supports [76] or by applying photocatalytic dehydrogenative coupling to convert CH_4 into higher hydrocarbons [77–80]. This was recently demonstrated for Au nanoparticles on ZnO nanosheets [78]. The study showed that this reaction is induced by the electron transfer between the photoexcited ZnO nanosheets and a surface-adsorbed CH_4 molecule. The reported quantum efficiency was 0.08%, which is comparable to that of natural photosynthesis [78]. These results indicate that changing product selectivity and producing high-value products at milder temperatures than those used in thermocatalytic operations can be done by plasmonic photocatalysis.
4. Since natural gas (CH_4) infrastructure already exists, production of CH_4 from CO_2 can be easily introduced in the existing infrastructure. CO_2 conversion to CH_4 not only cuts down CO_2 emissions, but it adds to the fuel storage for future use that is generated by the renewable source.

6.3.3 Challenges for Photothermal Catalytic CO_2 Reduction

Catalyst and support in photothermal catalysis play a dual role. Not only catalyst must be highly active and selective, both catalyst and support must be thermally stable. Furthermore in photothermal operation, catalyst and support must have high solar absorption capabilities to generate large amount of heat in an efficient manner. For nanoparticles, size and shape greatly affect plasmonic absorption in a plasmonic catalysis [81–83] and this makes the material stability even more important. It is known that an extensive exposure to heat causes sintering and loss in nanostructured morphology. Hence the use of nanomaterials for photo thermal catalysis requires control of these properties. Often this is achieved by the use of composite materials to optimize solar heat absorption and generation and catalytic activity and selectivity of the materials. This requires understanding the interactions among them for specific composite material. Support also plays an important role. Tan [84] showed that light activation can be achieved by both catalyst Au and support TiO_2 in Au/TiO_2 catalyst. With the use of different wavelength ranges of ultraviolet and visible light, the study highlighted the importance of understanding the pathways associated with photothermal CO_2 reactions to gain better insights into catalyst performance especially in terms of product selectivity.

Along with material fabrication, the design of a photothermal reactor also require special attention since conventional plug flow tube reactors are usually not suitable for this purpose. Novel reactors published by Palumbo et al. [85] and Meng et al. [54] are still not suitable due to the lack of a secondary heating system for deconvoluting photothermal heating and photo-induced charge transfer effects. In order to achieve this, other studies [86,87], heated the catalyst from the bottom while exposing the reactor top to continuous illumination. In yet another study [88], commercially available reactor for in situ spectroscopic analysis of oxide supported Au catalyst was used. Since in these studies, secondary heat sources were capable of heating the reactor up to 400°C–600°C, the authors were able to separate thermal catalysis from photocatalysis by minimizing thermal heating from the irradiation sources through the use of specific bandwidth sources and/or filters.

6.3.4 Future Potential

While as shown in Section 6.2, that thermal catalytic reduction of CO_2 using solar energy is reasonably well understood, CO_2 hydrogenation by photothermal catalysis requires more research for fundamental understanding. This is because first, catalytic processes for CO_2 hydrogenation and product selectivity are more complex and the effects of both thermal and light components of photothermal catalysis are not well understood separately. Furthermore better understanding of process chemistry and mechanisms involved during interactions between catalytic processes and light need to be better understood.

We have separated solar thermal and photothermal catalysis in this chapter. This is not needed in practice. Concurrent use of solar thermal technologies and plasmonic catalysis may complement each other. This complementary behavior may reduce the solar concentration required from the solar collector by the use of plasmonic materials in a solar thermal reactor. This can also reduce the severity of solar thermal system to achieve sufficient temperature.

Photothermal catalysis is truly an interdisciplinary activity. It involves

1. Infrastructure design for solar absorption; understanding of solar radiation consistency management of required space availability
2. Development of novel catalyst and support materials, particularly using nano-sized catalytic materials
3. Production of renewable hydrogen

4. Effective system design for CO_2 capture
5. Understanding of CO_2 conversion pathways to improve reaction rates and selectivity
6. Integration of products like CO, CH_4, and CH_3OH or higher hydrocarbons in the existing distribution network
7. Suitability analysis of photothermal CO_2 reduction in different geographical locations

Currently, this technology is not ready for commercialization, although it fits well in the use of renewable sources for CO_2 utilization. More progress is needed to make it commercially viable.

6.4 PHOTOCATALYSIS AND PHOTOELECTROCATALYSIS FOR CONVERSION OF CO_2

Photocatalytic and photo-electrocatalysis reduction of CO_2 using selected band of light in the ultraviolet or visible region can be carried out in several different ways:

1. With heterogeneous or homogeneous catalysts, generally semiconductors are capable of absorbing selected bandwidth of light and catalyze the reaction. Catalyst can also be separated from photosensitizer. Molecular homogeneous catalyst can also be heterogenized.
2. With heterogeneous photocatalysts in the presence of enzymes or biocatalysts in the solution to improve product selectivity.
3. With photo-electrocatalysis.
4. Using photoelectrodes with immobilized homogeneous catalysts on the electrodes.
5. Using photoelectrodes with enzymes or biocatalysts in the solution to improve product selectivity

In the section, we briefly assess potentials for each of these methods of photocatalysis and photoelectrocatalysis and briefly review progress made so far in these methods.

6.4.1 Heterogeneous and Homogeneous Photocatalytic Conversion of CO_2

Just like in natural photosynthesis, electron–hole pairs are generated when the heterogeneous photocatalysts are exposed to solar light. The photogenerated electrons induce CO_2 to undergo a redox reaction that results in CO, formic acid, and in some instances hydrocarbon and alcohol formations. Three important processes occur during the photocatalytic conversion of CO_2: (a) absorption of sunlight by the photocatalyst or catalyst and photosensitizer; (b) charge separation and transfer; and (c) catalytic reduction of CO_2 and oxidation of H_2O. Each procedure during the conversion of CO_2 is closely related to the photocatalysts. Until now, the photocatalysts have been mainly from semiconductor materials like TiO_2, which are abundant on earth and easy to obtain. The reaction products include CO, methane, formic acid, methanol, and perhaps C2 hydrocarbons.

Generally, there are two options to promote the photocatalytic reaction: (a) a photocatalyst acts as both light absorber and CO_2 reduction catalyst, or (b) a light-absorbing molecule (dye) or semiconductor is used to sensitize a CO_2 reduction catalyst. There are very few examples of system (a) many of them are molecular catalysts [89]. Examples of system (b) are more common and a range of catalyst and dye combinations have been reported [90]. Many efforts have been made to optimize the structure and composition of photocatalysts or integrate them with other functional units to construct multifunctional and more active catalysts by integrating them with other supporting network. For example, highly microporous metal-organic frameworks (MOF) are integrated with the photocatalysts to improve catalyst surface area and reactivity [91–93]. However, the performance of these complexes still do not satisfy practical requirements for CO_2 reduction to be commercially feasible [94]. The construction of multi-junctions randomly distributed on the surface of photocatalysts, improves interfacial electron–hole separation and migration. The heterojunction created by deposition of MnOx nanosheets and Pt nanoparticles on different facets of semiconductor TiO_2 has shown to improve the conversion efficiency of CO_2 [95].

A study [96] has also shown that two-dimensional nanosheets can reduce the separation between electrons and holes. This study synthesized a series of heterostructured composites by depositing different amounts of CdS on the surface of $BiVO_4$ nanosheets with variable thickness. The results showed that CO_2 reduction was enhanced by this CdS/$BiVO_4$ nanosheet composites and the content of CdS in the composite was responsible for the yield of CO and CH_4. In general, however, heterogeneous photocatalytic systems have low conversion efficiency, poor selectivity of products, competition for generation of H_2, and rapid recombination of photogenerated electrons and holes [97,98]. Furthermore, most heterogeneous photocatalytic systems use expensive and toxic transition metals which results in cost increase and waste disposal requirement.

To overcome the thermodynamic barriers of CO_2 reduction, molecular catalysts can be used to lower the overpotential by stabilizing the intermediate transition states between the linear CO_2 molecules and the intended product. CO_2 has multiple known binding modes to transition metal complexes [13]. The metal can then act as an inner sphere electron transfer agent to activate CO_2 for further transformation. With the choice of various metal centers and ligand structures, molecular catalysts are highly tunable to achieve intended properties such as fast kinetics and long-term stability. As shown by Kumar et al. [13], the performance of a molecular catalyst can be judged by a number of parameters that include:

1. Turnover number (TON) = moles intended products/moles catalyst
2. Catalytic selectivity (CS) = moles intended products/(moles H_2 + moles other products) and
3. Photochemical quantum yield (PHI) = (moles products/absorbed photons) × (number of electrons needed for conversion).

A homogeneous CO_2 photoreduction system consists of a molecular catalyst, light absorber, sacrificial electron donor, and/or electron relay. CO_2 can be reduced by molecular catalysts with or without the help of photosensitizers in homogeneous mode. The generic mechanism of the photocatalytic reduction of CO_2 (Eqs. 6.2–6.5) consists of a photosensitizer (P) capable of absorbing radiation in the ultraviolet or visible region and of the generation of an excited state (P^*). The excited state is reductively quenched by a sacrificial donor (D) generating a singly reduced photosensitizer (P^-) and oxidized donor (D^+). The choice of photosensitizer must be such that P^- is able to transfer an electron efficiently to the catalyst species (cat) to generate the reduced catalyst species (cat^-). In some cases, the photosensitizer and the catalyst are the same species. The cat^- is then able to bind CO_2 and proceed with the catalytic mechanism to release the intended products and regenerate cat.

$$P + h\nu \rightarrow P^* \tag{6.2}$$

$$P^* + D \rightarrow P^- + D \tag{6.3}$$

$$P^- + cat \rightarrow P + cat^- \tag{6.4}$$

$$cat^- + CO_2 \rightarrow cat + products \tag{6.5}$$

Common photosensitizers used in these systems include aromatics, e.g., p-terphenyl and phenazine, and polypyridine-coordinated transition metal complexes. Ruthenium(II) trisbipyridine ($[Ru(bipy)_3]^{2+}$) is the most often employed transition metal complex due to its strong visible-light absorption and high photostability [99]. The most common catalyst species include macrocycle complexes of Ni and Co, polypyridine Ru and Re catalysts, and suspended metal colloids. The macrocycle and polypyridine complexes are the most efficient. Lehn's $[ReX(bpy)(CO)_3]$-type catalyst ($X = Cl^-$, Br^-) being the best known and studied. Photocatalytic reactions do not involve electrical wiring to reduce the oxidized dye or photocatalyst, but a chemical reductant is required to allow for catalytic turnover. Typical reductants are triethanolamine (TEOA), Et3N, and ascorbic acid, which are consumed during the reaction and are, therefore, termed "sacrificial electron donors" The eventual goal is to replace these donors with a sustainable source of electrons such as water. Further details are given in several excellent reviews [97,98,100].

Conjugated metallomacrocycles such as corrins [101], corroles [102], porphyrins [102], and phthalocyanines [103] with a Co or Fe center have been shown to act as photocatalysts for CO_2 reduction. Such metallomacrocycles strongly absorb visible light and do not require the addition of a photosensitizer. These systems do, however, suffer from low PHI and low CS due to the significant production of H_2. While tetraazamacrocyclics such as Ni(cyclam) reduce CO_2 efficiently and selectively to CO electrocatalytically while adsorbed on an Hg electrode [13], for purely photocatalytic systems they tend to suffer from low PHI, CS, and TON [104,105]. When Co is used as the metal center, the photocatalytic properties are improved [106–108]. Proposed mechanism for CO and formate production from photocatalyzed reduction of CO_2 by tetraazamacrocycle transition metal complexes (M = Co, Ni) is illustrated in Figure 6.7.

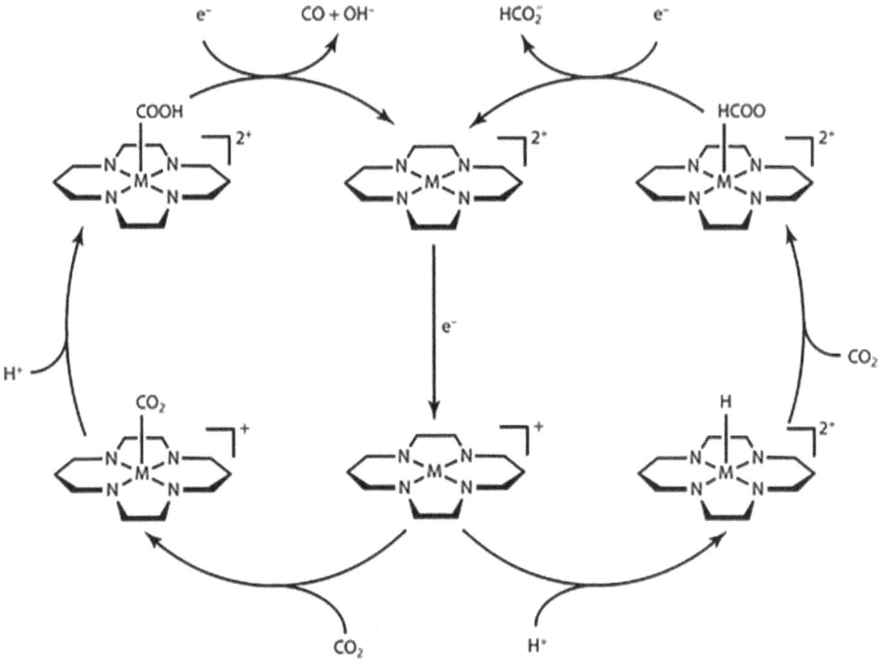

Kumar B, et al. 2012.
Annu. Rev. Phys. Chem. 63:541–69

FIGURE 6.7 Proposed mechanism for CO and formate production from photocatalyzed reduction of CO_2 by tetraazamacrocycle transition metal complexes (M = Co, Ni) [13].

Although the Re catalyst systems demonstrated remarkable PHI, they lacked extended absorption in the visible region. Ishitani and coworkers [109] addressed this issue by utilizing a bridging ligand to covalently attach a $[Ru(bipy)_3]^{2+}$-type photosensitizer, which absorbs strongly in the visible, to a $Re(bipy)(CO)_3X$ type catalyst to create a supramolecular dyad complex. This dyad exhibited significantly better performances (PHI = 0.12, TON = 170) relative to a simple, noncovalently linked mixture of the photosensitizer and catalyst (PHI = 0.062, TON = 101). Depending on the presence of water and the identity of sacrificial donor [71], system can produce CO or formate. This early work on molecular catalysts reviewed by Kumar et al. [13] needed significant improvement to be practical. If metal complexes involved expensive transition metals, improvements in catalytic activity and long-term stability are needed to make the system commercially viable. Better alternative is to use cheaper earth-abundant materials. The use of sacrificial donor is also expensive. Ideally, water should be used for electrons and hydrogen atoms for CO_2 reduction.

Molecular catalysts can be heterogenized. Porphyrins have highly delocalized electrons to form CdS/$BiVO_4$ composites a planar conjugated framework. This endows them with strong absorption in the visible light region and unique electronic redox characteristics. Moreover, the NH protons inside the ring are easily deprotonated and therefore exhibit remarkable coordination characteristics toward metal ions. Incorporation of porphyrin into MOF can further improve its photocatalytic performance. Wang et al. [110] demonstrated this by synthesizing an indium–porphyrin MOF framework. There are numerous other examples of heterogenized molecular catalyst examined in the literature. Ramaraj and Premkumar [111,112] absorbed metal phthalocyanines and porphyrins into a Nafion membrane and irradiated with visible light in $HClO_4$ solution with Et3N as electron donor. Sekizawa et al. [113] immobilized a dyad, consisting of a $[Ru(bpy)_3]^{2+}$ dye covalently linked to a catalytic $[RuCl_2(bpy)(CO)_2]$ unit, onto Ag-modified TaON particles via phosphonate anchoring groups. They also reported a Ru complex with phosphonate groups adsorbed on carbon nitride (C_3N_4), an organic polymer with semiconductor properties. Chambers et al. [114] functionalized a Rh CO_2 reduction catalyst with two carboxylic acid groups for incorporation into a metal-organic framework (MOF). The catalyst, $Cp^*Rh(bpydc)Cl_2$ (bpydc = 2,2′-bipyridine-5,5′-dicarboxylic acid) was inserted into a MOF with the formula $Zr_6(OH)_4(O)_4$(biphenyl-4,4′-dicarboxylate)$_6$. The system thermally dehydrogenated HCOOH at higher catalyst loadings to produce H_2 and CO_2. Increasing the amount of catalyst in the homogeneous system simply led to deactivation, with virtually no HCOOH or H_2 produced. The heterogeneous MOF system could be recycled up to six times or after 4 days of cumulative photolysis. Ha et al. [115] reported a system comprising a $[ReCl(bpy)(CO)_3]$ type catalyst and an organic dye co-adsorbed onto TiO_2 nanoparticles. Windle and Reisner [116] focused on the use of phosphonate anchoring groups to bind molecules tightly to the surface of metal oxide nanoparticles, in particular TiO_2. Phosphonate anchoring groups are a good candidate because their binding constants for TiO_2 are typically an order of magnitude greater than that of carboxylate groups and they are much less sensitive to water and high or low pH [117]. They also reported a system in which a $[ReL(bpy)(CO)_3]$ (L = 3-picoline or Br^-) photocatalyst was modified with phosphonate anchors and thus bound to TiO_2 nanoparticles.

In an effort to make solar-powered enzymes, Woolerton et al. [118] immobilized a carbon monoxide dehydrogenase onto dye-sensitized TiO_2. This is a viable strategy for photosensitizing an enzyme and achieving charge separation, whereas covalent attachment of a dye would be very challenging. The system produced CO at a turnover rate of $0.15\,s^{-1}$. The activity, however, decreased significantly during the first 4 hours of operation, indicating the fragility of the employed enzyme.

6.4.2 Enzyme Coupled to Photocatalysis

Product selectivity is a major issue with photocatalysis. This issue can be addressed by the use of enzymes in photocatalysis. Enzymes are biocatalysts renowned for their high efficiency and selectivity. In living cells, different enzymes often work together or in a specific order to catalyze multi-step biochemical reactions, playing crucial roles in the synthesis of natural products and metabolism [119]. Inspired by the biocatalytic reaction, enzymes including enzyme cascades were explored in vitro to complete the conversion of CO_2 to certain chemicals via a one-step or multi-step process.

Kuwabata et al. [120,121] demonstrated that CO_2 can be biocatalyzed into CH_3OH in a CO_2-saturated phosphate buffer solution, in which pyrroloquinoline quinone

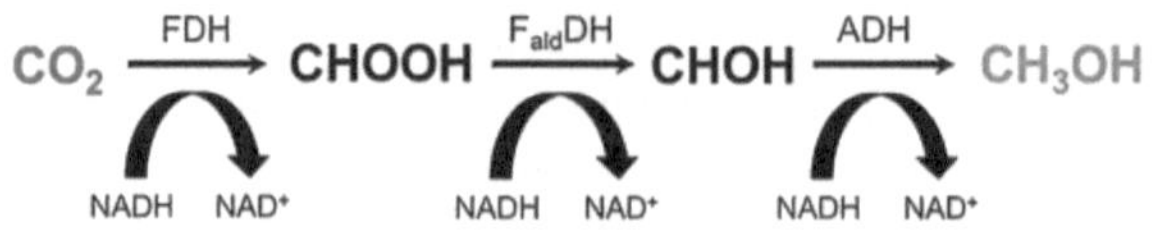

FIGURE 6.8 Biocatalytic transformation pathway of CO_2 to CH_3OH via stepwise reverse enzymatic catalysis by FDH, $F_{ald}DH$, and ADH [94].

(or methyl viologen) was used as an electron mediator, and formate dehydrogenase, formaldehyde dehydrogenase, and alcohol dehydrogenase were used as biocatalysts. The reduction of CO_2 to methanol was also presented by Obert and Dave [122] using three different dehydrogenases in three consequent reductions, in which reduced nicotinamide adenine dinucleotide (NADH) molecules were required at each step. As shown in Figure 6.8, such a multi-enzyme system was composed of three different dehydrogenases (FDH, F_{ald} DH, and ADH) that catalyze the conversion of CO_2 to CH_3OH in the presence of NADH. Each enzymatic step in the reduction cascade proceeds in the opposite direction of the natural (reversible) enzyme-catalyzed reaction and requires NADH as the electron donor for the reaction.

Singh et al. [123] observed that the formate dehydrogenase (ClFDH), formaldehyde dehydrogenase (BmFaldDH), and alcohol dehydrogenase (YADH) were from *Clostridium ljungdahlii*, *Burkholderia multivorans*, and *Saccharomyces cerevisiae*, respectively. A 500-fold increase in total turnover number can be obtained for the ClFDH–BmFaldDH–YADH cascade system compared to the *Candida boidinii* FDH–*Pseudomonas putida* FaldDH–YADH system. This implied that an enzyme cascade reaction can be optimized to get higher conversion efficiency. Furthermore, researchers noticed that the three dehydrogenases can not only be combined to convert CO_2 into methanol but also can be used individually to convert CO_2 to corresponding products, such as formate or formaldehyde. A barrier for the process is the expensive NAD(P)H. More efforts are being made to improve regeneration efficiency of NAD(P)H or improve its yield and reduce the production cost. The combination of enzymes and photocatalysts for CO_2 conversion has attracted increasing attention because it makes full use of the abundant energy supply of solar light and high specificity of enzyme catalysis [124–126]. These reactions can be conducted at mild conditions similar to the photosynthesis that occurs in plants or certain bacteria. The combination of electrodes with suitable biological enzymes can also minimize the requirement of high overpotentials to excite electro-catalytic reaction of CO_2 [127–129].

Photocatalysis coupled with enzymes provides a highly efficient, specific, and energy-saving strategy for CO_2 conversion and has attracted special attention by the researchers [130].

Yadav and coauthors [131] developed a graphene-based visible light active photocatalyst–FDH coupled system in which CO_2 was specifically converted to formic acid. In this study, chemically modified graphene (CCG) was covalently combined with the chromophore (multianthraquinone-substituted porphyrin, MAQSP) to form a new catalyst (CCGMAQSP). This catalyst enhanced the light-harvesting efficiency. The chromophore absorbed sunlight and acted as an electron donor.

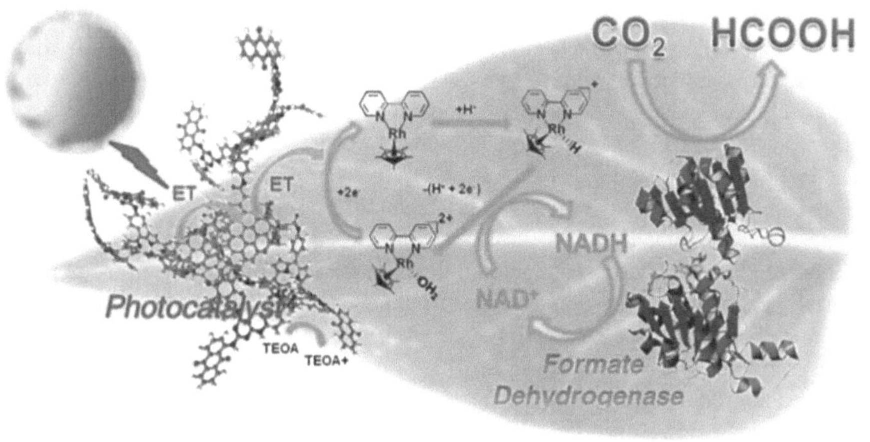

FIGURE 6.9 Graphene-based photocatalyst catalyzed artificial photosynthesis of formic acid from CO_2 under visible light [131,137].

The light-generated electrons are transferred to the organometallic rhodium complex through graphene (electron acceptor). The reduced rhodium complex further extracts H^+ from water. NAD^+ accepts electrons and H^+ to form NADH. Formate dehydrogenase converts CO_2 to formate in the presence of NADH, as shown in Figure 6.9.

In addition to rhodium complexes, methyl viologen (MV) was also used as an electron mediator. Kumar and coauthors [132] designed a photocatalyst of graphene oxide modified with cobalt metalized aminoporphyrin (GO-Co-ATPP) for conversion of CO_2 to formic acid under visible light.

Hybrid photocatalytic and enzymatic reaction which competes in the same environment has harmful effects on the stability and activity of enzymes. For example, FDH is divided into two types according to cofactor requirements: NADH-dependent FDH and metal-dependent FDH.

Although metal-containing FDHs have a higher catalytic activity for CO_2 reduction, these NADH-independent FDHs contain extremely unstable oxygen components, such as metal ions (tungsten or molybdenum), iron–sulfur clusters, and selenocysteine. The oxidation reaction of water may occur during photocatalysis to generate oxygen, which will affect the enzyme catalytic activity in the system, affect the stability of FDH, and then affect the final product of formic acid. The low compatibility of the photocatalysis and the biocatalysis in the system hindered its development.

In natural photosynthesis, thylakoids in chloroplast are employed to couple the photoreaction and the biological reaction system by which the enzymatic reaction is separated from the water oxidation reaction to protect enzymes from inactivation. In order to achieve cooperation between photocatalysis and biocatalysis and improve compatibility, Zhang et al. [133] developed an artificial thylakoid by decorating the inner wall of protamine–titania (PTi) microcapsules with cadmium sulfide quantum dots (CdS QDs), and coupled with biocatalysis to form an artificial photosynthesis system. Cds QDs absorb visible light and generate electrons and holes. The electrons are transferred to the outer surface of the capsule through the heterostructure of Cds QDs and amorphous TiO_2. Through the intermediate transfer of the rhodium complex, formate dehydrogenase converts CO_2 to formic acid in the presence of NADH. The size-selective capsule wall separates photocatalytic oxidation and enzymatic reduction of CO_2, thereby protecting the enzyme from inactivation that usually caused by photogenerated holes and active oxygen.

Most enzymes are powdered reagents, which makes them difficult to separate from the substrate and cannot be recycled. In order to address this issue, enzymes are often immobilized. For this purpose, zeolite imidazolate framework (ZIF) is a type of MOF material that possesses well-defined pore structure, excellent chemical–thermal stability, extremely high surface area, and other excellent properties [134]. Moreover, ZIF is easy to prepare and has little effect on enzyme activity because the preparation is usually conducted at mild conditions. It has become one of the common methods for enzyme immobilization [135]. Zhou et al. [136] combined ZIF and TCPP to construct a photocatalytic multi-enzyme cascade biomimetic carbon sequestration system (TCPP&FF@ZIF-8 (FF=FateDH and FaldDH)). TCPP was used as the photocatalyst and ZIF-8 as the multi-enzyme immobilized carrier for FateDH and FaldDH. The catalytic system was then used to absorb CO_2 and transform it to chemicals. The study showed that the system had excellent structural stability, light stability, and cycling stability.

Using different photocatalysts, enzymes, and cofactors, various products and yields can be obtained. Table 6.2 provides a simple comparison of the performance of different coupled photocatalytic/enzymatic CO_2 reduction systems in recent years.

TABLE 6.2
Performance Comparison of Different Coupled Photocatalytic/Enzymatic CO_2 Reduction Systems [137]

Photocatalyst	Enzyme	Cofactors	Efficiency[a]
CCG-IP	FateDH, FaldDH, ADH	NADH + $[Cp^*Rh(bpy)H_2O]^{2+}$Rh + TEOA	CH_3OH, 11.21 μM after 1 hour
CAN	FateDH, FaldDH, YADH	NADH + $[Cp^*Rh(bpy)H_2O]^{2+}$Rh + TEOA	CH_3OH, 0.21 mM min^{-1}
H_2TPPS	FDH, AldDH, ADH	MV^{2+}	CH_3OH, 6.8 μM after 100 minutes
C_{60} polymer film	FDH	NADH + TEOA	HCOOH, 239.46 μM after 2 hours
TiO_2	FDH	NADH	HCOOH, 1.634 mM after 4.5 hours
C_3N_4 (TPE-C_3N_4)	MAF-7@FDH	NADH + Rh + TEOA	HCOOH, 16.75 mM after 9 hours
CCGCMAQSP	FateDH, FaldDH, ADH	NADH + $[Cp^*Rh(bpy)H_2O]^{2+}$ Rh + TEOA	CH_3OH, 110.55 μM after 2 hours
CdS QDs+PTi	ClFDH	NADH + Rh	HCOOH, 1,500 μM h^{-1}

[a] The efficiency includes the products and rates.

The combination of photocatalysis and biocatalysis showed higher efficiency in CO_2 reduction than that of photocatalysis. The prominent problem in this approach is the interference between the photocatalysis and biocatalysis, resulting in corrosion of the photocatalyst and inactivation of the enzyme. However, the enzyme used in the biocatalysis is reversible and it is easy to perform the reverse reaction and charge reorganization [138]. To address this issue, as shown later, researchers adopted an electric field on the basis of photo-enzymatic catalysis, which can effectively promote charge separation and improve the conversion efficiency of CO_2.

6.4.3 Photoelectrocatalysis

The thermodynamic potentials for various CO_2 reduction products can be seen in Eqs. (6.6)–(6.10) (pH 7 in aqueous solution versus a normal hydrogen electrode (NHE), 25°C, 1 atm gas pressure, and 1 M for other solutes) [13]:

$$CO_2 + 2H^+ + 2e^- \rightarrow CO + H_2O \qquad E^\circ = -0.53\,V \tag{6.6}$$

$$CO_2 + 2H^+ + 2e^- \rightarrow HCO_2H \qquad E^\circ = -0.61\,V \tag{6.7}$$

$$CO_2 + 6H^+ + 6e^- \rightarrow CH_3OH + H_2O \qquad E^\circ = -0.38\,V \tag{6.8}$$

$$CO_2 + 8H^+ + 8e^- \rightarrow CH_4 + 2H_2O \qquad E^\circ = -0.24\,V \tag{6.9}$$

$$CO + e^- \rightarrow CO^{\bullet-} \qquad E^\circ = -1.90\,V \tag{6.10}$$

Although CO_2 has been shown to be reduced directly on metal surfaces, the overpotentials are either exceedingly high or the metal surfaces become poisoned and deactivated by the reduction products. In addition to thermodynamic considerations, there are also considerable kinetic challenges to the conversion of CO_2 to more complex products. Typically, multiple proton-coupled electron transfer (PCET) steps must be orchestrated with their own associated activation energies to overcome kinetic barriers to the forward reaction. A great deal of success has been achieved in the reduction of CO_2 to CO and formate. However, the multiple electron and proton transfers necessary to produce more useful products such as methane or methanol have only been demonstrated with low efficiency. To achieve success at efficient production of a CO_2 reduction product that can serve as a liquid fuel directly (i.e., methanol) would be a considerable challenge.

In a heterogeneous system, p-type semiconductor/liquid junctions are extensively studied as PV devices. The p-type semiconducting electrodes can act as photocathodes for photo-assisted CO_2 reduction. **There are** four different schemes of photo-assisted reduction of CO_2 using a semiconducting photocathode: (a) direct heterogeneous CO_2 reduction by a biased semiconductor photocathode, (b) heterogeneous CO_2 reduction by metal particles on a biased semiconductor photocathode, (c) homogeneous CO_2 reduction by a molecular catalyst through a semiconductor/molecular catalyst junction, and (d) heterogeneous CO_2 reduction by a molecular catalyst attached to the semiconductor photocathode surface [7,13]. Band energy diagrams for selected semiconductors are illustrated in Figure 6.10 [7].

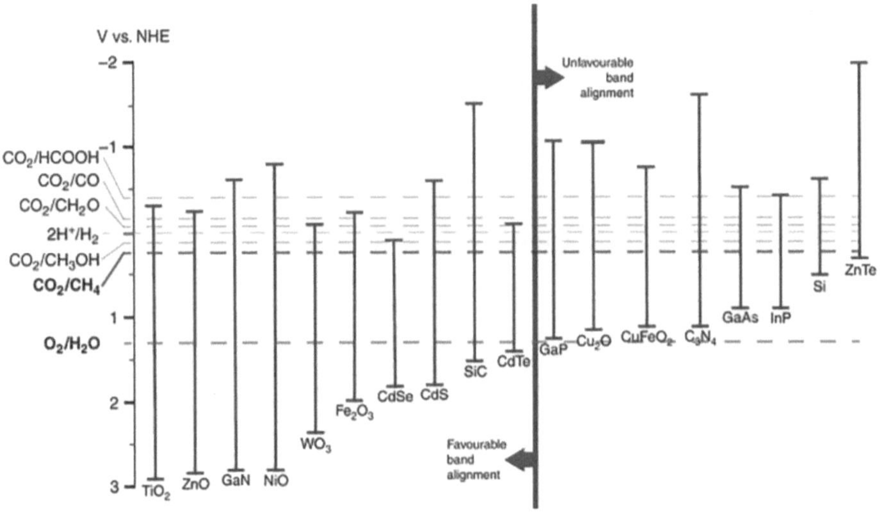

FIGURE 6.10 Band energy diagram of selected semiconductors. These materials are commonly used for photoelectrochemical water splitting and CO_2 reduction. Redox potentials of key CO_2 reduction reactions are also included. In principle, water splitting and CO_2 reduction can take place on the same semiconductor material if the conduction band energy level is aligned with, or more negative than, the energy level of the targeted CO_2 methanation reaction ($-0.24\ V_{NHE}$) and the valence band energy level is aligned with, or more positive than, the oxygen evolution reaction energy level ($1.23\ V_{NHE}$). This is indicated by the position of each material relative to the vertical bar dividing the figure [7].

Photoelectrocatalysis, which combines the advantages of photocatalysis and electrocatalysis, is considered to be an ideal strategy for the selective conversion of CO_2 into gaseous (such as CO, methane, etc.) and liquid products (such as formic acid, methanol, ethanol, etc.) under sunlight irradiation, and has, therefore, attracted great attention. A catalyst may be combined with a light-absorbing electrode, or a dye and catalyst are co-absorbed on an electrode. In this case, the driving force for the chemical reaction is obtained through light absorption and the electrons are delivered from an anode, where a sustainable oxidation reaction can occur, such as water oxidation. Thus, no (or only a small) electrochemical potential is required to perform the reduction of CO_2 at the cathode.

Photoelectrocatalysis makes the best use of solar energy to produce photoelectrons. The photogenerated electrons are transferred to the electrode surface under the action of an applied electric field, and finally obtained by CO_2 for catalytic reduction. The applied electric field can effectively facilitate charge separation in the photocatalytic process, promote electron migration, and significantly improve the intrinsic activity and energy efficiency of CO_2 molecules. The efficient utilization of solar energy in photoelectrocatalysis can effectively overcome the problem of high energy consumption in the electrocatalysis of CO_2.

Photoelectrode-based CO_2 reduction integrate the advantages of photosynthesis and electro-catalysis [139]. The process can be carried out in a number of different ways; photocathode–dark anode, photoanode–dark cathode, and photocathode–photoanode. The fact that each photoelectrode can consist of multiple absorber layers to better cover the solar spectrum complicates the process further. As shown later in Section 6.5, a sophisticated variant is the "buried junction" concept, where a solar cell is covered by one or more catalyst(s) (and possibly a protecting layer) and this whole assembly acts as a photoelectrode [140,141]. While electro-catalytic reduction of CO_2 has yet not reached a commercial scale, it is a very versatile process using different operating conditions and electro-catalysts with electricity as the main source of energy [142–144]. Experiments with different metal electrodes [145] have resulted in different products such as CO, CH_4, formic acid, formaldehyde, and methanol under different operating conditions. The standard potential required for all these products is similar to hydrogen evolution standard potential [146]. The hydrogen evolution reaction (HER) is very important during CO_2 electrocatalyst reduction in which H_2O is typically present as an electrolyte (and proton source). In order to make the process more effective, metal catalyst should be chosen such that desired selectivity of CO_2 reduction can be achieved at low overpotentials and high rates without reducing water simultaneously [147].

The direct conversion of CO_2 to methanol is important for two reasons; first, methanol can be used as raw materials for downstream operations to generate higher hydrocarbons, alcohols, acids, and fuels; second, methanol can provide required energy for the process. As shown in Table 6.3, different electrodes can be used to achieve methanol directly from CO_2 [147]. Canfield and Frese [206] proved that some semiconductors such as *n*-GaAs, *p*-InP, and *p*-GaAs have the ability to produce methanol directly from CO_2 although at extremely low current densities and faradaic efficiencies (FEs). Seshadri et al. [148] found that the pyridinium ion is a novel homogeneous electrocatalyst for CO_2 reduction to methanol at low overpotential. Recently, Pyridine has been widely explored in which it is used to

TABLE 6.3
CO_2 Electrochemical Reduction to Methanol [142]

Electrode	Type of Electrode	E vs. NHE (V)	Current Density (mA/cm²)	Faradaic Efficiency (%)	Electrolyte
p-InP	Semiconductor	−1.06	0.06	0.8	Sat. Na_2SO_4
n-GaAs			0.16	1.0	
p-GaAs			0.08	0.52	
CuO	Metal oxide	−1.3	6.9	28	0.5 M $KHCO_3$
RuO_2/TiO_2 nanotubes		−0.6	1	60	0.5 M $NaHCO_3$
Pt–Ru/C	Alloy	−0.06	0.4	7.5	Flow cell
n-GaP	Homogeneous catalyst	−0.06	0.27	90	10 mM pyridine at pH = 5.2
Pd		−0.51	0.04	30	0.5 M $NaClO_4$ with pyridine

act as cocatalyst to form the active pyridinium species in situ [142,149]. Generally, the one-electron reduction products of CO_2 show lower current density than the two-electron reduction products such as CO. Significant efforts are being made to increase current density and faradaic efficiency for one-step CO_2 conversion to methanol process.

Popić et al. [150] indicated that in case of CO_2 reduction on Ru modified by Cu and Cd adatoms, the production of methanol was dependent on the presence of adatoms at the surface of ruthenium. RuO_2 is a promising material to be used as an electrode for CO_2 reduction to methanol due to its high electrochemical stability and electrical conductivity. Qu et al. [151] prepared RuO_2/TiO_2 nanoparticles (NPs) and nanotubes (NTs) composite electrodes by loading of RuO_2 on TiO_2 nanoparticles and nanotubes, respectively. The results showed that the current efficiency of producing methanol from CO_2 was up to 60.5% on the RuO_2/TiO_2 NTs modified Pt electrode. This was significantly higher than the ones obtained for RuO_2 and RuO_2/TiO_2 NPs composite electrodes. The study concluded that nanotube structure gives better selectivity and efficiency of CO_2 electrochemical reduction process.

Hybrid system of a semiconductor light harvester and a complex of metal cocatalyst has received significant research interest. In this system, the water is considered the main source of electron donors and protons for the reduction of CO_2 at the surface of cathode. As shown in Figure 6.11, Zhao et al. [152] studied the full cell of photocathode with InP/Ru-complexes that was coupled with a TiO_2/Pt-based photoanode. In this full cell, in order to avoid the formate re-oxidation at the surface of photoanode, the proton exchange membrane was used as a separator. On the other hand, as shown in Figure 6.12, Arai et al. [153] constructed a wireless full cell for photoelectrochemical CO_2 reduction in which the system consists of the InP/Ru-complex as a hybrid photocathode and a photoanode of $SrTiO_3$. In this system, the redox reactions of CO_2 and H_2O occur via sunlight irradiation without applying any bias. The obtained results showed that the conversion efficiency from solar to chemical energy in these two full cells was 0.03% and 0.14% for TiO_2–InP/[RuCP] and $SrTiO_3$–InP/[RuCP], respectively. Barton et al. [141] successfully reduced CO_2 to methanol by using catalyzed p-GaP-based photoelectrochemical

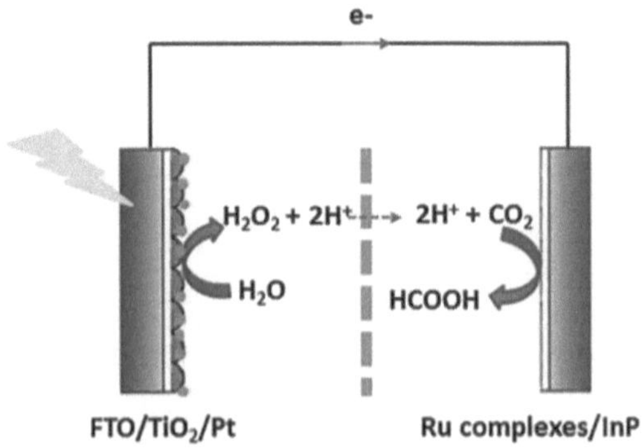

FIGURE 6.11 The two-compartment photoelectrochemical cell for CO_2 reduction [142].

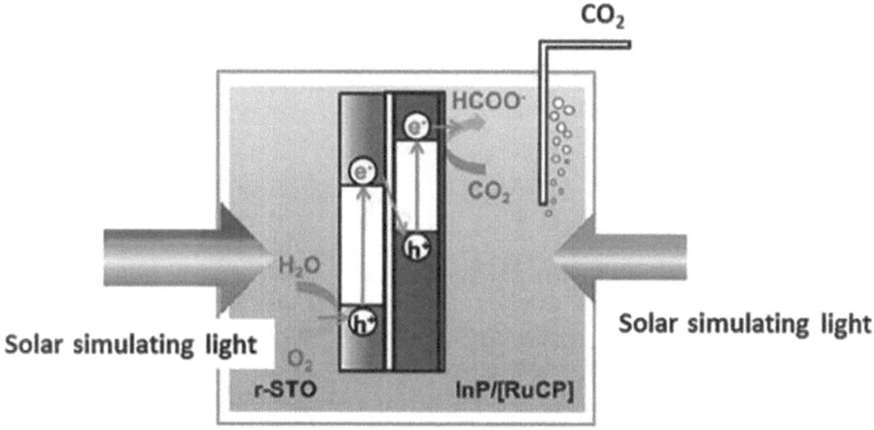

FIGURE 6.12 The one-compartment photoelectrochemical cell for CO_2 reduction [142].

cell in a process called chemical carbon mitigation. Chemical carbon mitigation term describes the photo-induced CO_2 conversion to methanol without the use of additional CO_2 generating power source. The methanol selectivity and CO_2 conversion in this study were found to be 100% and 95%, respectively. In order to promote rapid charge transfer and improve the performance of photoelectrocatalysis, Ding and coauthors [154] patterned a photocathode through photolithography to expose a third of the surface, which is an effective and robust Si–Bi interface formed by Bi^{3+}-assisted chemical etching of Si wafers and completed the reduction of CO_2. This method increased the current density and facilitated the reduction of CO_2 based on high product selectivity. Castro et al. [155] loaded different amounts of TiO_2 on the photoanode using a Cu plate as the photocathode to build a photoelectric chemical device, and combined this with an electrochemical filter-press cell. This device was employed to continuously convert CO_2 into alcohol with reducing energy consumption due to less external energy demand. Comparing the alcohol produced under different conditions, the TiO_2 photoanode system exhibited enhanced alcohol production and reduced energy consumption under ultraviolet light irradiation.

Different photocathodes have different light absorption capabilities, which essentially depend on the optical characteristics of the semiconductor. Table 6.4 lists and compares the performance of different photoelectrochemical systems of CO_2 reduction from the literature. The efficiency of CO_2 conversion is a criterion of photoelectric conversion efficiency.

Photoelectrocatalytic CO_2 reduction can also be achieved through layer-by-layer assembly: integrating a CO_2 reduction catalyst on top of a dye, which is adsorbed on a p-type semiconductor is a viable strategy. In this scenario, the dye layer absorbs visible light and the electrons are transferred to the catalyst. The oxidized dye is subsequently re-generated by hole transfer to the semiconductor, resulting in a cathodic current from CO_2 reduction. The dye layer may be molecular or a semiconducting material.

TABLE 6.4
Performance Comparison of Different Photoelectrochemical CO_2 Reduction Systems from Recent Literature [137]

Photocathode[a]	Condition[b]	Efficiency[c]
p+-n-n+-Si/TiO_2 + Cu/Ag	100 mW cm^2, 0.1 M $CsHCO_3$	C_2H_4, 10%–25%, 8 mA cm^2 at 0.4 V vs reversible hydrogen electrode (RHE) for 20 days
p-Si NWs + Sn	100 mW cm^2, 0.1 M $KHCO_3$	HCOOH, 88%, 18.9 μmol $h^{-1}cm^2$, 0.875 V vs RHE for 3 hours
CuO + Cu_2O	70 mW cm^2, 0.1 M $NaHCO_3$	CH_3OH, 95%, 85 mM at 0.2 V vs standard hydrogen electrode (SHE) after 1.5 hours
Si/GaN-NPhN4-Ru(CP)$^2{}_2{}^+$ RuCt	100 mW cm^2, 0.05 M $NaHCO_3$	HCOOH, 35%–64%, 1.1 mA cm^2 at 0.25 V vs RHE for 20 hours
p-n+-Si + SnO_2 NW	100 mW cm^2, 0.1 M $KHCO_3$	HCOOH, 59.2%, 18 mA cm^2 at 0.4 V vs RHE for 3 hours
Co_3O_4/CA + Ru(bpy)$_2$dppz	9 mW cm^2, 0.1 M $NaHCO_3$	HCOOH, 86%, 110 μmol $h^{-1}cm^2$ at 0.60 V vs normal hydrogen electrode (NHE) for 8 hours
FTO/TiO_2/Cu_2O + Ru-BNAH	100 mW cm^2, 0.1 M KCl	HCOOH, NA, 409.5 μmol at 0.9 V vs NHE after 8 hours
p-Si + Bi	50 mW cm^2, 0.5 M $KHCO_3$	HCOOH, 70%–95%, ~4 mA cm^2 at 0.32 V vs RHE for 7 hours
Fe_2O_3 NTs + Cu_2O	100 mW cm^2, 0.1 M $KHCO_3$	CH_3OH, 93%, 6 hours, 4.94 mmol L^1cm^2 at 1.3 V vs SCE for 6 hours
FTO/$CuFeO_2$ + CuO	100 mW cm^2, 0.1 M $NaHCO_3$	CH_3COOH, 80%, 142 μM at 0.4 V vs Ag/AgCl after 2 hours

[a] The configuration is described as "semiconductor + cocatalyst".
[b] The reaction conditions for photoelectrochemical (PEC) measurements include the light intensity of solar simulator and the electrolyte.
[c] The PEC efficiency parameters include the product, faradaic efficiency/photocurrent density/production rate or yield/stability at a certain working potential.

6.4.4 Electrolysis with Immobilized Molecular Catalysts

A number of studies reported in the literature have immobilized molecular catalysts on the electrodes to improve its CO_2 reduction ability. Meshitsuka et al. [156] immobilized Co and Ni metal phthalocyanines in benzene, on graphite electrode from an aqueous electrolyte solution. Phthalocyanines have poor solubility in common solvents and their planar aromatic structure allows for π–π stacking with graphite. The Co and Ni phthalocyanines display good electrocatalytic currents under CO_2. Lieber and Lewis [157] immobilized Co phthalocyanine onto a carbon cloth from THF. The catalyst showed excellent selectivity for the production of CO at an overpotential of 300 mV in pH 5 citrate buffer. The system yielded a TONCO = 105 at a turnover frequency (TOFCO) of 100 s^{-1}, while homogeneous Co phthalocyanine produced only TONCO = 4. Enyo et al. [158] functionalized a glassy carbon electrode with pyridine. The anchored pyridine was coordinated to the axial position of Co tetraphenyl porphyrin (CoTPP). The electrode displayed excellent selectivity and stability with FECO = 92% and TONCO = 107. H_2 generation was observed in the absence of CO_2.

Abruña et al. [159] electropolymerized a vinyl-functionalized Co tpy complex on Pt electrodes and in dimethylformamide (DMF) and it showed catalytic currents at 860 mV less negative potential than homogeneous $[Co(tpy)2]^{2+}$. Kaneko et al. [160] studied $[Co(tpy)2]^{2+}$ immobilized in a Nafion membrane on carbon electrodes in aqueous electrolyte solution. At −0.85 V vs NHE in pH 7 phosphate solution, FEHCOOH = 51% and FEH_2 = 13% was observed with a low TONHCOOH = 11. Fontecave et al. [161] functionalized a carbon electrode with terpyridine through diazonium coupling. The terpyridine was subsequently metallated with Co upon immersion in a solution of $CoCl_2$ in DMF. Electrolysis in DMF under a CO_2 atmosphere at a very negative potential (−1.73 V vs NHE) gave TONCO = 70.

Abruña and colleagues [162] reported a polymer film of [ReCl(4-vinyl-4′-methyl-2,2′-bipyridine)$(CO)_3$] on two types of light-absorbing electrodes, p-Si and polycrystalline p-WSe_2. Immobilization was achieved by cycling the electrode potential in an acetonitrile electrolyte solution of the Re complex. Sato, Arai, and colleagues [163] used a mixture of Ru complexes immobilized on an InP photoelectrode was coupled to a Pt–TiO_2 photoanode capable of water oxidation. The TiO_2 anode absorbs UV light and the InP harvests the longer wavelength visible light. The system was able to photo-catalytically reduce CO_2 to formate ($TONHCOO^-$ > 17) with no external bias and with electrons derived from photocatalytic water oxidation at the anode. The faradaic efficiency for formate was 70% with some CO and H_2 produced as by-products. Inoue and co-workers [164] reported photoelectron catalytic CO_2 reduction with a molecular dyad. A zinc porphyrin dye was covalently attached to a Re catalyst. A carboxylate group on the zinc porphyrin anchored the dyad onto p-type NiO electrodes. The faradaic efficiency for CO was 6.2% and TONCO = 10 over 50 hours. Co-adsorbing an additional zinc porphyrin dye, improved the TONCO by an order of magnitude but caused a four-fold drop in the faradaic efficiency for CO.

Lehn et al. [165] reported that [ReX(bpy)$(CO)_3$] (bpy = 2,2′-bipyridine; X = Cl^- or Br^-) complexes are selective photocatalysts for CO_2 reduction to CO. Subsequently [166], they demonstrated the electrocatalytic activity of the same complexes. Meyer et al. [167] reported on the heterogenization of this catalyst by electro-polymerizing [ReCl(CO)3(4-vinyl-4′- methyl-2,2′-bipyridine)] onto Pt electrodes. At an applied bias of −1.30 V vs NHE in CH_3CN, the Re-modified electrode showed good selectivity and electrocatalytic activity for the production of CO with FECO = 92% and TONCO = 516. The Re catalyst monomer showed a TONCO = 30 in homogeneous catalysis under comparable conditions. Kaneko et al. [168] studied immobilized [ReBr(bpy)$(CO)_3$] on graphite electrodes by casting a DMF/alcohol solution of the catalyst mixed with Nafion onto the surface of basal plane graphite. Electrolysis in pH 7 phosphate solution showed CO_2 reduction to a range of products, and the product distribution was dependent on the applied potential. FECO = 29, FEHCOOH = 48 with TONCO = 198, TONHCOOH = 148 were obtained under optimized conditions (between −1.05 and

−1.35 V vs NHE) with the major side product being H_2. Homogeneous [ReBr(bpy)$(CO)_3$] was less active, giving TONCO = 21 h^{-1} compared with 166 h^{-1} for the heterogenized catalyst.

A study [169] also reported that Mn analogs of the well-established [ReX(bpy)$(CO)_3$] (L = Br^-, CH_3CN) catalysts are active for electrocatalytic CO_2 reduction. This is important from a scale-up perspective as Mn is 1.3 million times more abundant than Re in the Earth's crust [170]. The Mn catalyst operates at a lower overpotential relative to the Re analog [169]. Cowan et al. [171] immobilized [MnBr(bpy)$(CO)_3$] in a Nafion membrane onto a glassy carbon electrode and produced CO and H_2 in a ratio of 1:2 at −1.17 V vs NHE in pH 7 phosphate solution. Addition of multi-walled carbon nanotubes increased the number of electro-active molecules in the membrane from 0.25% to 11%. The Nafion immobilized system produced CO at 100 mV less negative potentials than for homogeneous [MnBr(bpy)$(CO)_3$] (in CH_3CN/H_2O). Other CO_2 reduction systems include those using enzymes. Protein film electrochemical studies of a molybdenum- and tungsten-containing formate dehydrogenase have shown that this class of enzyme can reversibly inter-convert CO_2 and formate at the thermodynamic potential (i.e., in either direction with minimal overpotential) and with quantitative FE and selectivity [172]. Reversibility has also been demonstrated with a nickel-/iron-containing carbon monoxide dehydrogenase [173].

All these reported studies indicate that immobilization of molecular catalysts on electrodes work better than electrodes in the presence of homogeneous molecular catalysts. Significant research is being carried out to improve electrochemical performance with heterogenized molecular catalysts.

6.4.5 Photoelectrocatalysis with Biocatalysts (PEC)

Photoelectrochemical (PEC) cell combines photocatalysis, electrocatalysis, and bio-catalysis and maintains the optimal conditions of enzymes and improve conversion efficiency [174,175]. Generally, conducting wire is used to ensure the oriented transfer of reducing equivalents (primarily electrons, H^+) from the photoelectrode to biocatalyst. As mentioned earlier, a combination of photocatalysis with biocatalysis creates a competing environment for both types of catalyst resulting in the degradation of overall performance of combined system. In order to address this issue, Nam et al. [176] designed a two compartments photoelectrochemical (PEC) cell. In this cell, photocatalytic and biocatalytic reactions are separated in the anode and cathode compartments, respectively. As shown in Figure 6.13, in this PEC cell, photocatalytic water oxidation is carried out in an anode compartment with cobalt phosphate (Co-Pi) deposited hematite (Fe_2O_3) photoanode and formate dehydrogenase for NADH regeneration and CO_2 reduction are carried out in a cathode compartment. The cocatalyst (Co-Pi) in the photoelectrode can reduce the activation energy quickly, improve the quantum efficiency by promoting charge separation, and consume the photogenerated charges in time to improve the stability of the photoelectrode [177]. The compartment on right in Figure 6.13 carries out enzymatic reaction and thus, oxidation reaction of water is separated from enzymatic reaction and thus allowing the formate dehydrogenase to work at its optimal pH conditions.

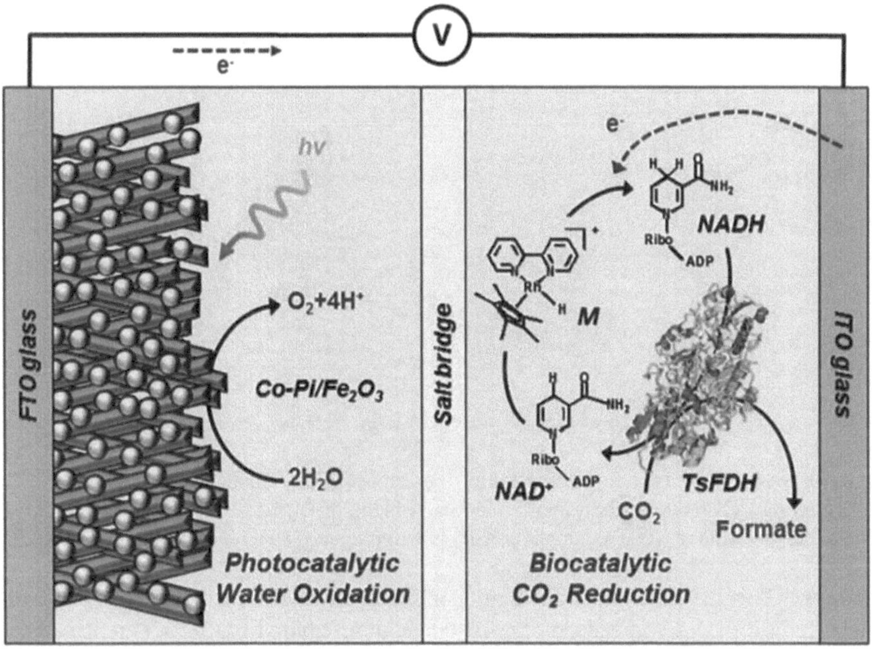

FIGURE 6.13 Schematic illustration of the biocatalytic PEC platform [120,137].

Generally, the performance of enzymes is affected by the reaction environment; particularly pH of the solution making the enzyme electrode unstable [178]. Eun-Gyu Choi et al. [179] studied the effect of pH on a coupled photoelectric–enzyme system. They found that RcFDH-driven CO_2 reduction was predominant at an acidic pH, whereas formate oxidation was favorable at basic conditions. By examining various pH values, they found that for their system, pH = 6.5 was optimum for CO_2 reduction to formic acid. For the stability and reusability of the enzyme, appropriate immobilization technique can be adopted. Lee et al. [128] reported that a tightly organized bio-photocathode (EC-PDA)-electrochemically synthesized polydopamine (PDA) film was copolymerized with FDH (E) and NADH (C) in which CO_2 can be reduced to formic acid with high selectivity. The PDA was chosen as the substrate for enzyme immobilization because of its excellent biocompatibility and charge transfer ability [180]. The PDA layer on the electrode can fulfill the requirement of electron transfer and enzyme stabilization and extend the service life of the enzyme [181]. A similar photoelectrochemical cell was constructed with the EC-incorporated PDA bioelectrode and cobalt phosphate/bismuth vanadate (CoPi/$BiVO_4$) photoanode in which the reduction of CO_2 can be achieved without external bias.

In another study, a PEC cell with Co-Pi/Fe_2O_3 photoanode and $BiFeO_3$ photocathode [182] drove surface charge accumulation and accelerated the transfer of electrons to the electrolyte, therefore resulting in an improvement in CO_2 conversion efficiency. The tandem PEC cell with an integrated enzyme cascade (TPIEC) system mimics the natural photosynthetic Z-scheme for the biocatalytic reduction of CO_2 to

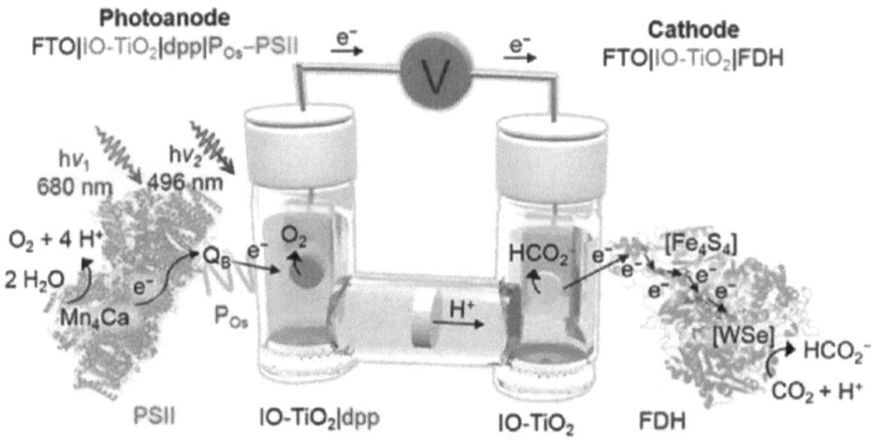

FIGURE 6.14 Schematic representation of the semi-artificial photosynthetic tandem PEC cell coupling CO_2 reduction to water oxidation. A blend of P_{Os} and PSII adsorbed on a dpp-sensitized photoanode (IO-TiO_2|dpp|P_{Os}-PSII) is wired to an IO-TiO_2|FDH cathode [137,185].

methanol. This device exhibited a significantly higher rate of methanol than those of other studies. In order to reduce cost, metal rhodium complex [183] was replaced by natural red as electron mediator and it carried out electron recycling between the electrode and NAD^+ [184]. Sokol et al. [185] adopted a semi-artificial design. As shown in Figure 6.14, in this design, a cathode containing formate dehydrogenase was connected to the photoanode containing photosynthetic water oxidase (photosystem II) to achieve the metabolic pathway of formic acid that is formed by light fixation of CO_2 in the absence of precious metal catalyst. This design demonstrated the feasibility of the non-metal catalysts for the conversion of CO_2 to formic acid and provides a novel method for CO_2 photoelectrochemical reduction. Table 6.5 lists the performance parameters of different photoelectrochemical/enzyme systems reported in the literature.

TABLE 6.5
Performance Comparison of different Coupled Photoelectrocatalytic/ Enzymatic CO_2 Reduction Systems [137]

Photoanode	Photocathode	Efficiency[a]
Co-Pi/Fe_2O_3	ITO/FDH	HCOOH, 6.4 μM h^{-1}
CoPi/$BiVO_4$	EC-PDA	HCOOH, FE: 99.18%
Co-Pi/oFe_2O_3	$BiFeO_3$-CcFDH/PcFaldDH/YADH	CH_3OH, 220 μM h^{-1}
FTO/IO-TiO_2/dPP/P_{Os}-PSII	FTO/IO-TiO_2/FDH	HCOOH, 0.185 μM cm^{-2}
FTO/FeOOH/$BiVO_4$	FTO/3D TiN-ClFDH	HCOOH, 0.78 μM h^{-1}, FE: 77.3%
TK/TiO_2	FDH-$CH_3V(CH_2)_9COOH$	HCOOH, 30.0 nM after 3 hours
Plain graphite rod	Pt-FDH	HCOOH, 15.49 μM mg Enzyme^{-1} min^{-1}

[a] The efficiency includes the product and rates.

In PEC cell, besides the use of noble metal mediator, the use of NADH is expensive. NADH easily forms enzymatically inactive dimers, NAD_2, resulting in the reduction of enzyme activity [186]. Over the last few decades [187–190], chemical, photochemical, and electrochemical regeneration of NADH has been developed. Electrochemically mediated electron injection into the enzyme not only bypasses the requirement of NADH but also simplifies product separation [191]. Amao and Shuto [192] demonstrated the use of viologen-modified FDH immobilized on an ITO electrode to electrochemically convert CO_2 to formic acid. This study showed evidence of electrons directly transferring to FDH active sites from electrodes and represents a new strategy for CO_2 reduction in the absence of NADH. Researchers have tried to use cheaper electron mediators, such as methyl viologen (MV^{2+}), instead of NADH. Miyatani et al. [193] developed a system using MV which combined the synthesis of formic acid from CO_2 (bicarbonate ion) with FDH and MV^{2+}, and photoreduction with ZnTMPyP as photosensor and TEOA as electron donor. Formic acid was successfully formed from bicarbonate ions and FDH in the absence of NADH. The study also demonstrated that the oxidation of formate to carbon dioxide does not occur readily with MV^{2+}. Therefore, the overall yield of formate would be preserved without loss from reoxidation [194]. Ishibashi et al. [195] showed that a photoelectrochemical system composed of TiO_2 nanoparticles as photocatalyst, MV^{2+} as electron carrier, and FDH as biocatalyst, resulted in successful reduction of CO_2 to formic acid without sacrificing reagent, external bias, and NADH.

6.5 PV/EC (OR PV+EC) CONCEPT

The transformation of carbon dioxide is an energy-intensive process that must involve inexpensive sources of energy and high energy efficiency while maintaining the lowest possible cost. Artificial photosynthetic systems, which are technological devices that utilize sunlight as a source of energy and water as a source of electrons to convert CO_2 into energy-dense organic compounds (fuels or other carbon-based feedstocks for the chemical industry), are attractive in that context. This can be achieved using a photovoltaic (PV) cell to provide photogenerated electrons and holes to an electrochemical cell (EC) for water oxidation at the anode and CO_2 reduction at the cathode. Only a few examples of such PV/EC systems have been reported, predominantly leading to high CO or formate selectivity [196–198], and only two such systems have led to high hydrocarbons or alcohols selectivity [199,200]. Among them, the most efficient PV/EC systems have been based on costly components: A record 13% solar-to-CO conversion was achieved using a GaInP/GaInAs/Ge photovoltaic cell [198], while a 3% solar- to-hydrocarbon efficiency was reported with an iridium oxide anode coupled to a four-terminal III-V/Si tandem cell [200].

The design of a cheap and high-efficiency PV/EC system indeed requires an integrative approach that takes the four following factors into account: (a) the development of robust CO_2 reduction (CO_2R) and oxygen evolution reaction (OER) electrocatalysts with low overpotentials and based on earth-abundant metals; (b) their operation in moderate pH conditions, limiting corrosion issues and electrolyte consumption and allowing for long-term operation; (c) their integration into an electrolyzer especially designed to maximize their efficiency and limit electrical energy losses; and (d) the final coupling of the electrolyzer to a low-cost PV system. This approach might result in lower current densities than currently reported using catalysts operating in highly basic media [201–203]. However, such current densities are sufficient to match the current densities provided by state-of-the-art perovskite PV cells [204].

While, there are several examples where PV-powered commercial electrolyzers have been used for hydrogen generation, very few of these setups exist for CO_2 reduction to an energy-dense product. The idea to power an electrolyzer by a PV device was first proposed by the patent of Smotkin et al. [205] for the water-splitting electrolyzer. Subsequently, a PV-powered electrolyzer has also been used to form syngas (CO and H_2) from CO_2 and water. For these different systems, the expressions for solar to chemical energy conversion efficiency are complicated because multiple products can form at the cathode and anode. In most cases, CO_2 photoelectrochemical reduction on photocathodes happens at high overpotentials, which further complicates this calculation.

Based on the successful H_2 evolution studies employing PV/EC systems, this configuration has attracted much attention also for CO_2 reduction. There are two types of PV/EC systems: (a) integrated systems, where the two functions are incorporated in the same unit, and (b) coupled ones, where regular PV panels are DC–DC connected to regular electrochemical cells. Some representative reported literature studies of CO_2 conversion on PV/EC are illustrated in Table 6.6.

In a study by Schreier et al. [198], atomic layer deposition of SnO_2 was performed on CuO nanowires for narrowing the product distribution of CO_2 reduction, thus yielding predominantly CO. The prepared catalyst was employed as both the cathode and anode for complete CO_2 electrolysis. In the resulting device, the electrodes were separated by a bipolar membrane, and a GaInP/GaInAs/Ge photovoltaic cell was used to drive the solar-driven splitting of CO_2 into CO and oxygen with a solar-to-CO efficiency of 13.4% and overall SFE of 14.4%. The operating current density, selectivity toward CO, and solar-to-CO efficiency remained almost stable during 5 hours of electrolysis. In another study, a CO_2 reduction system was integrated, achieving an average solar-to-CO efficiency of 13.9% and SFE (solar to fuel efficiency) of 15.6% with no appreciable performance degradation in 19 hours of operation [206].

Wang et al. [207] developed a two-step, redox-medium-assisted solar-driven CO_2 electroreduction system by incorporating a Zn/Zn(II) redox mediator that acted as the electron carrier during the photosynthesis. In the light reaction, the solar-driven oxygen evolution and Zn(II) reduction stored electrons in the Zn/Zn(II) medium. The carbon fixation released the stored electrons and led to an unassisted electrochemical reduction of CO_2. The energy diagram of each reaction part is shown in Figure 6.15a [207]. This redox-medium-assisted system enabled a solar-to-CO conversion efficiency of 15.6% under 1 Sun illumination. In addition, in a very recent study, solar-driven CO_2 reduction to CO with 19% solar-to-CO efficiency under 1 Sun illumination in a gas diffusion electrode (GDE) flow cell has been reported [208]. The use of a reverse assembled GDE (Figure 6.15b) prevented transition from a wetted to a flooded catalyst bed and allowed the device to operate stably for >150 hours with no loss in efficiency. The FE_{CO} and SFE over a 20 hours duration are shown in Figure 6.15c.

TABLE 6.6
Representative Studies on PV+EC CO_2 Conversion [6]

Light Absorber	Anode	Cathode	Illumination Conditions	Electrolyte	Products	Operation Point	FE (%)	SFE (%)	Maximum Test Time (Hour)
GaInP/GaInAs/Ge	CuO/SnO_2	CuO/SnO_2	1 Sun	Anolyte: CsOH, catholyte: $CsHCO_3$	CO	2.38–0.55 V vs RHE, 11.6 mA/cm^2	81	14.4	5
GaInP/GaInAs/Ge	Sr_2GaCoO_5	Ag	1 Sun	$NaNO_3$	CO	2.26 V, 3.54 mA/cm^2	85–89	15.6	19
Triple-junction GaAs (InGaP/GaAs/Ge) solar cell	Zn	Au	1 Sun	Anolyte: KOH with zinc acetate, catholyte: $KHCO_3$	CO	1.96 V, 10 mA/cm^2	~92	~16.9	24
GaInP/GaInAs/Ge	Ni or Pt	Ag/GDE	1 Sun	Anolyte: KOH; catholyte: KOH	CO	2.23–0.6 V vs RHE, 14.4 mA/cm^2	~100	19.1	150

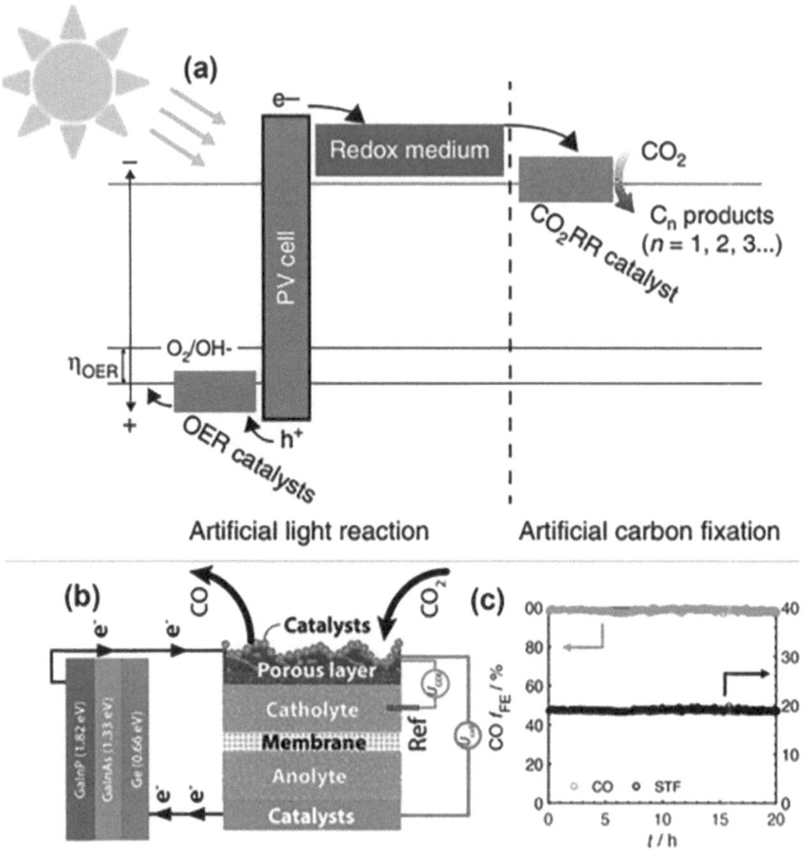

FIGURE 6.15 (a) Energy diagram of each part in a redox-medium-assisted system. (b) Illustration of a wire connection between the triple-junction cell and GDE cell and (c) CO faradaic efficiency and solar-to-fuel efficiency over 20 hours duration. [6]

In a practical PV/EC device, the PV cells with a much larger area will be required compared to the area of the EC component, because of the relatively low energy density of solar irradiation. Importantly, an EC component can be cost-effective when it is smaller and runs at higher current densities (such as those with GDE cells). There are some confusing reporting on current density and SFE, when PV cells with a much larger area are integrated with electrodes with much smaller area and the current density was normalized with smaller electrode areas leading to large current densities or SFEs. It is also possible to use large area inexpensive electrodes with smaller expensive PV cells. In these cases, the overpotential can be very small, because of operating at low current density. Finally, in some cases the SFE was obtained under irradiance significantly lower than 1 Sun, resulting in higher SFE values, which is unlikely to scale with the light intensity.

The above studies all reported CO as the main product with high SFEs (greater than 10%). As shown in Figure 6.16, the product distribution in these studies showed that CO is the most common product, accounting for 65.4% alone followed by

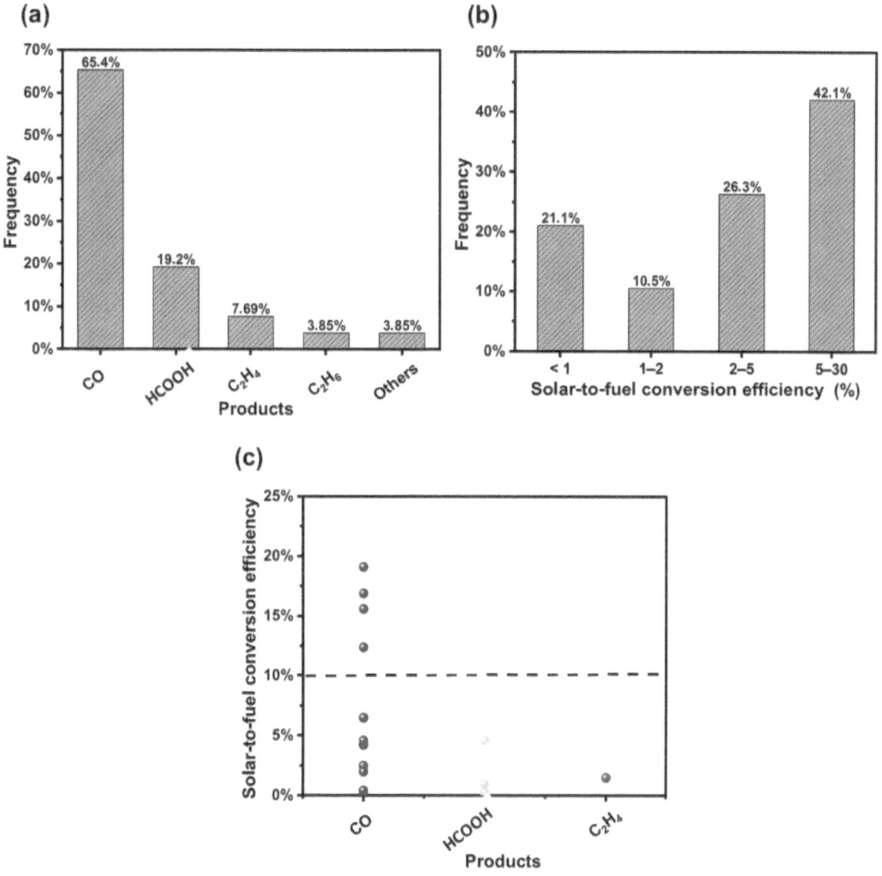

FIGURE 6.16 Statistical analysis of PV+EC CO_2 conversion studies: (a) product distribution, (b) SFE distribution, and (c) SFE distribution of different products [6].

HCOOH with 19.2%. Most of the studies reported SFE greater than 5%, accounting for 42.1%. Only 21.1% of the cases reported SFE less than 1%.

Huan et al. [209] used an electrolyzer with the same copper-based catalyst at both the anode and cathode and achieved CO_2 reduction to hydrocarbons (ethylene and ethane) with a 21% energy efficiency. Subsequent coupling of this system to a state-of-the-art perovskite PV minimodule demonstrated a 2.3% solar-to-hydrocarbons efficiency, setting a benchmark for an inexpensive all–earth-abundant PV/EC system.

The efficiency of a CO_2R/OER electrolyzer primarily depends on the activities of the catalysts, notably their ability to mediate both redox reactions with minimal overpotential losses. This is highly challenging as both anodic and cathodic reactions involve complex multielectron, multi-proton reactions, especially when hydrocarbons are targeted. Furthermore, resistivity issues and membrane potential contributions can also significantly influence energy losses. The use of a single-compartment cell with anode and cathode in close proximity effectively solve membrane potential and resistivity-related issues [196,199], but results in a gas mixture of the CO_2 reduction products and the anodically evolved oxygen. The lack of electrode separation also increases the chance of gas crossover that has a deleterious effect on the faradaic yield (FY) for CO_2 reduction. Huan et al. [209] developed a two-compartment electrolyzer containing an anion-exchange membrane between the anodic and cathodic compartments. The overall cell potential can thus be expressed as the sum of the equilibrium potential, the kinetic overpotentials of CO_2R and OER, and the resistive and concentration losses.

Thus far, copper is the only metal that has shown high selectivity for CO_2R to multi-carbon products, particularly when prepared by reduction of CuO materials [210–212]. In addition, literature has demonstrated that CuO can function as an efficient OER catalyst at the moderate pH conditions required for efficient CO_2R [213–216]. This strategy presents the additional advantage of limiting the poisoning of the cathode by redepositing the metal used for the anode [3]. In order to lower mass transfer losses (η_{conc}), Huan et al. [209] selected a dendritic nanostructured copper oxide material (DN-CuO) that is also highly efficient and stable OER catalyst [213]. This material presents both a macroporous structure, provided by cavities larger than 50 μm, and a mesoporous structure resulting from the dendritic structure constituting the walls. This unique morphology ensured an efficient mass transfer of reactants and products while preserving a high electrochemical surface area.

The design of the electrolyzer has a profound influence on the overall CO_2 reduction performance, as it not only affects the cell voltage but also the selectivity of CO_2R, the product separation, and the catalyst stability [217,218]. To reach the highest overall efficiency, Huan et al. [209] targeted minimizing both ohmic and mass transport losses. The most straightforward way to reduce the overall cell resistance is to lower the interelectrode distance and to use a concentrated electrolyte solution. However, such an approach faces two main limitations: (a) in a CO_2-saturated solution, a thin cathodic compartment favors formation and trapping of gas bubbles, thus strongly increasing the resistance of the cell, and (b) a concentrated bicarbonate solution increases the catalytic selectivity for proton reduction at the expense of CO_2 reduction [199,219,220]. Huan et al. [209] found that most optimal results were obtained using a 7 mm interelectrode distance and a 0.1 M cesium bicarbonate ($CsHCO_3$) CO_2-saturated (pH 6.8) solution as the cathodic electrolyte combined with a 0.2 M cesium carbonate (Cs_2CO_3) solution (pH 11) as the anodic electrolyte. Larger concentrations of the electrolytes, while decreasing the overall cell resistivity, lowered faradaic efficiency for CO_2 reduction. Similarly, the use of other alkali-metal cations resulted in an overall decrease in current for the same applied potential. Several other studies [221,222] have shown the beneficial influence of large alkali cations on both the selectivity for multi-carbon products in CO_2 reduction and on lowering the over- potential for water oxidation. Using HCO_3^- as the charge carrier for both the anodic and cathodic compartments together with an anion exchange membrane (AEM) allows for continuous operation of the system at high current densities: continuous CO_2 bubbling in the catholyte regenerates the diffused bicarbonate anions. In addition, the moderate pH difference between the anodic and cathodic compartments allows a minimal contribution of the pH gradient to the membrane potential.

The poor mass transfer of CO_2 to the cathode surface and low CO_2 solubility in electrolyte affects the efficiency of CO_2 conversion. In order to overcome these barriers Huan et al. [209] used a continuous-flow electrochemical cell in which the anolyte and catholyte are continuously flowing through the system. Constant saturation in CO_2 was ensured by continuously purging the catholyte with CO_2 in a separate compartment, which additionally continuously evacuated the reaction products. Huan et al. [209] also showed that continuous flow EC performed better than a standard H-type EC. Using Linear sweep voltammetry (LSV), they showed an excellent CO_2R electrocatalytic activity of DN-CuO. The anodic catalytic wave showed a similar onset potential for water oxidation [213], but accompanied with an increased current density, illustrating the beneficial influence of the electrolyzer system on the catalytic performance. Analysis of products indicated significant amounts of ethylene (C_2H_4) and ethane (C_2H_6) at applied potentials below −0.8 V vs. RHE. The highest selectivity for CO_2 reduction (vs. H_2 formation) and hydrocarbon production (vs. CO and HCOOH formation) was obtained at a cathode potential of −0.95 V vs. RHE. Among the CO_2 reduction products, C_2H_4 accounted for 57% (37% FY), C_2H_6 for 18% (12.8% FY), HCOOH for 11% (7% FY), and CO for 8% (5% FY). This is a very high selectivity and production rate for hydrocarbons at such a low cell potential [199,200] and corresponds to a record cell energy efficiency for hydrocarbons of 21%.

A PV/EC system has the advantage of relying on mature technologies benefiting from advanced experience of the industry innovations and continuously decreasing costs [223]. However, while the separation of light capture and catalysis within two different devices allows for independent optimization and better control of the performances and scalability, correct matching of the PV power output to the number and sizing of EC cells is challenging. When a solar cell and an electrolyzer are directly connected, the electrical circuit requires the operating current and voltage to be the same for the two devices, and their values are determined by the crossing point between the current–voltage curves for the two devices. More specifically, to reach the highest efficiency: (a) the operating point (current, voltage) of the device must be as close as possible to the maximum power point for solar-to-electric energy conversion, and (b) this working cell potential must correspond to the potential at which the highest selectivity for the products of interest, in this case hydrocarbons, are obtained.

Huan et al. [209] used triple cation perovskite solar cells stabilized with a photocured coating [224]. To provide a PV voltage and current density compatible with the electrolyzer conditions required for the most selective transformation of CO_2 to hydrocarbons, they built a minimodule constituted by two series of three perovskite solar cells connected in parallel. With this setup, stable currents were obtained and high selectivity for C2 hydrocarbons was preserved over more than 6 hours of operation time. When both systems were connected, stable current and potential were recorded at the electrolyzer terminals. Selectivity did not vary during the run, C_2H_4 and C_2H_6 being obtained as the main products with an average FY of 40.5% (34% for C_2H_4 and 6.5% for C_2H_6), together with CO and HCOOH in 4.8% and 6.4% FY, respectively. Concomitant hydrogen production was observed with 42.2% FY. The measured current densities and FY allowed determining a solar-to-hydrocarbon (ethylene and ethane) efficiency of 2.3%. This high efficiency constitutes a benchmark for solar-to-hydrocarbon products when using easily processable perovskite PV cells and earth-abundant metal catalysts.

6.6 PC, PEC AND PV/EC COMPARISONS

He and Janaky [6] compared the performance of photocatalytic, photoelectrocatalysis, and PV/EC concept under typical conditions. In this section, we summarize some of their important findings. The performance matrix used by them is illustrated in Table 6.7.

Activity, selectivity, and durability are usually the main three aspects to evaluate the performance of different solar-driven CO_2 conversion approaches. In PS/PC studies, product formation rate is the most commonly reported metric for evaluating the activity. In most cases, the formation rate is normalized over the weight of the catalyst or the geometrical area if it is immobilized on a substrate. Products, however, may vary in different catalytic systems. It is a worthwhile exercise to normalize the formation rate with the reaction stoichiometry (i.e., numbers of electrons transferred in the reaction); thus, comparisons can be reasonably made among different products. Quantum efficiency (QE) is another important component of the light-to-fuel efficiency, but it is not provided in all reported studies.

For the PEC approach, the faradaic efficiency (FE) is the most reported metric, often misleadingly interpreted as an activity descriptor. In addition, the onset potential of the reduction process (the potential/voltage at which the product detection measurement was carried out) and the corresponding normalized current density are also important metrics. Solar-to-fuel conversion efficiency (SFE) is a key metric, which is less reported in PEC systems. Strictly speaking, SFE is applicable only if no external bias is employed. Incident photon-to-current conversion efficiency (IPCE) and absorbed photon-to-current conversion efficiency (APCE) can also reflect the efficiency from different aspects [225].

In PV/EC-related systems, the performance metrics are similar to those of the PEC. The operation points (including voltage and current density) are usually provided, and SFE is more commonly reported. There are different performance metrics for PS/PC, PEC, and PV/EC systems, among which product distribution, light-to-fuel conversion efficiency, and maximum test time were selected as the indicators of selectivity,

TABLE 6.7
Summarized Performance Metrics for PS/PC, PEC, and PV/EC Systems [6]

PS/PC	PEC	PV + EC
Performance metrics formation rate	Formation rate (current density)	Formation rate (current density)
Conversion Turnover number (TON)	Potential/voltage	Potential/voltage
Selectivity	Selectivity (faradic efficiency, FE)	Selectivity (faradic efficiency, FE)
Quantum efficiency (QE)	Solar-to-fuel conversion efficiency (SFE)	Solar-to-fuel conversion efficiency (SFE)
Durability	Incident photon-to-current conversion efficiency (IPCE) Absorbed photon-to-current conversion efficiency (APCE) Durability	Durability

activity, and durability, respectively. The comparison study by He and Janaky [6] considered only the major products (selectivity > 30%) of a given study. For light-to-fuel efficiency the cases included in the product distribution analysis. For PS/PC studies, those investigations using sacrificial agents in the performance evaluation were excluded in the comparison of light-to-fuel efficiency, to ensure that water oxidation is the other half-reaction. For PEC studies, only studies without external bias were taken into account in the light-to-fuel conversion efficiency comparison. For PV/EC studies, SFE data were used. All the data for maximum time of run were included.

The comparison of product distribution is shown in Figure 6.17a. Overall, the products of PS/PC and PEC studies are more broadly distributed than those in PV/EC studies with a distinctly higher production of CO. The reason for this is that most of the PV/EC studies are proof of concept focusing on device fabrication or system validation, using commercial electrocatalysts, such as Ag [104] and Au [105] on which selective CO_2 reduction to CO has been widely reported. The gas products (mainly CO and CH_4) together account for nearly 80% in PS/PC studies, while they account for only about 40% in PEC studies, which might be associated with the factor that in PS/PC systems, both reduction and oxidation happen on the same particle, while in PEC systems they are spatially separated. The unfavored generation of liquid products in the PS/PC system may be plausibly further consumed by the photogenerated holes involving oxidation reaction conducted at the same particle surface. In addition, the frequency of HCOOH in PS/PC studies is lower than those of PEC and PV/EC plausibly because HCOOH, as one of the thermodynamically favored products, might be oxidized by the photogenerated holes or derived oxidizing intermediates, whereas this process is avoided to a great extent in PEC and PV/EC systems.

The activities of the three approaches are compared by light-to-fuel efficiencies, as shown in Figure 6.17b. The differences are striking! Most of the light-to-fuel efficiencies in PS studies (PC were not included in this analysis) are located in low-value ranges, with 29.0% located between 0% and 0.15% and 19.4% located between 0.15% and 0.4%, accounting for 48.4% together. For the PEC studies, the majority is less than 2%, accounting for 61.6%. While for that of PV/EC, the light-to-fuel efficiencies are more concentrated in the high-value range (≥4.5%). There is an obvious trend that PS studies frequently reported relatively lower light-to-fuel efficiencies while those of PEC are somewhat higher, and those of PV/EC are further improved.

For the comparison of durability (shown in Figure 6.17c), most of the PEC and PV/EC studies reported the longest measurement with the maximum test time less than 10 hours, accounting for more than 50%. Although fewer cases reported more than 50 hours of durability, some of those still have good stability. Moreover, the study found that PS/PC studies reported a higher frequency of more than 20 hours of durability. Generally, the instability in PS/PC systems is caused by photo corrosion resulting from the reduction/oxidation of the photoactive material by photogenerated electrons and holes. For that of PEC systems, not only photo corrosion but also electro corrosion and electrolyte degradation are considerable challenges for long-term durability. The lack of many long-duration PV/EC studies is somewhat surprising, because PV cells have a very long lifetime, while over 100–1,000 hours stability was also demonstrated for EC systems. The study recommended more work for the durability of the combined system.

For the sake of simplicity, He and Janaky [6] compared the most important descriptors behind these trends in Table 6.8. The reasons for these comparative performances are complex. It is therefore important to examine typical timescale of

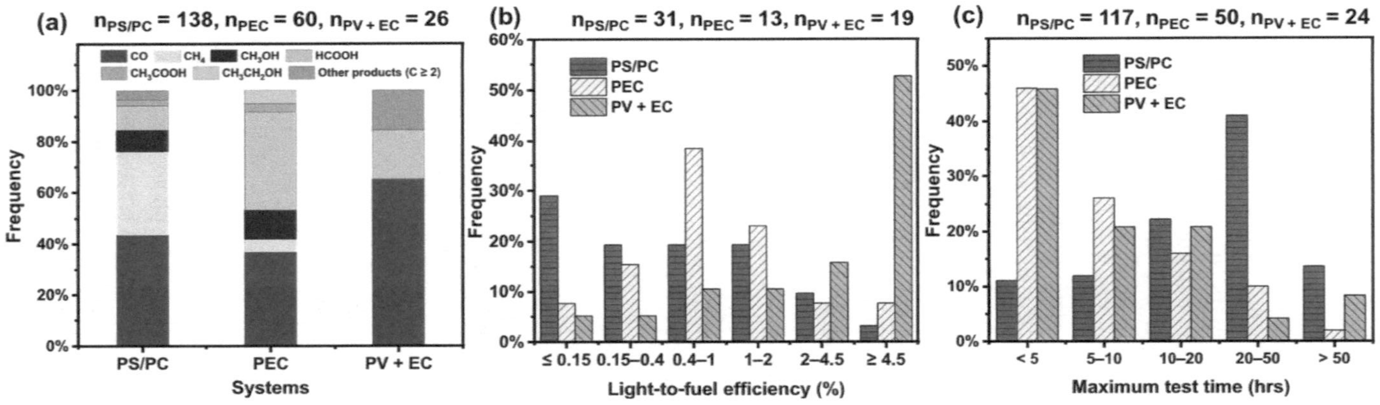

FIGURE 6.17 Comparisons of (a) product distribution, (b) light-to-fuel conversion efficiency, and (c) longest measurements in PS/PC, PEC, and PV/EC systems [6].

TABLE 6.8
Summary of the Differences among PS/PC, PEC, and PV + EC Systems from Six Aspects [6]

PS/PC System

1. One or more light absorbers are needed (see tandem and z-scheme configurations).
2. No need for carrier collection, but photogenerated holes and electrons need to reach the respective surface sites.
3. Both reactions proceed on the same particles. Preferably different sites for the two half-reactions. Back reactions are possible. The rates of the two half reactions have to match.
4. A high surface area is necessary to provide enough active sites for the reaction. High probability of surface recombination.
5. Stability: Intermediate stability, because of the presence of the solid/liquid interface.
6. Cost: Cheap experimental setup or device, but expensive multifunctional catalyst materials are needed.

PEC System

1. Either one (photocathode or photoanode) or two photoactive electrodes. The individual photoelectrodes can also be multicomponent.
2. Charge carrier trapping at defect sites at the electrode/electrolyte interface hinders charge carrier collection.
3. Slow charge carrier transfer to the substrate or mediator from the electrode surface, compared to the timescale of charge carrier recombination.
4. A high surface area is necessary to provide enough active sites for the reaction. High probability of surface recombination.
5. Very difficult to achieve reasonable stability, because of the presence of current flow and the electrode/electrolyte interface. Different protective coatings seem to ensure certain improvements.
6. More expensive and sophisticated cell designs are necessary, especially in the case of continuous flow processes. If a cocatalyst is employed, large amounts are needed because of the identical surface area of the light absorber and the electrochemical interface.

PV/EC System

1. Tailored photovoltaic cells can be designed (from single- to multijunction cells), to provide the necessary cell voltage.
2. Rapid charge carrier collection is achieved in the PV cell.
3. A separate electrochemical interface is responsible for the chemical reaction. Well-known stable and active electrocatalysts can be employed.
4. Nanostructured electrocatalysts can be used, without the detrimental surface recombination in the light absorber.
5. The stability is dictated only by the stability of the electrolyzer, as PV panels are stable for ages. Examples on the order of hundreds of hours are available.
6. Relatively expensive system cost. Much smaller electrochemically active area is needed (compared to the size of the PV), and thus less electrocatalysts, membranes, etc. have to be used. It is also possible to select high-performance PV cells with a smaller area (under concentrated light) and electrodes with a larger area.

photo-induced processes occurring in semiconductors and at semiconductor interfaces. Most importantly, the timescale of the chemical reactions (especially the CO_2 reduction reactions involving multielectron and multiproton transfer) is in the microsecond-to-second regime. This means that a substantially long (photo)electron lifetime is necessary for this process. Unfortunately, charge carrier recombination occurs at a much faster timescale (depending on the mechanism from sub-picosecond to microsecond). This mismatch already indicates that high light-to-fuel conversion efficiencies cannot be realistically expected from PS and PEC systems, unless cocatalysts can be found, which can properly "store" electrons. There is precedence in the literature, where complex PS/PC assemblies allowed charge carrier lifetime on the order of microseconds [226]. This lifetime enabled different redox reactions, although not those involving the transfer of multiple electrons and protons. The charge transfer is also often accompanied by corrosive processes inside the semiconductors which are typically faster than the CO_2 reduction reaction. This is not the case for the PV/EC method, where charge carriers are rapidly extracted from the PV cell (on the nanosecond–picosecond) timescale. Table 6.8 considers six parameters to compare PS/PC, PEC, and PV/EC processes. These are light absorption, charge carrier collection, charge transfer (reaction), nano-aspects, stability, and cost. The table outlines the behaviors of three processes for these six parameters.

Finally, He and Janaky [6] found that often different products formed via the different approaches even with similar catalysts. Taking Cu_xO-based catalysts as an example, several studies are listed in Table 6.9. There is a variety of products including CO, CH_4, C_2H_4, C_2H_6, HCOOH, CH_3COOH, and CH_3OH. He and Janaky [6] explained the difference based on difference in governing mechanisms in PS/PC and PEC systems and PV/EC system. The former is dictated by conduction band energy of the photocathode, while in latter it is dictated by electrode potential. He and Janaky [6] also pointed out that this difference may also allow more opportunities for PEC [227].

TABLE 6.9
Representative Studies on Solar-Driven CO_2 Conversion Using Cu_xO-Based Catalysts [6]

System	Catalyst	Main Product(s)	Formation Rate (mmol e−/g_{cat} h)	FE (%)	SFE (%)
PS/PC	Carbon quantum dots/Cu_2O	CH_3OH	0.336		
	Cathode				
PEC	Cu_2O–Cu	HCOOH		14	
		CH_3COOH		76	
	Cu	CH_4		47	
	Cu/Cu_2O	CH_3OH		53.6	
PV/EC	CuO	C_2H_4		34	
		C_2H_6		6.5	
	Cu_xO wire arrays	CO			2.5
		HCOOH			0.25

TABLE 6.10
Representative Studies on Solar-Driven CO_2 Conversion Using Au-Based Catalysts [6]

System	Catalyst	Main Product(s)	Selectivity (%)	FE (%)	SFE (%)
PS (Plasmonic catalysis)	Au	C_1–C_3 hydrocarbons	50% (C_{2+} hydrocarbons)		
PEC	Cathode				
	Au/Si	CO		91	
	Au/B doped g-C_3N_4	CH_3CH_2OH		47	
PV + EC	Au	CO		~92	15.6

The data in Table 6.10 for Au-based catalysts indicate that the same catalyst may play different roles in different scenarios resulting in different products. The results of the comparison done by He and Janaky [6] indicate that (a) only the minority of studies present all the factors, which fully describe the performance of a given system; (b) the CO_2 reduction products and their distribution are different in the different scenarios, and (c) at this stage only the PV/EC approach shows a performance (especially in terms of activity and durability), which can lead to industrial technologies in the near future. He and Janaky [6] also pointed out that there is a significant amount of research being carried out like for "doped photocatalysts" which will not result in making photocatalysis commercially viable. They also suggested that future study should confirm that CO_2 is the actual source of the carbon-containing product and full disclosure of all the factors (metrics) affecting the performance of the system should be made. They also conclude that only the PV/EC systems have the potential to become an industrial technology in the near and midterm future [228] due to rapid improvement in PCE, rapid cost decline for PV cells, and continuous progress in developing practical CO_2 electrolyzers [229] and gas diffusion porous electrodes to reduce mass transfer limitations and small turnover number at the active sites [230,231].

6.7 CLOSING PERSPECTIVES

CO_2 reduction can not only generate C1 hydrocarbons like CO, CH_4, and methanol but under suitable conditions, but it can also generate higher hydrocarbons. Product selectivity is, therefore, very important and in order to improve selectivity by photocatalysts, it is important to understand active sites at atomic level [232–238] and transformation mechanisms under practical conditions. For example, the [98] facet of a single-crystal Cu_2O particle is active for photo-driven CO_2 reduction to methanol, while the [90] facet is inert. The oxidation state of the active sites changes from Cu(I) toward Cu(II) because of CO_2 and H_2O co-adsorption and changes back to Cu(I) after CO_2 conversion under visible light illumination [236]. Biocatalysts are used to improve product selectivity, and in the case of biocatalysts, interactions between enzyme and active sites are important factors for the selectivity.

Immobilization of the catalysts not only combines the advantages of homogeneous and heterogeneous catalysis, but also allows easy separation of products from the catalyst and removes the limitations from the solubility of a homogeneous catalyst. The immobilization of catalysts can also prevent the formation of inactive species through dimerization and thereby enhance the activity and stability of the catalytic system [116]. Recent findings also demonstrate that immobilization can alter the kinetics of photo-generated catalytic intermediates to improve the system's activity [116]. Co-assembly of catalysts and dyes onto semiconductor surfaces provides pre-organization and overcomes diffusional limitations, which is important in light-driven systems. Furthermore, these materials can play an active role in some cases to provide long-lived charge separation via electron transfer into the conduction band [7,13,116], thereby promoting the activity and longevity of the system. In other cases, the material itself is the light absorber allowing for efficient electron transfer to an immobilized catalyst.

Immobilization of molecular catalysts is also a promising approach to establish more effective catalytic CO_2 conversion systems. The original work of [116,239] showed that heterogenization of a Co phthalocyanine resulted in a 25,000-fold enhanced activity compared to the corresponding homogenous system. In electrochemical systems, catalyst immobilization prevents a short circuit by diffusion of the reduced catalytic intermediates to the counter electrode. Turnover numbers and turnover frequencies can be more meaningfully determined when the catalyst is immobilized. Immobilization of molecular catalysts also makes more efficient use of the individual catalytic centers, as opposed to homogeneous electrocatalysis, where the vast majority of molecules are in the bulk solution. The use of porous electrode materials can significantly increase the surface area and therefore increases the current density for optimal electrocatalytic performance. Immobilization has led to a reduction in overpotential compared with homogeneous counterparts [19] and has allowed catalysis to move into aqueous solvent [240].

New homogeneous molecular catalysts can be immediate candidates for immobilization (heterogenization) for both photocatalysis and photoelectrocatalysis. New anchoring strategies and surfaces to which molecules are attached are being developed [7,13,116]. These may include changes in geometry and porosity of surface and the development of surfaces decorated with non-innocent chemical groups. Such a strategy may promote catalysis by creating a chemical environment around the molecule, mimicking the protein environment around an enzyme-active site. The electronics of semiconducting surfaces can be varied for efficient light harvesting and/or electron transfer.

For solar-driven photoelectrocatalysis, **hybrid photoelectrode assemblies** offer a platform to rationally design materials. Photoelectrocatalysis requires understanding of several complex requirements of photoelectrode which include (a) **Light absorption**: it is essential to harness a reasonable portion of the solar spectrum. (b) **Charge transport**: the semiconductor shall have high charge carrier mobility. The amount of bulk and surface traps should be minimized. (c) **Charge transfer kinetics**: facile charge transfer from the semiconductor to either the CO_2 molecules in the solution, or to a mediator (either immobilized or in the solution), is required. (d) **Stability and robustness towards chemical, electrochemical, and photo corrosion processes**.

These complex requirements indicate that no single material can satisfy all the issues. Often composite materials are required. In biological systems, one can find complex architectures with components that have precisely defined functionality and complementarity. As a bioinspired approach, some of the limitations of the individual components can be overcome by the rational design and assembly of hybrid PS/PC and PEC materials, where the different functionalities are decoupled. For example, $CsPbBr_3$@zeolitic imidazolate framework nanocomposites have been reported to exhibit enhanced CO_2 reduction activity due to the addition of zeolitic imidazolate framework with its original CO_2 capturing ability and the role of acting as a cocatalyst [241]. In addition, it was demonstrated by the example of Cu_2O that when a highly conductive scaffold is introduced into the photoelectrode, the charge carrier transport can be enhanced, and larger photocurrents can be harvested [242,243]. Similar trends have also been discovered in organic photoelectrodes [244]. This strategy greatly broadened the range of the catalyst and light absorber selection by reasonably combining the attractive features of each component.

The improvement in photoelectrocatalyst also requires enhancement in current density. This can be achieved by insertion of nanoscale materials and processes. The long-standing strategy for PEC to build thick electrode films (with film thickness of over several micrometers) [245], and there is no systematic study on the use of nanoparticles for PEC processes. For PS/PC, however, there is solid and coherent knowledge on the effect of catalyst size at the nanoscale [246]. The nanoscale effects on light absorption, charge carrier transport, band bending, and charge transfer in PEC cell are very important. Also, despite some very informative studies on Si micro- [247,248] and nanowires [249], there is a lack of knowledge whether there is an ideal morphology for PEC applications. In addition, the consequence of drastically increased surface/bulk ratio (which is the case for nanomaterials) on the contribution of surface functional groups to the materials' property has remained unknown.

Plasmonic catalysis has become an emerging avenue in CO_2 conversion by the promise of these particles to harness visible light as hot carriers and their intrinsic catalytic activity toward CO_2 reduction. Noble metal NPs, in particular, allow the integration of strong visible-wavelength plasmonic excitation with surface activation of CO_2 and therefore represent a novel and promising class of photocatalysts for CO_2 reduction [8]. The fundamental understanding of plasmon-assisted CO_2 reduction processes, however, is still lacking. In plasmonic CO_2 reduction reaction, catalytic activity, reaction pathways, and selectivity are expected to not only depend on the properties of the metal and metal/adsorbate interactions but also possibly be tuned by light excitation [250–252].

The integration of PV and EC technologies can also benefit from buried junction [253]. Buried junction use not only overcomes the instability issue in integrated PEC cells, but the PV components in the buried junction can also act as more efficient and stable light absorbers than most of the single semiconductors when in physical contact with the aqueous environment is avoided [254]. This strategy might provide a new direction toward the enhanced durability of PEC systems. More attention should also be given to the component development of PV/EC system. A recent study has demonstrated that the largest voltage losses may occur at the membrane or the membrane electrode assembly rather than at the catalyst layers in their flow cell [255]. A solid electrolyte may be a better alternative, because the formed liquid products are in a mixture with the dissolved salts in liquid electrolytes, requiring energy-intensive downstream separation. Furthermore, in the solid electrolyte, the generated cations (such as H^+) and anions (such as $HCOO^-$) are combined to form pure product solutions without mixing with other ions [234]. A reactor optimization for the effects of geometrical configuration, construction material, heat exchange, and mixing and flow characteristics [256,257] on CO_2 utilization is required.

High current density operation for both the PEC and PV/EC requires concentrated sunlight. With its use, the solar power input can be concentrated up to 500 Suns, which in combination with proper PV cells or photoelectrodes can allow current densities similar to those in industrial electrolyzers. For water splitting, 0.88 A/cm^2 current density was demonstrated with an STH efficiency of over 15%, under concentrated solar irradiation (up to 474 kW/m^2) [258]. Value-added anode processes are also worth more consideration. Most CO_2 reduction approaches couple cathodic CO_2 reduction with the anodic oxygen evolution reaction (OER), resulting in approximately 90% of the electricity input being consumed by the OER in the EC scenario [259]. This issue can be addressed by coupling other anodic oxidation reactions with less electricity needs and probably higher-value products. This strategy might also offer opportunities to the PC and PEC approaches as well, because there will be no need to cope with the difficult OER, and in fact, thermodynamically downhill processes can be designed.

In all different types of catalysis discussed here, the catalyst is the key component of different CO_2 reducing systems. A suitable catalyst can not only reduce energy consumption but also facilitate the generation and transfer of electrons. Photocatalysis and photoelectrocatalysis that adopt clean and sustainable solar energy to drive the conversion are of great interest, in which porphyrin-based macrocycles or their combination with other components present promising properties of light-capturing and charge transferring. Specific product generation from CO_2 greatly reduces the cost of subsequent product separation and purification. This has inspired researchers to pay particular attention to biocatalysts because of their high specificity and efficiency in catalyzing biochemical reactions. Among the aforementioned methods, enzyme coupled to photocatalysis and enzyme coupled to photoelectrocatalysis has integrated the two sides successfully, showing great potential in solar energy utilization and specific conversion of CO_2. They are worthy of more investigation to make bio-catalysis compatible with photocatalysis or photoelectrocatalysis. In general, continuous efforts are needed to improve the catalytic efficiency, conversion rate, and product selectivity.

While the above discussion indicates that electrochemical cells are very effective for CO_2 conversion to methanol and other chemicals, poor product selectivity and the low/high reaction temperatures are considered to be the main barriers in the heterogeneous CO_2 reduction process. Cheaper electrocatalyst with higher selectivity and lower overpotentials are still needed. Various heterogeneous electrocatalysts are selective, fast, and energy-efficient, but they are considered to be unstable catalysts. The use of renewable energy for power can facilitate process scaleup. More work is needed to reduce cost and develop new electrocatalytic materials that can be used to allow working at higher current densities without loss of faradaic efficiency. **Generally, the prospects to develop the successful commercial technologies for the efficient CO_2 conversion using solar energy are considered long term (>5 years out) with a bright future**.

REFERENCES

1. Pearlman, H., and C.-H. Chen 2012. Syngas production by thermochemical conversion of CO_2 and H_2O using a high-temperature heat pipe based reactor, Solar Hydrogen and Nanotechnology VII, edited by L. Vayssieres, *Proceedings of SPIE* Vol. 8469, 846900 ©2012 SPIE 000 code: 0277-786/12/$18. doi: 10.1117/12.930753.
2. Kraemer, S. 2017. Solar PACES, Technique uses solar thermal energy to split H_2O and CO_2 for jet fuel, a website report by Energy and Green Tech. (Npv., 2017).
3. Pullar, R. C., R. M. Novais, A. P. F. Caetano, M. A. Barreiros, S. Abanades, and F. A. C. Oliveira 2019. A review of solar thermochemical CO_2 splitting using ceria-based ceramics with designed morphologies and microstructures. *Frontiers in Chemistry* 7:601. https://www.frontiersin.org/article/10.3389/fchem.2019.00601 DOI=10.3389/fchem.2019.00601, ISSN: 2296-2646.
4. Aresta, M., A. Dibenedetto, and A. Angelini 2013. The use of solar energy can enhance the conversion of carbon dioxide into energy-rich products: Stepping towards artificial photosynthesis. *Philosophical Transactions of the Royal Society A* 371:20120111. http://dx.doi.org/10.1098/rsta.2012.0111.
5. Smestad, G. P., and A. Steinfeld 2012. Review: Photochemical and thermochemical production of solar fuels from H_2O and CO_2 using metal oxide catalysts. *Industrial & Engineering Chemistry Research* 51:11828–11840. doi: 10.1021/ie3007962.
6. He, J., and C. Janaḱy 2020. Recent advances in solar-driven carbon dioxide conversion: Expectations versus reality. *ACS Energy Letters* 5:1996–2014.
7. Ulmer, U., et al. 2019. Fundamentals and applications of photocatalytic CO_2 methanation. *Nature Communications* 10:3169. https://doi.org/10.1038/s41467-019-10996-2.
8. Wang, P., R. Dong, S. Guo, J. Zhao, Z.-M. Zhang, and T.-B. Lu 2020. Improving photosensitization for photochemical CO_2-to-CO conversion. *National Science Review* 7(9):1459–1467. https://doi.org/10.1093/nsr/nwaa112.
9. Teng, E., K. HaoTan, E. Lovell, R. J. Wong, J. Scott, and R. Amal 2017. A review on photo-thermal catalytic conversion of carbon dioxide. *Green Energy & Environment* 2(3):204–217.
10. Kareya, M. S., I. Mariam, A. A. Nesamma, and P. P. Jutur 2020. CO_2 sequestration by hybrid integrative photosynthesis (CO_2-SHIP): A green initiative for multi-product biorefineries. *Materials Science for Energy Technologies* 3:420–428.
11. Wang, H. 2020. Solar thermochemical fuel generation, an Intech open access paper. doi: 10.5772/intechopen.90767.
12. Henao, C. A., C. T. Maravelias, J. E. Miller, and R. A. Kemp 2008. Synthetic production of methanol using solar power, SAND2008–8149C, Sandai National Lab, Albuquerque, NM.
13. Kumar, B., M. Llorente, J. Froehlich, T. Dang, A. Sathrum, and C. P. Kubiak 2012. Photochemical and photoelectrochemical reduction of CO_2. *Annual Review of Physical Chemistry* 63:541–569. https://doi.org/10.1146/annurev-physchem-032511-143759.
14. Zhou, X., R. Liu, K. Sun, Y. Chen, E. Verlage, S. A. Francis, N. S. Lewis, and C. Xiang 2016. Solar-driven reduction of 1 atm of CO_2 to formate at 10% energy-conversion efficiency by use of a TiO_2 protected III-V tandem photoanode in conjunction with a bipolar membrane and a Pd/C cathode. *ACS Energy Letters* 1(4):764–770.
15. Grim, R. G., Z. Huang, M. T. Guarnieri, J. R. Ferrell, L. Tao, and J. A. Schaidle 2020. Transforming the carbon economy: Challenges and opportunities in the convergence of low-cost electricity and reductive CO_2 utilization. *Energy & Environmental Science* 13(2):472–494.
16. Nørskov, J. K., et al. 2019. Research needs towards sustainable production of fuels and chemicals; Energy X. Downloaded from: https://www.energy-x.eu/wpcontent/uploads/2019/09/Energy_X_Research-needs-report.pdf.

17. Barelli, L., G. Bidini, F. Gallorini, and S. Servili 2008. Hydrogen production through sorption-enhanced steam methane reforming and membrane technology: A review. *Energy* 33(4):554–570. doi: 10.1016/j.energy.2007.10.018.
18. Fan, J., and L. Zhu 2015. Performance analysis of a feasible technology for power and high-purity hydrogen production driven by methane fuel. *Applied Thermal Engineering* 75:103–114. doi: 10.1016/j.applthermaleng.2014.10.013.
19. Li, Y., N. Zhang, and R. Cai 2013. Low CO_2- emissions hybrid solar combined-cycle power system with methane membrane reforming. *Energy* 58:36–44. doi: 10.1016/j.energy.2013.02.005.
20. Klein, H. H., J. Karni, and R. Rubin 2009. Dry methane reforming without a metal catalyst in a directly irradiated solar particle reactor. *Journal of Solar Energy Engineering* 131(2):021001-1-021001-14. doi: 10.1115/1.3090823.
21. Edwards, J. H., et al. 2000. CSIRO's solar thermal-fossil energy hybrid technology for advanced power generation. In: *Proceedings of Solar Thermal 2000 10th SolarPACES International Symposium on Solar Thermal Concentrating Technologies*, Sydney, N.S.W., pp. 27–32.
22. Steinfeld, A., M. Brack, A. Meier, A. Weidenkaff, and D. Wuillemin 1998. A solar chemical reactor for co- production of zinc and synthesis gas. *Energy* 23(10):803–814. doi: 10.1016/ S0360-5442(98)00026-7.
23. Lorentzou, S., G. Karagiannakis, D. Dimitrakis, C. Pagkoura, A. Zygogianni, and A. G. Konstandopoulos 2015. Thermochemical redox cycles over Ce-based oxides. *Energy Procedia* 69:1800–1809. doi: 10.1016/j.egypro.2015.03.152.
24. Takacs, M. 2017. Splitting CO_2 with a ceria-based redox cycle in a solar-driven thermogravimetric analyser. *AIChE Journal* 63:1263–1271. doi: 10.1002/aic.15501.
25. Takacs, M., M. Hoes, M. Caduff, T. Cooper, J. R. Scheffe, and A. Steinfeld 2016. Oxygen nonstoichiometry, defect equilibria, and thermodynamic characterization of $LaMnO_3$ perovskites with Ca/Sr A-site and Al B-site doping. *Acta Materialia* 103:700–710. doi: 10.1016/j.actamat.2015.10.026.
26. Siegel, N. P., J. E. Miller, I. Ermanoski, R. B. Diver, and E. B. Stechel 2013. Factors affecting the efficiency of solar driven metal oxide thermochemical cycles. *Industrial & Engineering Chemistry Research* 52:3276–3286. doi: 10.1021/ie400193q.
27. Ackermann, S., J. R. Scheffe, and A. Steinfeld 2014a. Diffusion of oxygen in ceria at elevated temperatures and its application to H_2O/CO_2 splitting thermochemical redox cycles. *Journal of Physical Chemistry C* 118:5216–5225. doi: 10.1021/jp500755t.
28. Chueh, W. C., et al. 2010. High-flux solar-driven thermochemical dissociation of CO_2 and H_2O using nonstoichiometric ceria. *Science* 330:1797–1801. doi: 10.1126/science.1197834.
29. Chueh, W. C., and S. M. Haile 2010. A thermochemical study of ceria: Exploiting an old material for new modes of energy conversion and CO_2 mitigation. *Philosophical Transactions of the Royal Society A* 368:3269–3294. doi: 10.1098/rsta.2010.0114.
30. Haussener, S., and A. Steinfeld 2010. Effective heat and mass transport properties of anisotropic porous ceria for solar thermochemical fuel generation. In: *Conference Proceedings 2010 AIChE Annual Meeting, 10AIChE*, Salt Lake City, UT.
31. Haussener, S., and A. Steinfeld 2012. Effective heat and mass transport properties of anisotropic porous ceria for solar thermochemical fuel generation. *Materials* 5:192–209. doi: 10.3390/ma5010192.
32. Abanades, S., and G. Flamant 2006. Thermochemical hydrogen production from a two-step solar-driven water-splitting cycle based on cerium oxides. *Solar Energy* 80:1611–1623. doi: 10.1016/j.solener.2005.12.005.
33. Bhosale, R. R., G. Takalkar, P. Sutar, A. Kumar, F. Al Momani, and M. Khraisheh 2019. A decade of ceria based solar thermochemical H_2O/CO_2 splitting cycle. *International Journal of Hydrogen Energy* 44:34–60. doi: 10.1016/j.ijhydene.2018.04.08.

34. Mogensen, M., N. M. Sammes, and G. A. Tompsett 2000. Physical, chemical and electrochemical properties of pure and doped ceria. *Solid State Ionics* 129:63–94. doi: 10.1016/S0167-2738(99)00318-5.
35. Le Gal, A., S. Abanades, and G. Flamant 2011. CO_2 and H_2O splitting for thermochemical production of solar fuels using nonstoichiometric ceria and ceria/zirconia solid solutions. *Energy Fuels* 25:4836–4845. doi: 10.1021/ef200972r.
36. Wang, H., Y. Hao, and H. Kong 2015. Thermodynamic study on solar thermochemical fuel production with oxygen permeation membrane reactors. *International Journal of Energy Research* 39(13):1790–1799. doi: 10.1002/er.3335.
37. Zhu, L., Y. Lu, and S. Shen 2016. Solar fuel production at high temperatures using ceria as a dense membrane. *Energy* 104:53–63. doi: 10.1016/j. energy.2016.03.108.
38. Michalsky, R., D. Neuhaus, and A. Steinfeld 2015. Carbon dioxide reforming of methane using an isothermal redox membrane reactor. *Energy Technology* 3(7):784–789. doi: 10.1002/ente.201500065.
39. Tou, M., R. Michalsky, and A. Steinfeld 2017. Solar-driven thermochemical splitting of CO_2 and in situ separation of CO and O_2 across a ceria redox membrane reactor. *Joule* 1(1):146–154. doi: 10.1016/j.joule.2017.07.015.
40. Ozin, G. A. 2017. "One-pot" solar fuels. *Joule* 1(1):19–23. doi: 10.1016/j.joule.2017.08.010.
41. Wang, H., M. Liu, H. Kong, and Y. Hao 2019. Thermodynamic analysis on mid/low temperature solar methane steam reforming with hydrogen permeation membrane reactors. *Applied Thermal Engineering* 152:925–936. doi: 10.1016/j.applthermaleng.2018.03.030.
42. Mallapragada, D. S., and R. Agrawal 2014. Limiting and achievable efficiencies for solar thermal hydrogen production. *International Journal of Hydrogen Energy* 39(1):62–75. doi: 10.1016/ j.ijhydene.2013.10.075.
43. Said, S. A., D. S. Simakov, M. Waseeuddin, and Y. Román-Leshkov 2016. Solar molten salt heated membrane reformer for natural gas upgrading and hydrogen generation: A CFD model. *Solar Energy* 124:163–176. doi: 10.1016/j.solener.2015.11.038.
44. Zhang, L., et al. 2012. High CO_2 permeation flux enabled by highly interconnected three-dimensional ionic channels in selective CO_2 separation membranes. *Energy & Environmental Science* 5(8):8310–8317. doi: 10.1039/C2EE22045H.
45. Ceron, M.R., et al. 2018. Surpassing the conventional limitations of CO_2 separation membranes with hydroxide/ceramic dual-phase membranes. *Journal of Membrane Science* 567:191–198. doi: 10.1016/j.memsci.2018.09.028.
46. Wang, H., and Y. Hao 2017. Thermodynamic study of solar thermochemical methane steam reforming with alternating H_2 and CO_2 permeation membranes reactors. *Energy Procedia* 105:1980–1985. doi: 10.1016/j.egypro.2017.03.57022.
47. Blaber, M. G., M. D. Arnold, and M. J. Ford 2010. A review of the optical properties of alloys and intermetallics for plasmonics. *Journal of Physics: Condensed Matter* 22:143201.
48. Ndebeka-Bandou, C., F. Carosella, R. Ferreira, A. Wacker, and G. Bastard 2014. Free carrier absorption and inter-subband transitions in imperfect heterostructures. *Semiconductor Science and Technology* 29:023001.
49. Peng, X., and D. M. P. Mingos 2005. *Semiconductor Nanocrystals and Silicate Nanoparticles*, 118–119. Heidelberg, Germany: Springer-Verlag.
50. Robatjazi, H., et al. 2017. Plasmon-induced selective carbon dioxide conversion on earth-abundant aluminum–cuprous oxide antenna-reactor nanoparticles. *Nature Communications* 8:1–9.
51. González De La Cruz, G., and Y. G. Gurevich 1996. Electron and phonon thermal waves in semiconductors: An application to photothermal effects. *Journal of Applied Physics* 80:1726–1730.

52. Govorov, A. O., and H. H. Richardson 2007. Generating heat with metal nanoparticles. *Nano Today* **2**:30–38.
53. Gadzuk, J. W. 1983. Vibrational excitation in molecule—Surface collisions due to temporary negative molecular ion formation. *Journal of Chemical Physics* 79:6341.
54. Meng, X., et al. 2014. Photothermal conversion of CO_2 into CH_4 with H_2 over group VIII nanocatalysts: An alternative approach for solar fuel production. *Angewandte Chemie* 126:11662–11666.
55. Sastre, F., A. Puga, L. Liu, A. Corma, and H. Garcia 2014. Complete photocatalytic reduction of CO_2 to methane by H_2 under solar light irradiation. *Journal of the American Chemical Society* 136:6798–6801.
56. Jelle, A. A., et al. 2018. Highly efficient ambient temperature CO_2 photomethanation catalyzed by nanostructured RuO_2 on silicon photonic crystal support. *Advanced Energy Materials* 8:1702277.
57. O'Brien, P. G., et al. 2014. Photomethanation of gaseous CO_2 over Ru/silicon nanowire catalysts with visible and near-infrared photons. *Advanced Science* 1:1–7.
58. Ren, J., et al. 2016. Targeting activation of CO_2 and H_2 over Ru-loaded ultrathin layered double hydroxides to achieve efficient photothermal CO_2 methanation in flow-type system. *Advanced Energy Materials* 7:1601657.
59. O'Brien, P. G., et al. 2018. Enhanced photothermal reduction of gaseous CO_2 over silicon photonic crystal supported ruthenium at ambient temperature. *Energy & Environmental Science* 11:3443–3451.
60. Zhang, X., et al. 2018. Plasmon-enhanced catalysis: Distinguishing thermal and nonthermal effects. *Nano Letters* 18:1714–1723.
61. Zhang, X., et al. 2017. Product selectivity in plasmonic photocatalysis for carbon dioxide hydrogenation. *Nature Communications* 8:1–9.
62. Zhang, H., et al. 2016. Surface-plasmon-enhanced photodriven CO_2 reduction catalyzed by metal-organic-framework-derived iron nanoparticles encapsulated by ultrathin carbon layers. *Advanced Materials* 28:3703–3710.
63. Wang, C., et al. 2013. Visible light plasmonic heating of Au–ZnO for the catalytic reduction of CO_2. *Nanoscale* 5:6968–6974.
64. Jia, J. 2017. Photothermal catalyst engineering: Hydrogenation of gaseous CO_2 with high activity and tailored selectivity. *Advanced Science* 4:1700252.
65. Zhang, H., et al. 2016. Efficient visible-light-driven carbon dioxide reduction by a single-atom implanted metal-organic framework. *Angewandte Chemie International Edition* 100049:14522–14526.
66. Linic, S., U. Aslam, C. Boerigter, and M. Morabito 2015. Photochemical transformations on plasmonic metal nanoparticles. *Nature Materials* 14:567–576.
67. Christopher, P., H. Xin, A. Marimuthu, and S. Linic 2012. Singular characteristics and unique chemical bond activation mechanisms of photocatalytic reactions on plasmonic nanostructures. *Nature Materials* 11:1044–1050.
68. Rodriguez, J., et al. 2013. CO_2 hydrogenation on Au/TiC, Cu/TiC, and Ni/TiC catalysts: Production of CO, methanol, and methane. *Journal of Catalysis* 307:162–169.
69. Navarrete, A., et al. 2015. Novel windows for "solar commodities": A device for CO_2 reduction using plasmonic catalyst activation. *Faraday Discuss* 183:249–259.
70. Baffou, G., and R. Quidant 2014. Nanoplasmonics for chemistry. *Chemical Society Reviews* 43:3898–3907.
71. Li, Z., J. Zhao, and L. Ren 2012. Aqueous solution-chemical derived Ni–Al_2O_3 solar selective absorbing coatings. *Solar Energy Materials and Solar Cells* 105:90–95.
72. Buck, R., J. F. Muir, and R. E. Hogan 1991. Carbon dioxide reforming of methane in a solar volumetric receiver/reactor: the CAESAR project. *Solar Energy Materials* 24:449–463.

73. Toth, L. E. 1971. *Transition Metal Carbides and Nitrides 7.* New York: Academic Press.
74. Hoch, L. B., et al. 2016. Nanostructured indium oxide coated silicon nanowire arrays: A hybrid photothermal/photochemical approach to solar fuels. *ACS Nano* 10:9017–9025.
75. Busch, D. G., and W. Ho 1996. Direct observation of the crossover from single to multiple excitations in femtosecond surface photochemistry. *Physical Review Letters* 77:1338–1341.
76. Liu, L., et al. 2018. Sunlight-assisted hydrogenation of CO_2 into ethanol and C^{2+} hydrocarbons by sodium-promoted Co@C nanocomposites. *Applied Catalysis B* 235:186–196.
77. Chen, B., L. Wu, and C. Tung 2018. Photocatalytic activation of less reactive bonds and their functionalization via hydrogen-evolution cross-couplings. *Accounts of Chemical Research* 51:2512–2523.
78. Meng, L., Z. Chen, and Z. Ma 2018. Gold plasmon-induced photocatalytic dehydrogenative coupling of methane to ethane on polar oxide surfaces. *Energy & Environmental Science* 11:294–298.
79. Larionov, E., and M. M. Mastandrea 2017. Asymmetric visible-light photoredox cross-dehydrogenative coupling of aldehydes with xanthenes. *ACS Catalysis* 7:7008–7013.
80. Yi, H., et al. 2017. Photocatalytic dehydrogenative cross-coupling of alkenes with alcohols or azoles without external oxidant. *Angewandte Chemie International Edition* 56:1120–1124.
81. Jain, P. K., K. S. Lee, I. H. El-Sayed, and M. A. El-Sayed 2006. Calculated absorption and scattering properties of gold nanoparticles of different size, shape, and composition: Applications in biological imaging and biomedicine. *Journal of Physical Chemistry* B110:7238.
82. Link, S., and M. A. El-Sayed 1999. Size and temperature dependence of the plasmon absorption of colloidal gold nanoparticles. *Journal of Physical Chemistry B* 103:4212–4217.
83. Noguez, C. 2007. Surface plasmons on metal nanoparticles: The influence of shape and physical environment. *Journal of Physical Chemistry C* 111:3806–3819.
84. Tan, T. H., J. Scott, Y. H. Ng, R. A. Taylor, K.-F. Aguey-Zinsou, and R. Amal 2016. Understanding plasmon and band gap photoexcitation effects on the thermal-catalytic oxidation of ethanol by TiO_2-supported gold. *ACS Catalysis* 6:8021–8029.
85. Palumbo, R., M. Keunecke, S. M€oller, and Steinfeld 2004. Reflections on the design of solar thermal chemical reactors: Thoughts in transformation. *Energy* 29:727–744.
86. Westrich, T. A., K. A. Dahlberg, M. Kaviany, and J. W. Schwank 2011. High-temperature photocatalytic ethylene oxidation over TiO_2. *Journal of Physical Chemistry C* 115:16537–16543.
87. Tan, T. H., J. Scott, Y. H. Ng, R. A. Taylor, K.-F. Aguey-Zinsou, and R. Amal 2016. Understanding plasmon and band gap photoexcitation effects on the thermal-catalytic oxidation of ethanol by TiO_2-supported gold. *ACS Catalysis* 6:1870–1879.
88. Aniruddha, I. R., A. Upadhye, X. Zeng, H. J. Kim, I. Tejedor, M. A. Anderson and A. James 2015. Plasmon-enhanced reverse water gas shift reaction over oxide supported Au catalysts. *Catalysis Science & Technology* 5:2590–2601.
89. Taniguchi, I., B. Aurian-Blajeni, and J. O. M. Bockris 1984. The mediation of the photoelectrochemical reduction of carbon dioxide by ammonium ions. *Journal of Electroanalytical Chemistry* 161:385–388.
90. Taniguchi, I., B. Aurian-Blajeni, and J. O. M. Bockris 1983. Photo-aided reduction of carbon dioxide to carbon monoxide. *Journal of Electroanalytical Chemistry and Interfacial Electrochemistry* 157:179–182.

91. Zafrir, M., M. Ulman, Y. Zuckerman, and H. Halmann 1983. Photoelectrochemical reduction of carbon dioxide to formic acid, formaldehyde and methanol on p-gallium arsenide in an aqueous V(II)-V(III) chloride redox system. *Journal of Electroanalytical Chemistry* 159:373–389.
92. Bradley, M. G., T. Tysak, D. J. Graves, and N. A. Viachiopoulos 1983. Electrocatalytic reduction of carbon dioxide at illuminated p-type silicon semiconducting electrodes. *Journal of the Chemical Society* 7:349–5044.
93. Cabrera, C. R., and H. D. Abruña 1986. Electrocatalysis of CO_2 reduction at surface modified metallic and semiconducting electrodes. *Journal of Electroanalytical Chemistry* 209:101–745.
94. Arai, T., S. Sato, K. Uemura, T. Morikawa, T. Kajino, and T. Motohiro 2010. Photoelectrochemical reduction of CO_2 in water under visible-light irradiation by a p-type InP photocathode modified with an electropolymerized ruthenium complex. *Chemical Communications* 46:6944–4646.
95. Ikeda, S., A. Yamamoto, H. Noda, M. Maeda, and K. Ito 1993. Influence of surface treatment of the p-GaP photocathode on the photoelectrochemical reduction of carbon dioxide. *Bulletin of the Chemical Society of Japan* 66:2473–2477.
96. Bilgen, E. 2001. Solar hydrogen from photovoltaic-electrolyzer systems. *Energy Conversion and Management* 42:1047–1057.
97. Doherty, M. D., D. C. Grills, J. T. Muckerman, D. E. Polyansky, and E. Fujita 2010. Toward more efficient photochemical CO_2 reduction: Use of $scCO_2$ or photogenerated hydrides. *Coordination Chemistry Reviews* 254:2472–2482.
98. Morris, A. J., G. J. Meyer, and E. Fujita 2009. Molecular approaches to the photocatalytic reduction of carbon dioxide for solar fuels. *Accounts of Chemical Research* 42:1983–1994.
99. Durham, B., J. V. Caspar, J. K. Nagle, and T. J. Meyer 1982. Photochemistry of $Ru(bpy)_3^{2+}$. *Journal of the American Chemical Society* 104:4803–4810.
100. Takeda, H., and O. Ishitani 2010. Development of efficient photocatalytic systems for CO_2 reduction using mononuclear and multinuclear metal complexes based on mechanistic studies. *Coordination Chemistry Reviews* 254:346–354.
101. Grodkowski, J., and P. Neta 2000. Cobalt corrin catalyzed photoreduction of CO_2. *Journal of Physical Chemistry A* 104:1848–1853.
102. Grodkowski, J., P. Neta, E. Fujita, A. Mahammed, L. Simkhovich, and Z. Gross 2002. Reduction of cobalt and iron corroles and catalyzed reduction of CO_2. *Journal of Physical Chemistry A* 106:4772–4778.
103. Grodkowski, J., et al. 2000. Reduction of cobalt and iron phthalocyanines and the role of the reduced species in catalyzed photoreduction of CO_2. *Journal of Physical Chemistry A* 104:11332–11339.
104. Craig, C. A., L. O. Spreer, J. W. Otvos, and M. Calvin 1990. Photochemical reduction of carbon dioxide using nickel tetraazamacrocycles. *Journal of Physical Chemistry* 94:7957–7960.
105. Kimura, E., X. Bu, M. Shionoya, S. Wada, and S. Maruyama 1992. A new nickel(II) cyclam (cyclam = 1,4,8,11-tetraazacyclotetradecane) complex covalently attached to tris(1,10-phenanthroline)ruthenium(2+). A new candidate for the catalytic photoreduction of carbon dioxide. *Inorganic Chemistry* 31:4542–4546.
106. Tinnemans, A. H. A., T. P. M. Koster, D. H. M. W. Thewissen, and A. Mackor 1984. Tetraaza-macrocyclic cobalt(II) and nickel(II) complexes as electron-transfer agents in the photo(electro)chemical and electrochemical reduction of carbon dioxide. *Recueil des Travaux Chimiques des Pays-Bas* 103:288–295.
107. Matsuoka, S., et al. 1993. Efficient and selective electron mediation of cobalt complexes with cyclam and related macrocycles in the p-terphenyl-catalyzed photoreduction of carbon dioxide. *Journal of the American Chemical Society* 115:601–609.

108. Ogata, T., et al. 1995. Phenazine-photosensitized reduction of CO_2 mediated by a cobalt-cyclam complex through electron and hydrogen transfer. *Journal of Physical Chemistry* 99:11916–11922.
109. Gholamkhass, B., H. Mametsuka, K. Koike, T. Tanabe, M. Furue, and O. Ishitani 2005. Architecture of supramolecular metal complexes for photocatalytic CO_2 reduction: Ruthenium-rhenium bi- and tetranuclear complexes. *Inorganic Chemistry* 44:2326–2336.
110. Wang, S.-S., H.-H. Huang, M. Liu, S. Yao, S. Guo, J.-W. Wang, Z.-M. Zhang, and T.-B. Lu 2020. Encapsulation of single iron sites in a metal– porphyrin framework for high-performance photocatalytic CO_2 reduction. *Inorganic Chemistry* 59. doi: 10.1021/acs.inorgchem.0c00407.
111. Premkumar, J., and R. Ramaraj 1997. Photocatalytic reduction of carbon dioxide to formic acid at porphyrin and phthalocyanine adsorbed Nafion membranes. *Journal of Photochemistry and Photobiology A* 110:53.
112. Premkumar, J. R., and R. Ramaraj 1997. Photoreduction of carbon dioxide by metal phthalocyanine adsorbed Nafion membrane. *Chemical Communications*:343.
113. Sekizawa, K., K. Maeda, K. Domen, K. Koike, and O. Ishitani 2013. Artificial Z-scheme constructed with a supramolecular metal complex and semiconductor for the photocatalytic reduction of CO_2. *Journal of the American Chemical Society* 135:4596.
114. Chambers, M. B., X. Wang, N. Elgrishi, C. H. Hendon, A. Walsh, J. Bonnefoy, J. Canivet, E. A. Quadrelli, D. Farrusseng, C. Mellot- Draznieks, and M. Fontecave 2015. Photocatalytic carbon dioxide reduction with rhodium-based catalysts in solution and heterogenized within metal-organic frameworks. *ChemSusChem* 8:603.
115. Ha, E.-G., J.-A. Chang, S.-M. Byun, C. Pac, D.-M. Jang, J. Park, and S. O. Kang 2014. High-turnover visible-light photoreduction of CO_2 by a Re(i) complex stabilized on dye-sensitized TiO_2. *Chemical Communications* 50:4462.
116. Windle, C. D., and E. Reisner 2015. Heterogenised molecular catalysts for the reduction of CO_2 to fuels. *Chimia* 69(7/8):435–441. doi: 10.2533/chimia.2015.435.
117. Zhang, L., and J. M. Cole 2015. Anchoring groups for dye-sensitized solar cells. *ACS Applied Materials & Interfaces* 7:3427.
118. Woolerton, T. W., S. Sheard, E. Reisner, E. Pierce, S. W. Ragsdale, and F. A. Armstrong 2010. Efficient and clean photoreduction of CO_2 to CO by enzyme-modified TiO_2 nanoparticles using visible light. *Journal of the American Chemical Society* 132:2132.
119. Ardao, I., E. T. Hwang, and A.-P. Zeng 2013. Invitro multienzymatic reaction systems for biosynthesis. *Fundamentals and Application of New Bioproduction Systems* [A.-P. Zeng (ed.)]. Berlin/Heidelberg, Germany: Springer, pp. 153–184.
120. Kuwabata, S., R. Tsuda, K. Nishida, and H. Yoneyama 1993. Electrochemical conversion of carbon dioxide to methanol with use of enzymes as biocatalysts. *Chemistry Letters* 22:1631–1634.
121. Kuwabata, S., R. Tsuda, and H. Yoneyama 1994. Electrochemical conversion of carbon dioxide to methanol with the assistance of formate dehydrogenase and methanol dehydrogenase as biocatalysts. *Journal of the American Chemical Society* 116:5437–5443.
122. Obert, R., and B. C. Dave 1999. Enzymatic conversion of carbon dioxide to methanol: Enhanced methanol production in silica sol–gel matrices. *Journal of the American Chemical Society* 121:12192–12193.
123. Singh, R. K., et al. 2018. Insights into cell-free conversion of CO_2 to chemicals by a multienzyme cascade reaction. *ACS Catalysis* 8:11085–11093.
124. Appel, A. M., et al. 2013. Frontiers, opportunities, and challenges in biochemical and chemical catalysis of CO_2 fixation. *Chemical Reviews* 113:6621–6658.
125. Schlager, S., A. Dibenedetto, M. Aresta, D. H. Apaydin, L. M. Dumitru, H. Neugebauer, and N. S. Sariciftci 2017. Biocatalytic and bioelectrocatalytic approaches for the reduction of carbon dioxide using enzymes. *Energy Technology* 5:812–821.

126. Fu, Q., et al. 2018. Hybrid solar-to-methane conversion system with a Faradaic efficiency of up to 96%. *Nano Energy* 53:232–239.
127. Aresta, M., A. Dibenedetto, and C. Pastore 2005. Biotechnology to develop innovative syntheses using CO_2. *Environmental Chemistry Letters* 3:113–117.
128. Lee, S. Y., S. Y. Lim, D. Seo, J.-Y. Lee, and T. D. Chung 2016. Light-driven highly selective conversion of CO_2 to formate by electrosynthesized enzyme/cofactor thin film electrode. *Advanced Energy Materials* 6:1502207.
129. Kuk, S. K., et al. 2019. Continuous 3D titanium nitride nanoshell structure for solar-driven unbiased biocatalytic CO_2 reduction. *Advanced Energy Materials* 9:1900029.
130. Zhang, S., J. Shi, Y. Chen, Q. Huo, W. Li, Y. Wu, Y. Sun, Y. Zhang, X. Wang, and Z. Jiang 2020. Unraveling and manipulating of NADH oxidation by photogenerated holes. *ACS Catalysis* 10:4967–4972.
131. Yadav, R. K., J.-O. Baeg, G. H. Oh, N.-J. Park, K.-J. Kong, J. Kim, D. W. Hwang, and S. K. Biswas 2012. A photocatalyst–enzyme coupled artificial photosynthesis system for solar energy in production of formic acid from CO_2. *Journal of the American Chemical Society* 134:11455–11461.
132. Kumar, S., R. K. Yadav, K. Ram, A. Aguiar, J. Koh, and A. J. F. N. Sobral 2018. Graphene oxide modified cobalt metallated porphyrin photocatalyst for conversion of formic acid from carbon dioxide. *Journal of CO_2 Utilization* 27:107–114.
133. Zhang, S., J. Shi, Y. Sun, Y. Wu, Y. Zhang, Z. Cai, Y. Chen, C. You, P. Han, and Z. Jiang 2019. Artificial thylakoid for the coordinated photoenzymatic reduction of carbon dioxide. *ACS Catalysis* 9:3913–3925.
134. Jiang, H.-L., and Q. Xu 2011. Porous metal–organic frameworks as platforms for functional applications. *Chemical Communications* 47:3351–3370.
135. Nadar, S. S., L. Vaidya, and V. K. Rathod 2020. Enzyme embedded metal organic framework (enzyme– MOF): De novo approaches for immobilization. *International Journal of Biological Macromolecules* 149:861–876.
136. Zhou, J., S. Yu, H. Kang, R. He, Y. Ning, Y. Yu, M. Wang, and B. Chen 2020. Construction of multi- enzymecascade biomimetic carbon sequestration system based on photocatalytic coenzyme NADH regeneration. *Renewable Energy* 156:107–116.
137. Xu, L., Y. Xiu, F. Liu, Y. Liang, and S. Wang 2020. Research progress in conversion of CO_2 to valuable fuels. *Molecules* 25:3653. doi: 10.3390/molecules25163653. www.mdpi.com/journal/molecules.
138. K. Ma, O. Yehezkeli, E. Park, and J. N. Cha 2016. Enzyme mediated increase in methanol production from photoelectrochemical cells and CO_2. *ACS Catalysis* 6:6982–6986.
139. Le, M., M. Ren, Z. Zhang, P. T. Sprunger, R. L. Kurtz, and J. C. Flake 2011. Electrochemical reduction of CO_2 to CH_3OH at copper oxide surfaces. *Journal of the Electrochemical Society* 158(5):E45–E49.
140. Shironita, S., K. Karasuda, K. Sato, and M. Umeda 2013. Methanol generation by CO_2 reduction at a Pt–Ru/C electrocatalyst using a membrane electrode assembly. *Journal of Power Sources* 240:404–410.
141. Barton, E. E., D. M. Rampulla, and A. B. Bocarsly 2008. Selective solar-driven reduction of CO_2 to methanol using a catalyzed p-GaP based photoelectrochemical cell. *Journal of the American Chemical Society* 130(20):6342–6344.
142. Al-Saydeh, S., and Z. Javaid 2018. Carbon dioxide conversion to methanol: Opportunities and fundamental challenges. doi: 10.5772/intechopen.74779.
143. Jovanov, Z. P., et al. 2016. Opportunities and challenges in the electrocatalysis of CO_2 and CO reduction using bifunctional surfaces: A theoretical and experimental study of Au–Cd alloys. *Journal of Catalysis* 343, 215–231.

144. Kuhl, K. P., T. Hatsukade, E. R. Cave, D. N. Abram, J. Kibsgaard, and T. F. Jaramillo 2014. Electrocatalytic conversion of carbon dioxide to methane and methanol on transition metal surfaces. *Journal of American Chemical Society* 136(40):14107–14113.
145. White, R. E., C. G. Vayenas, and M. E. Gamboa-Aldeco 2009. *Modern Aspects of Electrochemistry*. New York: Springer.
146. Pradier, J. P., and C.-M. Pradier 2014. *Carbon Dioxide Chemistry: Environmental Issues*. Netherland: Elsevier.
147. Goeppert, A., M. Czaun, J.-P. Jones, G. K. S. Prakash, and G. A. Olah 2014. Recycling of carbon dioxide to methanol and derived products–closing the loop. *Chemical Society Reviews* 43(23):7995–8048.
148. Seshadri, G., C. Lin, and A. B. Bocarsly 1994. A new homogeneous electrocatalyst for the reduction of carbon dioxide to methanol at low overpotential. *Journal of Electroanalytical Chemistry* 372(1–2):145–150.
149. Morris, A. J., R. T. McGibbon, and A. B. Bocarsly 2011. Electrocatalytic carbon dioxide activation: The rate-determining step of Pyridinium-catalyzed CO_2 reduction. *ChemSusChem* 4(2):191–196.
150. Popić, J. P., M. L. Avramov-Ivić, and N. B. Vuković 1997. Reduction of carbon dioxide on ruthenium oxide and modified ruthenium oxide electrodes in 0.5 M $NaHCO_3$. *Journal of Electroanalytical Chemistry* 421(1):105–110.
151. Qu, J., X. Zhang, Y. Wang, and C. Xie 2005. Electrochemical reduction of CO_2 on RuO_2/TiO_2 nanotubes composite modified Pt electrode. *Electrochimica Acta* 50(16):3576–3580.
152. Zhao, J., X. Wang, Z. Xu, and J. S. C. Loo 2014. Hybrid catalysts for photoelectrochemical reduction of carbon dioxide: A prospective review on semiconductor/metal complex co-catalyst systems. *Journal of Materials Chemistry A* 2(37):15228–15233.
153. Arai, T., S. Sato, T. Kajino, and T. Morikawa 2013. Solar CO_2 reduction using H_2O by a semiconductor/metal-complex hybrid photocatalyst: Enhanced efficiency and demonstration of a wireless system using $SrTiO_3$ photoanodes. *Energy and Environmental Science* 6:1274–1282.
154. Ding, P., et al. 2019. Controlled chemical etching leads to efficient silicon–bismuth interface for photoelectrochemical CO_2 reduction to formate. *Materials Today Chemistry* 11:80–85.
155. Castro, S., J. Albo, and A. Irabien 2020. Continuous conversion of CO_2 to alcohols in a TiO_2 photoanode- driven photoelectrochemical system. *Journal of Chemical Technology & Biotechnology* 95:1876–1882.
156. Meshitsuka, S., M. Ichikawa, and K. Tamaru 1974. Electrocatalysis by metal phthalocyanines in the reduction of carbon dioxide. *Journal of the Chemical Society*:158.
157. Lieber, C. M., and N. S. Lewis 1984. Catalytic reduction of CO_2 at carbon electrodes modified with cobalt phthalocyanine. *Journal of the American Chemical Society* 106:5033.
158. Atoguchi, T., A. Aramata, A. Kazusaka, and M. Enyo 1991. Electrocatalytic activity of CoII TPP-pyridine complex modified carbon electrode for CO_2 reduction. *Journal of Electroanalytical Chemistry and Interfacial Electrochemistry* 318:309.
159. Guadalupe, A. R., D. A. Usifer, K. T. Potts, H. C. Hurrell, A.-E. Mogstad, and H. D. Abruna 1988. Novel chemical pathways and charge-transport dynamics of electrodes modified with electropolymerized layers of (Co(v-terpy)2)2+. *Journal of the American Chemical Society* 110:3462.
160. Yoshida, T., T. Iida, T. Shirasagi, R.-J. Lin, and M. Kaneko 1993. Electrocatalytic reduction of carbon dioxide in aqueous medium by bis(2,2′: 6′,2″-terpyridine)cobalt(II) complex incorporated into a coated polymer membrane. *Journal of Electroanalytical Chemistry* 344:355.

161. Elgrishi, N., S. Griveau, M. B. Chambers, F. Bedioui, and M. Fontecave 2015. Versatile functionalization of carbon electrodes with a polypyridine ligand: metallation and electrocatalytic H^+ and CO_2 reduction. *Chemical Communications* 51:2995.
162. Cabrera, C. R., and H. D. Abruna 1986. Blocking of recombination sites and photoassisted hydrogen evolution at surface-modified polycrystalline thin films of p-tungsten diselenide. *Journal of Electroanalytical Chemistry and Interfacial Electrochemistry* 209:101.
163. Sato, S., T. Arai, T. Morikawa, K. Uemura, T. M. Suzuki, H. Tanaka, and T. Kajino 2011. Selective CO_2 conversion to formate conjugated with H_2O oxidation utilizing semiconductor/complex hybrid photocatalysts. *Journal of the American Chemical Society* 133:15240.
164. Kou, Y., S. Nakatani, G. Sunagawa, Y. Tachikawa, D. Masui, T. Shimada, S. Takagi, D. A. Tryk, Y. Nabetani, H. Tachibana, and H. Inoue 2014. Visible light-induced reduction of carbon dioxide sensitized by a porphyrin-rhenium dyad metal complex on p-type semiconducting NiO as the reduction terminal end of an artificial photosynthetic system. *Journal of Catalysis* 310:57.
165. Hawecker, J., J.-M. Lehn, and R. Ziessel 1983. Efficient photochemical reduction of CO_2 to CO by visible light irradiation of systems containing $Re(bipy)(CO)_3X$ or $Ru(bipy)_3^{2+}$–Co^{2+} combinations as homogeneous catalysts. *Journal of the Chemical Society*:536.
166. Hawecker, J., J.-M. Lehn, and R. Ziessel 1984. Electrocatalytic reduction of carbon dioxide mediated by $Re(bipy)(CO)_3Cl$ (bipy = 2,2′-bipyridine). *Journal of the Chemical Society*:328.
167. O'Toole, T. R., L. D. Margerum, T. D. Westmoreland, W. J. Vining, R. W. Murray, and T. J. Meyer 1985. Electrocatalytic reduction of CO_2 at a chemically modified electrode. *Journal of the Chemical Society*:1416.
168. Yoshida, T., K. Tsutsumida, S. Teratani, K. Yasufuku, and M. Kaneko 1993. Electrocatalytic reduction of CO_2 in water by $[Re(bpy)(CO)_3Br]$ and $[Re(terpy)(CO)_3Br]$ complexes incorporated into coated nafion membrane (bpy = 2,2′-bipyridine; terpy = 2, 2′;6′,2″-terpyridine). *Journal of the Chemical Society*:631.
169. Bourrez, M., F. Molton, S. Chardon-Noblat, and A. Deronzier 2011. [Mn(bipyridyl)$(CO)_3$Br]: An abundant metal carbonyl complex as efficient electrocatalyst for CO_2 reduction. *Angewandte Chemie International Edition* 50:9903.
170. Smieja, J. M., M. D. Sampson, K. A. Grice, E. E. Benson, J. D. Froehlich, and C. P. Kubiak 2013. Manganese as a substitute for rhenium in CO2 reduction catalysts: the importance of acids. *Inorganic Chemistry* 52:2484.
171. Walsh, J. J., G. Neri, C. L. Smith, and A. J. Cowan 2014. Electrocatalytic CO_2 reduction with a membrane supported manganese catalyst in aqueous solution. *Chemical Communications* 50:12698.
172. Reda, T., C. M. Plugge, N. J. Abram, and J. Hirst 2008. Reversible interconversion of carbon dioxide and formate by an electroactive enzyme. *Proceedings of the National Academy of Sciences of the USA* 105:10654.
173. Parkin, A., J. Seravalli, K. A. Vincent, S. W. Ragsdale, and F. A. Armstrong 2007. Rapid and efficient electrocatalytic CO_2/CO interconversions by Carboxydothermus hydrogenoformans CO dehydrogenase I on an electrode. *Journal of the American Chemical Society* 129:10328.
174. King, P.W. 2013. Designing interfaces of hydrogenase–nanomaterial hybrids for efficient solar conversion. *Biochimica et Biophysica Acta - Bioenergetics* 1827:949–957.
175. Sakimoto, K. K., N. Kornienko, S. Cestellos-Blanco, J. Lim, C. Liu, and P. Yang 2018. Physical biology of the materials–microorganism interface. *Journal of the American Chemical Society* 140:1978–1985.

176. Nam, D. H., S. K. Kuk, H. Choe, S. Lee, J. W. Ko, E. J. Son, E.-G. Choi, Y. H. Kim, and C. B. Park 2016. Enzymatic photosynthesis of formate from carbon dioxide coupled with highly efficient photoelectrochemical regeneration of nicotinamide cofactors. *Green Chemistry* 18:5989–5993.
177. Yang, J., D. Wang, H. Han, and C. Li 2013. Roles of cocatalysts in photocatalysis and photoelectrocatalysis. *Accounts of Chemical Research* 46:1900–1909.
178. Cooney, M. J., V. Svoboda, C. Lau, G. Martin, and S. D. Minteer 2008. Enzyme catalysed biofuel cells. *Energy & Environmental Science* 1:320–337.
179. Choi, E.-G., Y.J. Yeon, K. Min, and Y. H. Kim 2018. Communication—CO_2 reduction to formate: An electro-enzymatic approach using a formate dehydrogenase from rhodobacter capsulatus. *Journal of the Electrochemical Society* 165:H446–H448.
180. Dreyer, D. R., D. J. Miller, B. D. Freeman, D. R. Paul, and C. W. Bielawski 2013. Perspectives on poly(dopamine). *Chemical Science* 4:3796–3802.
181. Shi, J., C. Yang, S. Zhang, X. Wang, Z. Jiang, W. Zhang, X. Song, Q. Ai, and C. Tian 2013. Polydopamine microcapsules with different wall structures prepared by a template-mediated method for enzyme immobilization. *ACS Applied Materials & Interfaces* 5:9991–9997.
182. Kuk, S. K., R. K. Singh, D. H. Nam, R. Singh, J.-K. Lee, and C. B. Park 2017. Photoelectrochemical reduction of carbon dioxide to methanol through a highly efficient enzyme cascade. *Angewandte Chemie International Edition* 56:3827–3832.
183. Zhang, L., N. Vilà, G.-W. Kohring, A. Walcarius, and M. Etienne 2017. Covalent immobilization of (2,20-bipyridyl)(pentamethylcyclopentadienyl)-rhodium complex on a porous carbon electrode for efficient electrocatalytic NADH regeneration. *ACS Catalysis* 7:4386–4394.
184. Srikanth, S., Y. Alvarez-Gallego, K. Vanbroekhoven, D. Pant 2017. Enzymatic electrosynthesis of formic acid through carbon dioxide reduction in a bioelectrochemical system: Effect of immobilization and carbonic anhydrase addition. *ChemPhysChem* 18:3174–3181.
185. Sokol, K. P., W. E. Robinson, A. R. Oliveira, J. Warnan, M. M. Nowaczyk, A. Ru, I. A. C. Pereira, and E. Reisner 2018. Photoreduction of CO_2 with a formate dehydrogenase driven by photosystem II using a semi-artificialz-scheme architecture. *Journal of the American Chemical Society* 140:16418–16422.
186. Ali, I., A. Gill, and S. Omanovic 2012. Direct electrochemical regeneration of the enzymatic cofactor 1,4-NADH employing nano-patterned glassy carbon/Pt and glassy carbon/Ni electrodes. *Chemical Engineering Journal* 188:173–180.
187. Xiu, Y., X. Zhang, Y. Feng, R. Wei, S. Wang, Y. Xia, M. Cao, and S. Wang 2020. Peptide-mediated porphyrin based hierarchical complexes for light-to-chemical conversion. *Nanoscale* 12:15201–15208.
188. Wang, S., D. Zhang, X. Zhang, D. Yu, X. Jiang, Z. Wang, M. Cao, Y. Xia, and H. Liu 2019. Short peptide-regulated aggregation of porphyrins for photoelectric conversion. *Sustainable Energy & Fuels* 3:529–538.
189. Wang, S., M. Li, A. J. Patil, S. Sun, L. Tian, D. Zhang, M. Cao, and S. Mann 2017. Design and construction of artificial photoresponsive protocells capable of converting day light to chemical energy. *Journal of Materials Chemistry A* 5:24612–24616.
190. Yadav, R. K., G. H. Oh, N.-J. Park, A. Kumar, K.-J. Kong, and J.-O. Baeg 2014. Highly selective solar- driven methanol from CO_2 by a photocatalyst/biocatalyst integrated system. *Journal of the American Chemical Society* 136:16728–16731.
191. Rabaey, K., and R. A. Rozendal 2010. Microbial electrosynthesis—Revisiting the electrical route for microbial production. *Nature Reviews Microbiology* 8:706–716.
192. Amao, Y., and N. Shuto 2014. Formate dehydrogenase–viologen-immobilized electrode for CO_2 conversion, for development of an artificial photosynthesis system. *Research on Chemical Intermediates* 40:3267–3276.

193. Miyatani, R., and Y. Amao 2004. Photochemical synthesis of formic acid from CO_2 with formate dehydrogenase and water-soluble zinc porphyrin. *Journal of Molecular Catalysis B: Enzymatic* 27:121–125.
194. Jayathilake, B. S., S. Bhattacharya, N. Vaidehi, and S. R. Narayanan 2019. Efficient and selective electrochemically driven enzyme-catalyzed reduction of carbon dioxide to formate using formate dehydrogenase and an artificial cofactor. *Accounts of Chemical Research* 52:676–685.
195. Ishibashi, T., M. Higashi, S. Ikeda, and Y. Amao 2019. Photoelectrochemical CO_2 reduction to formate with the sacrificial reagent free system of semiconductor photocatalysts and formate dehydrogenase. *ChemCatChem* 11:6227–6235.
196. Schreier, M., et al. 2015. Efficient photosynthesis of carbon monoxide from CO_2 using perovskite photovoltaics. *Nature Communications* 6:7326.
197. White, J. L., J. T. Herb, J. J. Kaczur, P. W. Majsztrik, and A. B. Bocarsly 2014. Photons to formate: Efficient electrochemical solar energy conversion via reduction of carbon dioxide. *Journal CO_2 Utilization* 7:1–5.
198. Schreier, M., F. Heroguel, L. Steier, S. Ahmad, J. S. Luterbacher, M. T. Mayer, J. Luo, and M. Gratzel 2017. Solar conversion of CO_2 to CO using Earth-abundant electrocatalysts prepared by atomic layer modification of CuO. *Nature Energy* 2:17087.
199. Ren, D., N. W. X. Loo, L. Gong, and B. S. Yeo 2017. Continuous production of ethylene from carbon dioxide and water using intermittent sunlight. *ACS Sustainable Chemistry & Engineering* 5:9191–9199.
200. Gurudayal, et al. 2017. Efficient solar-driven electrochemical CO_2 reduction to hydrocarbons and oxygenates. *Energy & Environmental Science* 10:2222–2230.
201. Ma, S., et al. 2016. One-step electrosynthesis of ethylene and ethanol from CO_2 in an alkaline electrolyzer. *Journal of Power Sources* 301:219–228.
202. Dinh, C.-T., et al. 2018. CO_2 electroreduction to ethylene via hydroxide-mediated copper catalysis at an abrupt interface. *Science* 360:783–787.
203. Gabardo, C. M., et al. 2018. Combined high alkalinity and pressurization enable efficient CO_2 electroreduction to CO. *Energy & Environmental Science* 11:2531–2539.
204. Correa-Baena, J.-P., et al. 2017. The rapid evolution of highly efficient perovskite solar cells. *Energy & Environmental Science* 10:710–727.
205. Smotkin, E., A. J. Bard, and M. A. Fox 1988. Multielectrode photoelectrochemical cell for unassisted photocatalysis and photosynthesis US 4793910, Gas Research Institute, Chicago, Il.;.
206. Zhou, L. Q., C. Ling, H. Zhou, X. Wang, J. Liao, G. K. Reddy, L. Deng, T. C. Peck, R. Zhang, M. S. Whittingham, C. Wang, C. Chu, Y. Yao, H. Jia 2019. A high-performance oxygen evolution catalyst in neutral-pH for sunlight-driven CO_2 reduction. *Nature Communications* 10(1):4081.
207. Wang, Y., J. Liu, Y. Wang, Y. Wang, G. Zheng 2018. Efficient solar-driven electrocatalytic CO_2 reduction in a redox-medium assisted system. *Nature Communications* 9(1):5003.
208. Cheng, W., M. H. Richter, I. Sullivan, D. M. Larson, C. Xiang, B. S. Brunschwig, H. A. Atwater, C. Xiang, B. S. Brunschwig, and H. A. Atwater 2020. CO_2 reduction to CO with 19% efficiency in a solar driven gas diffusion electrode flow cell under outdoor solar illumination. *ACS Energy Letters* 5(2):470–476.
209. Huan, T., D. Alves Dalla Corte, S. Lamaison, D. Karapinar, L. Lutz, N. Menguy, M. Foldyna, S.-H. Turren-Cruz, A. Hagfeldt, F. Bella, M. Fontecave, and V. Mougel 2019. Low-cost high-efficiency system for solar-driven conversion of CO_2 to hydrocarbons. *Proceedings of the National Academy of Sciences* 116. 201815412. doi:10.1073/pnas.1815412116.
210. Dutta, A., M. Rahaman, M. Mohos, A. Zanetti, and P. Broekmann 2017. Electrochemical CO_2 conversion using skeleton (sponge) type of Cu catalysts. *ACS Catalysis* 7:5431–5437.

211. Lum, Y., B. Yue, P. Lobaccaro, A. T. Bell, and J. W. Ager 2017. Optimizing C–C coupling on oxide-derived copper catalysts for electrochemical CO_2 reduction. *Journal of Physical Chemistry C* 121:14191–14203.
212. Mistry, H., et al. 2016. Highly selective plasma-activated copper catalysts for carbon dioxide reduction to ethylene. *Nature Communications* 7:12123.
213. Huan, T. N., et al. 2017. Nanostructured oxygen-evolving copper oxide electrocatalyst. *Angewandte Chemie International Edition* 56:4792–4796.
214. Joya, K. S., and H. J. M. de Groot 2016. Controlled surface-assembly of nanoscale leaf-type Cu- oxide electrocatalyst for high activity water oxidation. *ACS Catalysis* 6:1768–1771.
215. Liu, X., et al. 2016. Self-supported copper oxide electrocatalyst for water oxidation at low overpotential and confirmation of its robustness by Cu K-edge X-ray absorption spectroscopy. *Journal of Physical Chemistry C* 120:831–840.
216. Yu, F., F. Li, B. Zhang, H. Li, and L. Sun 2015. Efficient electrocatalytic water oxidation by a copper oxide thin film in borate buffer. *ACS Catalysis* 5:627–630.
217. Yang, W., K. Dastafkan, C. Jia, and C. Zhao 2018. Design of electrocatalysts and electro- chemical cells for carbon dioxide reduction reactions. *Advanced Materials Technologies* 3:1700377.
218. Weekes, D. M., D. A. Salvatore, A. Reyes, A. Huang, and C. P. Berlinguette 2018. Electrolytic CO_2 reduction in a flow cell. *Accounts of Chemical Research* 51:910–918.
219. Wuttig, A., M. Yaguchi, K. Motobayashi, M. Osawa, and Y. Surendranath 2016. Inhibited proton transfer enhances Au-catalyzed CO_2-to-fuels selectivity. *Proceedings of the National Academy of Sciences USA* 113:E4585–E4593.
220. Huan, T. N., et al. 2017. Electrochemical reduction of CO_2 catalyzed by Fe-N-C materials: A structure–selectivity study. *ACS Catalysis* 7:1520–1525.
221. Kang, Q., et al. 2017. Effect of interlayer spacing on the activity of layered manganese oxide bilayer catalysts for the oxygen evolution reaction. *Journal of the American Chemical Society* 139:1863–1870.
222. Zhu, L., J. Du, S. Zuo, and Z. Chen 2016. Cs(I) cation enhanced Cu(II) catalysis of water oxidation. *Inorganic Chemistry* 55:7135–7140.
223. McKone, J. R., et al. 2016. Translational science for energy and beyond. *Inorganic Chemistry* 55:9131–9143.
224. Bella, F., et al. 2016. Improving efficiency and stability of perovskite solar cells with photocurable fluoropolymers. *Science* 354:203–206.
225. Hao, E., and G. C. Schatz 2004. Electromagnetic fields around silver nanoparticles and dimers. *Journal of Chemical Physics* 120:357–366.
226. Kodaimati, M. S., K. P. McClelland, C. He, S. Lian, Y. Jiang, Z. Zhang, and E. A. Weiss 2018. Viewpoint: Challenges in colloidal photocatalysis and some strategies for addressing them. *Inorganic Chemistry* 57(7):3659–3670.
227. Beranek, R. 2019. Selectivity of chemical conversions: Do light driven photo electrocatalytic processes hold special promise? *Angewandte Chemie International Edition* 58(47):16724–16729.
228. Zhou, X., and C. Xiang 2018. Comparative analysis of solar-to-fuel conversion efficiency: A direct, one-step electrochemical CO_2 reduction reactor versus a two-step, cascade electrochemical CO_2 reduction reactor. *ACS Energy Letters* 3(8):1892–1897.
229. Higgins, D., C. Hahn, C. Xiang, T. F. Jaramillo, and A. Z. Weber 2019. Gas-diffusion electrodes for carbon dioxide reduction: A new paradigm. *ACS Energy Letters* 4(1):317–324.
230. Szczesny, J., A. Ruff, A. R. Oliveira, M. Pita, I. A. C. Pereira, A. L. De Lacey, and W. Schuhmann 2020. Electroenzymatic CO_2 fixation using redox polymer/enzyme-modified gas diffusion electrodes. *ACS Energy Letters* 5:321–327.

231. Liu, K., W. A. Smith, and T. Burdyny 2019. Introductory guide to assembling and operating gas diffusion electrodes for electrochemical CO_2 reduction. *ACS Energy Letters* 4(3):639–643.
232. Liu, J., B. Liu, Y. Ren, Y. Yuan, H. Zhao, H. Yang, and S. Liu 2019. Hydrogenated nanotubes/nanowires assembled from TiO_2 nanoflakes with exposed {111} facets: Excellent photo-catalytic CO_2 reduction activity and charge separation mechanism between (111) and (111) polar surfaces. *Journal of Materials Chemistry A* 7(24):14761–14775.
233. Wan, L., Q. Zhou, X. Wang, T. E. Wood, L. Wang, P. N. Duchesne, J. Guo, X. Yan, M. Xia, Y. F. Li, A. A. Jelle, U. Ulmer, J. Jia, T. Li, W. Sun, and G. A. Ozin 2019. Cu_2O nanocubes with mixed oxidation-state facets for (photo)catalytic hydrogenation of carbon dioxide. *Nature Catalysis* 2(10):889–898.
234. Gao, C., Q. Meng, K. Zhao, H. Yin, D. Wang, J. Guo, S. Zhao, L. Chang, M. He, Q. Li, H. Zhao, X. Huang, Y. Gao, and Z. Tang 2016. Co_3O_4 hexagonal platelets with controllable facets enabling highly efficient visible-light photocatalytic reduction of CO_2. *Advanced Materials* 28(30):6485–6490.
235. Chen, Q., X. Chen, M. Fang, J. Chen, Y. Li, Z. Xie, Q. Kuang, and L. Zheng 2019. Photo-induced Au-Pd alloying at TiO_2 {101} facets enables robust CO_2 photocatalytic reduction into hydrocarbon fuels. *Journal of Materials Chemistry A* 7(3):1334–1340.
236. Wu, Y. A., I. McNulty, C. Liu, K. C. Lau, Q. Liu, A. P. Paulikas, C. J. Sun, Z. Cai, J. R. Guest, Y. Ren, V. Stamenkovic, L. A. Curtiss, Y. Liu, and T. Rajh 2019. Facet-dependent active sites of a single Cu_2O particle photocatalyst for CO_2 reduction to methanol. *Nature Energy* 4(11):957–968.
237. Zhao, Y., Y. Wei, X. Wu, H. Zheng, Z. Zhao, J. Liu, and J. Li 2018. Graphene-wrapped Pt/TiO_2 photocatalysts with enhanced photogenerated charges separation and reactant adsorption for high selective photoreduction of CO_2 to CH_4. *Applied Catalysis B* 226:360–372.
238. Xia, C., P. Zhu, Q. Jiang, Y. Pan, W. Liang, E. Stavitski, H. N. Alshareef, and H. Wang 2019. Continuous production of pure liquid fuel solutions via electrocatalytic CO_2 reduction using solidelectrolyte devices. *Nature Energy* 4(9):776–785.
239. Yu, S., A. J. Wilson, G. Kumari, X. Zhang, and P. K. Jain 2017. Opportunities and challenges of solar-energy-driven carbon dioxide to fuel conversion with plasmonic catalysts. *ACS Energy Letters* 2(9):2058–2070.
240. Li, C., X. Tong, P. Yu, W. Du, J. Wu, H. Rao, and Z. M. Wang 2019. Carbon dioxide photo/electroreduction with cobalt. *Journal of Materials Chemistry A* 7(28):16622–16642.
241. Kong, Z. C., J. F. Liao, Y. J. Dong, Y. F. Xu, H. Y. Chen, D.-B. Kuang, and C. Y. Su 2018. Core@shell $CsPbBr_3$@zeolitic imidazolate framework nanocomposite for efficient photocatalytic CO_2 reduction. *ACS Energy Letters* 3(11):2656–2662.
242. Kecsenovity, E., B. Endrödi, P. S. Toth, Y. Zou, R. A. Dryfe, K. Rajeshwar, C. Janaky 2017. Enhanced photoelectrochemical performance of cuprous oxide/graphene nanohybrids. *Journal of the American Chemical Society* 139(19):6682–6692.
243. Kecsenovity, E., B. Endrodi, Z. Papa, K. Herná di, K. Rajeshwar, and C. Janaky 2016. Decoration of ultra-long carbon nanotubes with Cu_2O nanocrystals: A hybrid platform for enhanced photoelectrochemical CO_2 reduction. *Journal of Materials Chemistry A* 4(8):3139–3147.
244. Kormanyos, A., D. Hursán, and C. Janá ky 2018. Photoelectrochemical behavior of PEDOT/nanocarbon electrodes: Fundamentals and structure-property relationships. *Journal of Physical Chemistry C* 122(25):13682–13690.
245. Rajeshwar, K. 2001. *Electron Transfer at Semiconductor-Electrolyte Interfaces. Electron Transfer in Chemistry* [V. Balzani (ed.)]. Weinheim: WileyVCH.
246. Osterloh, F. E. 2013. Inorganic nanostructures for photoelectrochemical and photocatalytic water splitting. *Chemical Society Reviews* 42:2294–2320.

247. Boettcher, S. W., J. M. Spurgeon, M. C. Putnam, E. L. Warren, D. B. Turner-Evans, M. D. Kelzenberg, J. R. Maiolo, H. A. Atwater, and N. S. Lewis 2010. Energy-conversion properties of vapor-liquid-solid grown silicon wire-array photocathodes. *Science* 327(5962):185–187.
248. Warren, E. L., H. A. Atwater, and N. S. Lewis 2014. Silicon microwire arrays for solar energy-conversion applications. *Journal of Physical Chemistry C* 118(2):747–759.
249. Goodey, A. P., S. M. Eichfeld, K.-K. Lew, J. M. Redwing, and T. E. Mallouk 2007. Silicon nanowire array photelectrochemical cells. *Journal of the American Chemical Society* 129(41):12344–12345.
250. Panayotov, D. A., A. I. Frenkel, and J. R. Morris 2017. Catalysis and photocatalysis by nanoscale Au/TiO_2: Perspectives for renewable energy. *ACS Energy Letters* 2(5), 1223–1231.
251. Negrín-Montecelo, Y., M. Comesaña-Hermo, L. K. Khorashad, A. Sousa-Castillo, Z. Wanǵ M. Perez-Lorenzo, T. Liedl, A. O. Govorov, and M. A. Correa-Duarte 2020. Photophysical effects behind the efficiency of hot electron injection in plasmon-assisted catalysis: The joint role of morphology and composition. *ACS Energy Letters* 5(2):395–402.
252. Creel, E. B., E. R. Corson, J. Eichhorn, R. Kostecki, J. J. Urban, and B. D. McCloskey 2019. Directing selectivity of electrochemical carbon dioxide reduction using plasmonics. *ACS Energy Letters* 4(5):1098–1105.
253. Spitler, M. T., M. A. Modestino, T. G. Deutsch, C. X. Xiang, J. R. Durrant, D. V. Esposito, S. Haussener, S. Maldonado, I. D. Sharp, B. A. Parkinson, D. S. Ginley, F. A. Houle, T. Hannappel, N. R. Neale, D. G. Nocera, and P. C. McIntyre 2020. Practical challenges in the development of photoelectrochemical solar fuels production. *Sustainable Energy & Fuels* 4(3):985–995.
254. Xu, P., T. Huang, J. Huang, Y. Yan, and T. E. Mallouk 2018. DyeSensitized photoelectrochemical water oxidation through a buried junction. *Proceedings of the National Academy of Sciences of U. S. A.* 115(27):6946–6951.
255. Salvatore, D., and C. P. Berlinguette 2020. Voltage matters when reducing CO_2 in an electrochemical flow cell. *ACS Energy Letters* 5(1):215–220.
256. Castro, S., J. Albo, and A. Irabien 2018. Photoelectrochemical reactors for CO_2 utilization. *ACS Sustainable Chemistry & Engineering* 6(12):15877–15894.
257. Khan, A. A., and M. Tahir 2019. Recent advancements in engineering approach towards design of photo-reactors for selective photocatalytic CO_2 reduction to renewable fuels. *Journal of CO_2 Utilization* 29:205–239.
258. Tembhurne, S., F. Nandjou, and S. A. Haussener 2019. Thermally synergistic photo-electrochemical hydrogen generator operating under concentrated solar irradiation. *Nature Energy* 4(5):399–407.
259. Verma, S., S. Lu, and P. J. A. Kenis 2019. Co-electrolysis of CO_2 and glycerol as a pathway to carbon chemicals with improved technoeconomics due to low electricity consumption. *Nature Energy* 4(6):466–474.

7 Plasma-Activated Catalysis for CO_2 Conversion

7.1 INTRODUCTION

Plasma-activated catalysis involves plasma and catalysts operating together or in sequence to carry out a chemical reaction. In order to understand plasma-activated catalysis, we first need to understand plasma and how it can interact with the catalyst. Plasma is considered as the "fourth state of matter" after solid, liquid, and gas. With an increase in temperature, a solid material is transformed into liquid, gas, and finally plasma, which is an ionized gas. Only less than 1% of matter in the universe is solid, liquid, or gas largely because of the dominance of plasma phase in our solar system.

Plasma can be created in a number of different ways. It can be created by electrical discharges using direct, pulsed or alternating currents. This can include the use of electrodes made of dielectric materials. Plasma can also be created by radio frequency radiation and microwaves. Thus, the creation of plasma does not necessarily need high temperature.

Electromagnetic waves break down gas into ions, radicals and excited species along with some neutral species, atoms, and molecules. The degree by which electricity ionizes gas in plasma state is generally referred as degree of ionization. This degree can vary depending on the state of plasma from low, partial, or full. The fully ionized means 100% of plasma is in the ionic state. In most practical situations, low or partial ionization are rendered. Unlike neutral species, atoms and molecules, ions, electrons, radicals, and excited species are highly reactive, carry significant energy, and react with each other forming compounds and sets of new radicals and excited species. This reactive nature of plasma can be used for a number of potential applications [1,2]. In this chapter, we strictly focus on plasma-activated catalysis for conversion of carbon dioxide to chemicals and fuels.

Pressure and temperature conditions of plasma are important for their applications. Plasma can be at low pressure, atmospheric pressure, or high pressure. This pressure condition has a significant effect on the way chemical transformation can occur in the plasma state. Low pressure favors collisions between excited species and surface of the catalyst, thus making catalytic effect more noticeable. It reduces collisions of the excited species in the gas phase, thus reducing the possibility of gas-phase deactivation of the active species. Low and atmospheric pressures are also favorable conditions for the industrial operations. High pressure requires more energy consumption. For plasma-catalysis atmospheric pressure dielectric barrier discharge (DBD) is generally considered to offer the most favorable reaction conditions.

DOI: 10.1201/9781003229575-7

Another important variable for plasma is its temperature condition. Plasma temperature can be at local thermodynamic equilibrium or non-local thermodynamic equilibrium. Since plasma is a mixture of various types of ionic, active, and neutral species carrying different degree of vibrational, translational, rotational, and electronic freedom, in principle it can exhibit different temperatures. In general, electrons and active species have higher temperatures than neutral species because they are lighter and carry more motions. When all active and non-active species have the same temperature as the bulk gas, plasma is said to be in thermal equilibrium and this type of plasma is denoted as "thermal plasma." The equilibrium temperature in this type of plasma is generally greater than 10,000 K. In thermal plasma, reactive species are only created at very high temperatures which require high energy consumption.

When plasma is under non-equilibrium thermal condition, lighter electrons, ionic species, and radicals are generally at higher temperatures than heavier neutral species. Their low weight also makes them more mobile causing more motions of different types like vibrational, translational, rotational, etc., leading to more collisions among themselves, with neutral species and with the catalyst surface if present. Heavy neutral molecules are generally at the bulk gas temperature with low degree of motion. The high degree of motions of reactive species allows plasma to create reactive environment at low bulk gas temperature, generally below 1,000 K and they are thus called "non-thermal plasma."

Non-thermal plasma provides an environment of chemical transformation that is usually produced at high temperature in thermal catalysis. Within this plasma environment, there are temperature differences within different types of reactive and non-reactive species. Most active species like electrons and vibrationally excited molecules have generally the highest temperatures in that order. Ions, species with rotational degree of freedom, and neutral species have about the same and lower temperatures. When more active and lighter species like electrons and highly vibrated active species collide with heavier neutral species, because of their size and weight lose a little of their energy and continue to be active. Active species can recombine in the plasma phase to produce new compounds or new types of active species. Electric discharge also continues to provide new sets of electrons and active species and provide additional energy to existing electrons and active species [3,4]. Thus, non-thermal plasma environment continues to provide chemical transformation even in the gas phase. When the catalyst is present, electrons and active species bombard on active metals and inert support to allow further chemical transformation as long as active radicals are alive. In plasma catalysis, active radicals can also create new sets of chemical transformations generating new sets of products and reduce the role of coking. Porosity of the catalyst can also have a significant role in the effectiveness of reactive species to inhibit coking or improve selectivity of certain products.

In thermal catalysis, temperature is used to overcome activation barrier for stable molecules like CO_2. In DBD plasma, electrically charged active species overcomes this barrier with low power requirements. Unlike thermal catalysis, plasma-induced reactions can be easily started or turned off. Generally, plasma-catalysis system achieves stable operation within less than 30 minutes. Furthermore, required electricity for plasma can also be generated from surplus power from renewable sources like solar and wind energy. Such use can make plasma catalysis more of a green technology.

Extent of conversion, energy efficiency, and cost and yield of the desired product are generally considered to be performance parameters for plasma catalysis. The last parameter is only relevant when more than one product is formed such as in CO_2 hydrogenation or dry methane reforming. In these cases, often product selectivity for useful products like CO, methane and other lower hydrocarbon products and methanol are of importance. Other oxygenates such as formaldehyde, ethanol, acetaldehyde, dimethyl ether, and carboxylic acid, which can provide raw materials for more chemicals and fuels are also considered to be important [4-10]. For CO_2 dissociation where only one useful product, CO is formed, product selectivity is not relevant. Energy cost is important depending on the source of energy. If the electricity for CO_2 dissociation is supplied by fossil fuel, net CO_2 removal would be considerably reduced. Energy cost is very important for process scale-up and overall cost for industrial operations. In this chapter, we will mainly focus on conversion and energy efficiency. In plasma catalysis, the synergy between plasma and catalyst can significantly increase conversion efficiency without a reduction in energy efficiency.

For plasma catalysis to be most effective, it requires high electron density and a larger concentration of vibrationally excited molecules [11]. High electron density requires stronger electrical charge and more power. Often DBD plasma cannot provide both at the same time rendering its limitations in providing high conversion and energy efficiency.

High-temperature thermal plasma is normally not used for the CO_2 dissociation by plasma catalysis. In order to improve the performance of the low-temperature, atmospheric-pressure DBD plasma, often "warm plasma" is used. These types of plasma typically operate at low pressure like glow discharge and radio frequency discharge or at gas temperature exceeding 1,000 K. These different operating conditions can generate different types of dominant reactive species which can positively affect the conversion and energy efficiencies. Microwave and GA, or GAP are the most widely used warm plasmas. Like DBD, major contributions of these types of plasma are that they provide electron energy more targeted toward vibrational mode which is most energy efficient for CO_2 dissociation. Furthermore, the high gas and transitional temperature of 2,000–3,000 K of these plasmas compared to DBD plasma also contributes to the rate of chemical reactions and conversion efficiency. However, recent modeling studies have shown some harmful effects of higher temperature [12-14]. Since, these plasmas operate at somewhat lower pressure and higher temperature, they can also have some negative impacts on the cost. Because of their relevance to CO_2 conversion process, we mostly focus on plasma catalysis with DBD, MW, GA, and GAP plasma in this chapter. Performance of other types of plasma is strictly used for comparison.

7.2 PERSPECTIVES ON PLASMA-ACTIVATED CATALYSIS

Whitehead points out that plasma catalysis should be called plasma-activated catalysis [15]. Due to synergy between catalyst and plasma that exist during plasma catalysis, the overall performance of plasma catalysis not only depends on what is going on in the plasma phase and on the surface of the catalyst but also on the interactions between the ionic activities in the plasma and processes occurring on the catalyst external surface and within the catalyst pores. Electromagnetically driven ionization in the plasma targets and modifies an elementary set of reactions, which facilitates and modifies processes occurring on the catalyst surface. While plasma provides gas-phase environment that can activate stable CO_2 molecules, the overall transformation requires understanding of how plasma phase interacts with the catalyst. This process is complex, interactive, and time and scale dependent.

In conventional heterogeneous catalysis, reacting gas molecules are either physically or chemically adsorbed on the catalyst surface. Active catalytic materials are dispersed on the porous support to provide large surface area for the reaction and provide mechanical and thermal stability to the catalyst. The adsorbed active species like radicals either react on the catalyst surface to produce new products or react with gas-phase species. Reactions can occur among the active species and radicals in the gas phase or between adsorbed active species on the catalytic materials. Both of these processes are governed by different mechanisms (Langmuir-Hinshelwood vs. Eley-Rideal) [4,6]. The adsorption–desorption process is essential for catalytic reactions because radicals and active species are only formed at the catalyst surface. While similar principles apply for plasma catalysis except that in plasma catalysis electrically induced active ions, radicals also exist in the gas phase which can collide with other active or neutral species and create new radicals or compounds or lose their energy before being adsorbed on the catalyst surface [4,15,16]. The collision distance at 1 atm. in a DBD reactor is about 100 nm with time interval between two collision of about 1 ns. Active and neutral species in plasma catalysis, if survived, can also be adsorbed on the catalyst surface and produce new radicals or products.

In understanding plasma catalysis, it is important to know the lifetime of various gas-phase active species and radicals. For oxidation reactions, Whitehead [15] reports lifetime of two most active radicals, O^3P and OH to about 14 [17] and 20 μs [18]. Such small times between collision imply that in order for these radicals to interact with catalyst surface, they need to be very close to the surface. The gas-phase radicals–catalyst surface interactions can only occur if the radicals are formed in a very thin boundary layer (about 50 μm) near the catalyst surface. Thus, in order for plasma catalysis to be more effective reactor should be designed (such as a packed-bed reactor) to facilitate formation of active species very near catalyst surface. Catalyst porosity also plays an important role in making this happen. Porous catalyst with a thin layer of active materials coated on the surface can create active species within pores by having gas flow through it and having surface discharge propagating through the pores of the catalyst. Packed bed reactor inter-dispersed with dielectric materials can also help. The literature also points out that during reaction between CO_2 and methane in a DBD reactor, liquid film may be formed around catalyst which not only affects chemistry but also modifies the electric field [12].

As pointed out earlier, plasma catalysis is time and scale dependent. The previous paragraph articulated the importance of scale. The literature shows that interplay between active species and radicals created in the gas phase by plasma and species adsorbed on the catalyst surface must be analyzed together because what happens in one phase affects what happens in other phase and vice versa [19]. Since these processes are fast, it is important to analyze them in dynamic manner with on time recording of the plasma and catalyst behavior. Whitehead points out that conventional method of analyzing the performance of a plasma-catalytic reactor by periodic measurements do not give true understanding of all the phenomena occurring in the plasma-catalysis process [15]. He suggested the use of diffuse reflectance infrared Fourier transform spectroscopy (DRIFTS) for the dynamic measurements. Only with the knowledge of time dependent and microscale phenomena occurring in plasma and catalyst surface, significant improvement on the reactor performance can be achieved.

Many parameters affect the performance of a plasma-catalysis reactor. The active environments in a reactor include plasma phase, particularly near the catalysts surface, active catalyst materials and catalysts support. The active species and radicals created by plasma can react in all three phases simultaneously through constant process of adsorption and desorption. Besides chemical forces causing transformations, electric fields that may be created due to catalyst surface irregularities play an important role. This can not only enhance ionization but can also create localized hot spots accelerating reaction processes [20,21]. Kim et al. showed that during pulsed plasma discharge, active catalyst materials heat and cool faster than the support [20]. In a non-thermal plasma, there is a significant temperature inhomogeneity both in the plasma phase and on the catalyst surface for the reasons described above. This necessitates that modeling of plasma-catalytic reactor requires considerations of time, phase and position-dependent temperature inhomogeneity and their effects on various properties. Dielectric materials and the role they play on changing electric properties of discharge can also significantly affect the reactor performance.

As mentioned before, plasma catalysis is a complex process involving multiple scale and dimensions of gaseous and adsorbed species activities in multiple phases, accompanied by significant temperature variations and their effects on chemical and electrical properties of various phases. To tackle this task can be overwhelming. An engineering approach is to design a plasma-catalytic reactor and perform experiments under different operating conditions and discrete time levels and develop performance models for conversion and energy efficiency and desired product selectivity for the reactor with different degrees of sophistication and use these data and models for process optimization and reactor scale-up. Whitehead points out that while this approach is reasonable, it must be backed up with studies on basic understandings of various physical, chemical, catalytic, thermal, and electrical processes at molecular level both in plasma phase and on the catalyst surface occurring at different times and scales [15]. Time scale can be from pico seconds to minutes while distance scale can vary from nm to cm [20]. Only with that detailed understanding, significant improvements in conversion and energy efficiencies will be achieved to make plasma-catalysis process commercially viable. Full understanding of plasma catalysis will involve better understanding of synergy between plasma and catalyst phases.

7.3 SYNERGY BETWEEN PLASMA AND CATALYST

Whitehead points out that a combination of plasma and catalyst induces synergistic behavior that is not possible in thermal catalysis [15]. As mentioned above, the complex interactions between plasma and the catalyst are interactive, time and scale dependent, and nature of plasma and catalyst structure dependent. While the understanding of all the effects in a dynamic manner may be overwhelming task, the end targets for their role in CO_2 conversion and energy efficiency with added information on energy cost. In order to tackle the end targets, one need to understand synergy between plasma and catalyst. Some aspects of synergy between plasma and catalyst are stated below in Table 7.1 and graphically illustrated in Figure 7.1.

TABLE 7.1
Possible Catalyst-Plasma Synergistic Effects

Effects of Catalyst on the Plasma
a. Changing the distribution of active species
b. Generating micro discharges, particularly in the catalyst pores
c. Enhancing the electric field
d. Change in the discharge type
e. Pollutant concentration in the plasma
Effects of Plasma on Catalysts
a. Accelerating the selectivity
b. Improving the resistance of coke deposition
c. Lowering the activation barrier
d. Having more active sites
e. Changing reaction pathways, particularly surface reaction pathways
f. Generating photocatalytic effect, activation by photon irradiation
g. Enhancing adsorption and desorption probability at catalyst
h. Hot-spot formation
i. Higher catalyst surface area
j. Change in catalyst oxidation state
k. Reduction of metal oxide to metallic catalyst
l. Change in catalyst work function

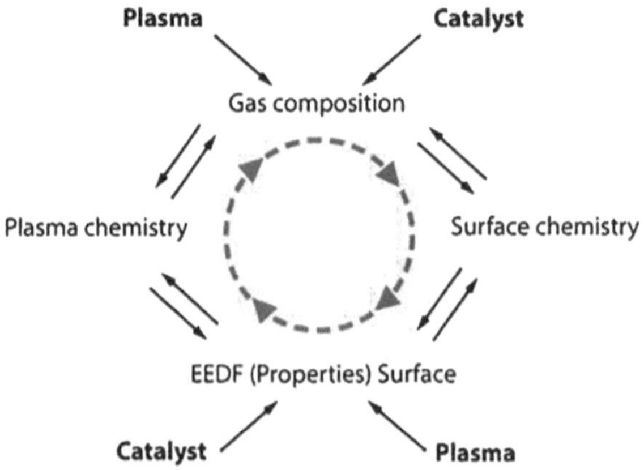

FIGURE 7.1 An overview of the possible effects of the catalyst on the plasma and vice versa, possibly leading to synergism in plasma catalysis [4].

The synergy is a result of four basic effects: physical, chemical, thermal, and electrical. The physical adsorption of reactive and non-reactive species can change the distribution and concentration of radicals and reactive species on the catalyst surface which in turn can affect the conversion of reactive species. Plasma can also physically change adsorption and desorption probability at the catalyst surface. Electrons produced by the discharge can change the behavior of adsorbed species and their impact on the catalyst surface may cause localized heating leading to desorption. Physical effects along with chemical effects can also change catalyst surface area by sintering and particle agglomeration. Synergy is also created by chemical effects by both plasma and the catalyst. Creations of a wide variety of reactive species by plasma bring more possibilities for their recombination, generations of new radicals, and new reaction pathways in the gas phase but particularly on the catalyst surface. Reactive species can accelerate both conversion and selectivity of the products. More vibrationally active species in plasma and active species on the catalyst surface can reduce activation barrier. Low activation barrier also allows a greater possibility for non-adiabatic barrier. Chemical effects can transform metal oxide catalysts to their metallic form and reduce coke formation on the surface and within pores. The electrical properties of the catalyst can also change the distribution and energy levels of gas-phase electrons and radicals as well as the binding energy of the adsorbed species.

Thermal and electric effects also result in plasma-catalyst synergy. Non-uniform temperature distribution among different reactive and non-reactive species may alter adsorption-desorption behavior of the species on the catalyst. They can also affect localized reaction rates in the plasma phase. The creation of hot spots on the active catalyst material can accelerate localized reaction rates and product selectivity. Hot spots can also affect local coke deposition rate. Different types of electrical discharges by different plasma can affect concentrations of different reactive species in plasma. Electric charges at the contact points of the catalyst can change reaction rate and selectivity. This is particularly important for dielectric materials whose conductivity can be a crucial parameter for the discharge properties [22]. There is a significant body of literature on the effects of dielectric materials on conversion and energy efficiency [23–29]. These studies have shown that alumina and quartz perform better than mullite and Pyrex, and that a thicker dielectric lead to a higher conversion and energy efficiency. The literature also shows that changing or doping high-voltage electrode with copper or gold gave 1.5 times better conversion results than iron, platinum, or palladium [29].

Strong electrical field can create micro discharge in the pores of the catalyst which can affect local reaction rates and paths. Catalyst surface roughness and geometric distortion can enhance electric field which in turn can change reacting environment. The presence of insulating surface and different discharge characteristics inside the pore of the catalyst compared to bulk due to microcharge formation in the pores can also cause variation in discharge types. Photons emitted by active plasma species can also activate catalyst. Novel approaches for discharge and reactor design are also taken to improve synergy. Efforts are also made to improve synergy by of micro plasma reactors [30]. A study has shown that a DBD/SOEC hybrid reactor increases conversion of CO_2 dissociation by a factor of four [31]. The use of pulsed power supply to improve nature of discharge has also been attempted [25]. The modeling of DBD reactor indicates that during CO_2 dissociation, electron impact excites CO_2 molecules resulting in its dissociation in CO and O [23,32-37].

While some physical, chemical, thermal, and electrical effects are well understood compared to others, it is often difficult to separate these effects. Some of these synergies should be examined separately to better handle their roles in overall conversion and energy efficiency, product selectivity, and energy cost. Due to synergistic effects, there is a constant interaction among gas-phase active and neutral species and various physical, chemical, thermal, and electrical properties of the catalyst, and this can vary with time because of their influence on and plasma and catalyst chemistry. There are also methods adopted to alter the synergy between plasma and catalysts. These involve either working on new catalyst preparation methods, changing catalyst compositions, or even changing the characteristics of catalyst by plasma treatment in the absence of the reaction. Numerous catalyst preparation methods that have been successful in altering catalyst performance during thermal catalysis can also be used for plasma catalysis [5]. A mixture of active metals can also have positive effects on the synergy. Tailored nano-catalysts can also work better than conventional catalysts. New catalysts can be targeted to specific conversion or product selectivity. Efforts are also made for activating catalyst at reduced temperatures resulting in significant reduction in the energy cost [5]. The method of catalyst, inert, and/or dielectric materials packing in the reactor can also alter synergy between plasma and catalyst.

Plasma catalysis can be carried out at low temperature because electrically induced vibrational excitation of CO_2 molecule can dissociate on the catalyst surface. The catalyst composition, structure, and thermal and electric behaviors can be altered by energetic ionic species. If catalyst has dielectric properties, it can further enhance electric field concentration both at the surface as well as in pores. For these reasons, catalyst pretreatment by plasma has been considered as a viable approach for altering synergy between plasma and catalyst. For thermal catalysis, catalysts are often pretreated by plasma which changes the metal dispersion on the catalyst [38–40]. Novel catalytic materials that involve nanoparticles, selective doping or low temperatures can be pretreated by plasma. Marinov et al. claimed that if catalyst is prepared and modified by plasma under controlled conditions, this modification process may be continued under the reaction conditions [41].

The literature shows that plasma pretreatment of the catalyst is fast and efficient and it can reduce active metal particle size, clean the catalyst surface, and eliminate the need for further calcination process which is important for reducing coke deposition on the catalyst surface and pores [5,42]. Physical, chemical, thermal, and electrical changes infused by plasma pretreatment can improve catalyst activity. Pretreatments can also alter catalyst physical, chemical, thermal, and electrical morphology and structure.

Neyts et al. point out that electronic properties of the catalyst can be changed by changing catalyst surface area and surface morphology both of which can be caused

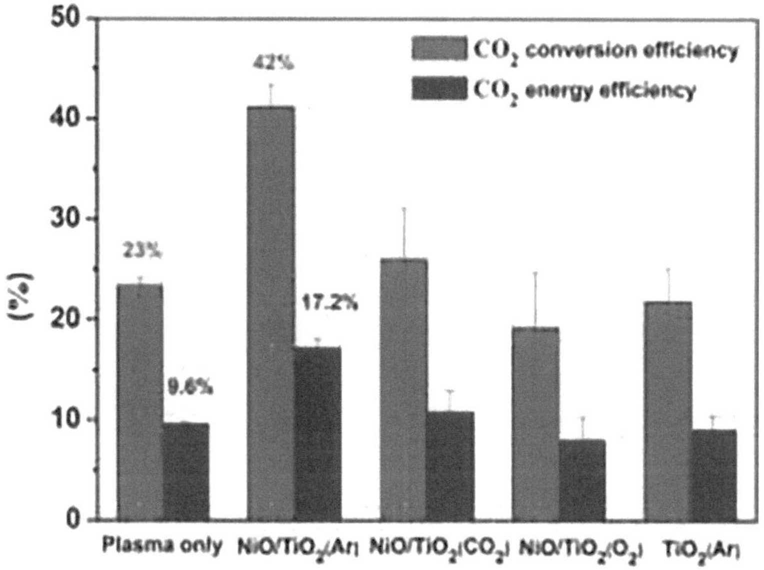

FIGURE 7.2 Effect of plasma-activated catalysis on CO_2 conversion and energy efficiency [51].

by physical modification [43]. The electronic properties are particularly affected for dielectric materials. Zhang et al. [44] and Witvrouwen et al. [45] indicated that plasma pretreatment increased the active surface area by producing better metal dispersion and wavy surface structure. These studies once again indicate closed links among physical, chemical, and electronic effects on the plasma.

Finally, plasma–catalyst interactions can also be altered by the insertions of inert gases like N_2, Ar and He in the plasma. There are mixed results reported on the roles of Ar and He on CO_2 conversion but in general, both improve CO_2 conversion with Ar doing little better than He [46–49]. N_2 plays a very important role on CO_2 conversion due to vibrationally excited N_2 molecules [50]. As shown in Figure 7.2, Chen [51] reported significant increase in both conversion and energy efficiency by the introduction of Ar in pulsed surface-wave sustained microwave discharge for TiO_2 supported NiO catalyst. Figure 7.2 also shows that different catalyst and different gas like O_2 gave poorer results.

When more than one gas plays role in plasma catalysis such as in CO_2 hydrogenation or dry reforming of CH_4 by CO_2, synergy can also affect the product selectivity and conversions of both reactants. Snoeckx and Bogaerts [4] illustrated synergy effects on both reactants for dry reforming of methane in Figure 7.3. The figure shows that a complex phenomenon originating from the interplay between the various plasma–catalyst interactions result in the combined effects that are larger than the sum of the two separate effects for both reactants. [52]. When adding a co-reactant (e.g. CH_4, H_2O, H_2), the catalyst also allows modifying the selectivity toward the targeted products.

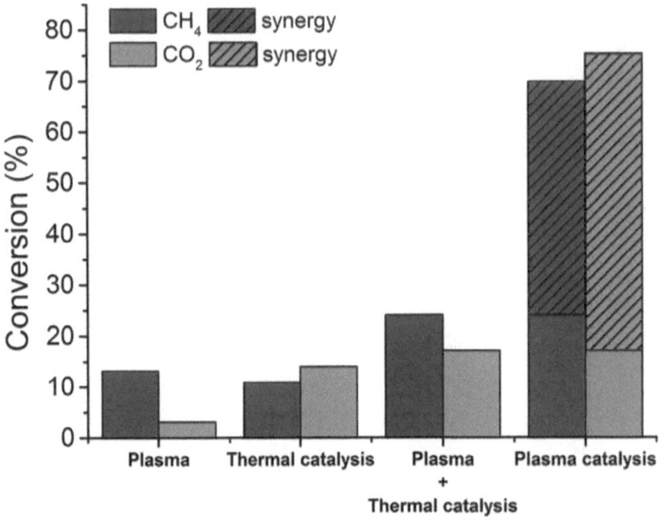

FIGURE 7.3 Demonstration of the synergy of plasma catalysis for the dry reforming of methane [4].

7.4 TYPES OF PLASMA SET-UPS USED FOR CO_2 CONVERSION

As mentioned in Section 7.1, based on the level and distribution of temperature within plasma phase, plasma can be categorized as (a) non-thermal plasma, (b) warm plasma, and (c) thermal plasma. Thermal plasma is at thermal equilibrium at very high temperatures (of the order of 10,000 K and higher). These types of plasma require significant thermal energy and generally are not suitable for the chemical transformation of the type discussed in this paper. Non-thermal plasma, on the other hand, has non-equilibrium temperature distribution, gas-phase temperature of around 1,000 K and contain a wide variety of active and neutral species like ions, electrons, radicals, stable atoms and gas molecules etc. Non-equilibrium temperature distribution and mixture of active and non-active species make them well suited for the chemical transformation such as one discussed here. One such non-thermal plasma is dielectric barrier discharge (DBD) which is very widely used for CO_2 transformation and it is further described in details here. DBD generally operates at atmospheric pressure which makes it very suitable for small as well as large-scale operations. While DBD has many attractive features, conversion and energy efficiency obtained in this type of reactor for CO_2 dissociation are limited and not at the level that plasma catalysis can be commercialized without significant improvement in its performance. Warm plasma has also been examined in significant details because it operates at about 2,000 K and at somewhat lower pressure. Most widely used warm plasma are microwave (MW) plasma and gliding arc (GA) and its companion guiding arc plasmatron (GAP) plasma. While there are many more variations to these non-thermal and warm plasma based on pressure level, discharge characteristics such as its level and duration, etc., these four plasmas (DBD, MW, GA and GAP) are most widely used for CO_2 dissociation. We will therefore mainly describe these ones in details here.

A. Dielectric Barrier Discharge (DBD) Plasma

A DBD is created by either two parallel or concentric cylindrical electrodes, one of which is covered with dielectric barrier made of glass, ceramics quartz or polymer. This barrier resides in between the electrodes, the 0dielectric barrier is a control mechanism for level of discharge and it prevents discharge from going into "thermal plasma" region (Figure 7.4a). When concentric cylindrical electrodes are used, dielectric barrier surrounds the inner cylinder with outer electrode wrapped around barrier in the form of foil or wired mesh. The gap between inner electrode and dielectric barrier is generally of the order of few millimeters and has flows through this gap under plug flow condition from one end to the other. The power is supplied by connection to one electrode while the other electrode is grounded. Generally, power has a voltage range of 1–100 kV and frequency range of few Hz to MHz. Most of the time DBD is operated at atmospheric pressure, although it can be operated in the pressure range of 0.1–1 atm. While the gap acts as a plasma region, generally only 1%–10% of it carries out plasma related non-equilibrium activities [23,53]. The rest of the gas is a carrier to active species and transports them through the discharge region. For the use of DBD in plasma catalysis, catalyst (and possible other inert or dielectric materials) is packed within the gap.

DBD is ignited by passing of the electric current through the gap which breaks down gas into active species like ions, radicals, electrons creating non-equilibrium situation. The temperature differences among various plasma species lead to non-homogeneous environment where large number

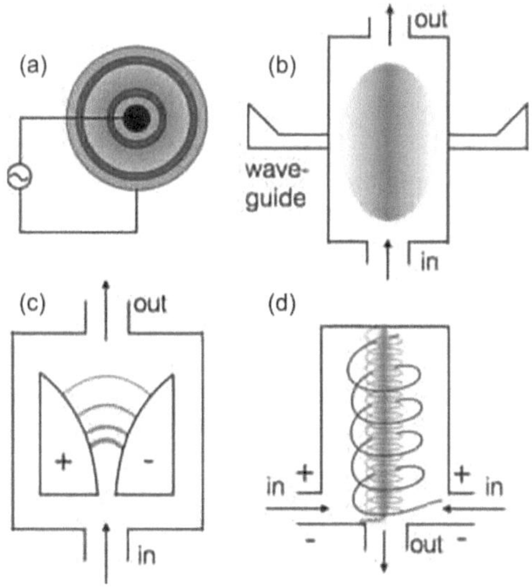

FIGURE 7.4 Schematic illustration of the three plasma reactors most often used for gas conversion applications, i.e., DBD (a), MW plasma (b), and GA discharge, in classical configuration (c) and cylindrical geometry, called GAP (d) [16].

of micro discharges are observed in a so called "filamentary mode". While the occurrence of filamentary mode to some extent depends on gas, CO_2 produces such a mode. The existence and behavior of micro discharges are discussed in great details in the literature [54,11]. For CO_2 conversion, DBD plasma (and reactor) has been most widely used because of it is simple to build, easy to operate with more normal operating conditions, and easy to scale up or down. The scale-up can be easily done by using a number of DBD in parallel. DBD is also easy to operate at an industrial scale [54].

B. Microwave (MW) Plasma

While DBD is created by passing of the electric current, MW plasma is generated by electromagnetic radiation in the frequency range of 2.45–0.915 GHz, although lower end can be extended up to 100 MHz [55]. MW plasmas can come in different forms such as electron cyclotron resonance plasmas, surface wave charges, cavity-induced plasmas, and free expanding atmospheric plasma torches, although surface wave plasma is most frequently used for CO_2 transformation. This type of discharge is created by crossing rectangular waveguide with MW radiation which is passed through transparent quartz tube walls (Figure 7.4b). This type of system is also called "guide-sulfatron" involving surface waves with wavelength of 915 MHz or 2.45 GHz, and this wave energy is transferred to the plasma. The wave energy is absorbed by the plasma by propagation of MW along the interface between quartz tube and the plasma column. While MW plasma operates in "thermal plasma" region with temperature of few 1,000 K at about 0.1 atm [56] and remains in this region for pressure between 0.1 and 1 atm, at low pressure (as low as few m Torr), it is in the non-equilibrium situations and creates reactive species like non-thermal plasmas. The efficiency of MW for CO_2 dissociation is greatly dependent upon pressure; at low pressure like 10 m Torr, they exhibit great energy efficiency. The efficiency also depends on the gas flow, and in the supersonic flow regime, efficiency as high as 90% has been reported [57].

The energy transfer in MW plasma for CO_2 dissociation depends significantly on the vibration energy possessed by CO_2 molecules [58]. MW is most efficient at low pressure because under these conditions, significant difference between an electron temperature of 1–2 eV and a gas temperature of 1,500 K exists and 95% of electron energy is transferred to the CO_2 molecules in the asymmetric vibrational mode [58–60]. High vibrational energy in CO_2 molecules leads to its dissociation [57].

C. Gliding Arc (GA) and Gliding Arc Plasmatron (GAP)

Gliding arc (GA) is a transient discharge which combines its operation in between thermal and non-thermal plasmas and, in this way, combines the advantages of both types of plasma systems (Figure 7.4c). Due to its transient nature, it goes through both quasi-equilibrium and non-equilibrium states with time and simultaneously transitions from thermal to non-thermal mechanisms of ionization. The periodic transformation from arc to strongly non-equilibrium discharge occurs at high currents. The conventional GA discharge passage goes through small electrode distance causing formation of arc to large electrode distance where arc extinguishes itself. The arc "glides" through narrow path to wider path of electrodes and thus its name "Gliding arc". This process generally results in low conversion of gas because a large fraction of gas does not pass-through arc discharge. The conversion can be improved by increasing the gas flow rate to several liters per minute.

Due to its transient nature, applied power and flow rate determine, whether GA can be operated in thermal or non-thermal region. During the beginning of arc, GA can also operate in transition regime (which leads to thermal and subsequent non-thermal) which is very efficient for CO_2 dissociation and in this regime in a revers vortex flow energy efficiency of 43% has been reported [61]. High temperature difference between arc and surrounding gas and high vibration energy by CO_2 appear to be the reasons for this high energy efficiency. In conventional GA, non-thermal plasma conditions can be achieved in nano seconds at high frequencies and low currents. This eliminates thermal plasma region after arc ignition. Because of its non-uniform and limited gas conversion, GA is not suitable for industrial operation. High gas flow is required to arc which further reduces gas residence time and hence gas conversion.

A new design of three-dimensional gliding arc plasmatron (GAP) operating between two cylindrical electrodes was designed to overcome gas flow limitations of GA. This is illustrated in Figure 7.4d. In this design, the grounded reactor outlet acts as anode and the inner cylinder acts as cathode. Gap is designed such that outlet is considerably smaller in diameter compared to the diameter of the reactor body such that tangential flow entering GAP cannot directly leave the reactor and it first flows in forward vortex form near the reactor walls, providing cooling to walls from arc heat generated at the reactor center. When this forward vortex reaches the upper end of the reactor, it has lost some energy due to frictional and inertial forces. When flowing downwards at the upper end, it moves in a smaller inner reverse vortex where it mixes with plasma arc resulting in more energy efficient conversion. Reverse vortex gas eventually leaves the reactor at gas outlet. Just like GA, arc is created at the shortest interelectrode distance and it glides and expands until it reaches the upper part of reactor. The arc rotates around axis and stabilizes at the center in about 1 ms. Unlike in Ga, however, in GAP the inner reverse vortex gas passes through stabilized arc allowing a large fraction of gas converted to the product. Although this design with reverse vortex flow allows larger gas flow exposure to arc than GA design, it still limits gas-arc exposure time.

Besides the above three main discharge types for CO_2 conversion, some research is also being carried out with other plasma types, which include pulse discharge [62,63], radio frequency (RF) discharges, several different atmospheric pressure glow discharges (APGD), corona discharges, spark discharges, and nanosecond pulsed discharges. They are not studied as widely as other three for CO_2 conversion, hence they will not be described here. They are, however, extensively described in the literature [4,5,16].

7.5 ROLE OF PLASMA CHEMISTRY AND REACTOR DESIGN CONSIDERATIONS

A. Role of plasma chemistry

What makes plasma-catalyst chemistry different from conventional heterogeneous catalysis is that in the latter case, gas phase is only activated by thermal forces but most of the chemical transformation occurs on the catalyst surface. In the former case, on the other hand, gas-phase activation and ionization are profoundly affected by the electrical discharges and they play significant role on activation and ionization of neutral species which can react among themselves along with their interactions with the catalyst surface. When gas phase is exposed to electrical discharges, neutral molecules first go into rotatory motion, followed by vibration leading to electronic excitation. This electronic excitation results in bond dissociation which ultimately result into gas ionization forming active species, electrons and radicals which initiate reactions among themselves or form new products by interactions with catalyst surface.

The gas molecules' dissociation and formation of new products in plasma chemistry depend on average electron energy or its equivalent electron temperature. This mean average energy is generally correlated to the reduced electric field of the plasma defined as a ratio of electric charge E divided by the concentration of neutral species n. This ratio is called Townsend, Td; with 1 Td equivalent to 10–21 V m^2. 100 Td is equivalent to 2 eV electron temperature [4,16]. The amount of electrical energy contributing to various forms of molecular motions stated above depends on the level of reduced electric field. This relationship is illustrated in Figure 7.5. As shown, at low reduced electric field most of energy goes into rotational and vibrational motions of molecules leading to electronic excitations. High level of electronic excitations leading to ionization and molecular dissociation occurs at a reduced electric field of 100 Td and higher where DBD plasma mostly operate (greater than around 200 Td). The figure also shows that MW and GA plasma excite electrons at lower Td. Thus, the value of reduced electric field has a wide implication on which channel energy is transferred. Furthermore, this relationship depends on the nature of discharge. For reduced electric charge greater than 200 Td, about 70%–80% of charge goes into electronic excitation with about 5% each going into dissociation and ionization channels. At low E/n of about 50%, 90% of charge goes into molecules vibrational movements. This also means that most dissociation in DBD occurs by less energy efficient electronic impacts and collisions. While in MW and GA (GAP), dissociation is a result of more energy-efficient asymmetric vibrational motions of molecules. Plasma chemistry is also affected by the nature of gas. Different gases like CH_4, H_2O, and H_2 and some inert gases like Ar, He, and N_2 produce different levels of channel distribution as a function of E/n [64]. It is, therefore, important to know what levels of reduced electric field is required for given gas composition to carry out required level of electronic excitation followed by dissociation and ionization.

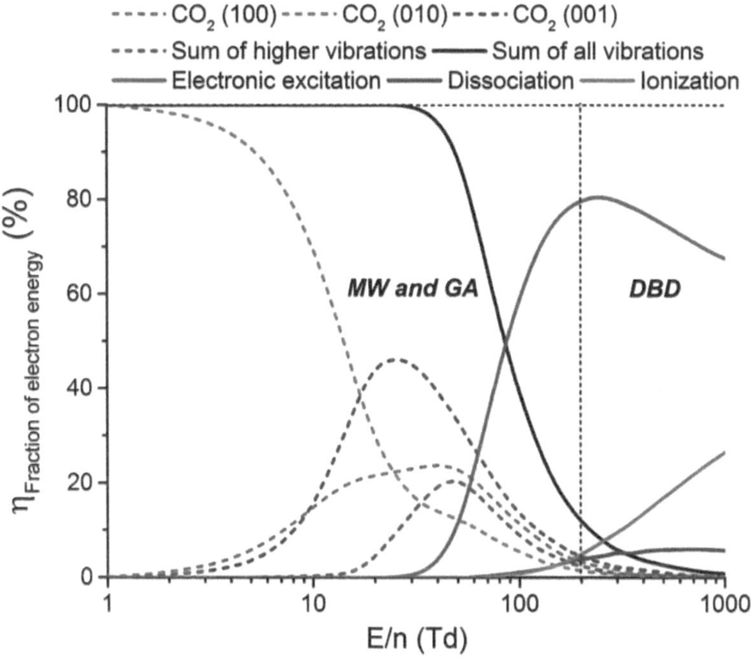

FIGURE 7.5 The fraction of electron energy transferred to different channels of excitation as a function of the reduced electric field (E/n) [4].

Plasma chemistry is a little more complex than what is described above and illustrated by Figure 7.5. While it is important to know where electric charge is contributing in these different levels of molecular movements, formation of electrons, molecular dissociation, and ultimate ionization, since all of these species are in constant motion and state of internal collisions and interactions with the catalyst surface, translation energy of the species may bring some species back to ground state without reaction. As mentioned earlier, the lifetime of some of the excited active species may be short and may not play role in interactions with the catalyst, particularly in DBD plasma catalysis [65]. This has implications on how plasma-catalyst reactor is designed. Furthermore, the literature indicates [2] that most important state of molecule in plasma chemistry is its vibrational mode which carries a temperature of about 1,000–4,000 K and initiates dissociation in homogeneous phase [2] and induces chemical adsorption on the catalyst surface [24,66–71], both of which result in homogeneous plasma chemistry and heterogeneous catalytic chemistry. The latter is most important to take full advantage of plasma-catalyst synergy.

B. Performance parameters and reactor design consideration

The performance of CO_2 dissociation by plasma-activated catalysis is generally evaluated by three important parameters; conversion, energy efficiency and specific energy input. Traditionally, to characterize the process efficiency, two main parameters are *conversion* and *energy* efficiency. Besides these two, specific energy input (SEI) is also important variable. Here we define these parameters based on information provided in numerous review articles [3–10,16].

The conversion efficiency (χ) and energy efficiency (η) of CO_2 are defined as follows:

$$X = \left(\text{moles of } CO_2 \text{ input} - \text{moles of } CO_2 \text{ output}\right) / \text{moles of } CO_2 \text{ input} \quad (7.1)$$

$$\eta = \chi * 2.9\,\text{eV}/\text{SEI} \quad (7.2)$$

The specific energy input SEI (expressed in J/cm^3 or kJ/L) is defined as the plasma power divided by the gas flow rate, and this is the dominant determining factor for the conversion and energy efficiency in a plasma process [2]. Thus, specific energy input (SEI) or the electrical energy deposited per mole of gas is:

$$\text{SEI} = \text{plasma power}/\text{molar flow rate} \quad (7.3)$$

Besides these three parameters, yield of product selectivity is also an important performance variable when more than one product is formed such as in CO_2 hydrogenation or dry reforming. In the case of CO_2 dissociation, when only one product is formed, selectivity parameter is not needed. These three definitions can be further defined in a number of different ways and units as illustrated in the literature [4]. These different definitions take into account presence of other gases, product selectivity and different sets of conversion units. Besides these variables, energy cost is also an important variable for large scale operations. An evaluation of this variable, however, requires source of energy, its current unit cost along with the value for energy efficiency.

Due to synergy between plasma and catalyst, design of the reactor system is very important in order to take maximum advantage of synergy. As mentioned earlier, catalyst properties can be changed by pretreatment of catalyst with plasma. Once catalyst is fully prepared, the reaction between plasma and catalyst can be carried out in two different way; One approach is to create plasma in the first reactor followed by exposure of plasma to the catalyst in the second reactor. This is called two-stage reactors approach.

Earlier discussion on the details of plasma chemistry, however, indicated that lifetime of some of the active species created by plasma may be too short and they may react in the gas phase or die before they have a chance to interact with the catalyst. We pointed out that in order to take maximum advantage of synergy between plasma and catalyst, reactive species need to be created very near catalyst surface so that they can interact with species and vacancies on the catalyst surface. In two-stage plasma catalysis, often known as post-plasma catalysis, only long-lived vibration-excited species, radicals and ionized molecules along with intermediate products can react with the catalyst [2,65,72]. This limits the maximum interactions for all reactive species created by plasma and the catalyst. The reactions among long lived active species and intermediate products with catalyst can also be accomplished by multiple plasma-catalyst hybrid reactors in series.

The second approach is to have a hybrid plasma-catalyst packed-bed reactor where the plasma (i.e. active species) is created close to the catalyst surface so that synergy between plasma and catalyst is fully exploited. This approach is well accepted by most researchers. In this approach, not only active species created by plasma interact with species adsorbed on the catalyst surface most effectively, plasma can also influence acid-base nature of the supports, create more dispersed and uniform distribution of active metals along with infusing changes in the morphology, texture and microstructure of nanoparticle contained within the active materials. It can also modify the nature of the active metal support interface. The packed-bed approach also allows the best path to overcome activation barrier at low temperature that prevails in non-thermal or warm plasmas. Close interactions between reactive and adsorbed species can also enhance conversion [3–5,16].

The objective of most plasma-catalytic process is to take advantages of physical, chemical, thermal and electrical effects in ways to optimize synergy. This also means improved conversion without losing energy efficiency [4,5,16,57,73]. This is particularly important for creating a viable industrial process. In order to gain most from physical, chemical, thermal, and electrical effects for conversion and energy efficiency, proximity of plasma and catalyst is essential, particularly because these four effects are not completely separable and they reinforce each other [3–5,16,59,60,74,75]. Another advantage of packed-bed reactor is that it can be easily inserted within the path of most types of plasma discharges including the four we discussed earlier.

One other advantage of packed-bed reactor is that if one reactor does not give necessary conversion or desired product selectivity, additional reactor(s) can be used in series to improve the performance. Packed-bed reactor can also be set up in parallel to increase flowrate of gas and scale up the process. Packed-bed reactor also offers flexibility in the use of different size and shape of packing to alter openings in between catalyst particles that can affect boundary layer around the catalyst and make better use of electrical properties like micro discharge at the contacts of various particles. Packed bed can be full or partial, or it can also consist of catalyst with other cheaper, dielectric or inert materials to save the catalyst cost. The literature suggests barium titanate or quartz wool. Better conversion of CO_2 is also obtained by pre-prepared homogeneous distribution of catalyst in the bed for full exposure to CO_2 and active species [4,5,16,75].

DBD rector operates in filamentary mode in the absence of packing [2]. In this case, microcharges extend across discharge gap between the electrodes with lifetime of tens of nano seconds. Packing reduces open volume and size of the gaps both of which can enhance microcharges and their role in chemical activity [60]. Once again this is a case of maximizing roles of physical, chemical, thermal and electrical phenomena on synergy. When the packed bed is full and contains dielectric materials, surface charges dominate and weak lower amplitude pulses prevail in the interstices between the particles [76–80]. The inert particles like barium titanate can change surface charges to improve CO_2 dissociation rate by enhancing electron energy for better transformation process [76]. Dielectric materials can reduce electron density which can promote oxygen atoms recombination at the expense of recombination of CO and O to reform CO_2 [3–5,77,79,80]. Hot spots created on active metal can further accelerate reaction and change calcination process to reduce coking. Microcharges created in catalyst pores can also have an effect on coking. Thus, there are many options to improve conversion with packed-bed reactor by many physical, chemical, thermal and electrical phenomena occurring simultaneously to enhance synergy.

Catalyst can also be coated on the surface of the electrodes. This is called in-plasma catalysis. This provides the best method for plasma-catalyst interactions. Electric field and electron density can be enhanced by more touching of the catalyst or dielectric beads. This causes stronger and faster interactions with the plasma. There is a positive contribution from the packing to the conversion when using smaller sizes packing materials due to electric field enhancement at the contact points [81], but also a negative contribution resulting from the lower residence time of the gases because of the volume reduction at a given flow rate. Overall, this can give either a positive or negative effect depending on the nature of material inserted into the voids. Furthermore, as size decreases it is important to sustain a discharge as the breakdown voltage for plasma action strongly increases [81]. Michielsen et al. suggest that all these observations have wider implications for plasma catalysis as the conversion obtained is not just a function of the active sites on the catalyst which is deposited on a support material but also on how that support material is packed into the reactor [82]. This conclusion emphasizes the important part played by the mechanical construction of a plasma-catalytic reactor and how it is packed. For example, Tu and Whitehead noted that for dry reforming of methane, using a DBD reactor, only partially filled reactor with a catalyst doubles the CH_4 conversion and H_2 yield [80].

Although MWs also have a simple geometry, due to the high gas temperature inside the discharge zone (1,000–2,000 K compared to 300–400 K for DBD), catalysts are often placed downstream, due to their low thermal stability. Finally, GA discharges have rather complex geometries and the same higher gas temperatures, thus catalysts are typically introduced downstream, although the use of a spouted bed has also been reported [83]. If these MW and GA discharges can be operated at slightly lower temperatures ($\leq$1,000 K), this would yield new possibilities. It would also open up the way for using sufficiently thermally stable catalysts inside the discharge zone, and at the same time it would allow thermal activation of catalysts inside the discharge zone by the plasma. Various reactor options illustrated by Chen et al. are shown in Figure 7.6 [3].

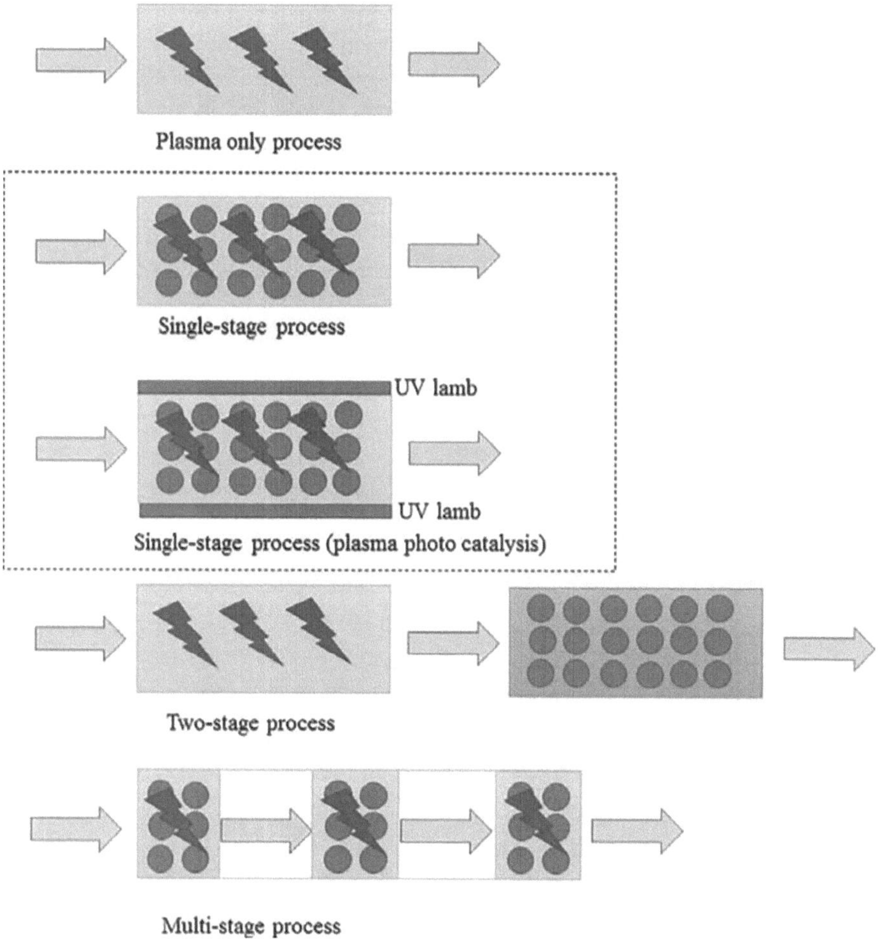

FIGURE 7.6 Schematic diagram of different plasma-catalyst configurations according to the catalyst bed position and number. (1) Plasma alone without a catalyst or with catalyst, (2) catalyst downstream of the discharge, and (3) multistage catalyst-plasma together. These examples mainly apply to a DBD. For MW and GA discharges, the catalyst is most commonly placed in the downstream region [3].

7.6 EFFECTIVENESS OF VARIOUS TYPES OF PLASMA FOR CO_2 DISSOCIATION

A. CO_2 dissociation chemistry in plasma catalysis

As mentioned in earlier section, the biggest contribution of electric charge in plasma is to put molecules in vibrational mode at low temperature (gas temperature in DBD reactor is generally around 100–200 K). This vibrational force on CO_2 can efficiently dissociate CO_2 molecules, although carbon-oxygen bond is very stable and it requires about 5.5 eV to break it.

Dissociation created by direct electronic impact generally requires greater than 7 eV energy. The vibrational mode provides more efficient way of CO_2 dissociation. As mentioned earlier, at low normalized E/n, electric charge puts molecules in a vibrational mode which is followed by the collisions between two vibrational level molecules. Such vibrational-vibrational collisions (V-V collisions) lead to higher vibrational levels molecules by so called ladder climbing approach. Eventually, highly vibrated molecules lead to dissociation of CO_2 molecules. The difference in energy required by direct electronic activation-dissociation process compared to ladder climbing step wise vibrational level increase followed by dissociation process is illustrated in Figure 7.7. The later approach requires a theoretical value of 5.5 eV compared to former process which requires greater than 7 eV [84].

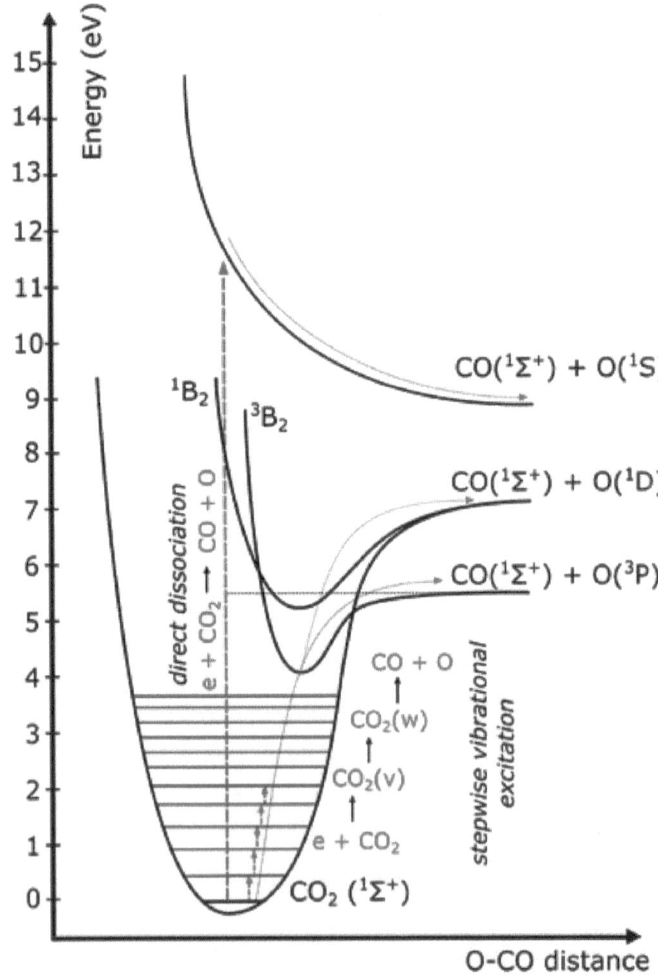

FIGURE 7.7 Schematic of some CO_2 electronic and vibrational levels, illustrating that much more energy is needed for direct electronic excitation–dissociation than for stepwise vibrational excitation, *i.e.*, the so-called ladder-climbing process [84].

The difference in direct electron impact driven CO_2 dissociation versus stepwise vibrational excitement driven approach is manifested in DBD versus MW or GA plasma; the former uses direct electron impact [23], and therefore, less energy efficient, while the latter two use more energy-efficient vibrational excitement approach [2,85].

Another way to demonstrate these two different mechanisms is to examine their effects on the kinetic mechanism of dissociation. The dissociation of a CO_2 molecule in plasma can be represented by the following overall reaction [2,86]:

$$CO_2 \rightarrow CO + 1/2\,O_2^-, \qquad \Delta H = 2.9\,eV/molecule \tag{7.4}$$

When dissociation occurs by direct electron-impact dissociation, the dissociation reaction takes path

$$CO_2 \rightarrow CO + O, \qquad \Delta H = 5.5\,eV/molecule \tag{7.5}$$

which as shown by Chen et al. [3], often accompanied by the further recombination of atomic O:

$$M + O + O \rightarrow O_2 + M\,(\text{Misa particle}) \tag{7.6}$$

On the other hand, vibrationally excited CO_2 molecules may also undergo decomposition via the collisions with atomic O:

$$CO_2 + O^{\text{Vibrat.}} \rightarrow CO + O_2, \qquad \Delta H = 0.3\,eV/molecule \tag{7.7}$$

as well as with the plasma electrons:

$$e + CO_2^{\text{vibr}} \rightarrow CO + 1/2\,O_2^- \left(\text{the energy required is} << 1\,eV\right) \tag{7.8}$$

Note that both Eqs. (7.7) and (7.8) are more energy efficient than Eq. (7.5).

While the vibrational force driven CO_2 dissociation is more efficient than electron impact driven CO_2 dissociation in plasma chemistry, situation for CO_2 dissociation in the presence of catalyst changes significantly. Chen et al. [3] pointed out that both theoretical and experimental literature studies indicate that oxygen vacancies on the catalyst surface play an important role in CO_2 adsorption, activation and dissociation [73,75,87–89]. In plasma catalysis, O atom created in gas-phase dissociation can be adsorbed on the vacant site on the catalyst. More importantly, the energetic electrons supplied by the plasma enhance the dissociative electron attachment of CO_2 at the catalyst surface leading to CO_2 dissociation into CO and O. This is followed by movement of CO molecule by desorption or to another site and remaining bridging O atom fills in the vacancy. The oxygen vacancy can also be filled or regenerated via the recombination on the surface of a bridging oxygen atom with a gaseous oxygen atom. Such oxygen vacancy regeneration maintains the equilibrium of the active sites in the catalyst and

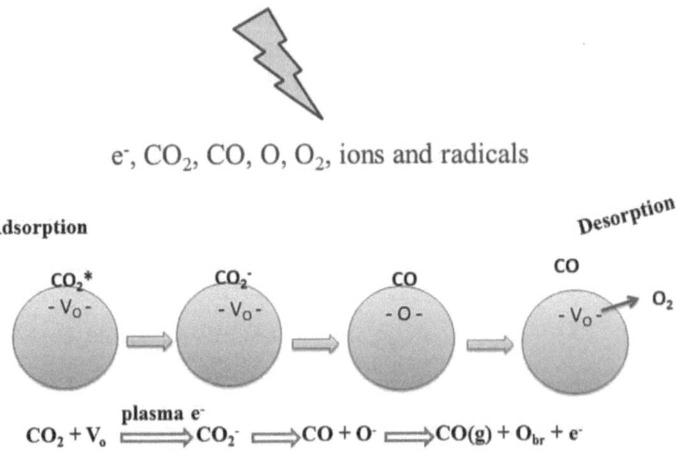

FIGURE 7.8 Schematic mechanism of plasma-assisted catalytic process for CO_2 conversion [3].

controls the CO_2 conversion [73]. In the case of photocatalysts, once plasma can generate electrons of very similar energy (3–4 eV) to the photons, oxygen vacancy can be regenerated by oxidizing the surface O^{2-} anions using holes, followed by releasing O_2 [89].

The CO_2 dissociation mechanism on catalyst surface in the presence of plasma is graphically illustrated in Figure 7.8 [3].

B. Effectiveness of DBD plasma

For CO_2 dissociation, non-thermal DBD plasma is most widely used and studied because it is easy to operate and scale-up and due to its low temperature and atmospheric pressure operation [23,90–92]. The temperature, however, somewhat depends on the applied electric power. It is also perceived as the best plasma for industrial-scale operation. Duan et al. [30] and Mei et al. [93] have shown that without any carrier gas, CO_2 can be conveniently dissociated with a conversion efficiency of about 30% with a power density of 14.75 W/cm^2 in DBD.

DBD plasma is normally used in a packed-bed reactor mode to obtain the best results. A number of design parameters that affect the performance of packed-bed DBD reactor include level of power and applied frequency, gas flow rate, discharge length, discharge gap, and reactor temperature and pressure. The materials used for catalyst, electrodes, dielectric material, and the nature of their packings, size, and shape are also important parameters for DBD reactor performance. The nature of inert (Ar, He, N_2) or other (O_2, H_2, H_2O, CH_4) gases also affects reactor performance [4,5,16]. These operating variables have been tested to improve values of CO_2 conversion and energy efficiency [4,5,16].

As mentioned earlier, it is important to optimize synergy among physical, chemical, thermal, and electrical effects created by the reactor design and process variables mentioned above. For CO_2 dissociation, three important performance parameters described earlier are conversion, energy efficiency and specific energy input, SEI. To that end, there are several obvious deductions based on literature results. Higher power leads higher conversion due to higher electron density, while higher gas flow reduces residence time and thereby reduces the conversion. Tighter packing reduces void space and increases gas velocity, which once again reduces conversion. Larger discharge length, or length of the packed bed increases residence time and therefore increases conversion. Larger gas flow and more micro discharges in the bed give rise to better energy efficiency [94]. Unless synergy is optimized, generally high conversion is accompanied by low energy efficiency. For example, the literature shows that the CO_2 conversion of 30% is accompanied by the energy efficiency of 5%–10% [4,23].

The literature also shows that there are a number of ways to improve energy efficiency and conversion [25,89,95]. Electric power in burst mode can increase energy efficiency up to 20% [25]. The nature of catalyst and dielectric materials can have significant effects on DBD reactor performance. Simultaneous increase in conversion and energy efficiency by a factor of two have been reported using ZrO_2 and TiO_2 catalyst with $BaTiO_3$ (dielectric materials) packing [89,96]. Another study showed that the use of Ni/SiO_2 catalyst with $BaTiO_3$ improved conversion from 19% to 23.5% [95]. The best combination of a conversion of 37.8% and an energy efficiency of 6.4% was obtained at 60 W power. The use of photocatalytic catalyst can have an additional effect on conversion and energy efficiency [89].

As mentioned earlier, for optimizing synergy, physical, chemical, thermal, and electrical interactions between plasma and packing materials (catalyst or other inert dielectric materials) are very important. These effects include physical and chemical adsorption, hot-spot formation and its effect on conversion, microdischarges and their effects on both conversion and energy efficiency by modifications of catalyst structure or reaction pathways. For example, Ray and Subrahmanyam found that the insertion of CeO_2 in discharge length enhanced conversion due to formation of oxygen valency defects to stabilized produced atomic oxygen [97]. The literature [4,82,89,96,98] also shows that packing effect (with or without dielectric materials) on conversion and energy efficiency is not always positive as shown in Figure 7.9 [82]. Enhancement in electrical effects and reduction in resident time counter acts on conversion. This figure is a classic example that shows that in a packed bed, physical, chemical, and electrical effects can counteract with each other and interactions among them do not always manifest into enhanced performance. The effects of applied power, discharge gap, packing material, and packing size and shape can affect the overall reactor performance differently in different situations.

The literature [4] shows that SEI is the most dominant factor for conversion and energy efficiency [23,89,90,92], and while conversion increases with SEI, energy efficiency decreases, particularly for SEI greater than 10 eV per molecule. Different combinations of power and gas flow for the same

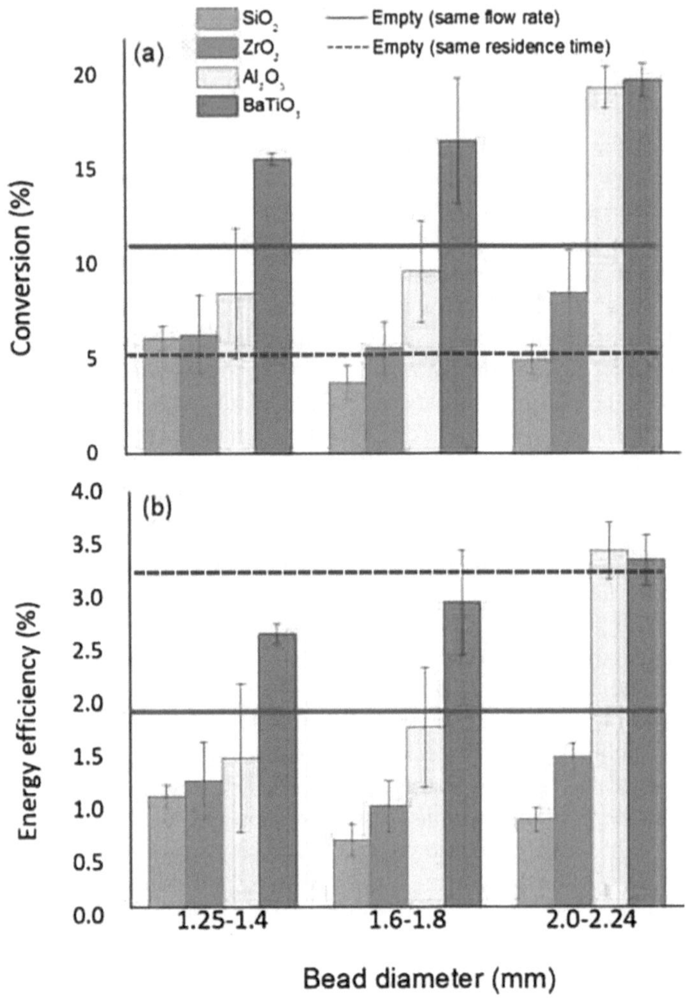

FIGURE 7.9 CO_2 conversion (a) and energy efficiency (b) in a DBD, with and without dielectric packing, for four different packing materials (legend) and three different bead sizes (x-axis), in the case of a DBD reactor with an Al_2O_3 dielectric barrier, 4.5 mm gap, stainless steel outer electrode, applied frequency of 23.5 kHz, 10 W input power, and 50 mL/min gas flow rate [16]. The error bars are defined based on 12 gas chromatography measurements. Comparison is also made with the results of an unpacked reactor, at the same flow rate (solid line) and the same residence time (but much higher flow rate of 192 mL/min; dashed line).

SEI can results in a different conversion because conversion is more dependent on gas flow and residence time than power input [23,25]. Low power and low gas flow are desirable for both conversion and energy efficiency. The literature also shows that while conversion as high as 40% and energy efficiency as high as 23% (for pulsed power) are reported, most data for CO_2 dissociation lie below the energy efficiency of 15% and the conversion of 40%.

Yu et al. showed that discharge behavior and conversion are more affected by the discharge gap than length for fixed SEI [92]. For each system, there is a minimum discharge gap above which low conversion and energy efficiency are obtained. There are mixed results on the effects of applied frequency on conversion and energy efficiency. While Aerts et al. [23] found negligible effect of applied frequency on both conversion and efficiency, Ozkan et al. [24] found both conversion and energy efficiency decreased with an increase in frequency from 15 to 30kHz. This discrepancy was explained on the basis that power input dictates optimum discharge frequency [99].

The literature shows mixed results on the effects of gas temperature on conversion [25,26,90]. While one study indicates a slight increase in conversion (26%–28.5%) with increase in gas temperature from 303 to 443 K [25], the other study indicates the possibility of an increase in recombination reaction for CO and O (to CO_2) and therefore, lower conversion with an increase in wall temperature [90]. Yet another study suggests that cooling of the reactor results in conversion increase from 0.5% to 3% and it should be done by higher gas flow rate [26].

There is a general agreement in the literature that conductivity of dielectric materials affect discharge properties [22,23]. Alumina and quartz perform better than mullite and pyrex [24], and thicker dielectric materials give higher conversion and energy efficiency [25]. Many novel dielectric materials have been tested in the literature [77–79], and one study [26] indicates that coating high-voltage electrodes by either Cu or Au gives 1.5 times more conversion than Fe, Pt or Pd. The literature [26,33,46] indicates that apparent conversion increases and energy efficiency decreases with insertion of inert gases like Ar and He, while N_2 addition shows a completely different behavior [47].

The literature indicates that novel DBD reactor designs like microplasma reactor or DBD/SOEC hybrid reactors with pulsed power supply can lead to increase in conversion [25,30,31]. The latter concept can increase the conversion by a factor of four. Modeling efforts suggest that in a DBD reactor, CO_2 dissociation is dominated by electron impact process [1,23,33–37]. Snoeckx and Bogaerts conclude that while DBD is scalable, easy to operate, and provide reasonable 40% conversion, its energy efficiency needs to be increased by a factor of four [4].

C. Effectiveness of MW plasma

The best way to understand effectiveness of MW plasma is to examine the effect of reduced electric field E/n in Figure 7.5 on various channels of excitation. Unlike DBD, MW plasma is very effective for creating high levels of vibrational excitation through a ladder-climbing approach (outlined in the previous section) at low reduced electric field [13]. These along with high electron density allows them to achieve very high energy efficiency [58]. The literature indicates that while glow discharge can give 30% efficiency at 7kV voltage and RF can give 90% CO_2 conversion at 1kW, MW discharge can give 90% conversion at merely 100 W power [100–102].

The requirement for MW to reach high conversion and energy efficiency is to operate at reduced pressures [57–60,99]. While earlier studies reported 80%–90% energy efficiency with 10%–20% conversion, in supersonic and sub sonic flows at a reduced pressure of up to few 100 mbar, these numbers on energy efficiency and conversion are not repeated in recent studies [57,58]. Van Rooij et al. and Bongers et al. [59,99] reported 10%–20% energy efficiency with 40%–50% conversion. Silva et al., on the other hand, reported a lower energy efficiency (10%) but much higher conversion (50%–80%) [60]. Other studies at atmospheric pressure reported in the literature indicate poorer performance of (20/10) or (50/5) energy efficiency/conversion mix. This level of energy efficiency is somewhat poorer than GA numbers of 30%–40% going as high as 60% [61,103–106].

Design variables important for MW are power, gas flow rate and its nature, temperature, reactor geometry, and composition of gas mixture. In line with the results shown in Figure 7.5, Bongers et al. recommended a reduced electric field of 20–50 Td with low gas temperature and reverse vortex gas flow [99]. High gas flow and reduced pressure are also desirable. MW can obtain an energy efficiency of 20%, which is twice the DBD efficiency, at atmospheric pressure [102]. Sulfaguide can generate an energy efficiency of 90% [102]. The effect of gas temperature on performance for MW is complex. The most desired temperature range is 1,000–2,000 K, and for low E/n and reduced pressure, 2,000 K gas temperature has been observed. For these gas temperatures, temperature of electrons and vibrationally excited species can have temperatures as high as 3,000–8,000 K (1–5 eV) [59,107].

Just like for DBD, SEI is an important parameter for MW, and similar types of trade-off for energy efficiency and conversion exist [59,99,102,108,109]. While conversion increases with SEI, energy efficiency decreases above SEI of 0.1–1 eV. At an SEI of 0.5 eV, an increasing pressure changes flow regime from diffuse to contracted causing thermal excursion up to 14,000 K [59] leading to thermal equilibrium and resulting decline in energy efficiency [102]. The reported effects of gas composition variation by inert gases are mixed [48–50]. Some observed no effect of Ar [48], some observed better conversion but lower energy efficiency with Ar than He [49], while others [50] indicated improved conversion with N_2. A recent study [53] on dissociation of carbon dioxide in the presence of TiO_2 supported NiO catalyst in a pulsed surface-wave sustained microwave discharge with three different gases: O_2, Ar, and CO_2 described earlier indicated near doubling of conversion and energy efficiency in the presence of Ar compared to plasma-only case. These results were described earlier in Figure 7.2.

MW has been studied with post-discharge catalyst packing of Rh/TiO_2 with little success [102]. A study using plasma-pretreated TiO_2-supported NiO catalysts, however, improved the energy efficiency to 17%, a factor of two higher than plasma-only case [73]. The general conclusion is that although present reported maximum values of energy efficiency are in the range of 45%–50%, MW plasma is capable of exceeding 60% energy efficiency [2,57,110].

D. Effectiveness of GA and GAP plasmas

There are two important parameters in the use of various types of plasma for CO_2 dissociation; pressure and mechanism for CO_2 dissociation. Generally atmospheric pressure and dissociation process by vibrational excitation are preferred [106]. Unlike MW and DBD, GA and GAP offer both of these advantages. As shown in Figure 7.5, GA and GAP provide high degree of vibrational excitation at low reduced electric field, E/n (less than 100 Td). Both GA and GAP can be operated at atmospheric pressure.

The differences between GA and GAP are their percentage of gas in contact with discharge and the nature of flow for CO_2 gas. GA relies on two-dimensional electrode blades, and its design has flowrate and residence time limitations with only about 20% of gas has contact with discharge [61,111,112]. This limits the theoretical maximum conversion to be about 20% depending on geometry [61,111,112]. GAP uses cylindrical electrodes and designed such that it allows reverse (inner) vortex flow to have higher fraction of gas in contact with discharge. This allows GAP to have maximum theoretical conversion efficiency of around 40% (twice that of GA). In practice, however, except for one study [113], the reported conversion is below 15%. Energy efficiency is in the range of 40%–50% [92,106,114] going as high as 65% [106,115] except for Indarto et al. [113] who used N_2 and reported less than 5% energy efficiency with 35% conversion. In general, GA and GAP plasmas offer energy efficiency 3–4 times higher than DBD plasma.

SEI is also an important parameter for GA and GAP with its optimal range of 0.1–1 eV per molecule. It shows trade-off between conversion and energy efficiency [61,103,104,113]. The effect of power varies between GA and GAP. In GA, higher power leads to higher conversion but lower energy efficiency [61]. GAP operates in two low and high current regimes, each showing opposite effects. Low current gives high energy efficiency and low conversion and vice versa for high current regime [104]. In both regimes, low gas flow and reverse vortex flow give higher conversions and higher energy efficiencies at higher SEI [104,106]. RVF can be regulated by adjusting reactor geometry [52,104,106]. One study showed an optimum conversion with respect to gas flow [115]. The conversion in GA can be increased by reducing interelectrode distance. Opposite results are obtained with an increase in interelectrode distance [61].

The effect of gas outlet diameter (anode) on GAP performance is illustrated in Figure 7.10. Smaller the outlet diameter, higher the reverse vortex flow resulting in better performance as indicated in Figure 7.10. In both GA and GAP, fraction of gas passage through the arc is limited which results in limited conversion [16,104].

As mentioned before, major reason for improved performances by MW and GA (and GAP) compared to DBD is prevailing dissociation mechanism. In the former cases, it is vibrational kinetics, while in the latter case, it is direct electron impact dissociation [2,4,13,116]. In both cases, conversion can also be improved by limiting recombination of CO and O reaction by

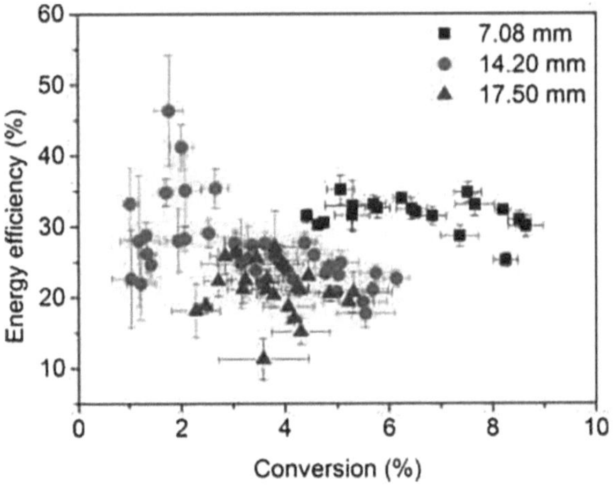

FIGURE 7.10 Energy efficiency vs. CO_2 conversion in a GAP for three different configurations with different anode diameters (cf. legend) and different combinations of power and gas flow rate [104].

either removing O using scavenging materials or chemicals or using membranes [114]. The latter approach is demonstrated by Mori et al. using hybrid DBD/SOEC reactor [117]. The literature shows that addition of O_2 has a negative effect on conversion [113]. The addition of N_2 can increase apparent conversion by a factor of two, but it also produces undesirable NOx.

E. Comparison of performances of different types of plasmas for CO_2 dissociation

Numerous other types of plasma namely corona discharges [106,118–122], glow discharges [100,123,124], non-self-sustained discharges [125], capillary discharges [126], and nanosecond pulsed discharges [127] have been investigated in the literature with the performance of energy efficiency/conversion of 40/15 same as that of DBD, except in one study [125] of non-self-sustained discharge, similar ratio was 50/30. The study by Wen and Jiang [121] showed that high surface area and strong adsorption on catalyst can improve performance, while the study by Brock et al. [100] showed that Rh-coated reactor has the highest activity for the CO_2 dissociation in a fan-type AC glow discharge plasma reactor using 2.5% CO_2 in He. Table 7.2 summarizes a comparison of literature data for various types of plasma as reported by Chen et al. [3].

Figure 7.11 shows a comparison of conversion and energy efficiency data for different types of plasma reported by Snoeckx and Bogaertz [4]. This comparison shows that there is an energy efficiency target of 60% based on making plasma catalysis competitive with electrochemical water splitting and solar thermochemical conversion [1,16]. This target can be achieved for MW and GA (GAP) plasma, but it is factor of four higher than what is possible for DBD. Conversion efficiency of 20% by all plasmas remains too low

TABLE 7.2
Summary of Literature Data on CO_2 Splitting [3]

Plasma Type	Comments	Gas Mixture	Catalyst	χ (%)	η (%)	SEI (eV/Molecule)
DBD		CO_2	—	17	9	5.8
DBD		CO_2	—	30	1	87
DBD		CO_2	γ-Al_2O_3	20	4.9	12
DBD		CO_2	$BaTiO_3$	38	17	6.5
DBD		CO_2	—	18	4	13
DBD	Low flow rate	CO_2	—	14	8	5.2
DBD	10% CO_2 in the gas mixture	CO_2–H_2O–Ar	Ni/γ-Al_2O_3	36	23	4.5
DBD		CO_2	—	35	2	50.8
DBD		CO_2	—	28.2	11.1	7.4
DBD		CO_2–N_2	—	4.5	4.5	2.9
DBD		CO_2	CaO	39.2	7.1	16
DBD		CO_2	—	20	10.4	5.6
DBD		CO_2	ZrO_2	2.9	9.6	9.6
DBD		CO_2	Ni/SiO_2 + $BaTiO_3$	23.5	2.31	29.5
DBD		CO_2	CeO_2 (2 mm)	10.6	27.6	1.11
DBD		CO_2	TiO_2 (3–4 mm)	8.2	15.54	1.53
Glow		CO_2–Ar	Rh-coated	30	1.4	62
RF		CO_2	—	20	3	19
MW		CO_2–Ar	—	10	20	1.4
MW	Supersonic flow	CO_2	—	10	90	0.3
MW		CO_2	NiO/TiO_2	42	18	7.0
MW		CO_2–N_2	—	80	6	39
MW		CO_2	—	20	20	2.9
MW		CO_2	—	12	45	0.8
MW	CO_2:H_2O = 1:1	CO_2–H_2O	—	12	8.7	4
MW		CO_2	NiO/TiO_2	45	56	2.3
MW		CO_2	—			
Corona		CO_2	—	11	2	16
Gliding arc		CO_2	—	4.6	43	0.3
Gliding arc		CO_2	—	15	19	2.3
Gliding arc		CO_2	—	10	34	0.85

for commercialization, although MW shows possibility of achieving 40% conversion and higher (up to 90%) with energy efficiency up to 40%. While a wide variety of both conversions and energy efficiencies achievable with MW discharges for the conversion of pure CO_2, at very high conversion and energy efficiencies, MW plasma would operate in thermal plasma region.

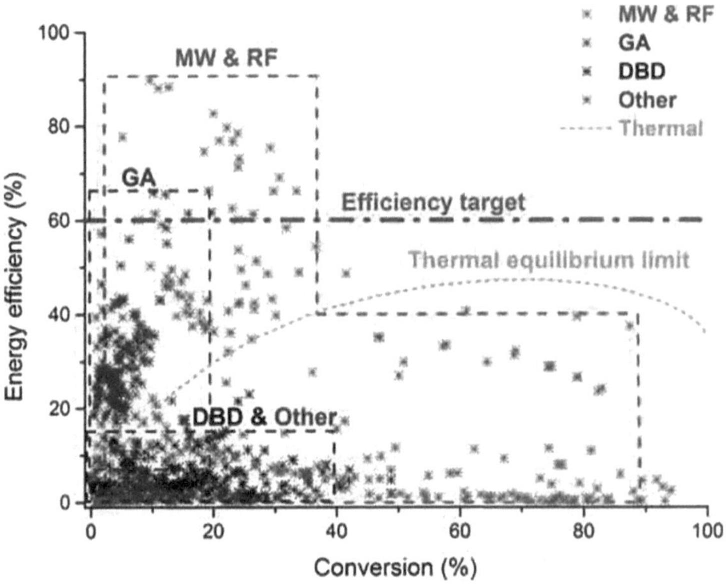

FIGURE 7.11 Comparison of all data collected from the literature for CO_2 splitting in the different plasma types, showing the energy efficiency as a function of conversion. The thermal equilibrium limit and the 60% efficiency target are also indicated. (Adapted from Refs. [4,16].)

7.7 ARTIFICIAL PHOTOSYNTHESIS

Many processes such as ammonia production emitting CO_2 also emit H_2O as a side product, and this facilitates exploration of artificial photosynthesis (reaction between CO_2 and H_2O) to convert CO_2. Unfortunately, like all other reactions involving CO_2, reaction between CO_2 and H_2O is highly endothermic and can be represented as

$$CO_2(g)+H_2O(g)=CO(g)+O_2(g)+H_2(g) \qquad \Delta H=525\,kJ/mol \qquad (7.9)$$

In order to overcome high activation barrier of this reaction, high temperature is required. Even at 2,000 K, the CO_2 conversion by this reaction is only 1.5% [4]. Furthermore, CO produced by this reaction can easily recombine with H_2O by reverse water gas shift reaction to reproduce CO_2 by the following reaction:

$$CO+H_2O=CO_2+H_2 \qquad \Delta H=-41\,kJ/mol \qquad (7.10)$$

As shown in this book, attempts have been made to implement artificial photosynthesis by high-temperature electrolysis or solar energy-based photocatalysis; however, these approaches require either significant energy or they produce low efficiency. The reaction is an ideal candidate for plasma catalysis if it can be activated at low temperature with the use of electrons and activated species created by plasma. Just like CO_2 dissociation, most promising plasma discharge candidates appear to be non-thermal DBD [128,129] or warm plasmas like MW [74,130–132] and GA (or GAP) [61,103], surface discharge [133], and a negative DC corona discharge [134].

The advantages for using non-thermal or warm plasmas for artificial photosynthesis are (a) the reaction can be carried out at low temperatures, (b) pressure can be atmospheric or below atmospheric and (c) the process can be carried out in a decentralized manner near the source of CO_2 at various scales. Syngas produced from this reaction can be used for other hydrocarbons, fuels or fuel additives by F-T and other well know syntheses. The literature [129-131,133–137] indicates that the reaction can produce H_2O_2, methane, oxalic acid, formic acid, dimethyl ether, methanol, ethanol, acetylene, and propadiene, but in very small amounts to be commercially justifiable. The main products appear to be syngas. The observed behavior of various products and literature data indicate that except perhaps for negative DC corona charge, oxidative pathway for CO_2/H_2O conversion and production of oxygenated hydrocarbons is not possible in one step in the absence of catalyst.

Most literature studies for artificial photosynthesis used DBD, MW, or GA plasma, except one [128] which used ferroelectric pellets packed-bed reactor. The studies [61,103,130–135] show that for plasma-only (no catalyst) cases, addition of water harms CO_2 conversion and increases energy cost. The studies also show that MW and GA plasmas are more efficient than DBD, and they allow more water because they operate at higher (order of 1,000 K) temperatures. The best results for all plasma, however, can only be obtained by inserting catalyst in the plasma discharge.

Guo et al. [134] found that while surface wave sustained discharge operating at 915 MHz in a pulse regime, the highest yield of CO and H_2 were only 8% and 4% respectively. The use of NiO/TiO_2 catalyst in DBD with insulating barrier improved CO_2 conversion from 23% (plasma only) to 43% with syngas as the sole product. Ma et al. [138] studied DBD reactor at 105°C, in the presence of Ar and found that syngas ratio can be adjusted between 0 and 0.86. The addition of 10% Ni/Al_2O_3 catalyst increased CO_2 conversion and addition of 20% catalyst increased H_2O conversion up to 14.7%. However, at 30% catalyst, significant carbon deposition occurred. In some of these studies on Ni catalyst, the products also contained some carbon nanofibers and methane.

In another study [94], the use of unreduced and reduced $Ni/g\text{-}Al_2O_3$ catalyst in a DBD reactor was examined. The results showed that compared to the syngas ratio of 0.55–0.18 in plasma-only case, unreduced $NiO/g\text{-}Al_2O_3$ catalyst gave the syngas ratio (H_2/CO) of 0.95–0.45 and partially reduced catalyst gave a syngas ratio of 0.66–0.35. The presence of catalyst also produced small amounts of methane, methanol, C_2H_2, propadiene, and carbon nanofibers. High gas flow gave higher H_2/CO ratios. Chen et al. [74] found that addition of 10% H_2O to CO_2 in MW plasma increased conversion from 23% to 31% with lower energy cost from 30.2 to 22.4 eV per molecule. The addition of NiO/TiO_2 catalyst further increased conversion to 48% with further reduction of energy cost to 14.5 eV per molecule. Hoeben et al. [139] noted that by application of pulsed corona discharges at high energy density in a CO_2 atmosphere over a water film, methane formation was obtained under mild reaction conditions. Dissociation of CO_2 and H_2O seems to induce CO hydrogenation chemistry at the corona-energized high-voltage wires. The study noted that more improvement of the conversion efficiency is needed by application of a nickel-based low-temperature plasma-aided catalyst.

Under right set of operating conditions, artificial photosynthesis can also produce methanol by the following reaction:

$$CO_2(g) + 2H_2O(g) \rightarrow CH_3OH(g) + 3/2O_2(g) \qquad \Delta H = +676\,kJ/mol \quad (7.11)$$

The reaction requires very high temperature and carries very low energy efficiency. The conversion of 60% and the energy efficiency of 40% at 3,300 K and nearly 100% conversion and 25% energy efficiency at 5,000 K can be obtained. Literature studies [94,131,134] also indicate that the best way to increase methanol production from CO_2/H_2O mixture is to use mixture of catalyst like $Cu/ZnO/ZrO_2/Al_2O_3/SiO_2$, CUO/ ZnO/ZrO_2, Cu/ZnO promoted with Pd and Ga, and Pd/ZnO or Pd/SiO_2 promoted with Ga. Eliasson et al. [135] examined $CuO/ZnO/Al_2O_3$ catalyst for CO_2/H_2O in DBD discharge and found that methanol yield and selectivity increased by more than factor of 10 compared to plasma-only case.

All the literature on reforming [140] indicates that the best way to process CO_2/H_2O mixture is to combine it with O_2 and CH_4 and process as a mixture of $CO_2/H_2O/O_2/CH_4$ to induce tri-reforming reactions which will result in the best syngas yield of 1.5–2.0. This is the most desirable syngas ratio for F-T and other related syntheses. Snoeckx and Bogaerts [4] collected all the literature data for the energy cost as a function of conversion and they indicated that MW plasma offers lowest energy cost and highest conversion. **The general conclusion from the literature data is that except for MW and perhaps GAP, in the presence of catalyst, artificial photosynthesis has too high energy cost and not high enough conversion to be commercially viable.**

7.8 CO_2 HYDROGENATION

As mentioned repeatedly in this book, major opportunity for CO_2 utilization lies in its conversion to chemicals and fuels. Some of the chemicals and fuels shown in Table 7.3 offer good prospects because of their global demands.

The most common method for transformation of carbon dioxide into chemicals and fuels is the hydrogenation. Hydrogen is a high energetic material that can be used to reduce carbon dioxide [42]. The products obtained vary depending on the catalyst used and the pressure and temperatures conditions. Different products that can be obtained from hydrogenation are briefly illustrated in Table 7.4.

TABLE 7.3
Some Potential Products from Carbon Dioxide as Raw Material [141]

Product	Reactants	Potential of CO_2 Reduction (Ton CO_2/Ton of Product)	Global Production (MT/yr)
Urea	NH_3 and CO_2	0.75	198.4
Polycarbonate	CO_2, Propylene oxide	0.5	3.6
Methanol	CO_2 and H_2	1.375	75
Dimethyl carbonate	CH_3OH and CO_2	1.467	0.24
Dimethyl ether	CO_2 and H_2	1.913	6.3

TABLE 7.4
Different Products Obtained from Carbon Dioxide Hydrogenation [142]

Reaction	Products
$CO_2 + H_2$	CO
	HCOOH
	CH_4
	$+R_2NH \rightarrow HCONR_2$
	Hydrocarbons
	CH_3OCH_3
	CH_3OH
	Higher alcohols

From a thermodynamic point of view, the reactions generating the products described in Table 7.4 have a positive change of enthalpy and they are endothermic. The use of homogeneous and heterogeneous catalysts is widely extended to perform these reactions. Homogenous catalytic systems usually show higher catalytic activity than heterogeneous catalysts. On the other hand, heterogeneous catalysts are better from the point of view of the reactor design, separation, handling, stability, and reusability of the catalyst. These advantages usually reduce the operation costs, which makes them the best option. The products formed from hydrogenation can be further transformed to a variety of fuels and chemicals by well-known FT, methanol, and other oxygenated syntheses.

Three main products generated by the hydrogenation of CO_2 are CO, CH_4, or CH_3OH depending on operating conditions and the stoichiometry of the reaction. Figure 7.12 summarizes the paths followed in a DBD plasma reactor for selectivity of CO, CH_4, or CH_3OH, as a function of the CO_2 hydrogenation. These paths indicate

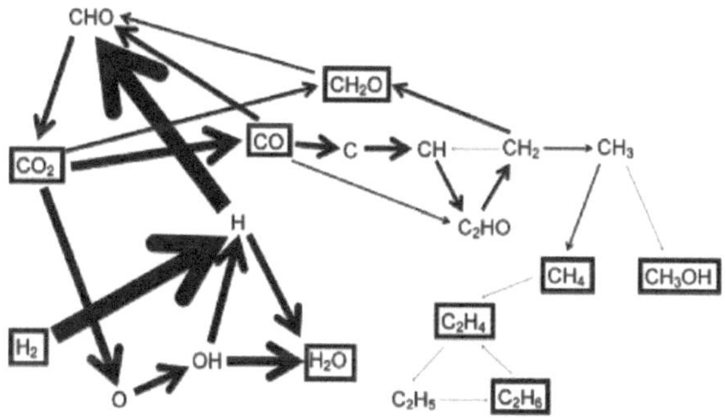

FIGURE 7.12 Dominant reaction pathways for the plasma-based conversion (without catalysts) of CO_2 and H_2 into various products, in a 50/50 CO_2/H_2 gas mixture. The thickness of the arrow lines is proportional to the rates of the net reactions. The stable molecules are indicated with black rectangles [143].

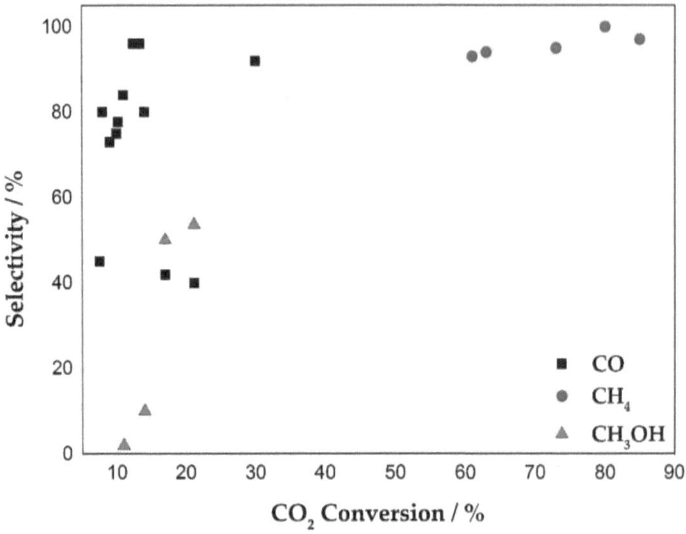

FIGURE 7.13 Overview of selectivity into CO, CH_4, and CH_3OH, as a function of CO_2 conversion, based on all reports available in literature about plasma catalysis (all in DBD reactors) [143].

that the main products are CO and CH_4 for direct hydrogenation of CO_2 in plasma catalysis, while the selectivity of CH_3OH is relatively low. With the right set of catalysts, possibilities for olefins and other gasoline hydrocarbons also exist. Some insights from heterogeneous catalysis, especially the combination of metal catalysts, metal oxide catalysts, and zeolite catalysts can be helpful to develop a methodology for plasma catalytic hydrogenation of CO_2. Based on the literature data for DBD reactor, an overview of product selectivity as a function of CO_2 conversion is illustrated in Figure 7.13. The figure clearly indicates that at low conversion, CO is a dominant product, while at high CO_2 conversion, methane is the dominant product.

7.8.1 Aspects of CO_2 Hydrogenation Mechanisms

A. CO production

CO is produced during CO_2 hydrogenation by reverse water gas shift reaction. The mechanisms suggested for reverse water-gas shift (RWGS) reaction fall into two main categories: (a) the redox pathway and (b) the formate decomposition mechanism. The redox mechanism is most often described over a Cu-based catalyst, the metallic Cu^0 atoms are partially oxidized to form Cu_2O and carbon monoxide and the oxidized copper is later reduced by the hydrogen present in the system, reforming it to its metallic state accompanied by the formation of a water molecule. In contrast to the redox mechanism in which the hydrogen molecule does not actively interact with the carbon dioxide molecule, the formate decomposition pathway suggests that the hydrogen molecule first hydrogenates the carbon dioxide into a formate intermediate. It is then from this intermediate that the cleaving of the carbon-oxygen double bond occurs to release a CO molecule. The formate-mediated route is also reported to occur for the methanation pathway, but has often been debated regarding the extent of its contribution toward product formation [144–147].

B. Methane production

Just like CO formation, mechanistic pathways for methane production fall into two different pathways: (a) direct hydrogenation of CO_2 or (b) CH_4 formation via a CO intermediate, with majority of the findings suggesting the latter process is the more probable methanation route.

A number of studies [148-151] have examined details of methanation via CO intermediate using different experimental techniques. These studies resulted in two different conclusions. Marwood et al. [149] proposed the formation of formate through a carbonate species, which was subsequently hydrogenated into an adsorbed CO species (CO_{ads}) (illustrated in Figure 7.14), which has been suggested to be the rate determining step [150]. In contrast, Jacquemin et al. [151] reported instead that CO_{ads} is a product of the direct dissociation of CO_2 without mentioning the detection (and role) of formate species.

While it was acknowledged that both formate and CO are present under the reaction conditions commonly applied in the mechanistic studies of CO_2 methanation, Vesselli et al. [153] demonstrated that formate was purely a spectator species and that the reaction proceeded via the direct hydrogenation of the C–O bond of CO_2. This conclusion arose mainly from the findings that formate has high reaction barriers for further hydrogenation for its transformation into CO. Further study [92] concluded that the mechanism involving CO without direct participation of the formate species being the most energetically plausible.

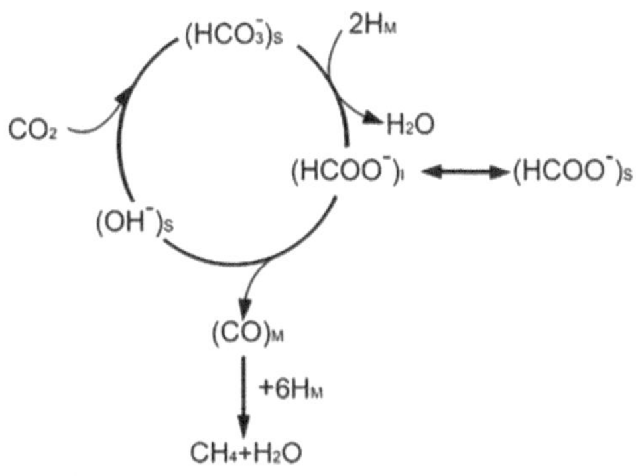

FIGURE 7.14 Proposed CO_2 methanation mechanism involving the formation of formate through a carbonate species. S = support, M = metal, I = metal-support interface [152].

C. Methanol synthesis

Methanol has advantage over CO and methane because it is in liquid form at room temperature. Furthermore, much like CO, it can also serve as an important starting constituent for olefins and aromatics [154–157]. Methanol synthesis, however, is usually less selective than methane and CO under low-pressure conditions. High pressures (50–100 bars) are often required to suppress CO formation via the RWGS pathway.

Arguments on the CO_2 to methanol reaction pathway are also dominated by the formate route and the CO route factions. The pathway suggested for the latter case proposes that CO_2 first undergoes RWGS-like hydrogenation before the CO formed is sequentially hydrogenated into formyl (HCO), formaldehyde (H_2CO), methoxy (H_3CO), and finally methanol [158]. The study by Grabow and Mavrikakis [158], however, suggested the formate pathway being more likely, whereby the hydrogenation of CO_2 into methanol occurs via the following sequence: $CO_2^* \rightarrow HCOO^* \rightarrow HCOOH^* \rightarrow CH_3O_2^* \rightarrow CH_2O^* \rightarrow CH_3O^* \rightarrow CH_3OH^*$ where the symbol * indicates an adsorbed species.

Further studies [159] indicated that catalyst plays an important role along with feed gas composition, temperature, and pressure in the reaction mechanism due to different sites at which key reactions occur. For CO_2 conversion, transition metals such as platinum, copper, cobalt, ruthenium, rhodium, and nickel are among the most active for CO_2 hydrogenation. A bimettalic catalyst can be used for improving selectivity. For example, on combining platinum and cobalt a greater tendency toward CO production was reported [160], whereas varying the composition of a cobalt-iron bimetallic catalyst influenced the resulting alkane/alcohol selectivity [161]. More recently, a nickel-gallium catalyst [35] was developed and shown to be capable of higher selectivity toward methanol production.

7.8.2 Methane and CO Productions

As shown in Figure 4.12, major reactions during hydrogenation of CO_2 produce CO and CH_4. Methane is produced by the highly exothermic Sabatier reaction:

$$CO_2 + 4H_2 \rightarrow CH_4 + 2H_2O \qquad (\Delta H = -165\,kJ/mol) \tag{7.12}$$

which is accompanied by several parallel and side reactions such as reverse WGSR, Boudouard and Bosch reactions, and CO hydrogenation reaction. The prevailing reactions during thermal catalysis in mechanistic terms are illustrated in Figure 7.15. While at low and intermediate temperatures, CO_2 conversion is low and CO dominates product distribution, at high temperatures, CO_2 conversion is high and methane selectivity dominates. As shown in Figure 7.15, methane is produced by the adsorption of CO_2 and CO produced at low temperatures on the catalyst surface. According to the literature, a catalyst can behave as either an acid or a base of Lewis. Thus, as shown in Figure 7.15, CO_2 can be adsorbed on a Lewis base site through the carbon, while oxygen can be adsorbed to a Lewis acid site, leading to the production of a

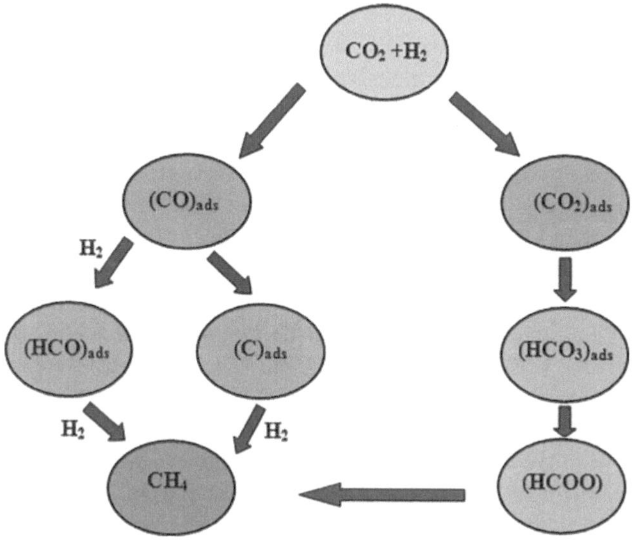

FIGURE 7.15 Global reaction mechanisms of the CO_2 methanation. (Adapted from Fechete [162].)

water molecule. In the presence of H_2 and H atoms, both adsorbed CO_2 and CO are converted to HCO atoms. Furthermore, CO is reduced to C. Both adsorbed HCO and C species are further converted to CH_4. HCO can also be converted to HCOO which can be transformed to methane as well. On a Lewis acid catalyst, and at an atmospheric pressure, methane formation is dependent upon temperature. At temperature below 200°C, no methane is produced; at temperature above 350°C, RWGS becomes dominant. Thus at these intermediate temperatures, with low CO_2 conversion, CO is the dominant product (see Figure 7.13). This also results in syngas production, resulting in the reduction of methane selectivity. Water blinds Lewis site, and its removal requires high temperature.

While thermal catalysis at temperatures higher than 450°C produces higher CO_2 to CH_4 conversion of 80% and methane selectivity at 95%, high temperatures lead to short catalyst life due to coking. It is therefore desirable to reduce working temperature during CO_2 hydrogenation. The use of non-thermal or warm plasma can reduce working temperature to remove water (100°C–160°C) and produce high methane selectivity at lower temperatures. The low temperature can also result in reduction of coking and increased catalyst life. DBD catalytic reactors also require very low consumption of electricity power, less than 12 kJ/mole of CH_4 and, at the same time, they meet all the requirements for a large-volume methane production. As shown below, the nature of catalyst plays a significant role in the effectiveness of the plasma.

The literature data on plasma catalysis for CO_2 hydrogenation are divided in two parts; effectiveness of plasma alone and effectiveness in the presence of catalyst. de Bie et al. [163] performed an extensive computational study on the hydrogenation of CO_2 in a DBD, using a one dimensional (1D) fluid model. The H_2/CO_2 mixing ratio was varied in the entire range from 1/9 to 9/1. The most abundant products predicted by the model were CO, H_2O, and CH_4, with small amounts of formaldehyde, C_2H_6, O_2, and methanol. The study concluded that a CO_2/H_2 mixture in a DBD-only set-up is not suitable for the production of value-added chemicals. Hayashi et al. [133] investigated both the effects of H_2O and H_2 as an additive gas for the plasma-based conversion of CO_2 using surface discharge. A 1/1 mixing ratio was used at atmospheric pressure. While the conversion and yield results for CO_2/H_2 were higher than for the CO_2/H_2O case, they were still too low. These results also indicated that conversions in CO_2 hydrogenation are about a factor of 2–3 lower (and the energy costs the same factor higher) than for DRM and pure CO_2 splitting. In the absence of catalyst, Zeng and Tu [164] found the CO_2 conversion for DBD plasma to be 7.5%, and the main products to be CO, H_2O and CH_4. The CO and CH_4 selectivity were 46% and 8%, respectively. Nizio et al. [165] found that with DBD at 90°C and H_2/CO_2 ratio of 4/1 without a catalyst, the CO_2 conversion was around 5%, without any methanation taking place. With the use of available data, Snoeckx and Boegarts [4] presented unfavorable energy cost for plasma-only case. **Thus, the limited data available in the literature indicate that the hydrogenation of CO_2 for plasma-only cases is not meaningful. Catalyst is essential to take advantage of plasma. Furthermore, MW is more energy efficient compared to DBD.**

In plasma catalysis, catalyst structure and composition have an effect on the kinetics of the reaction. A literature survey [4] indicates that most widely examined and successful catalyst is nickel with different types of supports and promoters for plasma catalysis. These include nickel-based catalysts supported on ceria-zirconia-mixed oxides ($Ce_xZr_{1-x}O_2$) and also hydrotalcite, zeolites, and mesoporous ceria-zirconia composites. Zr promotion has been found to be successful. TiO_2 support has also been examined to take advantage of photo-catalytic effect. These materials show different porosity and a large specific surface area as well as a good thermal stability. Some catalysts were doped with cerium oxide to provide the storage and mobility of oxygen. Ni-Ce gave good thermal and electrical conductivity both in the adiabatic and isothermal conditions. Most catalytic studies have used DBD plasma, although some used MW plasma. The electrical energy supplied to the catalyst via a high voltage sinusoidal current (14 kV) produces streamers, which are responsible for the positive or negative polarization of catalytic sites. The polarization leads to the adsorption of reactants and desorption of the reaction products (mainly water) at low temperatures (<200°C); without polarization, the temperatures are higher (between 300°C and 420°C).

As mentioned earlier, plasma-activated DBD catalytic reactors require usually only very low consumption of electricity power, and at the same time, they meet all the requirements for large volume methane production. Non-thermal plasma which produces a variety of active species such as electrons, ions and radicals has been applied to the methanation of CO [4]. Song et al. [166] investigated the production of synthesis gas with the use of a Ni catalyst packed in dielectric barrier discharge (DBD) reactor, and reported that the CO selectivity considerably increased with the application of plasma. Oshima et al. [167] reported the effect of electric field on the RWGS reaction over different La-ZrO_2 supported catalysts at low temperatures.

Literature studies indicate that it is possible to convert CO_2 into methane within the range of 110°C–420°C, and even at temperatures between 110°C and 270°C, without external heating, i.e. under adiabatic conditions (note that the methanation reaction is exothermic, $\Delta H = -165$ kJ/mol). Amouroux et al. [168] studied a process for the carbon dioxide reduction to methane by a DBD plasma-activated catalyst. They showed that under adiabatic conditions, a DBD plasma was able to greatly improve the conversion of CO_2 at low temperatures. Fan et al. [169] studied CO_2 methanation over Ni/$MgAl_2O_4$ catalysts prepared by different decomposition methods. They obtained smaller Ni particles with enhanced metal-support interaction and unique structure through DBD decomposition, which facilitated the hydrogenation of CO_2 to CH_4. The DBD plasma decomposition was found to be beneficial for improvement in the catalytic activity. Chen et al. [74] added a Ni/TiO_2 catalyst to a surface-wave MW plasma. For CO_2/H_2, conversion was enhanced by a factor of 2 by the catalyst, up to 28%, which was, however, still lower than the value for the pure CO_2 case (41%), and also no methanol or CH_4 formation was observed. A summary of DBD application with Ni catalyst is also reported by Debek et al. [170]. They also reported possible paths for catalyst development.

In the presence of the DBD plasma, the Zr-promoted catalysts show higher activity at lower temperatures, i.e., around 70% CO_2 conversion was measured at 240°C–250°C for both HTNi-Zr and HTNi-ZrCe. The non-doped HTNi catalyst maintains its activity (around 80% conversion with almost 100% selectivity to methane) within the entire temperature window. Even in the presence of a DBD plasma, the highest availability of zero-valent Ni^0 sites of intermediate crystal size might be responsible for the enhanced activity evidenced for this HTNi catalyst. There is no general agreement about the type of basic sites leading to enhanced catalytic activity [114,171]. Even in the presence of plasma, it seems that low- or medium-strength basic sites are still needed in order to boost the methanation reaction. On the contrary, the presence of too strong basic sites introduced by means of Ce and Zr doping does not seem to be beneficial for the overall reaction mechanism. TEM and X-ray diffraction provide no evidence any important change in the morphology and structure of the catalyst upon plasma methanation. Crystal sizes remained practically unchanged.

Most detailed studies on CO_2 hydrogenation with DBD reactor were carried out by Mikhail et al. [172], Ahmed et al. [173], Ge et al. [174], Nizio et al. [175], and Jwa et al. [136]. Mikhail et al. [172] examined a hybrid plasma catalytic system for CO_2 methanation based on the combination of a DBD plasma and Ni/CeZrOx (15NiCZ5842) catalyst. The hybrid plasma catalytic process was active at a low temperature (<270°C) on the selective conversion of CO_2 into methane. In the temperature range of 200°C–300°C, and in the presence of plasma, CO_2 conversion reached 73%–75% with 100% selectivity. The optimum reaction temperature for conversion and selectivity was found to be between 230°C and 270°C. This study also showed that the level of voltage made significant difference on the effect of temperature on methane yield. The study also showed that in plasma configuration worked better than post plasma configuration. Ahmed et al. [173] showed that the high methane yields were obtained at temperature as low as 150°C. The study also showed 20 times

CO_2 conversion activity in presence of plasma compared to thermal catalysis and five times more methane selectivity compared to plasma-only case at 150°C.

The methanation of CO_2 in a DBD packed with Ni/zeolite pellets was investigated by Jwa et al. [136]. Thermal and plasma-assisted catalytic hydrogenations were compared with a varying nickel loading for a temperature range of 180°C–360°C for a stoichiometric 4:1 ratio of H_2/CO_2. For the thermal catalytic hydrogenation case, a conversion of 96% was observed at 360°C, while for the plasma-assisted hydrogenation, the same conversion was already reached at 260°C. Ge et al. [174] examined the effects on the low-temperature activity of an Ni-Ce catalyst and reaction performance in plasma-assisted methanation. The catalyst was 3DOM Ni–Ce catalyst with a pore diameter of 146.6 ± 8.4 nm synthesized through inverse replica of PMMA templates. A methanation reaction test in the thermal fixed-bed reactor showed that plasma reduction generated higher low-temperature activity than the thermal method. The plasma catalysis gave a relatively stable and higher CO_2 conversion capacity; CO_2 conversion was still 82% even when GHSV reached 50,000 h^{-1}.

Nizio et al. [175] examined low-temperature hybrid packed-bed plasma-catalytic methanation over Ni-Ce-Zr hydrotalcite-derived catalysts (Ni-$Ce_xZr_{1-x}O_2$). H_2/CO_2 ratio was 4/1. While methanation by thermal catalysis required temperature higher than 320°C, CO_2 conversions as high as 80% were measured in the presence of the DBD plasma, even at very low reaction temperatures, i.e., 110°C under adiabatic conditions. Furthermore, hydrogenation was very selective (up to 99% at temperature around 260°C) to methane with CO yield lower than 2%.

While as shown above, Ni catalyst with various supports or promoters has been widely examined in plasma catalysis, few other metals have also been examined including Cu, Mn, Cu-Mn, and Ru. Zeng et al. [171] and Zeng and Tu [164] found that among Cu, Mn, and Cu-Mn mixture on Al_2O_3 support, Mn performed best with the highest CO_2 conversion (higher than 10%), close to 80% selectivity and yield for CO and close to 8% selectivity and yield for CH_4, with energy efficiency of 35% for CH_4 and 55% for CO. The best results were obtained at a H_2/CO_2 ratio of 4/1. Reverse water-gas shift reaction (RWGS) and carbon dioxide methanation were found dominant in the plasma CO_2 hydrogenation process with conversion to CO dominating at intermediate temperature while conversion to methane dominating at high temperatures. The best results for CO production were obtained with Mn/y-Al_2O_3 catalyst which enhanced the yield of CO by 114% and the energy efficiency of CO production by 116% compared to the plasma reaction without a catalyst. The temperatures were considerably lower than ones used for thermal catalysis. Another study [176] indicated that the plasma-generated C (or CO) and H_2 reacting into methane (or oxygenates) is also facilitated by multicomponent systems ($Cu/ZnO/ZrO_2/Al_2O_3/SiO_2$) that show good performance for $CO/CO_2/H_2$ mixtures which is generated in situ during the plasma-based conversion of CO_2/H_2. Another study [177] showed that noble metal $Ru/TiO_2/Al_2O_3$ catalyst at a very low H_2/CO_2 ratio of 1/3 and a low temperature of 240°C gave a CO_2 conversion of 78% and a methane selectivity of 93%. There may be some photo-catalytic effects with this catalyst due to TiO_2, and the catalyst may be expensive for commercial purposes. Overall, it appears that Ni-based catalyst has the best commercial potential. A comparison of the performance of CO_2 hydrogenation

TABLE 7.5
Comparison of the Performance of CO_2 Hydrogenation over Ni-Supported on Various Supports in DBD Reactors under NTP Conditions (after [178])

Catalysts	DBD Power (W)	WHSV (mL/gcat h)	NTP Only	Catalyst Thermal[a]	NTP + Catalyst	CH_4 Selectivity (%)	CH_4 Yield (%)
15Ni–CeO_2/Al_2O_3	15–40	40,000	3	4 (at 250°C)	70	96	67
15Ni/$Ce_{0.1}Zr_{0.9}O_2$	1–3	50,000	—	0 (at 280°C)	80	99.7	79.8
15Ni–TiO_2/Al_2O_3	—	1,100	13	5 (at 220°C)	50	—	—
15Ni/UiO66	1–3	30,000	5	5 (at 200°C)	85	99	84.2
15Ni/CZ/SBA-15	—	20,000	—	<1 (at 200°C)	80	99	79
15NiLa/Na–BETA	1–3	23,007	10	0 (at 200°C)	84	97	81
NiCe/Cs–USY	35	40,000	<5	20 (at 250°C)	79	98	77

[a] Catalyst (thermal): CO_2 conversion achieved by catalysts activated under thermal conditions at different temperatures for comparison.

by Ni catalyst on different types of supports in a DBD reactor under NTP conditions is illustrated in Table 7.5 [178].

7.8.3 Methanol Production

In the production of methanol by CO_2 hydrogenation, the catalysts used play an essential role in the selectivity of the reaction and the conversion achieved. In the 70s more active catalysts operated at 50–100 bar and 200°C–300°C. The production of methanol from carbon dioxide can be performed in one single reaction. However, the conversion is increased when an intermediate step is included in order to produce first carbon monoxide which then reacts again with hydrogen to obtain methanol. These reactions are:

$$CO_2 + H_2 \Leftrightarrow CO + H_2O; \qquad (\Delta H = +41.19\,kJ/mol) \tag{7.13}$$

$$CO + 2H_2 \Leftrightarrow CH_3OH; \qquad (\Delta H = -90.70\,kJ/mol) \tag{7.14}$$

While many catalysts such as Cu, Ni, Ru, Pd, Mo, or CeO_2 have been examined, copper-based catalysts have shown the most success. Cu can be mixed different supports, but ZnO/Cu catalysts have been the most successful combination for selectivity because the synergy created between both materials, and both are cheap and abundant materials, and their chemistry has been widely studied.

The production of methanol using plasma catalysis has been examined by various studies. Wang et al. [179] studied the use of a novel (Water electrode design) to produce methanol from CO_2/H_2 mixture at atmospheric pressure and room temperature (~30°C) with and without catalyst. While two catalysts; Cu/γ-Al_2O_3 or Pt/γ-Al_2O_3 were examined, the maximum methanol yield of 11.3% and the methanol selectivity of 53.7% were achieved over the Cu/γ-Al_2O_3 catalyst with a CO_2 conversion of 21.2% in the plasma process, while no reaction occurred at ambient conditions without using plasma. The study showed that the reaction performance of the plasma hydrogenation process was strongly dependent on the design and structure of the plasma reactors and the catalysts, while the influence of H_2/CO_2 molar ratio on the reaction was less critical. Copper catalysts performed better than platinum catalysts. Ronda-Lloret et al. [180] examined CO_2 hydrogenation at atmospheric pressure and low temperature using supported cobalt oxide catalysts. The study showed that packing a catalyst in a DBD plasma enhances the conversion and narrows the product distribution of CO_2 hydrogenation. The study found that the basicity of the MgO support enhances the conversion of CO_2 compared to more acidic supports (γ-Al_2O_3). Both types of supports promote the production of CO as the main product. Although CO is the main product, the catalysts with higher cobalt metal loadings and a good cobalt oxide dispersion favor methanol production. The most active catalyst was 15 wt% Co_xO_y/MgO, which converted 33% CO_2 and 24% H_2 near room temperature and at atmospheric pressure. This catalyst also gave 10% methanol yield.

CO_2 hydrogenation in a DBD reactor at 8 bar pressure and H_2/CO ratio 3/1 without and with CuO/ZnO/Al_2O_3 catalyst was examined by Eliasson et al. [135]. While in the absence of catalyst, experiments produced mainly CO and H_2O with 3%–4% methane and 0.4%–0.5% methanol, methanol yield increased by tenfold and selectivity by 10- to 20-fold in the presence of catalyst. Optimization of the system using low power and high pressure improved methanol selectivity over methane selectivity. Plasma changed the temperature for the maximum catalyst activity from 220°C to 110°C; however, low yield (1%) and high power used made the results not viable at industrial scale.

Two studies [181,182] also examined methanol formation via CO_2 hydrogen using MW discharge. Maya [181] mainly obtained CO and water at 1–2 Torr pressure; secondary products like acetylene, methane, methanol, ethylene, formaldehyde and formic acid were obtained in small amounts when H_2/CO_2 ratio exceeded 1/1. Dela Fuente et al. [182] examined effects of gas flow, H_2/CO_2 mixing ratio and SEI on plasma-only surface wave MW plasma reactor and found that the best performance was for H_2/CO_2 ratio 3 which resulted in 82% CO_2 conversion at energy cost of 28 eV per converted CO_2 molecule. Very small amounts of ethylene and methanol were also produced. The higher conversion was relegated to higher electron density and temperature. The study of Kano et al. [183] with RF impulse discharge at 1–10 Torr pressure found major products to be CO, H_2O and methane with some methanol, formaldehyde and formic acid.

7.9 DRY REFORMING OF METHANE

Dry methane reforming (DRM) has drawn most attention as viable CO_2 utilization technology because it may have one of the greatest commercial potentials. DMR also involves two major carbon compounds (CO_2 and CH_4) that are causing significant environmental issues and treat them simultaneously. The overall DRM reaction is:

$$CH_4 + CO_2 \rightarrow 2H_2 + 2CO \qquad (\Delta H = 247\,\text{kJ/mol}) \tag{7.15}$$

The products of this reaction are the main components of syngas (H_2 and CO), which can be converted to the synthetic fuels as well as H_2 carrier via well-established C1 chemistry.

Conventionally, the H_2/CO ratio of 1/1 from DRM is more suitable for the Fischer-Tropsch synthesis than other methane reforming reactions [140]. Contrary to pure CO_2 splitting, DRM can yield a wide variety of gaseous and liquid products. This can complicate calculations of energy efficiency which must take into account all the products. Energy efficiency can also be replaced by thermal efficiency based on higher or lower heating values of the products. Besides conversions of CO_2 and methane, product selectivity toward syngas and some hydrocarbons are emphasized in the literature.

As shown in Eq. (7.15), DRM is a highly endothermic reaction requiring significant thermal energy to overcome activation barrier. In thermal catalysis, DRM requires high temperature (greater than 900 K) to carry out the reaction. These high temperature leads to significant coke formation and resulting catalyst activation. An extensive review of thermal catalysis for DRM is reviewed by Shah and Gardner [184]. The paper indicates the importance of lowering temperature for this reaction. Plasma catalysis is well suited for this purpose.

Sheng et al. [185] explains the role plasma can play in overcoming activation barrier through energy diagram shown in Figure 7.16. The Gibbs free energy required to overcome activation barrier is related to required heat by the relation

$$\Delta H = T\Delta S + \Delta G \tag{7.16}$$

In thermal catalysis, at least 900 K is required to have negative Gibbs free energy. This temperature is achieved by infusing large thermal energy presumably by combustion of materials that produced CO_2 or by any other means. The exchange of heat

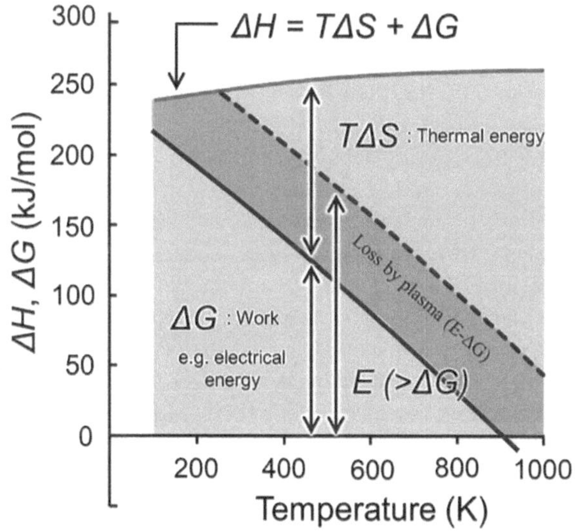

FIGURE 7.16 Energy diagram of DMR. (Adapted from Sheng et al. [185].)

between source and the DMR catalyst bed is not always most efficient and the process operates as Sheng et al. [185] described in the "heat transfer limiting regime".

Gibbs free energy barrier can also be overcome by electric energy of plasma, and this in turn can satisfy Eq. (7.16) at low temperature. Electrical energy accelerates electrons whose energy is transferred to the molecules to initiate DMR at much lower temperature than thermal catalysis. Electronic collision process is independent of reaction temperature if gas density does not change significantly. Meanwhile, a part of the electrical energy is converted to heat. Electrical energy consumed by non-thermal plasma (E) is depicted in the dashed line in Figure 7.16; inevitably, E is greater than ΔG at a fixed temperature. Although heat generated by non-thermal plasma is considered as energy loss (i.e., $E-\Delta G$), both excited species and heat are utilized via endothermic DMR, which enables efficient use of electrical energy without *heat transfer limitation*. As Sheng et al. [185] point out, electrification has the greatest advantage that the energy transfer and the control are independent of temperature gradient. As for other CO_2 conversion applications described earlier a number of non-thermal, warm, and other types of plasma can be used to carry out DRM. Since for dry reforming of methane, DBD, MW and GA plasmas are most widely used, in this chapter we will mainly focus on the roles of these plasmas on DRM. It should be noted that while electric discharge can over activation barrier at low temperature through electron energy transfer, the kinetics of the reaction may still require moderately high temperatures to accelerate the rate of the reaction. These temperatures are still considerable lower than what are required for thermal catalysis and these low temperatures inhibit the coke formation thus allowing the catalyst to have an extended life.

7.9.1 Role of Plasma Pretreatment of the Catalyst for DRM

We previously discussed role of plasma pretreatment on altering synergy between plasma and catalyst. In thermal catalysis, coke formation is a major challenge for DRM reaction. Plasma pretreatment can reduce coke formation during the dry reforming reaction. A possible mechanism for this speculated by Puliyalil et al. [186] is illustrated in Figure 7.17 [187]. In this mechanism, trapping of electrons from plasma on the metallic particles creates a thin plasma sheath around it [188]. This sheath causes strong electrostatic repulsive forces on catalyst which can result in elongation and distortion of catalyst particles on the surface [189]. Metallic particles can also be reduced due to nucleation and slower crystal growth caused by plasma electron bombardment. All these morphological changes can result in the increase in the catalyst active surface area. Plasma treatment can also increase the concentration of chemisorbed O_2, and doping of heteroatoms on the catalysts surface which can also improve the activity of the catalyst [190,191]. Electronic band structure and surface state of the materials can be changed by surface defects and doping of heteroatoms created by interactions between active species of plasma and catalyst [43,192].

Plasma treatment is used for various oxidation-reduction reactions. Reduction of NiO to Ni is well known [193,194]. Benrabbah et al. [195] showed that H_2 pretreatment of Ce-Zr promoted Ni catalyst for CO_2 methanation at lower temperature and with low H_2 consumption. Similarly in another study [196], Ir/Al_2O_3 catalyst was reduced to metallic Ir by Ar plasma treatment to improve catalytic activity for DRM.

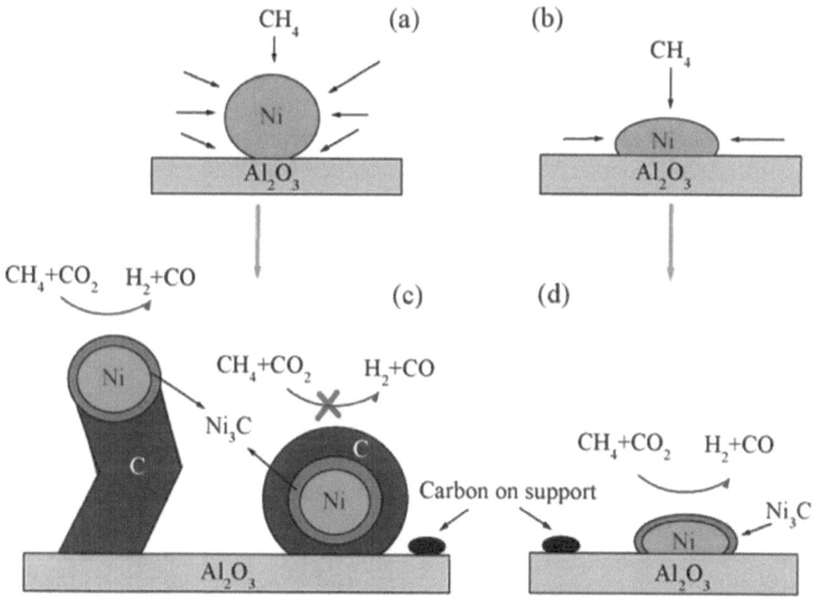

FIGURE 7.17 Schematic representation of CO_2/CH_4 reforming over (a and c) NiAl–C (calcinated); and (b and d) NiAl-PC (plasma treated prior to calcination) samples [186].

However, as pointed out by Puliyalil et al. [186], plasma pretreatments may not necessarily induce the same chemical effects to different metallic species.

7.9.2 Thermal versus Plasma Catalysis for DRM

As mention in Section 7.3, plasma-catalyst synergy plays an important role in altering reactions mechanisms during plasma catalysis. For example, excited CO_2 molecule dissociates in CO and O and O dissociated by this process plays an important role on the oxidation process on the catalyst surface. Similar process is followed for reduction process by excited H_2O or H_2 molecules. Ni catalysts have been extensively used in reforming processes including dry reforming. It is possible that during dry reforming, deposited carbon from Ni catalysts can be removed by plasma excited CO_2 and H_2O [197]. Plasma excited species can also create different pathways on the catalyst surface. For example, excited CO_2 can oxidize Ni/Al_2O_3 catalyst to for NiO layer which can drive oxidation-reduction cycle. High NiO on Ni catalyst has been observed when electron energy is high. The formation of a thin layer of NiO, 20 μm depth, external to catalyst has been observed in DBD plasma [185]. This leads to formation of fine amorphous carbon filaments only at the external surface with no coke in the pores of the catalyst where no interactions between DBD and the catalyst occurs. The NiO layer absorbs more oxygen which is required for thermal equilibrium inducing oxidation-reduction cycle and resulting new pathways such as methane dehydrogenation which prevents coke deposition during DRM reaction. This synergistic phenomena between plasma and catalyst enhances reaction compared to what is observed in thermal catalysis.

As pointed out in Section 7.3, the interactions between plasma and catalyst mainly occur in a thin layer of external surface of the catalyst. Catalyst pore size can largely reside in the micrometer range. For catalyst pores less than 10 μm, gas ionization by plasma cannot occur, and for pore size less than 2 nm, standard Paschen-type gas breakdown is not possible. Plasma-catalyst in the external surface is propagated from particle to particle contacts. However, electron density in this narrow filamentary channels is much smaller 10^{14} cm^{-3} [54,198,199] than a standard molecule density of 10^{19} cm^{-3}. This small concentration of active species cannot significantly change CH_4/CO_2 reaction in the gas phase, but they can cause significant changes on the reaction at the catalyst surface, if they are adsorbed. For this reason, as indicated earlier, the hetero-phase interactions between DBD and the external pellet surface provide the most important reaction sites, and this means that active species in the plasma should be created near the catalyst surface.

As pointed out in Section 7.3, the effects a catalyst has on plasma are somewhat different from those it has on a thermal catalytic reaction. In plasma catalysis, catalytic reactions can occur even at room temperature. Active species created by plasma are highly electrophilic and easily adsorbed on the catalyst surface. They can react among themselves, with other active species in the gas phase or desorb away. Chemisorption of active species is faster and more efficient than ground state molecules [200], and these species can also react with active species in the gas phase by ER mechanisms. Thus in plasma catalysis, both ER and LH mechanisms are simultaneously active even at low temperatures. This is more complex than what happens in thermal catalysis. A comparison of thermal and plasma-catalysis processes is graphically illustrated in Figure 7.18 [20].

The dissociation mechanism on the surface of an active material can be explained differently in the presence and absence of plasma. When a CO_2 molecule approaches the surface of NiO/TiO_2 catalyst, it gets adsorbed at the oxygen vacancies and undergoes dissociative electron attachment (DEA) [87]. DEA is defined as the low-energy electron-induced formation of a negative metastable ion (in the present example: CO_2), which undergoes subsequent dissociation. On the other hand, when plasma

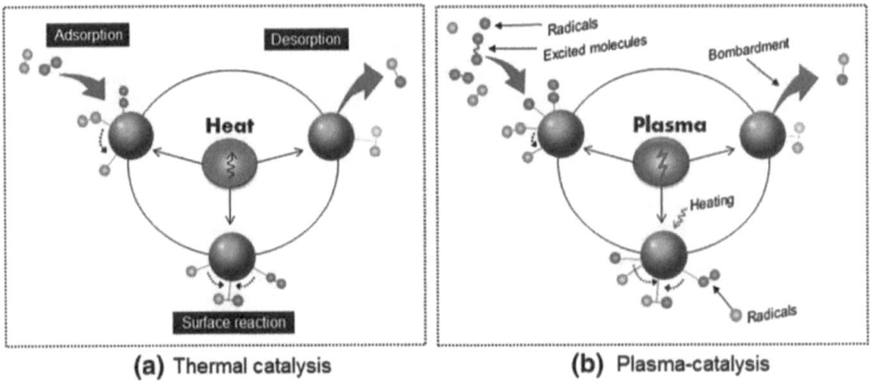

FIGURE 7.18 Schematic comparison of thermal and plasma catalytic surface processes. [186].

is introduced along with the catalyst, the dissociation rate tends to increase [23]. As shown in Section 7.3, this phenomenon results in enhancement of CO_2 conversion. The results, however, could be different for different catalysts.

While synergy occurring in plasma catalysis is complex and not well understood at the fundamental level, plasma-catalyst synergistic effects during CH_4 activation have been demonstrated in the past. Nozaki et al. [199] examined the effects of plasma activation of CH_4 in a DBD reactor with and without a catalyst packing. In a Ni/SiO_2 catalyst-only process, there was no CH_4 conversion at temperatures below 400°C and a spontaneous increase in the conversion rate was observed at temperatures around 600°C. When the catalyst was coupled with plasma, strong synergistic effects yielded very high conversion rates (roughly >50% increase) in the temperature range between 400°C and 600°C.

Plasma catalysis offers some advantages over thermal catalysis. Plasma-assisted DRM performed by Scapinello et al. [201] resulted in H_2/CO molar ratio equal to 1.1, while thermal catalytic DRM performed by Theofanidis et al. [202] resulted in $H_2/CO = 1$. A higher amount of CO_2 (40% more) per mol of produced syngas is consumed in the plasma-assisted DRM process, as compared with the combined DRM and SRM thermal catalytic process. Concurrently, a higher amount of H_2O (62% more) per mol of produced syngas is consumed in the latter case. Plasma-assisted DRM is more sustainable than the combined DRM and SRM thermal catalytic process as it increases CO_2 capture; it reduces H_2O scarcity by utilizing more efficiently H_2O input streams, and it is compatible with renewable energy technologies from which the electricity needed for the plasma initiation may be harvested. The plasma-assisted process may be more easily integrated with existing processes since it is relatively more compact than the thermal catalytic one. Moreover, except for electricity, no utilities are required since the cold utility demand is satisfied by the refrigerant, while, in the thermal catalytic processes, cooling water is required. Short start-up/shut-down periods are also counted as unique technology features compared to the thermal catalytic process.

In the study done by Kamashima et al. [197], dynamic comparison of thermal and plasma-catalysis DRM was performed by pulse reforming technique. CH_4/CO_2 ratio was changed from 0.5 to 1.5, and pressure was kept at 5 kPa. The specific energy input of discharge power was 1.2 eV per molecule. Other operating parameters are described by Kamashima et al. [197]. The results of this study are illustrated in Figure 7.19 [197], and they indicate that CH_4 conversion and H_2 yield were monotonically increased with the CH_4/CO_2 ratio and in all cases, plasma catalysis performed better than thermal catalysis. The syngas ratio (H_2/CO) also increased with methane flowrate.

Kamashima et al. [197] explained the results shown in Figure 7.19 by models presented in Figure 7.20. This figure indicates that there are two simultaneous routes for CH_4 conversion. Route (I) is a reforming path leading to syngas: CH_4 is chemisorbed as CH_x^* on metallic sites, and this in turn is oxidized by CO_2^* to form CH_xO^* before complete dehydrogenations to C* occurs producing syngas. In CO producing route (II), there is a thin C* rich layer on Ni surface and CH_4 almost irreversibly dehydrogenates toward carbon atom, and then C*-rich layer is oxidized slowly by CO_2^* (Reaction 7.17) to produce CO. When the CH_4/CO_2 ratio exceeds 1.0, CH_4 prefers to dehydrogenate to solid carbon through route (II) due to the low proportion of CO_2.

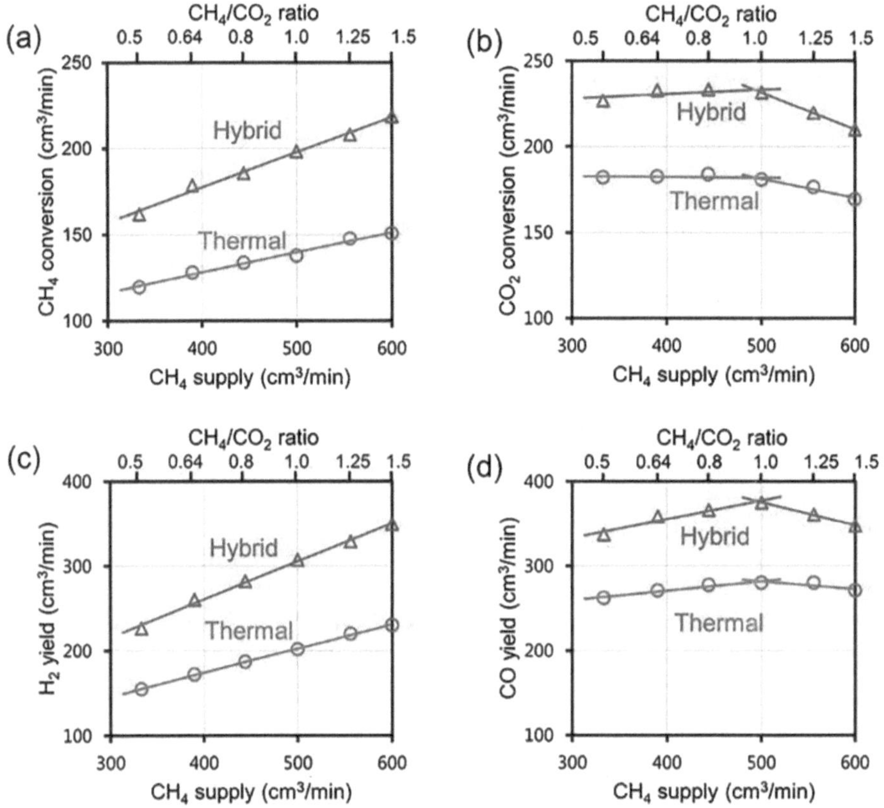

FIGURE 7.19 Effects of CH_4/CO_2 ratio on DMR at ca. 600°C: (a) CH_4 conversion, (b) CO_2 conversion, (c) H_2 yield, and (d) CO yield [185].

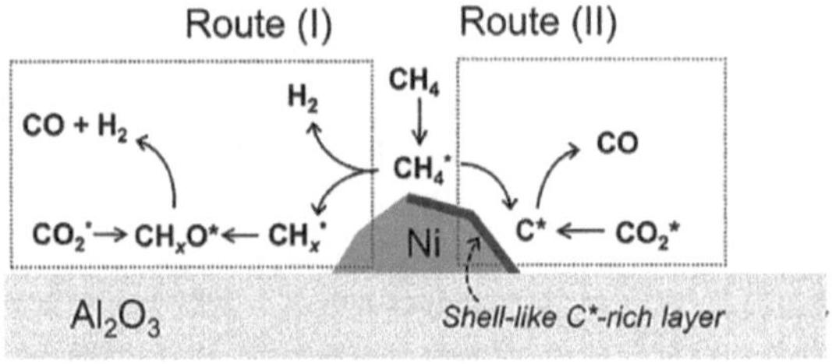

FIGURE 7.20 Two simultaneous routes for CH_4 conversion [185].

Subsequently, a non-negligible amount of solid carbon is produced, and CO_2 conversion and CO yield turned to proportionally decrease.

$$C + CO_2 \rightarrow CO + CO \tag{7.17}$$

$$CH_4 \rightarrow CH_4^* \rightarrow CH_x^* + (4 - x)/2\,H_2 \qquad (x = 0 - 3) \tag{7.18}$$

$$H_2 + CO_2 \rightarrow CO + H_2O \tag{7.19}$$

$$C + H_2O \rightarrow H_2 + CO \tag{7.20}$$

Compared with thermal reforming, both CH_4 conversion and H_2 yield are clearly promoted in hybrid reforming (Figure 7.19), and the main pathway of CH_4 conversion and H_2 yield could be simply described as CH_4 dehydrogenation (Reaction 7.18). Kamashima et al. [197] proposed that CH_4 dehydrogenation was enhanced by the synergistic effect of DBD and catalyst. Other related studies [203,204] have indicated that vibrational excitation of CH_4 molecules and low-energy impact by long-lived species play important roles in dehydrogenation. The reaction mechanism of plasma-enabled catalysis can be explained by the Langmuir-Hinshelwood (LH) reaction scheme.

Similarly, the CO_2 conversion and CO yield are also promoted in hybrid reforming compared to thermal reforming (Figure 7.19). H_2O is simultaneously produced as a by-product by reverse water gas shift (RWGS) reaction (7.19). Plasma-activated H_2O promotes reaction with adsorbed carbon and creates additional pathways (Eq. 7.20) to syngas (H_2 and CO). The CO_2 conversion and CO yield were promoted in the hybrid reforming, illustrating that the reverse-Boudouard reaction is enhanced by DBD. The reaction between plasma- activated CO_2 and adsorbed carbon increases CO yield. The same result are obtained in the de-coking period [205]. Thus, the presence of adsorbed carbon creates key pathways for plasma-induced synergistic effect. Consequently, plasma-activated CO_2 and H_2O promote surface reaction and increase CO and H_2 yield. Figure 7.19 clearly shows that the slope of each line increased in hybrid reforming compared with thermal reforming, attributing to the non-thermal plasma-excited species. The increase of slope can be further explained by the promoted overall reaction order, which plays the key role in the estimation of the rate-determining step [206].

7.9.2.1 Limitations of DBD Reactor

While plasma catalysis performs better than thermal catalysis, literature studies also indicate limitations of DBD reactor for DRM. While the highest conversion for DBD has been reported as 66% [207], the energy cost has been four times higher (18 eV per molecule) than set target of 4.27 eV per molecule [208]. These numbers do depend on the operating conditions.

Other studies [209,210] show that low pressures seem to favor the reactant conversion, whereas the syngas ratio does not show any pressure dependence. On the other hand, the selectivity of both CO and light hydrocarbons is reported to increase with rising pressure. Zhou et al. [209] observed slightly higher conversions and product yields upon increasing the temperature, with the syngas ratio being independent of the temperature. For a temperature range of about 300–873 K, Zhang et al. [211] and Goujard et al. [212] reported the conversions and the hydrocarbon selectivity increase with temperature, while the H_2 and CO selectivity decreased. There are, however, mixed results on temperature effects between these two studies which are explained on the basis of shift in reaction mechanism at 673 K [212], There are also mixed results on the effects of discharge gap and use of multiple reactors on the conversion

[207–213]. However, both studies found increase in selectivity when applying multi stage ionization. One study showed [214] better conversion with porous electrodes. The nature of metal [215,216] also has effect on oxygenated synthesis, and pulsed power gave better conversion [217].

There are other operating parameters that can affect DBD performance. Higher CO_2 flow rate (or content) can prevent backward reaction, i.e., $CH_3+H \rightarrow CH_4$, and promote reaction $O+H/OH \rightarrow OH/H_2O$ due to higher oxygen content resulting in higher conversion [208]. However, higher gas flow rate also decreases residence time which can counteract above effect. The literature also found that a larger number of filaments with lower energy are better for conversion and energy cost than smaller number of filaments with higher energy [208]. The effects of adding inert gases like He, Ar, and N_2 in gas are mixed. Two studies [218,219] reported that addition of He changes velocity distributions function of electrons resulting in changes in electron reaction rates, energy losses and energy cost. The highest conversion of 84.2% predicted by a model results in an energy cost of 30.1 eV per molecule. The effects of conversion on energy cost are illustrated by Snoeckx and Bogaerts [4]. In general, energy costs for DBD are very high for its commercial use.

7.9.3 Selectivity Improvements in Plasma Activated DRM

One of the accepted approaches for CO_2 utilization is to reduce it with H_2 inside a catalyst embedded plasma reactor, producing liquid fuels such as CH_3OH [135]. By introducing a commercial $CuO/ZnO/Al_2O_3$ catalyst inside the discharge zone, CH_3OH yield increased up to ten times compared to the one obtained in catalyst free discharge at a gas temperature of 100°C. Furthermore, the selectivity increased up to 20%.

As shown earlier, methanol selectivity during CO_2 hydrogenation is very low with high H_2 consumption. DRM reaction is more interesting compared to CO_2 dissociation and hydrogenation because it can produce many oxygenated products in one step with the right type of catalyst. In a recent study [220], V^{5+}/Al_2O_3 catalyst was embedded in the discharge zone along with glass balls and $BaTiO_3$. Major reaction by-products obtained include H_2, CO, CH_4, C_3H_6, and HCHO with a 100% C_2H_6 conversion and a high HCHO selectivity (11.4%). The improvement in selectivity was attributed to the synergistic effects of active vanadium catalyst, ferroelectric $BaTiO_3$ and plasma activation. The process allows a successful utilization of CO_2 for the production of HCHO, a necessary chemical in industry used for wood processing, textile manufacturing and the production of formaldehyde resins, fertilizers, chelating agents, and polyhydric alcohols [221].

As mentioned earlier, another useful metal for conversion of CO_2 to alcohols is Cu. This metal adsorbs CO_2 as COO species on the catalyst surface which is then reduced to HCOO which can be easily converted to alcohols. This was validated by Zhao et al. [222] in the selective synthesis of ethanol from CO_2 and water vapor with the assistance of commercially available $Cu/ZnO/Al_2O_3$ catalyst packed in a negative corona reactor. It is assumed that the active Cu species on the surface are partially oxidized in the plasma, which provides better selectivity toward ethanol compared to other competing products such as methanol. A study [223] showed that addition of more CO_2 in a DBD reactor can increase CH_4 conversion. Dissociation of CO_2 species in active radical such as O1D can pick up hydrogen from CH_4 to generate another active radical CH_3. This active radical can react with other hydrocarbon radicals, atomic oxygen or OH radical to form higher hydrocarbons or liquid oxygenates. The material used for electrode in DBD reactor can also make difference. In a study [216], selectivity toward carboxylic acid was doubled by changing electrode material from steel to Ni or Cu. Another study [225] showed that different zeolites in a DBD reactor can give different levels of C_4 hydrocarbons in the order zeolite HY > zeolite NaA > zeolite NaY > fleece.

Plasma discharge power can also affect product selectivity. Puliyalil [186] reports that higher power decreases lower hydrocarbons and alcohols because of their activation and conversion to higher hydrocarbons and alcohols. However, very high power can result into breakdown of these higher hydrocarbons into basic compounds like coke, hydrogen, and CO. Catalyst properties can also significantly change with the discharge power. Mehta et al. [5] report selectivity to liquid and gas products by various catalysts as shown in Figure 7.21.

In DRM, it is important to have high concentration of CO_2 to provide oxidative environment for higher conversion of CH_4. As shown by Zhang et al. [226] for La_2O_3/g-Al_2O_3 catalyst in a pulse corona reactor, this changed the nature of hydrocarbon production. As CO_2 content in the feed was increased by a factor of 4, C_2H_2 selectivity decreased by about 35% along with a significant increase in the formation of C_2H_6 and C_2H_4 (5%–26% and 7%–16%, respectively). However, for higher concentrations of CO_2, the selectivity toward higher hydrocarbons or alcohols tends to decrease with a simultaneous increase in the selectivity toward CO or CO_2 in the outlet. Furthermore, the influence of Pd doping on La_2O_3/g-Al_2O_3 in the plasma catalytic conversion of CH_4 over CO_2 was studied. It was revealed that traces of Pd on La_2O_3/g-Al_2O_3 increased the C2 selectivity up to 70% and along with a high C_2H_4 (about 65%) in the product mixture. On the other hand, La_2O_3/g-Al_2O_3 catalyst without Pd gave very high C_2H_2 (76%) yield with C_2H_4 yield below 12%.

In thermal catalysis, Cu is widely used as a promoter to improve the activity and selectivity of Fe_2O_3 catalysts toward alcohols [227]. The same catalyst selection strategy has been applied in a two stage plasma catalytic conversion process [228]. The comparison of catalyst performance in plasma revealed that CH_3OH selectivity was 10.5% lower when Fe_2O_3/CP was used instead of the CuO modified catalyst. Product selectivity also depends on the configuration of the reactor design. A study [229] was done to evaluate effects of various catalyst packing strategy on the performance of A DBD reactor for partial oxidation of methane over Fe_2O_3-CuO/g-Al_2O_3 catalyst. The study showed that CH_4 conversion and maximum CH_3OH yield were dependent upon whether reaction was carried out in one stage or two stage. Furthermore different products were formed in two different reactor configurations [186].

Snoecks et al. [230] point out that currently four different approaches for plasma-based liquefaction are being investigated. The 'direct gas-phase plasma methane liquefaction' is the conversion of pure methane into hydrogen gas and liquid hydrocarbon chains originating from the remaining CH_2 blocks. The 'oxidative plasma liquefaction' methods can be subdivided in an indirect and a direct approach. The indirect approach converts methane into syngas together with an oxidant, such as

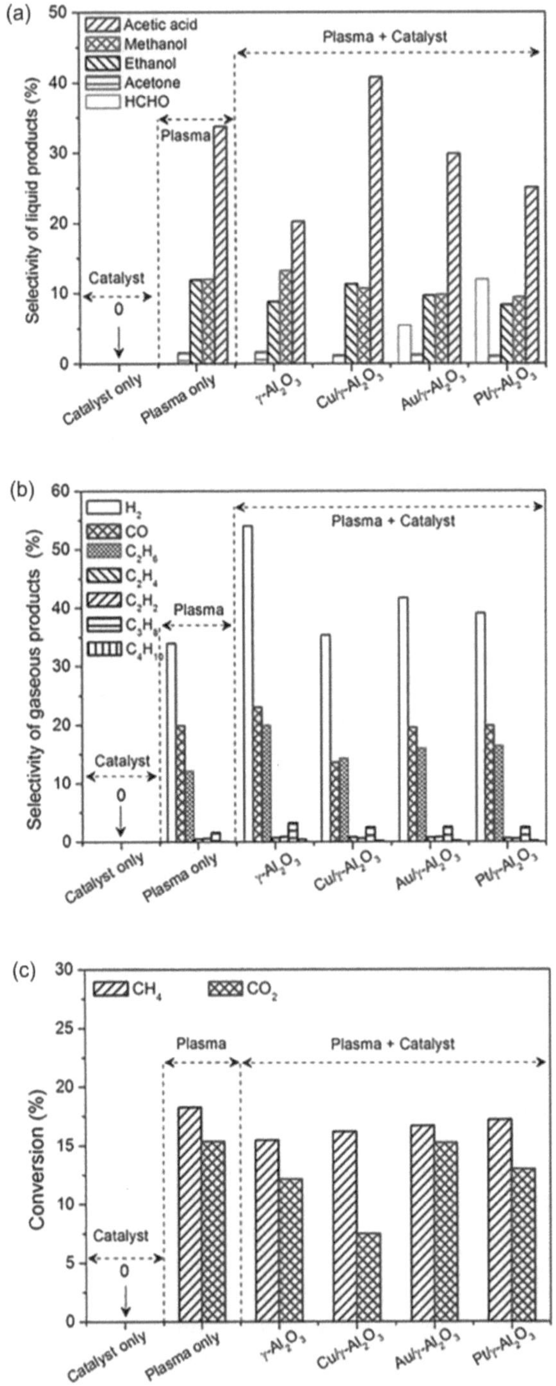

FIGURE 7.21 Selectivities to liquid products (a) and gaseous products (b) in plasma–catalytic DRM performed at ~30°C and 1 atm using a water-cooled DBD reactor. (c) Overall conversions of CH_4 and CO_2. [5,224].

O_2, CO_2, or H_2O, known as partial oxidation of methane (POX), dry reforming of methane (DRM) and steam methane reforming (SMR), respectively. The syngas is then further processed into liquids using Fischer-Tropsch or methanol synthesis. The direct approach, on the other hand, tries to convert methane with the same oxidants into oxygenated liquid products, such as alcohols and aldehydes, in one step. Finally, recently a new approach 'direct two phase plasma-assisted liquefaction' has emerged, which—as its name suggests—aspires the direct liquefaction of methane, through its incorporation into a second phase, namely existing liquid hydrocarbons.

7.9.4 DRM Using Other Plasma Reactors

Snoeckx and Bogaertz [4] presented an excellent review on the performance of DRM under different plasma discharges and presented a summary of all the published data in terms of energy efficiency versus conversion as shown in Figure 7.22. Based on their analysis of literature data, following summary statements can be made regarding effectiveness of various types plasma for DRM.

1. While conversion of dry reforming to syngas by DBD plasma catalysis is facilitated with the Ni-based and other catalysts at low temperatures, its energy cost is five times more expensive than what is required for its commercial viability. This statement also applies to most favorable packed-bed reactor configuration. Just as for CO_2 dissociation, this is a result of less energy efficient electron impact driven conversion than more energy

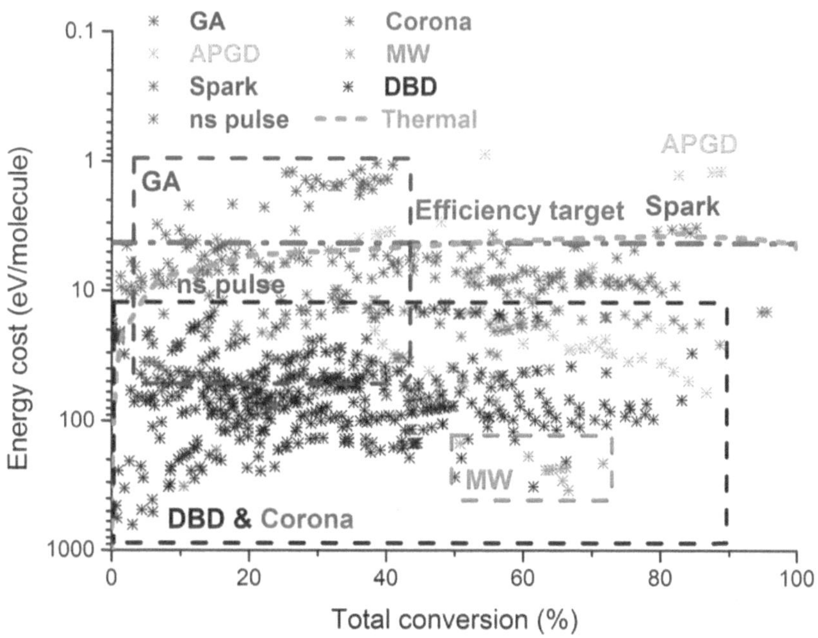

FIGURE 7.22 Comparison of all the data collected from the literature for DRM regarding the different plasma types, showing the energy cost as a function of the conversion [4].

efficient vibration mode conversion. DBD can be more useful if a catalyst can be found to use it for one step conversion to oxygenated products.

2. There is lack of data for DRM with MW plasma in the literature. Since this plasma was very effective for CO_2 dissociation, research should be carried out to evaluate its effectiveness for DRM. MW plasma with catalyst can be energy efficient and can give high conversions.
3. GA plasma has ability to exceed target energy efficiency even when conversion is 40%. Different catalysts should be tested to improve conversion. GA operates in non-equilibrium transient regime. More efforts should also be more focused on improved GAP plasma.
4. While spark discharge and nanosecond pulse discharge energy cost ranges from 3 eV per molecule (target number) to 10 eV per molecule, conversion of 40%–50% for nanosecond discharge and 85%–95% for spark discharge are favorable. The operation of these discharges, particularly that of spark discharge, should be optimized with additional studies.
5. The literature indicates that the best results so far are obtained with APGD plasma. They can get conversion in the range of 90% and energy cost as low as 1.2 eV per molecule. Further study of this plasma for DRM should be the highest priority. Furthermore more efforts should be made to examine selectivity of DRM reactions for APGD and GA (or GAP) plasmas. The results of energy cost versus conversion for APGD plasma reported by Snoeckx and Bogaerts [4] are shown in Figure 7.23.
6. Table 7.6 presents an overview of the best solar-to-fuel efficiencies, along with the reported conversions reported by Snoeckx and Bogaerts [4]. In this table, for some plasma types, two or three 'best values' are listed, as some conditions lead to the best conversion, while others lead to the best efficiency; Processes reaching efficiencies below 10% are considered inefficient, while those between 10% and 15% are considered promising, above 15% very promising and values above 20% might already be cost

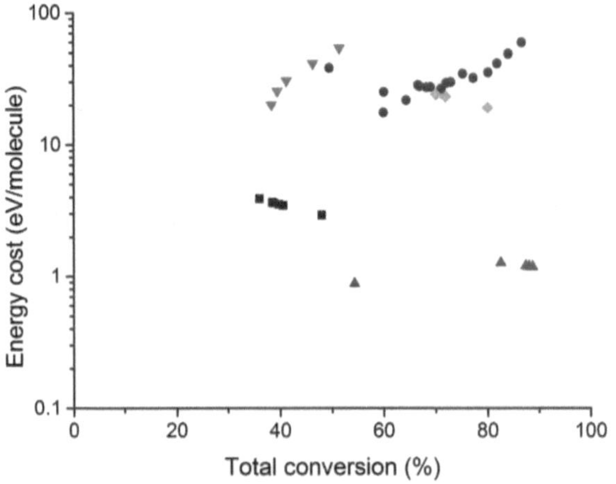

FIGURE 7.23 Best results with APGD plasma for energy cost and conversion for DRM [4].

TABLE 7.6
CO_2 Conversion and Energy Efficiency for Various Types of Plasma for CO_2 Dissociation and DRM Reported by Snoeckx and Bogaerts [4]

	CO_2 Splitting		Dry Reforming of Methane	
Plasma	χ_{CO_2} (%)	Solar to Fuel Efficiency (%)[a]	χ_{CH_4}/χ_{CO_2} (%)	Solar to Fuel Efficiency (%)[a]
DBD	25.8	5.8	7/3	4.2
	42.0	1.7	88/78	1.8
MW	9.7	22.5		
(1980)	35.9	17.1		
	87.4	9.4		
MW	9.7	12.6		
(2010)	29.5	10.8		
	82.9		71/69	11.6
GA	12.2	16.4	13/9	22.1
			41/36	15.5
			45/34	11.1
APGD	50.0	6.75	94/77	23.0
Ns-pulsed	7.1	2.5	61/50	14.5
Corona	6.1	1.6	23/36	12.2
Spark			87/83	14.7

[a] Using PV efficiency of 25%. [231–233]

competitive. Snoeckx and Bogaerts [4] point out that these data apply to the production of the syngas components CO and H_2; when considering the direct oxidative pathway to produce liquid fuels, lower values can be cost competitive

7. While DBD has been the most popular plasma discharge for plasma catalysis, an evaluation of all aspects of plasma-catalysis performances for all four basic CO_2 conversion reactions appears to indicate that the best potentials for commercialization of this technology are in MW, GAP, and APGD plasmas. The use of these plasmas will require novel thoughts on catalyst and reactor designs.

7.10 COMPARISON OF PLASMA-ACTIVATED CATALYSIS WITH OTHER CO_2 CONVERSION STRATEGIES

It is important to make some assessment of commercial potential of plasma-catalysis technology and compare them with other technologies examined in the previous chapters. Major positive attributes of plasma catalysis are its (a) high reactivity, (b) low-temperature operation, (c) flexibility with respect to scale and operation, (d) ability to connect easily to the sources of CO_2, and (e) ability to use renewable sources like wind, solar, hydro, tidal waves, etc. as well as low carbon nuclear sources for required electric power. Plasma catalysis is also capable of changing reaction paths to obtain organic oxygenates from CO_2 in a single reaction step. A comparison of plasma catalysis to other technologies considered in this book for CO_2 conversion to useful chemicals, materials and fuels or fuel additives developed by Snoeckx and Bogaerts [4] is illustrated in the following Table 7.7. This table considers nine parameters for comparison among various evolving technologies. We briefly summarize analysis of these nine parameters with the inclusion of major thoughts reported by Snoeckx and Bogaerts [4].

The use of rare earth metals is one of the key disadvantages of most technologies listed in the table, except for biochemical and plasma-chemical conversion. While rare earth materials provide high activity levels for many traditional and novel CO_2 conversion technologies, they are too expensive and not at bountiful for full commercialization of these technologies. Other technologies must find ways to replace them with other less expensive and bountiful materials to achieve improved performance.

TABLE 7.7
Overview of Traditional Thermal Catalysis and the Different Emerging Technologies, Indicating Their Distinctive Key Advantages and Disadvantages [4]

	Traditional Catalysis (MW Heating)	Electro-Chemical	Solar Thermo-Chemical	Photo-Chemical	Bio-Chemical	Plasma Chemical
Use of rare earth materials	Yes	Yes	Yes	Yes	No	No
Renewable energy	Indirect	Indirect	Direct	Direct[a]	Direct	Indirect
Turnkey process	No	No[b]	NA	Yes	No	Yes
Conversion and yield	High	High	High	Low	Medium	High
Separation step	Yes	Yes[c]	No	Yes	Yes[d]	Yes[e]
Oxygenated products	Yes	Yes	No	Yes	Yes	Yes
Investment cost	Low	Low	High	Low	High/Low	Low
Operating cost	High (Low for MW Heating)	Low	Low	Low	High	Low
Overall flexibility	Low	Medium	Low	Low	Low	High

[a] Bio- and photochemical processes can also rely on indirect renewable energy when they are coupled with artificial lighting.

[b] Electrochemical cells are turnkey, but generally the cells need to operate at elevated temperatures and the cells are sensitive to on/off fluctuations.

[c] The need for post-reaction separation for the electrochemical conversion highly depends on the process and cell type used.

[d] Biochemical CO_2 conversion requires very energy-intensive post-reaction separation and processing steps.

[e] The need for post-reaction separation for plasma technology highly depends on the process.

As mentioned earlier, plasma-chemical conversion can use renewable sources of power but only in an indirect way. The best way is to use excess power generated from solar, wind, hydro, or tidal waves to run plasma reactor. Excess power from nuclear reactor can also be used. Since some of the renewable powers are intermittent, power can be drawn from utility grid, microgrid, or mini grids. Power can also be obtained in a standalone operation as long as there is storage for other needs. As shown in the table, some of the other technologies can use renewable power and heat directly. This does mean that energy efficiency for plasma catalysis can be lower than the ones using renewable energy directly.

As discussed earlier, plasma catalysis requires no pre-heating or long stabilization times (about 30 minutes). This is not the case for traditional catalysis. Also, plasma reactor can be turned off with the flick of a switch, since no sensitive cool-down times are required and there is no risk of damaging the reactors with repeated on–off cycles. Electrochemical cells, on the other hand, operate at high temperatures and suffer performance if turned on and off too frequently. The nutrients, light, temperature, and mixing in biochemical processes need to be looked after on a continuous basis for good performance of the bio reactors.

Except for photo and biochemical technologies, most other technologies are capable of delivering high conversions and yields. Product separation is a major issue with most technologies because they can be expensive and energy consuming. Solar thermochemical and electrochemical conversion (depending on the process and cell type) are the only two technologies capable of generating separated product streams, i.e., separated CO and O_2 streams in the case of pure CO_2 splitting. This is an important advantage, since the separation of CO and O_2 is rather energy intensive at this point. Membrane technology may be a convenient solution in some cases.

Currently, product separation can be a major disadvantage of plasma technology, particularly for DRM, since it can produce multiple products including hydrocarbons, acids and alcohols and other oxygenated species. If the major product is syngas, it can be easily processed by F-T or methanol syntheses. The required syngas ratio can be tailored by operating conditions of plasma technology. The problem of product separation in DRM can be partially alleviated if the value-added liquid oxygenated products like acids, alcohols, aldehydes, and other oxygenated compounds can be produced by direct one-step oxidative process instead of current indirect two-step process [4,186]. The investment and operating costs are in general considered to be low for most technologies. The main exceptions are solar thermochemical conversion, which has a high investment cost for concentrating the solar energy and specially designed high-temperature solar reactors. For biochemical conversion, both investment and operating costs can be high, depending on the bioreactor type. Furthermore, the plasma-chemical conversion is a highly modular technology, which is not dependent on an economy of scale and thus allows for local on-demand production capabilities.

Finally, flexibility is an important issue for any CO_2 conversion technology to valuable products. The plasma-chemical conversion has a tremendous advantage here, due to its feed flexibility (CO_2, CO_2/CH_4, CO_2/H_2, and CO_2/H_2O), its energy source flexibility (solar, wind, hydro, nuclear power), its operation flexibility (instant on/off, power scalability) and its flexibility of scale. None of the other technologies possess this unique combination of features required for successful worldwide implementation as a CCU technique for broad base of CO_2 conversion reactions and sources of energy.

As shown in this chapter, both CO_2 dissociation and DRM give reasonable efficiency (60%) and high conversion (90%) in some types of plasma. Furthermore, the production of liquids in a one-step process, through a direct oxidative pathway (in combination with catalysts), has shown some success. This method will reduce the energy efficiency target and make it more competitive (i.e. by at least a factor of 2–3 in the case of methanol). CO_2 hydrogenation to produce CO or methane has shown significant success. More work is needed to produce methanol by this path. Plasma catalysis can be used at variable locations, and it can use raw materials with varying composition.

Snoeckx and Bogaerts [4] present a comparison of plasma-driven methanol production with traditional systems. For small-scale production, the costs associated with a plasma reactor may be significantly lower than that of a two-step steam-reforming to methanol synthesis plant. Compared to the thermal catalysis, fast turn on/off time and room temperature operation are significant advantageous of plasma-assisted conversion processes as well, which can reduce the energy consumption to a larger extent. Nevertheless, one can question the feasibility of plasma catalysis as the process is operated at atmospheric pressure with limited gas inlet flows. Such a drawback may be overcome by utilizing the parallel reactor concept for DBD as introduced by Kogelschatz for large-scale ozone synthesis [54]. The industrial-scale reactor thus constructed will consist of bundles of small-scale reactors arranged in parallel. Such reactors allow operations at higher gas inlet flows, incorporation of the catalyst and higher energy input. Current market methanol production costs are on the approximate level of 0.1–0.3$ per liter [54]. In summary, plasma technology fares very well in this comparison, with its main disadvantages being: (a) its current need for post-reaction separation processes, (b) the fact that the energy efficiency is dependent on the reactor type, and (c) the need for plasma catalysis to improve the product yield and selectivity particularly for oxygenated products.

REFERENCES

1. Bogaerts, A, E. Neyts, R. Gijbels, and J. Van der Mullen 2002. Gas discharge plasmas and their applications. *Spectrochimica Acta Part B* 57:609–658.
2. Fridman, A. 2008. *Plasma Chemistry*. New York: Cambridge University Press.
3. Chen, G., L. Wang, T. Godfroid, and R. Snyders 2018. Progress in plasma assisted catalysis for carbon dioxide reduction, an Intech paper. doi: 10.5772/intechopen.80798.
4. Snoeckx, R., and A. Bogaerts 2017. Plasma technology- a novel solution for CO_2 conversion? Review article. *Chemical Society Reviews* 46:5805.
5. Mehta, P, P. Barboun, D. Go, J. Hicks, and W. Schneider 2019. Catalysis enabled by plasma activation of strong chemical bonds: A review. *ACS Energy Letters* 4:1115–1133. http://pubs.acs.org/journal/aelccp.
6. Brune, L, A. Ozkan, E. Genty, T. Visart de Bocarmé, and F. Reniers 2018. Dry reforming of methane via plasmacatalysis: Influence of the catalyst nature supported on alumina in a packed-bed DBD configuration. *Journal of Physics D: Applied Physics* 51:234002 (14 pp). https://doi.org/10.1088/1361-6463/aac047.

7. Evangelos, D., S. Marco, and D. Georgios Stefanidis 2017. Investigating the plasma-assisted and thermal catalytic dry methane reforming for syngas production: Process design, simulation and evaluation. *Energies* 10:1429. doi: 10.3390/en10091429.
8. Sheng, Z., S. Kameshima, K. Sakata, and T. Nozaki 2018. Plasma-enabled dry methane reforming: An intech open access paper. doi: 10.57772/intechopen.80523.
9. Chen, G., N. Britun, T. Godfroid, M. Ogletree, and R. Snyders 2017. Role of plasma catalysis in the microwave plasma-assisted conversion of CO_2, an intech open access paper. doi: 10.57772/67874.
10. Liu, M., Y. Yi, L. Wang, H. Guo, and A. Bogaerts 2019. Hydrogenation of carbon dioxide to value-added chemicals by heterogeneous catalysis and plasma catalysis. *Catalysts* 9:275. doi: 10.3390/catal9030275.
11. Fridman, A., A. Chirokov, and A. Gutsol 2005. Non-thermal atmospheric pressure discharges. *Journal of Physics D: Applied Physics* 38:R1–R24.
12. Christensen, P. A., A. B. Ali, Z. Mashhadani, M. A. Carroll, and P. A. Martin 2018. The production of ketene and C_5O_2 from CO_2, N_2 and CH_4 in a nonthermal plasma catalysed by earth-abundant elements: An in-situ FTIR study. *Plasma Chemistry and Plasma Processing* 38(3):461–484.
13. Kozák, T., and A. Bogaerts 2014. Splitting of CO_2 by vibrational excitation in non-equilibrium plasmas: A reaction kinetics model. *Plasma Sources Science and Technology* 23:45004.
14. Kozák, T., and A. Bogaerts 2015. Evaluation of the energy efficiency of CO_2 conversion in microwave discharges using a reaction kinetics model. *Plasma Sources Science and Technology*, 24:15024.
15. Whitehead, J. 2019. Plasma catalysis: Is it a question of scale? *Frontiers of Chemical Science and Engineering* 13(2):264–273. https://doi.org/10.1007/s11705-019-1794-3.
16. Bogaerts, A., and C. Neyts Erik 2018. Plasma technology: An emerging technology for energy storage. *ACS Energy Letters* 3:1013–1027.
17. Stere, C. E., W. Adress, R. Burch, S. Chansai, A. Goguet, W. G. Graham, and C. Hardacre 2015. Probing a non-thermal plasma activated heterogeneously catalyzed reaction using in situ DRIFTS- MS. *ACS Catalysis* 5(2):956–964.
18. Jia, Z., and A. Rousseau 2016. Sorbent track: Quantitative monitoring of adsorbed VOCs under in- situ plasma exposure. *Scientific Reports* 6(1):31888.
19. Azzolina-Jury, F., and F. Thibault-Starzyk 2017. Mechanism of low pressure plasma-assisted CO_2 hydrogenation over Ni-USY by microsecond time-resolved FTIR spectroscopy. *Topics in Catalysis* 60(19–20):1709–1721.
20. Kim, H. H., Y. Teramoto, A. Ogata, H. Takagi, and T. Nanba 2016. Plasma catalysis for environmental treatment and energy applications. *Plasma Chemistry and Plasma Processing* 36(1):45–72.
21. Neyts, E. C. 2016. Plasma-surface interactions in plasma catalysis. *Plasma Chemistry and Plasma Processing* 36(1):185–212.
22. Belov, I., S. Paulussen, and A. Bogaerts 2016. Appearance of a conductive carbonaceous coating in a CO_2 dielectric barrier discharge and its influence on the electrical properties and the conversion efficiency. *Plasma Sources Science and Technology* 25:15023.
23. Aerts, R., W. Somers, and A. Bogaerts 2015. Carbon dioxide splitting in a dielectric barrier discharge plasma: A combined experimental and computational study. *ChemSusChem* 8:702–716.
24. Smith, R. R. 2004. Preference for vibrational over translational energy in a gas-surface reaction. *Science* 304:992–995.
25. Ozkan, A., A. Bogaerts, and F. Reniers 2017. Routes to increase the conversion and the energy efficiency in the splitting of CO_2 by a dielectric barrier discharge. *Journal of Physics D: Applied Physics* 50:84004.

26. Wang, J., G. Xia, A. Huang, S. L. Suib, Y. Hayashi, and H. Matsumoto 1999. CO_2 decomposition using glow discharge plasmas. *Journal of Catalysis* 185:152–159.
27. Li, R., Q. Tang, S. Yin, and T. Sato 2007. Investigation of dielectric barrier discharge dependence on permittivity of barrier materials. *Applied Physics Letters* 90(13):131502.
28. Li, R., Q. Tang, S. Yin, and T. Sato 2006. Plasma catalysis for CO_2 decomposition by using different dielectric materials. *Fuel Processing Technology* 87:617–622.
29. Wang, S., Y. Zhang, X. Liu, and X. Wang 2012. Enhancement of CO_2 conversion rate and conversion efficiency by homogeneous discharges. *Plasma Chemistry and Plasma Processing* 32:979–989. doi: 10.1007/s11090-012-9386-8.
30. Duan, X., Y. Li, W. Ge, and B. Wang 2015. Degradation of CO_2 through dielectric barrier discharge microplasma. *Greenhouse Gases: Science and Technology* 5:131–140.
31. Tagawa, Y., S. Mori, M. Suzuki, I. Yamanaka, T. Obara, J. Ryu, and Y. Kato 2011. Synergistic decomposition of CO_2 by hybridization of a dielectric barrier discharge reactor and a solid oxide electrolyser cell. *Kagaku Kogaku Ronbunshu* 37:114–119.
32. Bogaerts, A., A. Berthelot, S. Heijkers, St. Kolev, R. Snoeckx, S. Sun, G. Trenchev, K. Van Laer, and W. Wang 2017. CO_2 conversion by plasma technology: insights from modeling the plasma chemistry and plasma reactor design. *Plasma Sources Science and Technology* 26:063001.
33. Lindon, M. A., and E. E. Scime 2014. CO_2 dissociation using the Versatile atmospheric dielectric barrier discharge experiment (VADER) *Frontiers in Physics Plasma Physics* 2:1–13.
34. Aerts, R., T. Martens, and A. Bogaerts 2012. Influence of vibrational states on CO_2 splitting by dielectric barrier discharges. *Journal of Physical Chemistry C* 116(44):23257–23273.
35. Koelman, P., S. Heijkers, S. T. Mousavi, W. Graef, D. Mihailova, T. Kozak, A. Bogaerts, and J. van Dijk 2017. A comprehensive chemical model for the splitting of CO_2 in non-equilibrium plasmas. *Plasma Processes and Polymers* 14:1600155.
36. Ponduri, S., M. M. Becker, S. Welzel, M. C. M. Van De Sanden, D. Loffhagen, and R. Engeln 2016. Fluid modelling of CO_2 dissociation in a dielectric barrier discharge. *Journal of Applied Physics* 119(9):093301.
37. Bogaerts, A., C. De Bie, R. Snoeckx, and T. Kozák 2016. Plasma based CO_2 and CH_4 conversion: A modeling perspective. *Plasma Processes and Polymers*:1–21.
38. Glonek, K., A. Wroblewska, E. Makuch, B. Ulejczyk, K. Krawczyk, R. J. Wrobel, Z. C. Koren, and B. Michalkiewicz 2017. Oxidation of limonene using activated carbon modified in dielectric barrier discharge plasma. *Applied Surface Science* 420:873–881.
39. Liu, L., C. H. Zheng, S. H. Wu, X. Gao, M. J. Ni, and K. F. Cen 2017. Manganese cerium oxide catalysts prepared by non-thermal plasma for NO oxidation: Effect of O_2 in discharge atmosphere. *Applied Surface Science* 416:78–85.
40. Wang, Z., Y. Zhang, E. C. Neyts, X. X. Cao, X. S. Zhang, B. W. L. Jang, and C. J. Liu 2018. Catalyst preparation with plasmas: How does it work? *ACS Catalysis* 8(3):2093–2110.
41. Marinov, D., O. Guaitella, T. de los Arcos, A. von Keudell, and A. Rousseau 2014. Adsorption and reactivity of nitrogen atoms on silica surface under plasma exposure. *Journal of Physics. D, Applied Physics* 47(47):475204.
42. Witvrouwen, T., S. Paulussen, and B. Sels 2012. The use of non-equilibrium plasmas for the synthesis of heterogeneous catalysts. *Plasma Processes and Polymers* 9:750–760.
43. Neyts, E. C., K. Ostrikov, M. K. Sunkara, and A. Bogaerts 2015. Plasma catalysis: Synergistic effects at the nanoscale. *Chemical Reviews* 115:13408–13446.
44. Zhang, Y., J. Li, G. An, and X. He 2010. *Sensors and Actuators B* 144:43–48.
45. Witvrouwen, T., S. Paulussen, and B. Sels 2012. *Plasma Processes and Polymers* 9:750–760.

46. Ramakers, M., I. Michielsen, R. Aerts, V. Meynen, and A. Bogaerts 2015. Effect of argon or helium on the CO_2 conversion in a dielectric barrier discharge. *Plasma Processes and Polymers* 12:755–763.
47. Snoeckx, R., S. Heijkers, K. Van Wesenbeeck, S. Lenaerts, and A. Bogaerts 2016. CO_2 conversion in a dielectric barrier discharge plasma: N_2 in the mix as a helping hand or problematic impurity? *Energy & Environmental Science* 9:30–39.
48. Spencer, L. F. 2012. The study of CO_2 conversion in a microwave plasma/catalyst system, The University of Michigan.
49. Tsuji, M., T. Tanoue, K. Nakano, and Y. Nishimura 2001. Decomposition of CO_2 into CO and O in a microwave-excited discharge flow of CO_2/He or CO_2/Ar mixtures. *Chemistry Letters*:22–23.
50. Heijkers, S., R. Snoeckx, T. Kozák, T. Silva, T. Godfroid, N. Britun, R. Snyders, and A. Bogaerts 2015. CO_2 conversion in a microwave plasma reactor in the presence of N_2: Elucidating the role of vibrational levels. *Journal of Physical Chemistry C* 119(23):12815–12828.
51. Chen, G. 2015. A study of microwave plasma assisted CO_2 conversion by plasma catalysis, Ph.D. thesis, Universite de Mons and Universite Libre de Bruxelles, Brussels, Belgium.
52. Liu, J. L., H. W. Park, W. J. Chung, and D. W. Park 2016. High-efficient conversion of CO_2 in AC pulsed tornado gliding arc plasma. *Plasma Chemistry and Plasma Processing* 36:437–449.
53. Bogaerts, A., W. Wang, A. Berthelot, and V. Guerra 2016. Modeling plasma-based CO_2 conversion: Crucial role of the dissociation cross section. *Plasma Sources Science and Technology* 25:55016.
54. Kogelschatz, U. 2003. Dielectric-barrier discharges: Their history, discharge physics, and industrial applications. *Plasma Chemistry and Plasma Processing* 23:1–46.
55. Lebedev, Y. A. 2010. Microwave discharges: Generation and diagnostics. *Journal of Physics: Conference Series* 257:012016.
56. Bogaerts, A., E. Neyts, R. Gijbels, and J. Van der Mullen 2002. Gas discharge plasmas and their applications (review). *Spectrochimica Acta Part B* 57:609–658.
57. Asisov, R. I., A. K. Vakar, V. K. Jivotov, M. F. Krotov, O. A. Zinoviev, B. V. Potapkin, A. A. Rusanov, V. D. Rusanov, and A. Fridman 1983. Non-equilibrium plasma-_chemical process of CO_2 decomposition in a supersonic microwave discharge. *Proceedings of the USSR Academy of Sciences* 271:94–97.
58. Rusanov, V. D., A. A. Fridman, and G. V. Sholin 1981. The physics of a chemically active plasma with nonequilibrium vibrational excitation of molecules. *Uspekhi Fizicheskikh Nauk* 134:185–235.
59. Rooij van, G. J., D. C. M. van den Bekerom, N. den Harder, T. Minea, G. Berden, W. W. Bongers, R. Engeln, M. F. Graswinckel, E. Zoethout, and M. C. M. van de Sanden 2015. Taming microwave plasma to beat thermodynamics in CO_2 dissociation. *Faraday Discussions* 183:233–248.
60. Silva, T., N. Britun, T. Godfroid, and R. Snyders 2014. Optical characterization of a microwave pulsed discharge used for dissociation of CO_2. *Plasma Sources Science and Technology* 23:025009.
61. Nunnally, T., K. Gutsol, A. Rabinovich, A. Fridman, A. Gutsol, and A. Kemoun 2011. Dissociation of CO_2 in a low current gliding arc plasmatron. *Journal of Physics D: Applied Physics* 44:274009.
62. Sunka, P., V. Babický, M. Clupek, P. Lukes, M. Simek, J. Schmidt, and M. Cernák 1999. Generation of chemically active species by electrical discharges in water. *Plasma Sources Science and Technology* 8:258–265.
63. Locke, B. R., and S. M. Thagard 2012. Analysis and review of chemical reactions and transport processes in pulsed electrical discharge plasma formed directly in liquid water. *Plasma Chemistry and Plasma Processing* 32:875–917.

64. Janeco, A., N. R. Pinhao, and V. Guerra 2015. Electron kinetics in $He/CH_4/CO_2$ mixtures used for methane conversion. *Journal of Physical Chemistry C* 119: 109–120.
65. Whitehead, J. C. 2016. Plasma–catalysis: The known knowns, the known unknowns and the unknown unknowns. *Journal of Physics D: Applied Physics* 49:243001.
66. Rettner, C. T., and H. Stein 1987. Effect of vibrational energy on the dissociative chemisorption of N_2 on Fe (111). *Journal of Chemical Physics* 87:770–771.
67. Holmblad, P. M., J. Wambach, and I. Chorkendorff 1995. Molecular beam study of dissociative sticking of methane on Ni (100). *Journal of Chemical Physics* 102:8255–8263.
68. Romm, L., G. Katz, R. Kosloff, and M. Asscher 1997. Dissociative chemisorption of N_2 on Ru (001) enhanced by vibrational and kinetic energy: Molecular beam experiments and quantum mechanical calculations. *Journal of Physical Chemistry B* 101:2213–2217.
69. Murphy, M. J., J. F. Skelly, A. Hodgson, and B. Hammer 1999. Inverted vibrational distributions from N_2 recombination At Ru (001): Evidence for a metastable molecular chemisorption well. *Journal of Chemical Physics* 110:6954–6962.
70. Luntz, A. C. 2000. A simple model for associative desorption and dissociative chemisorption. *Journal of Chemical Physics* 113:6901–6905.
71. Diekhöner, L., H. Mortensen, A. Baurichter, E. Jensen, V. V. Petrunin, and A. C. Luntz 2001. N_2 dissociative adsorption on Ru (0001): The role of energy loss. *Journal of Chemical Physics* 115:9028–9035.
72. Whitehead, J. C. 2010. Plasma catalysis: A solution for environmental problems. *Pure and Applied Chemistry* 82:1329–1336.
73. Chen, G., V. Georgieva, T. Godfroid, R. Snyders, and M.-P. Delplancke-Ogletree 2016. Plasma assisted catalytic decomposition of CO_2. *Applied Catalysis B: Environmental* 190:115–124.
74. Chen, G., N. Britun, T. Godfroid, V. Georgieva, R. Snyders, and M.-P. Delplancke-Ogletree 2017. An overview of CO_2 conversion in a microwave discharge: The role of plasma-catalysis. *Journal of Physics D: Applied Physics* 50:084001.
75. Chen, G., T. Godfroid, V. Georgieva, N. Britun, M.-P. Delplancke-Ogletree, and R. Snyders 2017. Plasma-catalytic conversion of CO_2 and CO_2/H_2O in a surface-wave sustained microwave discharge. *Applied Catalysis B: Environmental* 214:114–125.
76. Tu, X., H. J. Gallon, and J. C. Whitehead 2011. Electrical and spectroscopic diagnostics of a single-stage plasma-catalysis system: Effect of packing With TiO_2. *Journal of Physics D: Applied Physics* 44:482003.
77. Gallon, H. J., X. Tu, and J. C. Whitehead 2012. Effects of reactor packing materials on H_2 production by CO_2 reforming of CH_4 in a dielectric barrier discharge. *Plasma Processes and Polymers* 9:90–97.
78. Wang, W., H. H. Kim, K. Van Laer, and A. Bogaerts 2018. Streamer propagation in a packed bed plasma reactor for plasma catalysis applications. *Chemical Engineering Journal* 334:2467–2479.
79. Tu, X., H. J. Gallon, M. V. Twigg, P. A. Gorry, and J. C. Whitehead 2011. Dry reforming of methane over a Ni/Al_2O_3 catalyst in a coaxial dielectric barrier discharge reactor. *Journal of Physics D: Applied Physics* 44:274007.
80. Tu, X., and J. Whitehead 2012. Plasma-catalytic dry reforming of methane in an atmospheric dielectric barrier discharge: Under-standing the synergistic effect at low temperature. *Applied Catalysis B* 125:439–448.
81. Butterworth, T., R. Elder, and R. Allen 2016. Effects of particle size on CO_2 reduction and discharge characteristics in a packed bed plasma reactor. *Chemical Engineering Journal* 293:55–67.

82. Michielsen, I., Y. Uytdenhouwen, J. Pype, B. Michielsen, J. Mertens, F. Reniers, V. Meynen, and A. Bogaerts 2017. CO_2 dissociation in a packed bed DBD reactor: First steps towards a better understanding of plasma catalysis. *Chemical Engineering Journal* 326:477–488.
83. Lee, H., and H. Sekiguchi 2011. Plasma–catalytic hybrid system using spouted bed with a gliding arc discharge: CH_4 reforming as a model reaction. *Journal of Physics D: Applied Physics* 44:274008.
84. Hibert, C., I. Gaurand, O. Motret, and J. M. O. H. Pouvesle 1999. (X) measurements by resonant absorption spectroscopy in a pulsed dielectric barrier discharge. *Journal of Applied Physics* 85(10):7070–7075.
85. Ozkan, A., et al. 2016. The influence of power and frequency on the filamentary behavior of a flowing DBD—Application to the splitting of CO_2. *Plasma Sources Science and Technology* 25:025013.
86. Chung, W. C., and M. B. Chang 2016. Review of catalysis and plasma performance on dry reforming of CH_4 and possible synergistic effects. *Renewable and Sustainable Energy Reviews* 62:13–31.
87. Lee, J., D. C. Sorescu, and X. Deng 2011. Electron-induced dissociation of CO_2 on TiO_2. *Journal of the American Chemical Society* 133:10066–10069.
88. Liu, L., and Y. Li 2014. Understanding the reaction mechanism of photocatalytic reduction of CO_2 with H_2O on TiO_2- based photocatalysts: A review. *Aerosol and Air Quality Research* 14:453–469.
89. Mei, D., X. Zhu, C. Wu, B. Ashford, P. T. Williams, and X. Tu 2016. Plasma- photocatalytic conversion of CO_2 at low temperatures: Understanding the synergistic effect of plasma- catalysis. *Applied Catalysis B: Environmental* 182:525–532.
90. Brehmer, F., S. Welzel, R. M. C. M. Van De Sanden, and R. Engeln 2014. Co and byproduct formation during CO_2 reduction in dielectric barrier discharges. *Journal of Applied Physics* 116:123303.
91. Paulussen, S., et al. 2010. Conversion of carbon dioxide to value-added chemicals in atmospheric pressure dielectric barrier discharges. *Plasma Sources Science and Technology* 19:034015.
92. Yu, Q., M. Kong, T. Liu, J. Fei, and X. Zheng 2012. Characteristics of the decomposition of CO_2 in a dielectric packed-bed plasma reactor. *Plasma Chemistry and Plasma Processing* 32:153–163.
93. Mei, D., Y. L. He, S. Liu, J. Yan, and X. Tu 2016. Optimization of CO_2 conversion in a cylindrical dielectric barrier discharge reactor using design of experiments. *Plasma Processes and Polymers* 13:544–556.
94. Mahammadunnisa, S., L. Reddy, D. Ray, C. Subrahmanyam, and J. C. Whitehead 2013. CO_2 reduction to syngas and carbon nanofibres by plasma-assisted in situ decomposition of water. *International Journal of Greenhouse Gas Control* 16:361–363.
95. Zhang, K., G. Zhang, X. Liu, A. N. Phan, and K. Luo 2017. A study on CO_2 decomposition to CO and O_2 by the combination of catalysis and dielectric- barrier discharges at low temperatures and ambient pressure. *Industrial and Engineering Chemistry Research* 56:3204–3216.
96. Van Laer, K., and A. Bogaerts 2015. Improving the conversion and energy efficiency of carbon dioxide splitting in a zirconia packed dielectric barrier discharge reactor. *Energy Technology* 3:1038–1044.
97. Ray, D., and C. Subrahmanyam 2016. CO_2 decomposition in a packed DBD plasma reactor: Influence of packing materials. *RSC Advances* 6:39492–39499.
98. Duan, X., Z. Hu, Y. Li, and B. Wang 2015. Effect of dielectric packing materials on the decomposition of carbon dioxide using DBD microplasma reactor. *AIChE Journal* 61:898–903.

99. Bongers, W., H. Bouwmeester, B. Wolf, F. Peeters, S. Welzel, D. van den Bekerom, N. den Harder, A. Goede, M. Graswinckel, P. W. Groen, J. Kopecki, M. Leins, G. van Rooij, A. Schulz, M. Walker, and R. van de Sanden 2017. Plasma-driven dissociation of CO_2 for fuel synthesis. *Plasma Processes and Polymers* 14:1600126.
100. Brock, S. L., M. Marquez, S. L. Suib, Y. Hayashi, and H. Matsumoto 1998. Plasma decomposition of CO_2 in the presence of metal catalysts. *Journal of Catalysis* 180:225–233.
101. Spencer, L. F., and A. D. Gallimore 2011. Efficiency of CO_2 dissociation in a radio-frequency discharge. *Plasma Chemistry and Plasma Processing* 31:79–89.
102. Spencer, L. F., and A. D. Gallimore 2013. CO_2 dissociation in an atmospheric pressure plasma/catalyst system: A study of efficiency. *Plasma Sources Science and Technology* 22:015019.
103. Indarto, A., D. R. Yang, J. W. Choi, H. Lee, and H. K. Song 2007. Gliding arc plasma processing of CO_2 conversion. *Journal of Hazardous Materials* 146:309–315.
104. Ramakers, M., G. Trenchev, S. Heijkers, W. Wang, and A. Bogaerts 2017. Gliding arc plasmatron: Providing a novel method for CO_2 conversion. *ChemSusChem* 10:2642–2652.
105. Wang, W., D. Mei, X. Tu, and A. Bogaerts 2017. Gliding arc plasma for CO_2 conversion: Better insights by a combined experimental and modelling approach. *Chemical Engineering Journal* 330:11–25.
106. Sun, S. R., H. X. Wang, D. H. Mei, X. Tu, and A. Bogaerts 2017. CO_2 conversion in a gliding arc plasma: Performance improvement based on chemical reaction modeling. *Journal of CO_2 Utilization* 17:220–234.
107. Berthelot, A., and A. Bogaerts 2017. Modeling of CO_2 plasma: Effect of uncertainties in the plasma chemistry. *Journal of Physical Chemistry C* 121:8236–8251.
108. Spencer, L. F., and A. D. Gallimore 2010. Efficiency of CO_2 dissociation in a radio-frequency discharge. *Plasma Chemistry and Plasma Processing* 31:79–89.
109. Vesel, A., M. Mozetic, A. Drenik, and M. Balat-Pichelin 2011. Dissociation of CO_2 molecules in microwave plasma. *Chemical Physics* 382:127–131.
110. Asisov, R. I., A. A. Fridman, V. K. Givotov, E. G. Krasheninnikov, B. I. Patrushev, B. V. Potapkin, V. D. Rusanov, and M. F. Krotov 1983. Carbon dioxide dissociation in non-equilibrium plasma, *5th International Symposium on Plasma Chemistry*, p. 52.
111. Rusu, I., and J. M. Cormier 2003. On a possible mechanism of the methane steam reforming in a gliding arc reactor. *Chemical Engineering Journal* 91:23–31.
112. Ouni, F., A. Khacef, and J. M. Cormier 2006. Effect of oxygen on methane steam reforming in a sliding discharge reactor. *Chemical Engineering & Technolog* 29(5):604–609.
113. Indarto, A., J. W. Choi, H. Lee, and H. K. Song 2006. Conversion of CO_2 by gliding arc plasma. *Environmental Engineering Science* 23(6):1033–1043.
114. Aerts, R., R. Snoeckx, and A. Bogaerts 2014. In-situ chemical trapping of oxygen after the splitting of carbon dioxide by plasma. *Plasma Processes and Polymers* 11:985–992.
115. Kim, S. C., M. S. Lim, and Y. N. Chun 2014. Reduction characteristics of carbon dioxide using a plasmatron. *Plasma Chemistry and Plasma Processing* 34(1):125–143.
116. Bogaerts, A., T. Kozák, K. Van Laer, and R. Snoeckx 2015. Plasma-based conversion of CO_2: Current status and future challenges. *Faraday Discussions* 183:217–232.
117. Mori, S., N. Matsuura, L. L. Tun, and M. Suzuki 2016. Direct synthesis of carbon nanotubes from only CO_2 by a hybrid reactor of dielectric barrier discharge and solid oxide electrolyser cell. *Plasma Chemistry and Plasma Processing* 36:231–239.
118. Mikoviny, T., M. Kocan, S. Matejcik, N. J. Mason, and J. D. Skalny 2004. Experimental study of negative corona discharge in pure carbon dioxide and its mixtures with oxygen. *Journal of Physics D: Applied Physics* 37(1):64–73.

119. Xu, W., M. W. Li, G. H. Xu, and Y. L. Tian 2004. Decomposition of CO_2 using DC corona discharge at atmospheric pressure. *Japanese Journal of Applied Physics* 43:8310–8311.
120. Morvova, M., F. Hanic, and I. Morva 2000. Plasma technologies for reducing CO_2 emissions from combustion exhaust with toxic admixtures to utilisable products. *Journal of Thermal Analysis* 61(1):273–287.
121. Wen, Y., and X. Jiang 2001. Decomposition of CO_2 using pulsed corona discharges combined with catalyst. *Plasma Chemistry and Plasma Processing* 21:665–678.
122. Maezono, I., and J. S. Chang 1990. Reduction of CO_2 from combustion gases by DC corona torches. *IEEE Transactions on Industry Applications* 26(4):651–655.
123. Suib, S. L., S. L. Brock, M. Marquez, J. Luo, H. Matsumoto, and Y. Hayashi 1998. Efficient catalytic plasma activation of CO_2, NO and H_2O. *Journal of Physical Chemistry B* 102:9661–9666.
124. Brock, S. L., T. Shimojo, M. Marquez, C. Marun, S. L. Suib, H. Matsumoto, and Y. Hayashi 1999. Factors Influencing the decomposition of CO_2 in AC Fan-type plasma reactors: Frequency, waveform, and concentration effects. *Journal of Catalysis* 184:123–133.
125. Andreev, S. N., V. V. Zakharov, V. N. Ochkin, and S. Y. Savinov 2004. Plasma-chemical $CO(_2)$ decomposition in a non-self-sustained discharge with a controlled electronic component of plasma. *Spectrochimica Acta Part A* 60:3361–3369.
126. Mori, S., A. Yamamoto, and M. Suzuki 2006. Characterization of a capillary plasma reactor for carbon dioxide decomposition. *Plasma Sources Science and Technology* 15:609–613.
127. Bak, M. S., S. K. Im, and M. Cappelli 2015. Nanosecond-pulsed discharge plasma splitting of carbon dioxide. *IEEE Transactions on Plasma Science* 43:1002–1007.
128. Futamura, S., and H. Kabashima 2004. Synthesis gas production from CO_2 and H_2O with nonthermal plasma. *Studies in Surface Science and Catalysis* 153:119–124.
129. Snoeckx, R., A. Ozkan, F. Reniers, and A. Bogaerts 2017. The quest for value-added products from carbon dioxide and water in a dielectric barrier discharge: a chemical kinetics study. *ChemSusChem* 10:409–424.
130. Ihara, T., M. Kiboku, and Y. Iriyama 1994. Plasma reduction of CO_2 with H2O for the Formation of Organic Compounds. *Bulletin of the Chemical Society of Japan* 67:312–314.
131. Ihara, T., T. Ouro, T. Ochiai, M. Kiboku, and Y. Iriyama 1996. Ihara, T., T. Ouro, T. Ochiai, M. Kiboku, and Y. Iriyama 1996. Formation of Methanol by Microwave-Plasma Reduction of CO_2 with H_2O. *Bulletin of the Chemical Society of Japan* 69:241–244.
132. Chen, G., T. Silva, V. Georgieva, T. Godfroid, N. Britun, R. Snyders, and M. P. Delplancke-Ogletree 2015. Simultaneous dissociation of CO_2 and H_2O to a syngas in a surface wave microwave discharge. *International Journal of Hydrogen Energy* 40:3789–3796.
133. Hayashi, N., T. Yamakawa, and S. Baba 2006. Effect of additive gases on synthesis of organic compounds from carbon dioxide using non-thermal plasma produced by atmospheric surface discharges. Vacuum 80:1299–1304.
134. Guo, L., X. Ma, Y. Xia, X. Xiang, and X. Wu 2015. A novel method of production of ethanol by carbon dioxide with steam. *Fuel* 158:843–847.
135. Eliasson, B., U. Kogelschatz, B. Xue, and L.-M. Zhou 1998. Hydrogenation of Carbon Dioxide to Methanol with a Discharge-Activated Catalyst. *Industrial & Engineering Chemistry Research* 37:3350–3357.
136. Jwa, E, S. B. Lee, H. W. Lee, and Y. S. Mok 2013. Plasma-assisted catalytic methanation of CO and CO_2 over Ni–zeolite catalysts. *Fuel Processing Technology* 108:89–93.

137. Studt, F., I. Sharafutdinov, F. Abild-Pedersen, C. F. Elkjær, J. S. Hummelshøj, S. Dahl, I. Chorkendorff, and J. K. Nørskov 2014. Discovery of a Ni-Ga catalyst for carbon dioxide reduction to methanol. *Nature Chemistry* 6:320–324.
138. Ma, X., et al. 2019. Plasma assisted catalytic conversion of CO_2 and H_2O over Ni/Al_2O_3 in a DBD reactor. *Plasma Chem Plasma Process* 39:109–124. https://doi.org/10.1007/s11090-018-9931-1.
139. Hoeben, W. F. L. M., E. J. M. Heesch, van F. J. C. M. Beckers, W. Boekhoven, and A. J. M. Pemen 2015. Plasma-driven water assisted CO_2 methanation. *IEEE Transactions on Plasma Science* 43(6):1954–1958.
140. Shah, Y. T. 2017. *Chemical Energy from Natural and Synthetic Gas*. New York, NY: CRC Press.
141. Huang, R. J., et al. 2014. High secondary aerosol contribution to particulate pollution during haze events in China. *Nature* 514(7521):218–222. doi: 10.1038/nature13774.
142. Wang, W., S. Wang, X. Maa, and J. Gong 2011. Recent advances in catalytic hydrogenation of carbon dioxide. *Chemical Society Reviews* 40:3703–3727.
143. Liu, M., Y. Yi, L. Wang, H. Guo, and A. Bogaerts 2019. Hydrogenation of carbon dioxide to value-added chemicals by heterogeneous catalysis and plasma catalysis, an MDPI open access article, *Catalysts* 9:275. doi: 10.3390/catal9030275. www.mdpi.com/journal/catalysts.
144. Kung, H. H. 1989. *Transition Metal Oxides*, vol. 45, 1st edn, Surface chemistry and catalysis. Amsterdam, The Netherlands: Elsevier.
145. Goguet, A., F. C. Meunier, D. Tibiletti, J. P. Breen, and R. Burch 2004. *Journal of Physical Chemistry B* 108:20240–20246.
146. Davis, W., and M. Martín 2014. *Journal of Cleaner Production* 80:252–261.
147. Aldana, P. U., et al. 2013. *Catalysis Today* 215:201–e207.
148. Peebles, D., D. Goodman, J. White 1983. *Journal of Physical Chemistry* 87:4378–e4387.
149. Marwood, M., R. Doepper, and A. Renken 1997. In-situ surface and gas phasre analysis for kinetic studies under transient conditions-The catalytic hydrogenation of CO_2. *Applied Catalysis A., General* 151:223–246.
150. Prairie, M. R., A. Renken, J. G. Highfield, K. R. Thampi, and M. Gr€atzel 1991. A fourier transform infrared spectroscopic study of CO_2 methanation on supported Ruthenium. *Journal of Catalysis* 129:130–144.
151. Jacquemin, M, A. Beuls, and P. Ruiz 2010. Catalytic production of methane from CO_2 and H_2 at low temperature: Insight on the reaction mechanism. Catalysis Today 157:462–466.
152. Quin, Z., Y. Zhou, Y. Jiang, Z. Liu, and H. Ji 2017. Recent advances in heterogeneous catalytic hydrogenation of CO_2 to methane, an open access intech paper. doi: 10.5772/65407.
153. Vesselli, E., et al. 2009. Hydrogen-assisted transformation of CO_2 on nickel: The role of formate and carbon monoxide. *Journal of Physical Chemistry Letters* 1:402–406.
154. Olah, G. A., G. S. Prakash, and A. Goeppert 2011. Anthropogenic chemical carbon cycle for a sustainable future. *Journal of the American Chemical Society* 133:12881–12898.
155. Ortelli, E, J. Wambach, and A. Wokaun 2001. Methanol synthesis reactions over a CuZr based catalyst investigated using periodic variations of reactant concentrations. *Applied Catalysis A: General* 216:227–241.
156. Jadhav, S. G., P. D. Vaidya, B. M. Bhanage, and J. B. Joshi 2014. Catalytic carbon dioxide hydrogenation to methanol: A review of recent studies. *Chemical Engineering Research and Design* 92:2557–2567.
157. Grabow, L, and M. Mavrikaki 2011. Mechanism of Methanol Synthesis on Cu through CO_2 and CO Hydrogenation. *ACS Catalysis* 1:365–384.
158. Graciani, J., et al. 2014. Highly active copper-ceria and copper-ceria-titania catalysts for methanol synthesis from CO_2. *Science* 345:546–550.

159. Studt, F., et al. 2015. The mechanism of CO and CO_2 hydrogenation to methanol over Cu-based catalysts. *ChemCatChem* 7:1105–1111.
160. Alayoglu, S., et al. 2011. CO_2 hydrogenation studies on Co and CoPt bimetallic nanoparticles under reaction conditions using TEM, XPS and NEXAFS. *Topics in Catalysis* 54:778.
161. Gnanamani, M. K., et al. 2016. Hydrogenation of carbon dioxide over Co–Fe bimetallic catalysts. *ACS Catalysis* 6:913–927.
162. Nizio, M. 2016. Plasma catalytic process for CO_2 methanation. *Catalysis*. Université Pierre et Marie Curie – Paris VI. English. NNT: 2016PA066607.
163. De Bie, C., J. van Dijk, and A. Bogaerts 2016. CO_2 hydrogenation in a dielectric barrier discharge plasma revealed. *Journal of Physical Chemistry C* 120:25210–25224.
164. Zeng, Y., and X. Tu 2015. Plasma-catalytic CO_2 hydrogenation at low temperatures. *IEEE Transactions on Plasma Science* 44:1–7.
165. Nizio, M., A. Albarazi, S. Cavadias, J. Amouroux, M. E. Galvez, and P. Da Costa 2016. Hybrid plasma-catalytic methanation of CO_2 at low temperature over ceria zirconia supported Ni catalysts. *International Journal of Hydrogen Energy* 41:11584–11592.
166. Song, H., J.-W. Choi, S. Yue, H. Lee, and B.-K. Na 2004. Synthesis gas production via dielectric barrier discharge over Ni/γ-Al_2O_3 catalyst. *Catalysis Today* 89:27–33. doi: 10.1016/j.cattod.2003.11.009.
167. Oshima, K., T. Shinagawa, Y. Nogami, R. Manabe, S. Ogo, and Y. Sekine 2014. Low temperature catalytic reverse water gas shift reaction assisted by an electric field. *Catalysis Today* 232:27–32.
168. Amouroux, J., and S. Cavadias 2017. Electro catalytic reduction of carbon dioxide under plasma DBD process. *Journal of Physics D: Applied Physics* 50. doi: 10.1088/1361-6463/aa8b56.
169. Fan, Z., S. Kaihang, N. Rui, Z. Binran, and C.-J. Liu 2015. Improved activity of Ni/$MgAl_2O_4$ for CO_2 methanation by the plasma decomposition. *Journal of Energy Chemistry* 24. doi: 10.1016/j.jechem.2015.09.004.
170. Dębeka, R., F. Azzolina-Jurya, A. Traverta, and F. Maugéa. A review on plasma-catalytic methanation of carbon dioxide – Looking for an efficient catalyst. https://www.sciencedirect.com/science/article/pii/S1364032119306355. Manuscript_e90b722ee1805ea7a1048009dc4781bb.
171. Zheng, X., S. Tan, L. Dong, S. Li, and H. Chen 2015. Plasma-assisted catalytic dry reforming of methane: highly catalytic performance of nickel ferrite nanoparticles imbedded in silica. *Journal of Power Sources* 274:286–294.
172. Mikhail, M., B. Wang, R. Jalain, S. Cavadias, M. Tatoulian, S. Ognier, M. Galvez, and P. Da Costa 2018. Plasma-catalytic hybrid process for CO_2 methanation: Optimization of operation parameters. *Reaction Kinetics, Mechanisms and Catalysis* 126. doi: 10.1007/s11144-018-1508-8.
173. Ahmad, F., E. Lovell, H. Masood, P. J. Cullen, K. Ostrikov, J. Scott, and R. Amal 2020. Low-temperature CO_2 methanation: Synergistic effects in plasma-Ni hybrid catalytic system. *ACS Sustainable Chemistry & Engineering*. doi: 10.1021/acssuschemeng.9b06180.
174. Ge, Y., T. He, D. Han, G. Li, R. Zhao, and J. Wu 2019. Plasma-assisted CO_2 methanation: Effects on the low-temperature activity of an Ni–Ce catalyst and reaction performance. *Royal Society Open Science*. https://doi.org/10.1098/rsos.190750
175. Nizio, M., R. Benrabbah, M. Krzak, R. Debek, M. Motak, S. Cavadias, M. Galvez, and P. Da Costa 2016. Low temperature hybrid plasma-catalytic methanation over Ni-Ce-Zr hydrotalcite-derived catalysts. *Catalysis Communications* 83. doi: 10.1016/j.catcom.2016.04.023.
176. Li, W., H. Wang, X. Jiang, J. Zhu, Z. Liu, X. Guo, and C. Song 2018. A short review of recent advances in CO_2 hydrogenation to hydrocarbons over heterogeneous catalysts. *RSC Advances* 8:7651.

177. Tsiotsias, A. I., N. D. Charisiou, I. V. Yentekakis, and M. A. Goula 2021. Bimetallic Ni-based catalysts for CO_2 methanation: A review. *Nanomaterials (Basel)* 11(1): 28. doi: 10.3390/nano11010028.
178. Xu, S., H. Chen, C. Hardacre, and X. Fan 2021. Non-thermal plasma catalysis for CO_2 conversion and catalyst design for the process. *Journal of Physics D: Applied Physics* 54:233001.
179. Wang, L., Y. Yi, H. Guo, and X. Tu 2017. Atmospheric pressure and room temperature synthesis of methanol through plasma-catalytic hydrogenation of CO_2. *ACS Catalysis* 8. doi: 10.1021/acscatal.7b02733.
180. Ronda-Lloret, M., Y. Wang, P. Oulego, G. Rothenberg, X. Tu, and N. Raveendran Shiju 2020. CO_2 hydrogenation at atmospheric pressure and low temperature using plasma-enhanced catalysis over supported cobalt oxide catalysts. *ACS Sustainable Chemistry & Engineering* 47:17397–17407. https://doi.org/10.1021/acssuschemeng.0c05565.
181. Maya, L. 2000. Plasma-assisted reduction of carbon dioxide in the gas phase. *Journal of Vacuum Science and Technology A* 18:285–287.
182. de la Fuente, J. F., S. H. Moreno, A. I. Stankiewicz, and G. D. Stefanidis 2016. A new methodology for the reduction of vibrational kinetics in non-equilibrium microwave plasma: application to CO_2 dissociation. *International Journal of Hydrogen Energy*:1–11.
183. Kano, M., G. Satoh, and S. Iizuka 2012. Reforming of carbon dioxide to methane and methanol by electric impulse low-pressure discharge with hydrogen. *Plasma Chemistry and Plasma Processing* 32:177–185.
184. Shah, Y. T., and T. Gardner 2014. Dry reforming of hydrocarbon feedstocks. *Catalysis Reviews: Science and Engineering* 54:476–536. New York: CRC Press, Taylor and Francis.
185. Sheng, Z., S. Kameshima, K. Sakata, and T. Nozaki 2018. Plasma-enabled dry methane reforming, an Intech open access paper. doi: 10.5772/intechopen.80523.
186. Puliyalil, H., D. L. Jurkovi´c, V. D. B. C. Dasireddy, and B. Likozar 2018. A review of plasma-assisted catalytic conversion of gaseous carbon dioxide and methane into value added platform chemicals and fuels. *RSC Advances* 8:27481. doi: 10.1039/c8ra03146k.
187. Zhu, X., P. Huo, Y. Zhang, D. Cheng, and C. Liu 2008. Structure and reactivity of plasma treated Ni/Al_2O_3 catalyst or CO_2 reforming of methane. Applied Catalysis B 81:132–140.
188. Filipi˘c, G., and U. Cvelbar 2012. Copper oxide nanowires: A review of growth. *Nanotechnology* 23:194001.
189. Taghvaei, H, M. Heravi, and M. R. Rahimpour 2017. Synthesis of supported nanocatalysts via novel non-thermal plasma methods and its application in catalytic processes. *Plasma Processes and Polymers*:1600204.
190. Ding, D., Z.-L. Song, Z.-Q. Cheng, W.-N. Liu, X.-K. Nie, X. Bian, Z. Chen, and W. Tan 2014. Plasma-assisted nitrogen doping of graphene-encapsulated Pt nanocrystals as efficient fuel cell catalysts. *Journal of Materials Chemistry A* 2:472–477.
191. Liang, J, Y. Jiao, M. Jaroniec, and S. Z. Qiao 2012. Rücktitelbild: Sulfur and nitrogen dual-doped mesoporous graphene electrocatalyst for oxygen reduction with synergistically enhanced performance. *Angewandte Chemie International Edition* 51:11496–11500.
192. Kim, D. H., G. S. Han, W. M. Seong, J.-W. Lee, B. J. Kim, N.-G. Park, K. S. Hong, S. Lee, and H. S. Jung 2015. Niobium doping effects on TiO_2 mesoscopic electron transport layer-based perovskite solar cells. *ChemSusChem* 8:2392–2398.
193. Li, Y, Wei, Z., and Y. Wang 2014. Ni/MgO catalyst prepared via dielectric-barrier discharge plasma with improved catalytic performance for carbon dioxide reforming of methane. *Frontiers of Chemical Science and Engineering* 8:133–140.

194. Fang, X., J. Lian, K. Nie, X. Zhang, Y. Dai, X. Xu, X. Wang, W. Liu, C. Li, and W. Zhou 2016. Dry reforming of methane on active and coke resistant $Ni/Y_2Zr_2O_7$ catalysts treated by dielectric barrier discharge plasma. *Journal of Energy Chemistry* 25:825–831.
195. Benrabbah, R, C. Cavaniol, H. Liu, S. Ognier, S. Cavadias, M. E. G´alvez, and P. Da Costa 2017. Plasma DBD activated ceria-zirconia-promoted Ni-catalysts for plasma catalytic CO_2 hydrogenation at low temperature. *Catalysis Communications* 89:73–76.
196. Zhao, Y., Y. xiang Pan, Y. Xie, and C. jun Liu 2008. Carbon dioxide reforming of methane over glow discharge plasma-reduced Ir/Al_2O_3 catalyst. *Catalysis Communications*:1558–1562.
197. Kameshima, S., K. Tamura, R. Mizukami, T. Yamazaki, and T. Nozaki 2017. Parametric analysis of plasma-assisted pulsed dry methane reforming over Ni/Al_2O_3 catalyst. *Plasma Processes and Polymers* 14:1600096. doi: 10.1002/ppap.201600096.
198. Nozaki, T., Y. Miyazaki, Y. Unno, and K. Okazaki 2001. Energy distribution and heat transfer mechanisms in atmospheric pressure non-equilibrium plasmas. *Journal of Physics D: Applied Physics* 34:3383.
199. Nozaki, T., Y. Unno, and K. Okazaki 2002. Thermal structure of atmospheric pressure non- equilibrium plasmas. *Plasma Sources Science and Technology* 11:431. doi: 10.1088/0963-0252/11/4/310. http://dx.doi.org/110.5772/intechopen.80523
200. Ashik, U. P. M, and W. M. Wan Daud 2015. Probing the differential methane decomposition behaviors of n-Ni/SiO_2, n-Fe/SiO_2 and n-Co/SiO_2 catalysts prepared by co-precipitation cum modified Stober method. *RSC Advances* 5:46735–46748.
201. Scapinello, M., L. M. Martini, G. Dilecce, and P. Tosi 2016. Conversion of CH_4/CO_2 by a nanosecond repetitively pulsed discharge. *Journal of Physics D: Applied Physics* 49:75602.
202. Theofanidis, S. A., V. V. Galvita, H. Poelman, and G. B. Marin 2015. Enhanced carbon-resistant dry reforming Fe-Ni catalyst: Role of Fe. *ACS Catalysis* 5:3028–3039.
203. Dombrowski, E., E. Peterson, D. Del Sesto, and A. L. Utz 2015. Precursor-mediated reactivity of vibrationally hot molecules: Methane activation on Ir (111). *Catalysis Today* 244:10–18. doi: 10.1016/j. cattod.2014.10.025.
204. Nozaki, T., N. Muto, S. Kado, and K. Okazaki 2004. Dissociation of vibrationally excited methane on Ni catalyst: Part 1. Application to methane steam reforming. *Catalysis Today* 89:57–65. doi: 10.1016/j.cattod.2003.11.040.
205. Kameshima, S., K. Tamura, Y. Ishibashi, and T. Nozaki 2015. Pulsed dry methane reforming in plasma-enhanced catalytic reaction. *Catalysis Today* 256:67–75. doi: 10.1016/j.cattod.2015.05.011.
206. Nozaki, T., H. Tsukijihara, W. Fukui, and K. Okazaki 2007. Kinetic analysis of the catalyst and nonthermal plasma hybrid reaction for methane steam reforming. *Energy & Fuels* 21(5):2525–2530. doi: 10.1021/ef070117.
207. Wang, Q., B.-H. Yan, Y. Jin, and Y. Cheng 2009. Investigation of dry reforming of methane in a dielectric barrier discharge reactor. *Plasma Chemistry and Plasma Processing* 29:217–228.
208. Snoeckx, R., Y. X. Zeng, X. Tu, and A. Bogaerts 2015. Plasma-based dry reforming: improving the conversion and energy efficiency in a dielectric barrier discharge. *RSC Advances* 5:29799–29808.
209. Zhou, L. M., B. Xue, U. Kogelschatz, and B. Eliasson 1998. Nonequilibrium plasma reforming of greenhouse gases to synthesis gas. *Energy Fuels* 12:1191–1199.
210. Liu, C. J., B. Xue, B. Eliasson, F. He, Y. Li, and G. H. Xu 2001. Methane conversion to higher hydrocarbons in the presence of carbon dioxide using dielectric-barrier discharge plasmas. *Plasma Chemistry and Plasma Processing* 21:301–310.

211. Zhang, X., and Cha, M. S. 2013. Electron-induced dry reforming of methane in a temperature-controlled dielectric barrier discharge reactor. *Journal of Physics D: Applied Physics* 46:415205.
212. Goujard, V., J.-M. Tatibouet, and C. Batiot-Dupeyrat 2009. Use of a non-thermal plasma for the production of synthesis gas from biogas. Applied Catalysis A 353:228–235.
213. Li, Y., C. J. Liu, B. Eliasson, and Y. Wang 2002. Synthesis of oxygenates and higher hydrocarbons directly from methane and carbon dioxide using dielectric-barrier discharges: product distribution. Energy Fuels 16:864–870.
214. Rico, V. J., Hueso, J. L, Cotrino, J., and A. R. Gonzalez-Elipe 2010. Evaluation of different dielectric barrier discharge plasma configurations as an alternative technology for green C_1 chemistry in the carbon dioxide reforming of methane and the direct decomposition of methanol. *Journal of Physical Chemistry A* 114:4009–4016.
215. Li, Y., G. Xu, C. Liu, B. Eliasson, and B. Xue 2001. CO-generation of syngas and higher hydrocarbons from CO_2 and CH_4 using dielectric-barrier discharge: Effect of electrode materials. Energy Fuels 15:299–302.
216. Scapinello, M., L. M. Martini, and P. Tosi 2014. CO_2 hydrogenation by CH_4 in a dielectric barrier discharge: Catalytic effects of nickel and copper. *Plasma Processes and Polymers* 11:624–628.
217. Song, H., H. Lee, J.-W. Choi, and B. Na 2004. Effect of electrical pulse forms on the CO_2 reforming of methane using atmospheric dielectric barrier discharge. *Plasma Chemistry and Plasma Processing* 2004 24:57–72.
218. Janeco, A., N. R. Pinhao, and Guerra, V. 2015. Electron kinetics in He/CH_4/CO_2 mixtures used for methane conversion. *Journal of Physical Chemistry C* 119:109–120.
219. Goujard, V., J.-M. Tatibouet, and C. Batiot-Dupeyrat 2011. Carbon dioxide reforming of methane using a dielectric barrier discharge reactor: Effect of helium dilution and kinetic model. *Plasma Chemistry and Plasma Processing* 31:315–325.
220. G´omez-Ram´ırez, A., V. J. Rico, J. Cotrino, A. R. Gonz´alez- Elipe, and R. M. Lambert 2014. Low temperature production of formaldehyde from carbon dioxide and ethane by plasma-assisted catalysis in a ferroelectrically moderated dielectric barrier discharge reactor. *ACS Catalysis* 4:402–408.
221. Reuss, G., W. Disteldorf, A. O. Gamer, and A. Hilt 2000. *Ullmann's Encyclopedia of Industrial Chemistry*, pp. 1–34, Weinheim, Germany: Wiley-VCH Verlag GmbH & Co. KGaA.
222. Zhao, B., Y. Liu, Z. Zhu, H. Guo, and X. Ma 2018. Highly selective conversion of CO_2 into ethanol on Cu/ZnO/Al_2O_3 catalyst with the assistance of plasma. *Journal of CO_2 Utilisation* 24:34–39.
223. Amin, N. 2006. Co-generation of synthesis gas and C_{2+} hydrocarbons from methane and carbon dioxide in a hybrid catalytic-plasma reactor: A review. Fuel 85:577–592.
224. Wang, L., Y. Yi, C. Wu, H. Guo, and X. Tu 2017. One-step reforming of CO_2 and CH_4 into high- value liquid chemicals and fuels at room temperature by plasma-driven catalysis. *Angewandte Chemie International Edition* 56:13679–13683.
225. Zhang, K., B. Eliasson, and U. Kogelschatz 2002. Direct conversion of greenhouse gases to synthesis gas and C_4 hydrocarbons over zeolite HY promoted by a dielectric-barrier discharge. Industrial & Engineering Chemistry Research 41:1462–1468.
226. Zhang, X., B. Dai, A. Zhu, W. Gong, and C. Liu 2002. *Catalysis Today* 72:223–227.
227. Sun, C., D. Mao, L. Han, and J. Yu 2016. Effect of preparation method on performance of Cu–Fe/SiO_2 catalysts for higher alcohols synthesis from syngas. *RSC Advances* 6:55233–55239.
228. Chen, L., X. Zhang, L. Huang, and L. Lei 2010. Post-plasma catalysis for methane partial oxidation to methanol: role of the copper-promoted iron oxide catalyst. Chemical Engineering & Technology 33:2073–2081.

229. Chen, L., X. Zhang, L. Huang, and L. Lei 2010. Application of in-plasma catalysis and post-plasma catalysis for methane partial oxidation to methanol over a Fe_2O_3-CuO/γ-Al_2O_3 catalyst. Journal of Natural Gas Chemistry 19:628–637.
230. Snoeckx, R., A. Rabinovich, D. Dobrynin, A. Bogaerts, and A. Fridman 2016. Plasma based liquefaction of methane: The road from hydrogen production to direct methane liquefaction-review. Wiley online library for natural sciences, plasma processes and polymers. https://doi.org/10.1002/ppap.201600115.
231. Locke, B. R., and K.-Y. Shih 2011. *Plasma Sources Science and Technology* 20:34006.
232. Bruggeman, P., and C. Leys 2009. *Journal of Physics D: Applied Physics* 42:53001.
233. Malik, M. A., A. Ghaffar, and S. A. Malik 2001. *Plasma Sources Science and Technology* 10:82–91.
234. Zhang, J. Q., Y. J. Yang, J. S. Zhang, and Q. Liu 2002. Study on the conversion of CH_4 and CO_2 using a pulsed microwave plasma under atmospheric pressure. *Acta Chimica Sinica* 60:1973–1980.

Index